Big Cats

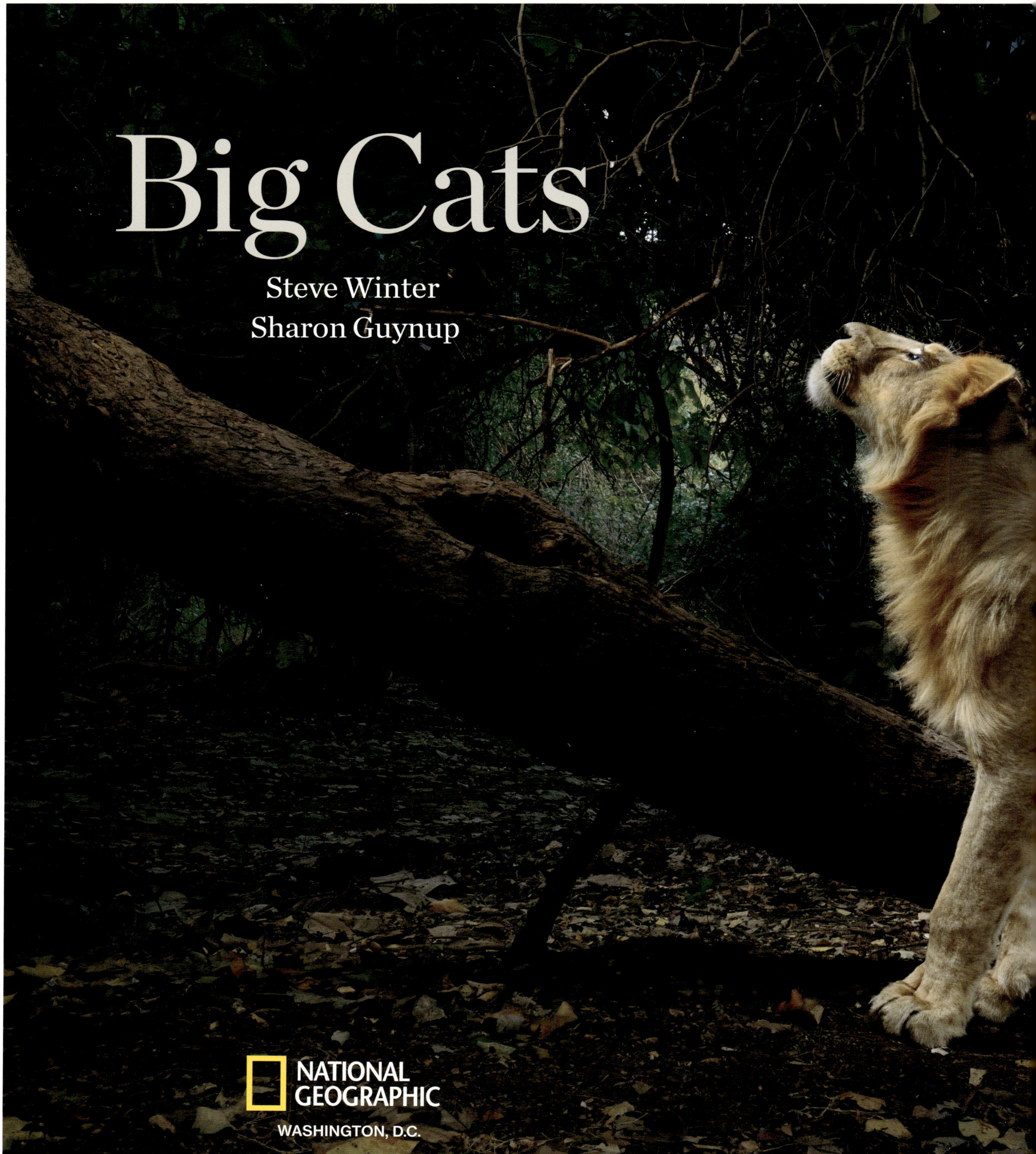

Big Cats

Steve Winter
Sharon Guynup

NATIONAL GEOGRAPHIC
WASHINGTON, D.C.

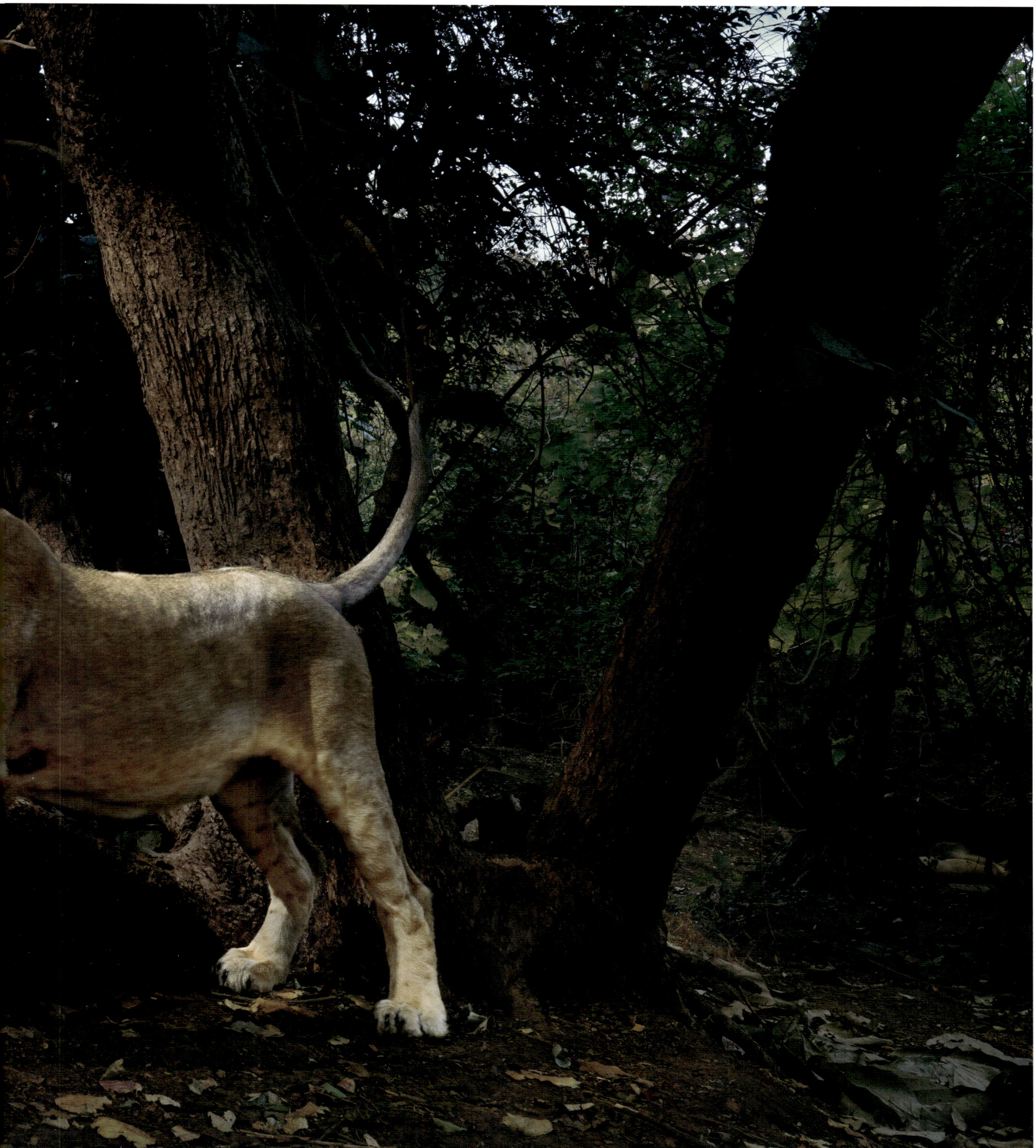

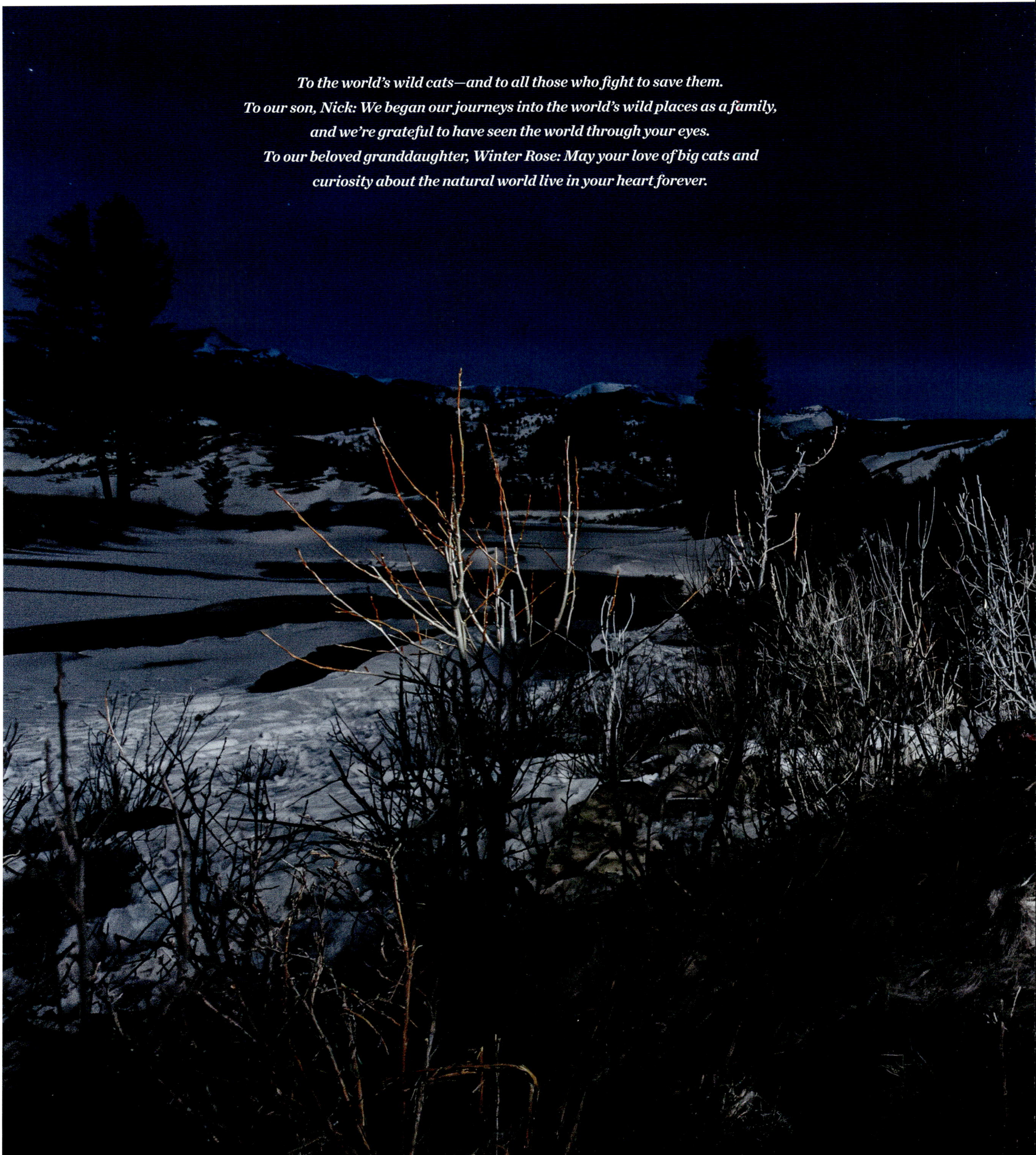

To the world's wild cats—and to all those who fight to save them.

To our son, Nick: We began our journeys into the world's wild places as a family, and we're grateful to have seen the world through your eyes.

To our beloved granddaughter, Winter Rose: May your love of big cats and curiosity about the natural world live in your heart forever.

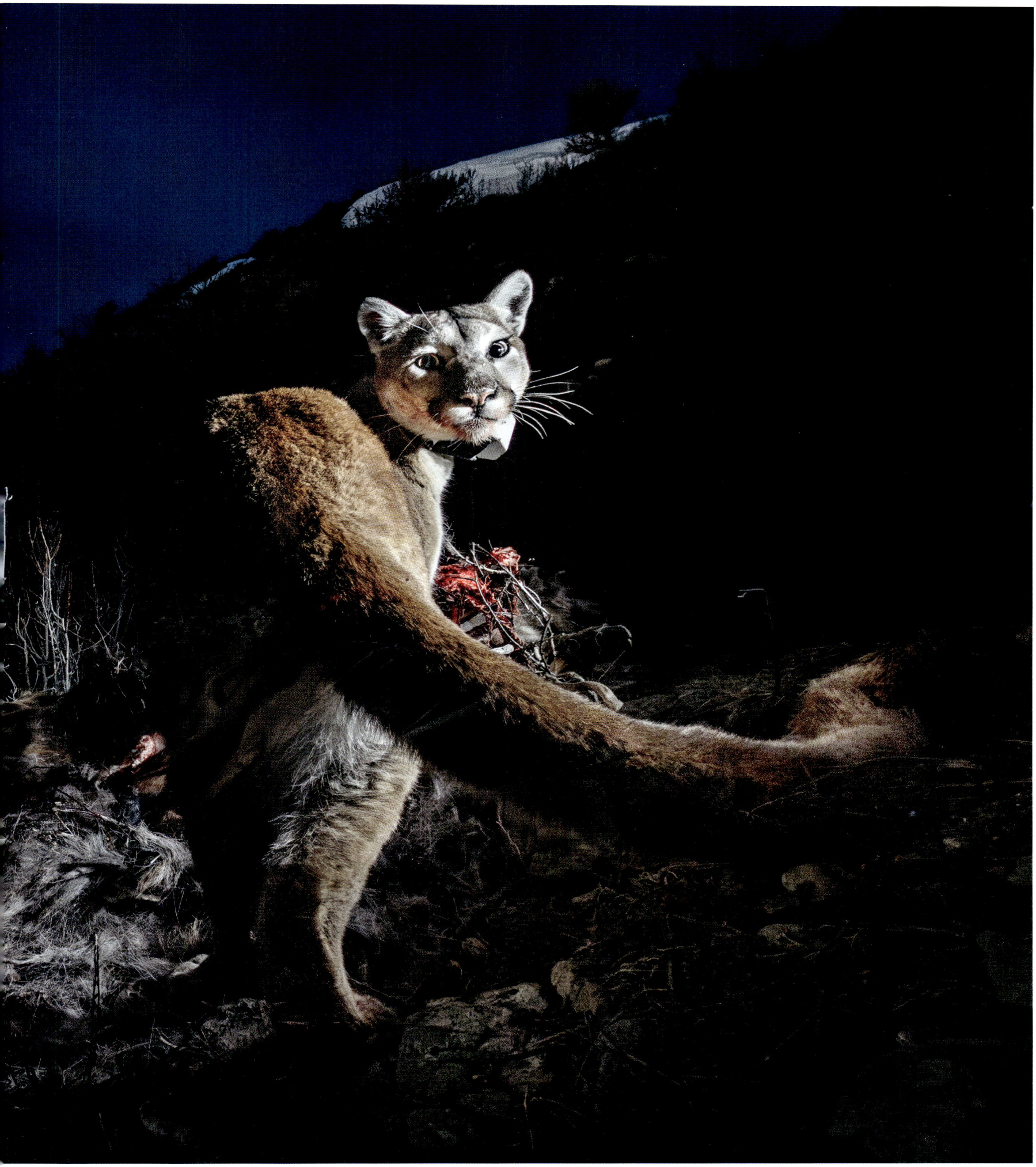

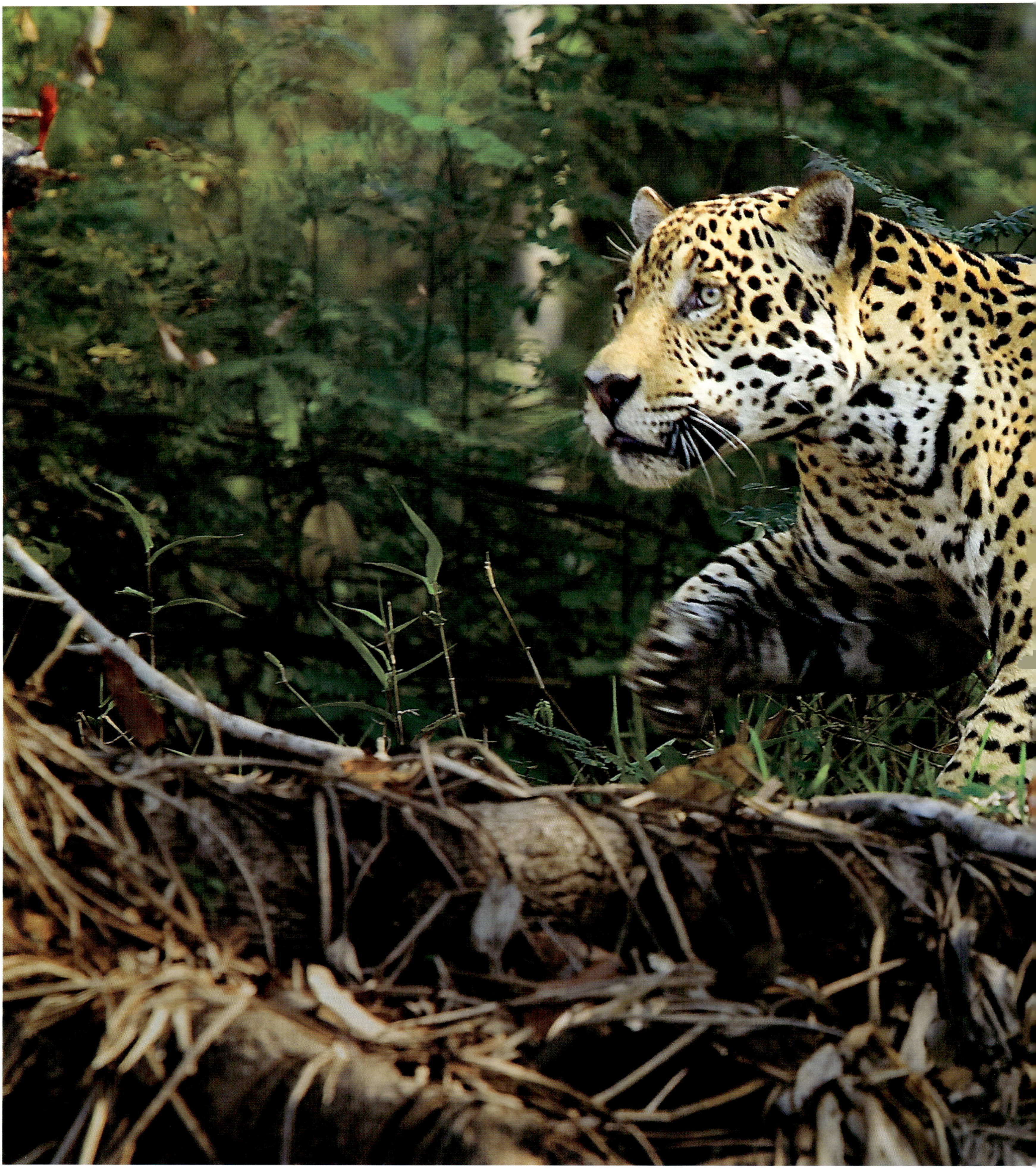

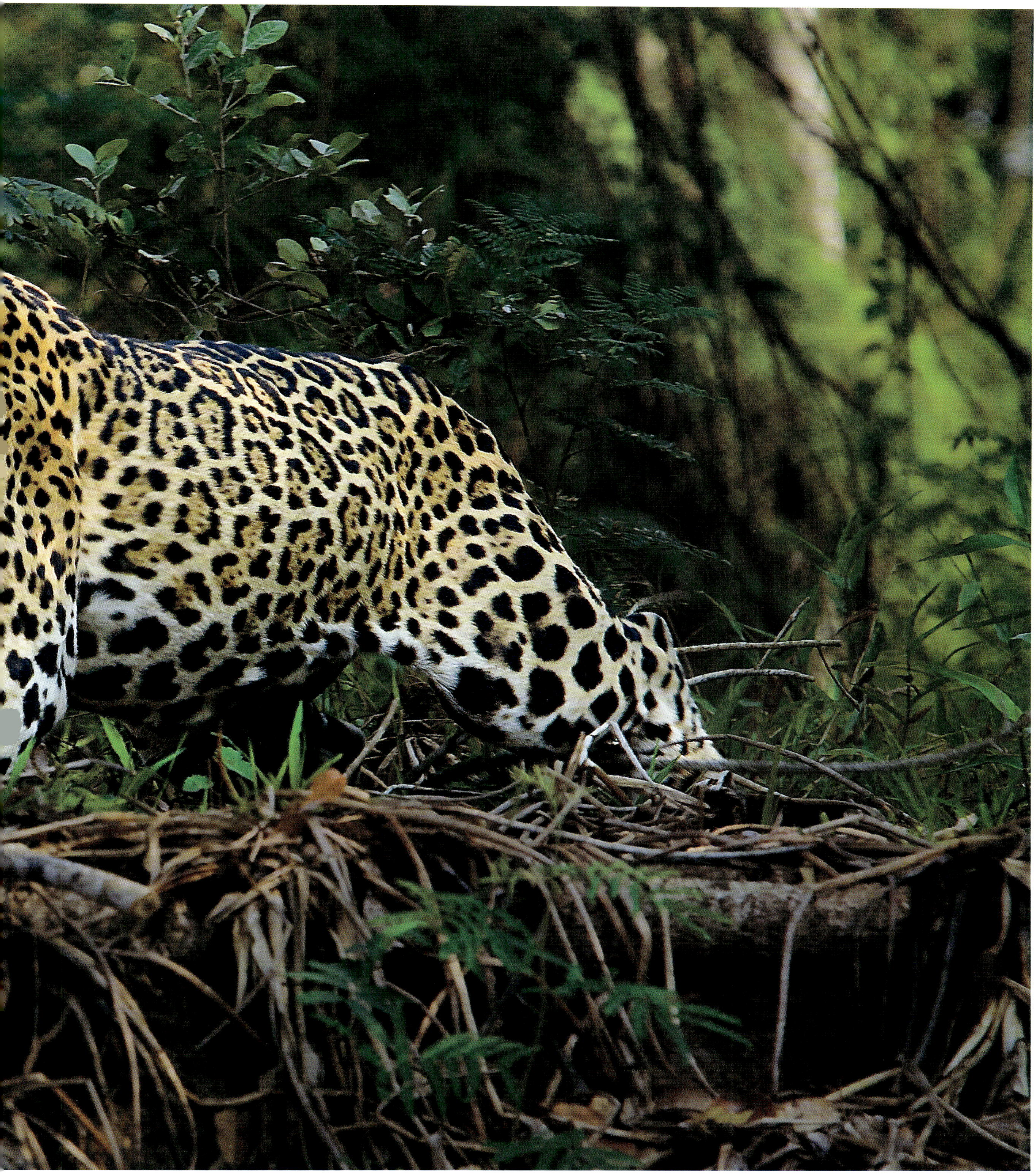

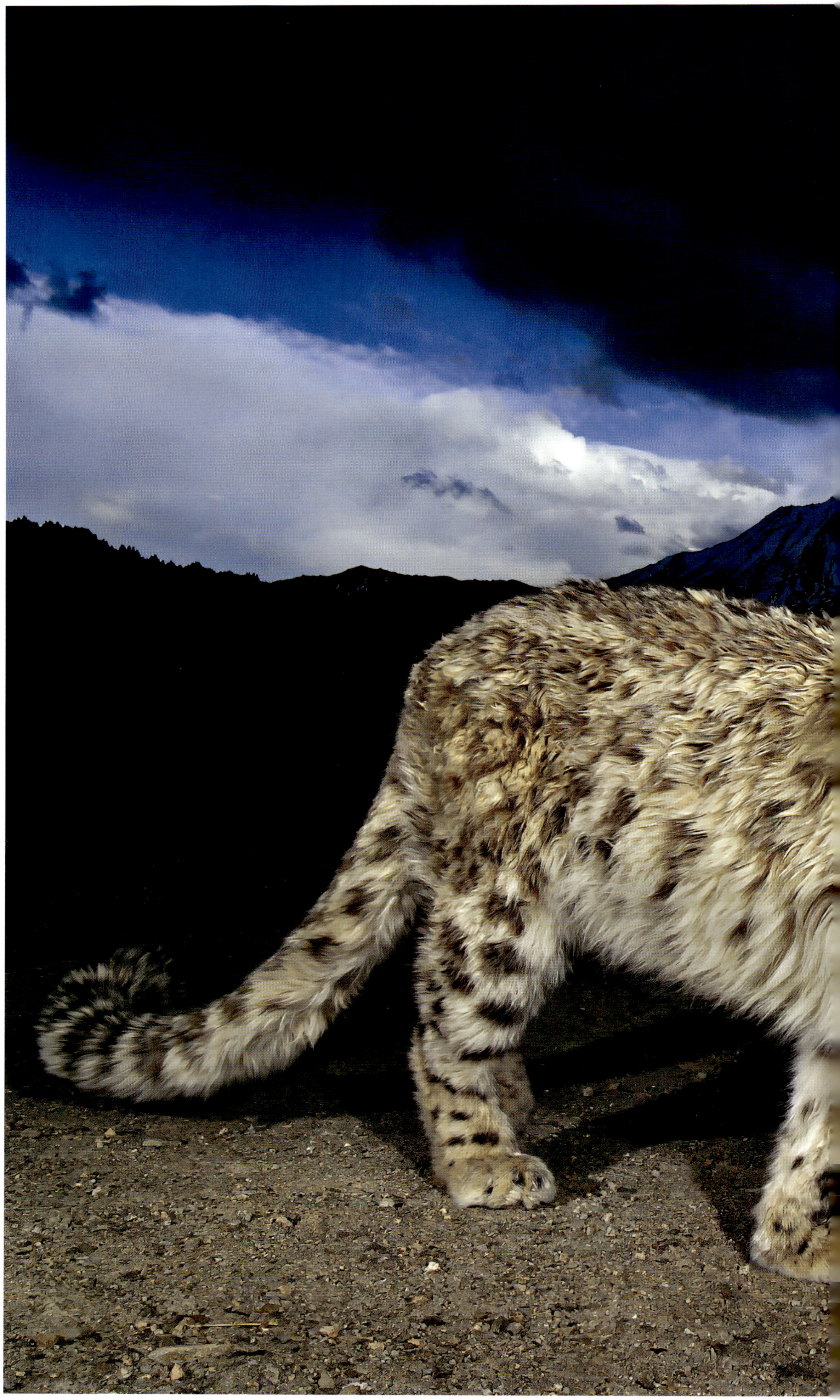

This snow leopard triggered a remote camera on a trail at 16,000 feet in the Indian Himalaya. Scientists helped me place my camera traps in places where these elusive cats mark their territory.

PAGES 2–3: A young male lion leaves a scent-mark message for other lions at a scratch tree in India's Gir National Park.

PAGES 4–5: The flash of a hidden camera diverts a Wyoming cougar from its kill. These largely nocturnal cats are so secretive that I need remote camera traps to illuminate their lives.

PAGES 6–7: A jaguar hunts along a riverbank in the Brazilian Pantanal. This cat, a powerful swimmer, ambushes and eats any of 85 different prey animals on land and in water.

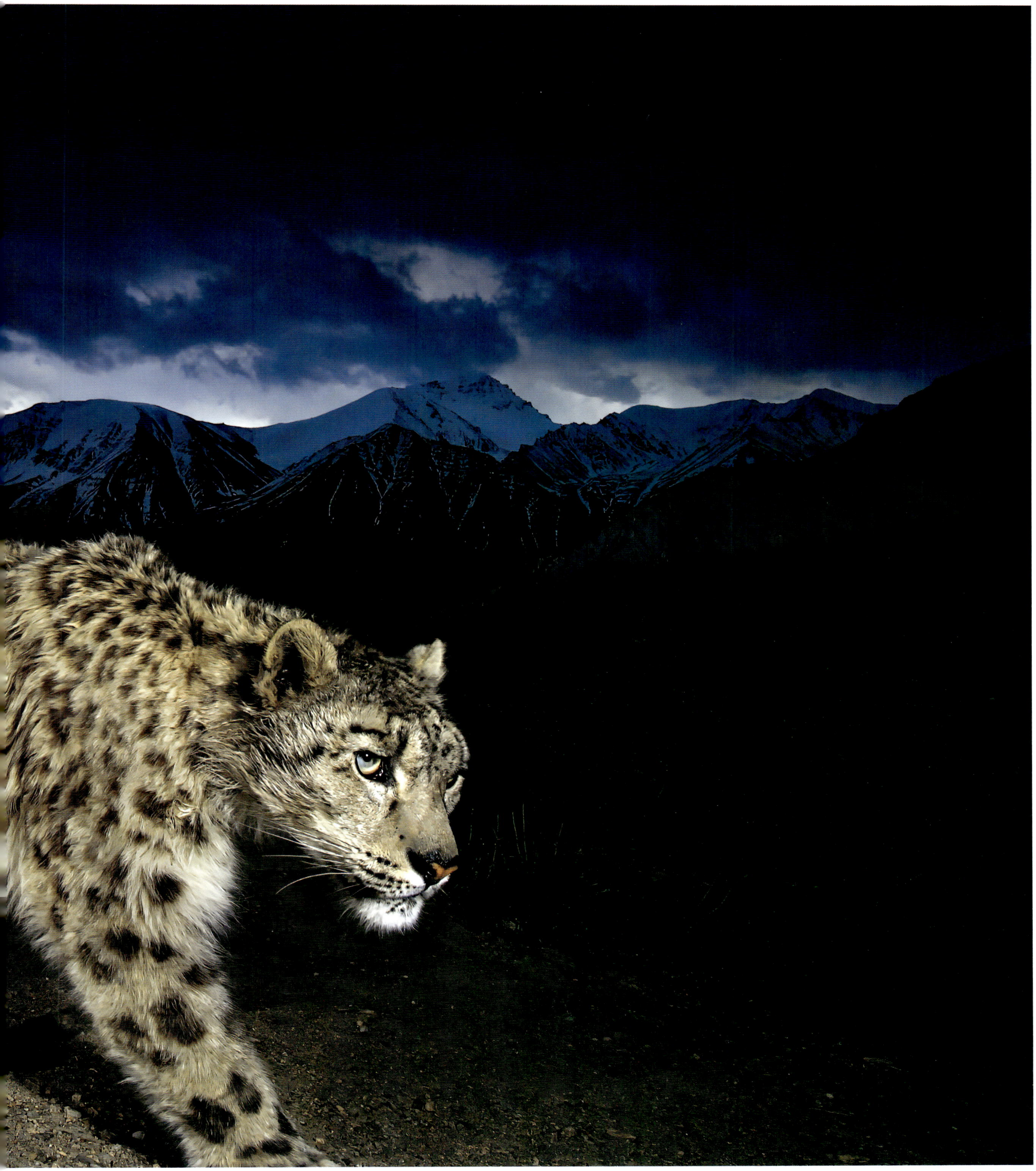

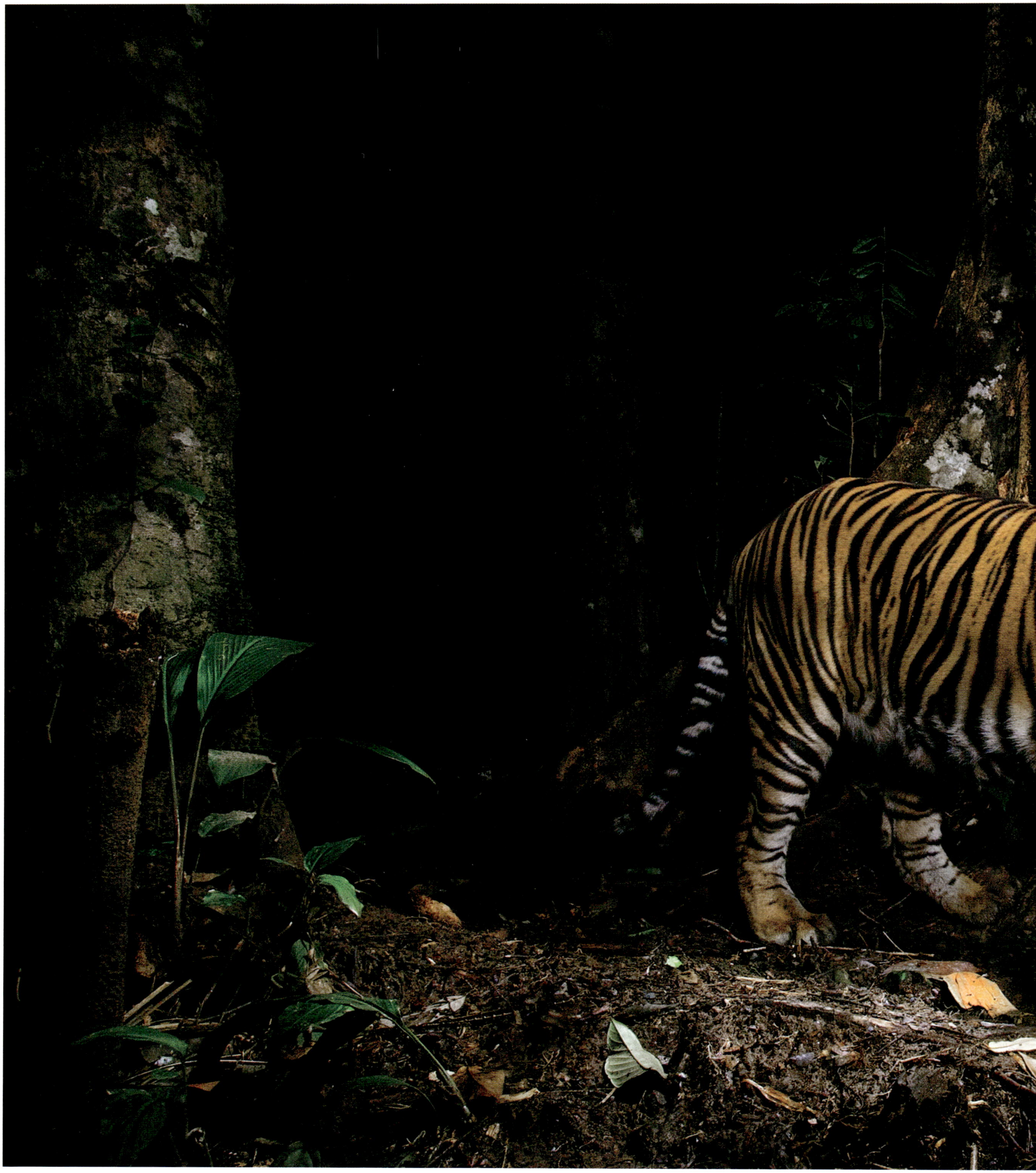

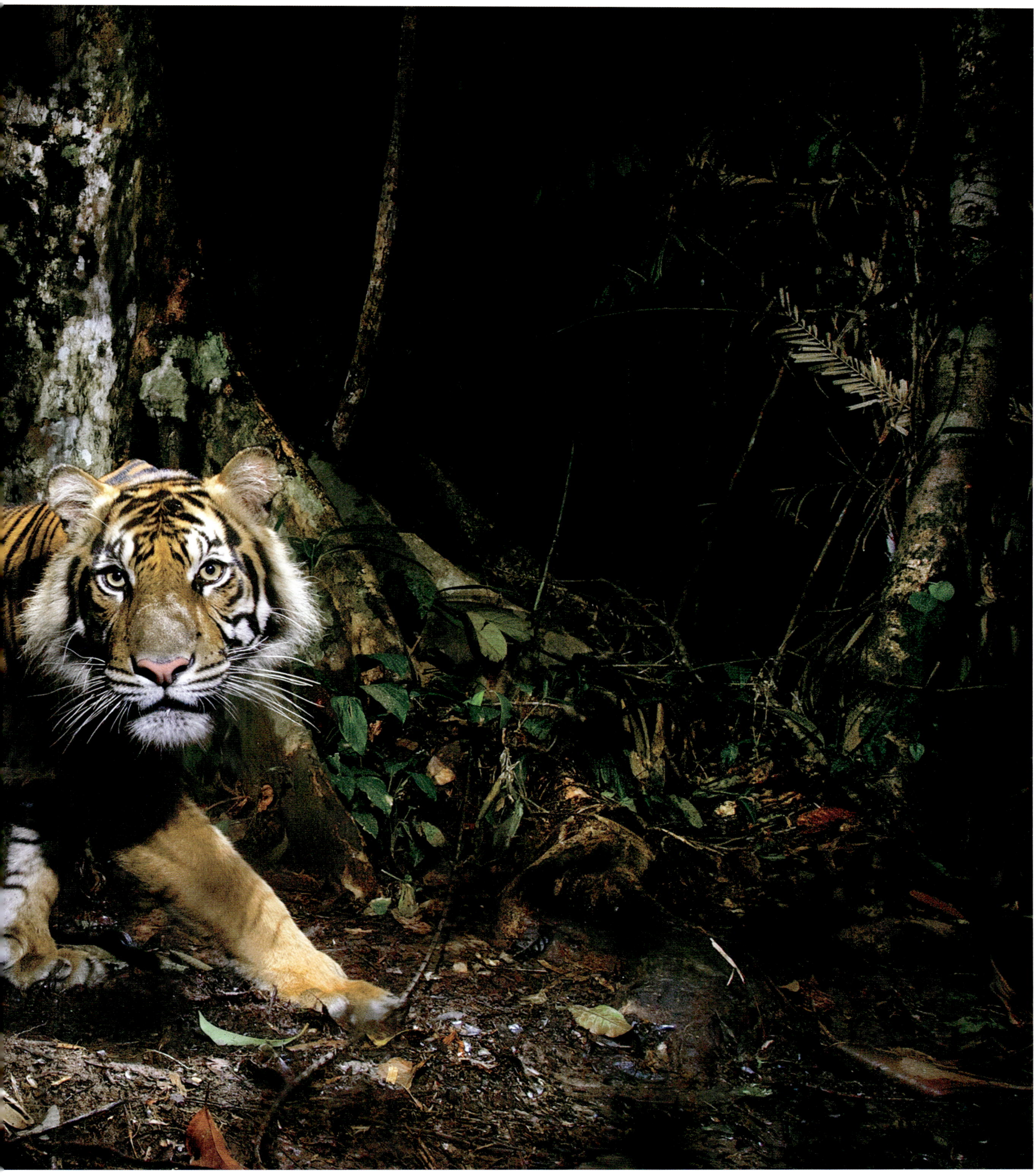

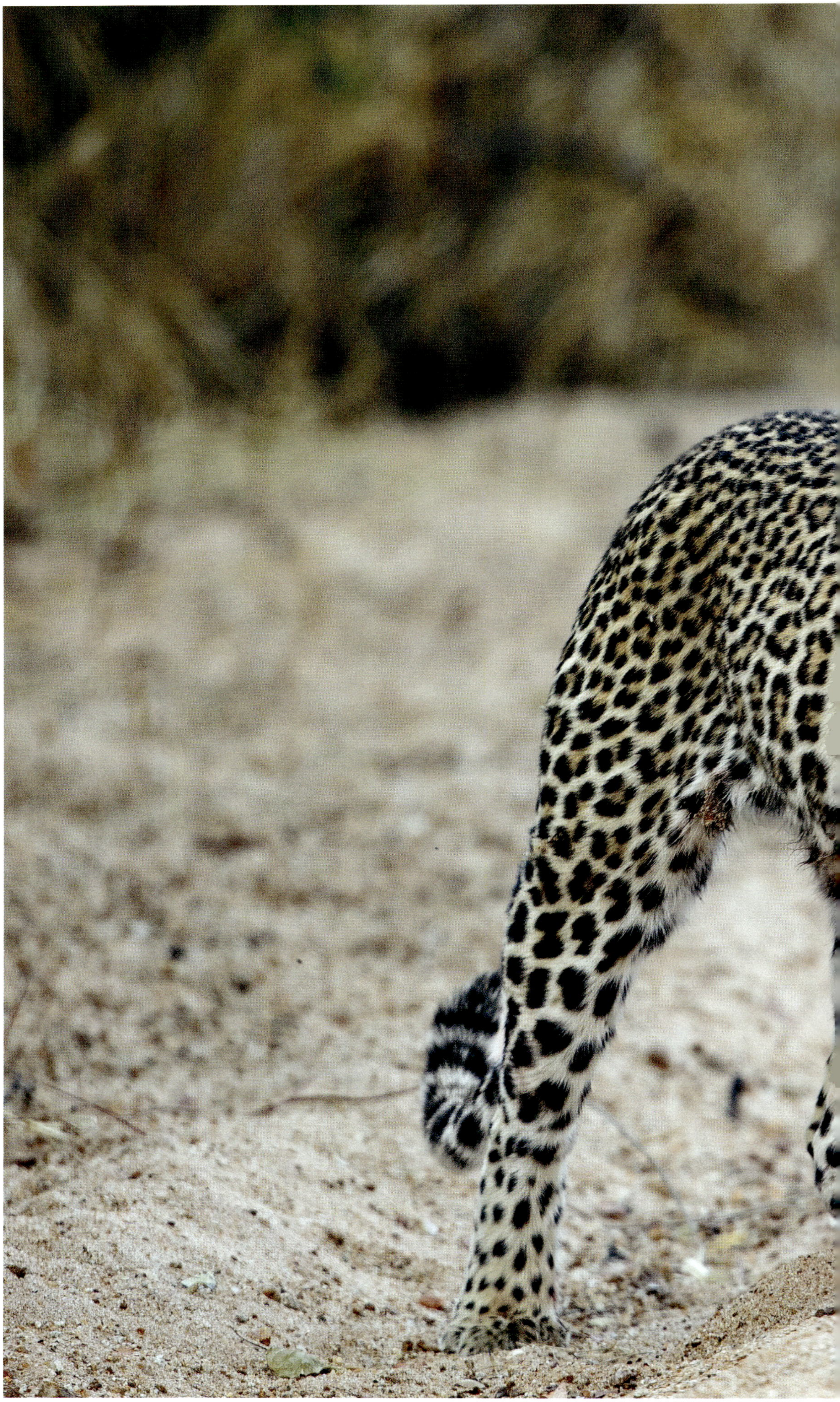

A leopardess walks with her cub in Sabi Sand Nature Reserve, South Africa. Leopards are the most persecuted big cat, but in this reserve, they are protected.

PAGES 10–11: A male Sumatran tiger peers at a camera trap he triggered in the forests of northern Sumatra. Due to lost habitat and poaching, these critically endangered tigers could be the next subspecies to go extinct.

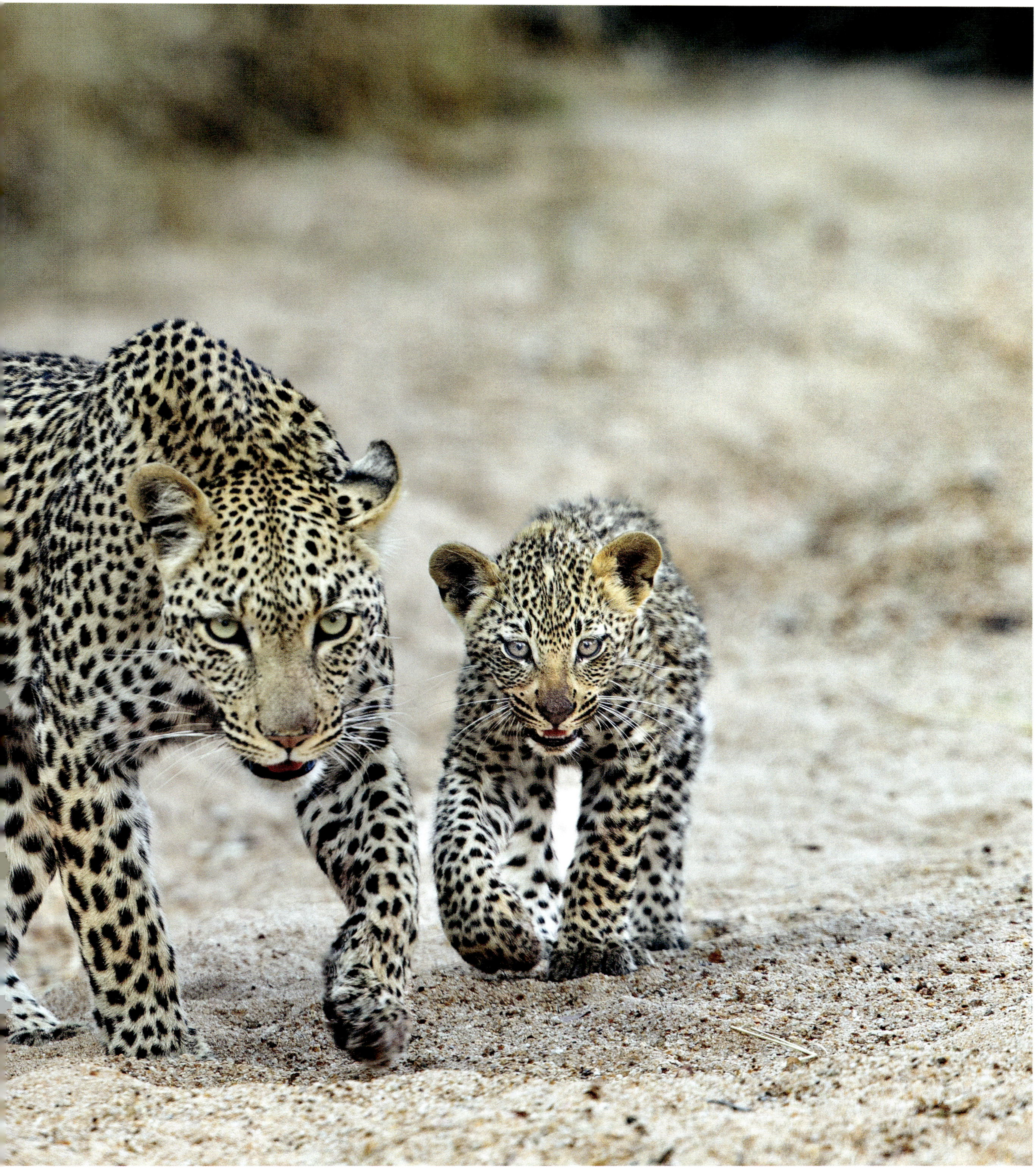

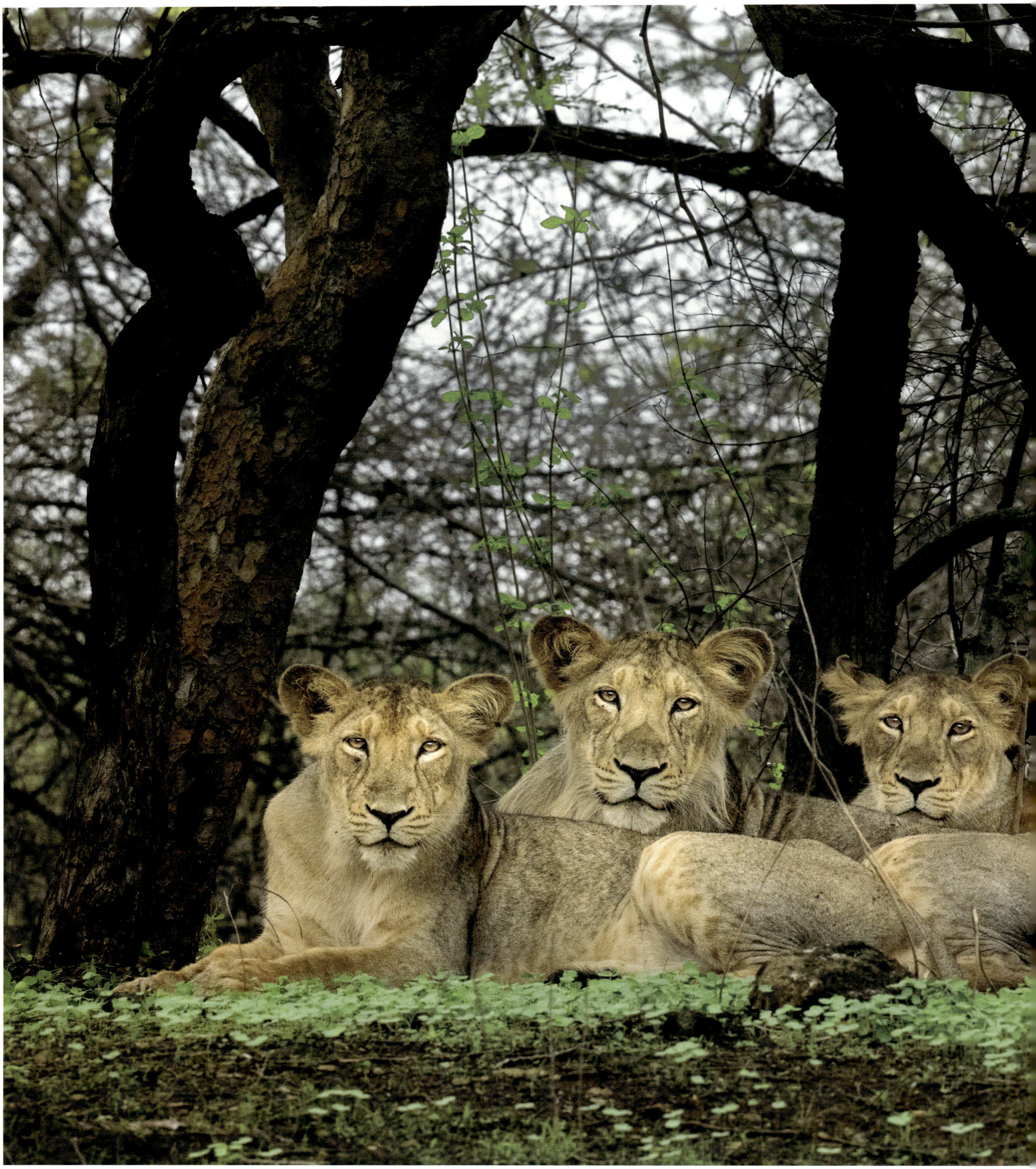

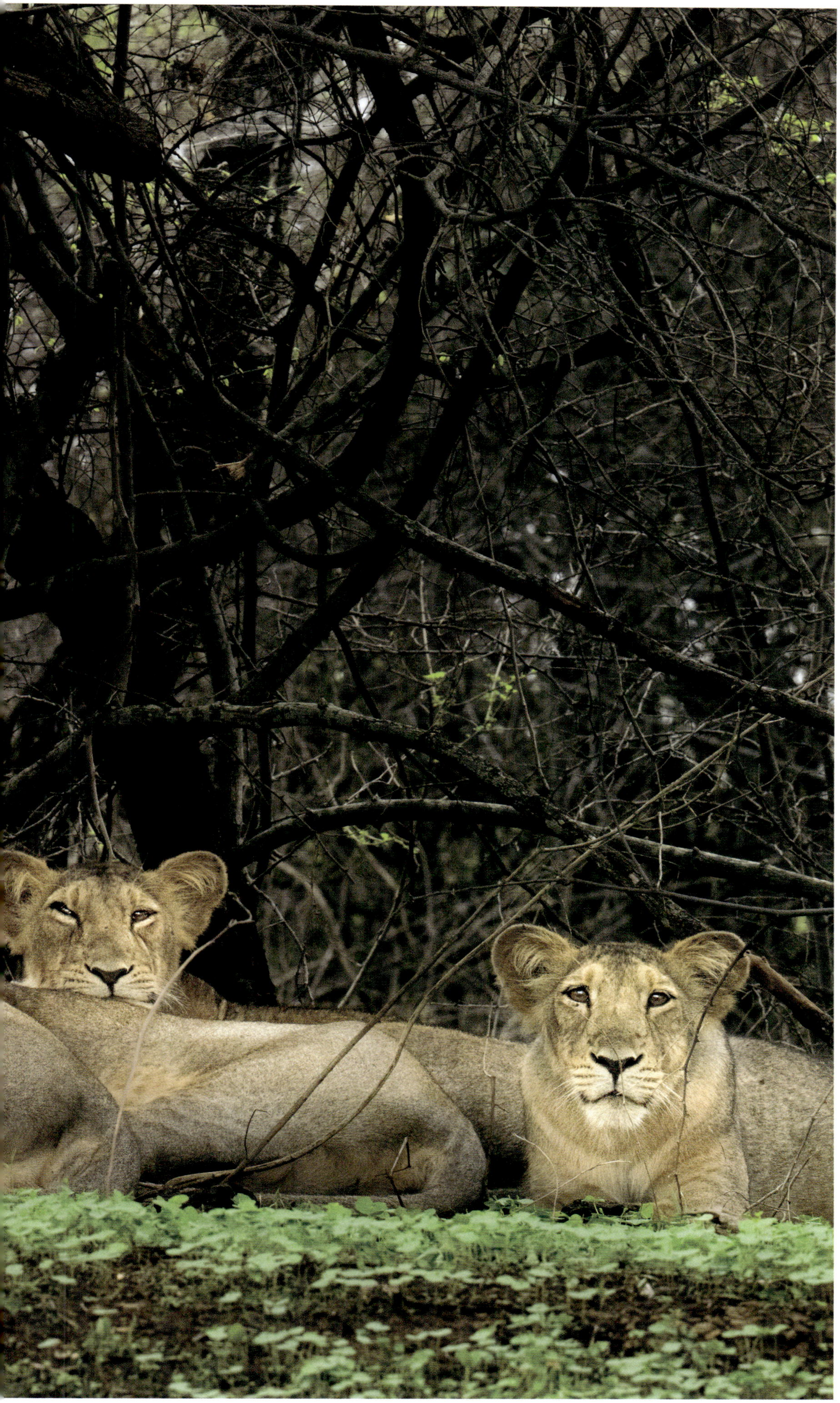

These young cats, photographed in India's Gir National Park, are among the world's last Asiatic lions. The population of this little-known subspecies has rebounded from a few dozen a century ago to nearly 900 today – but they need more space in a human-dominated landscape.

Contents

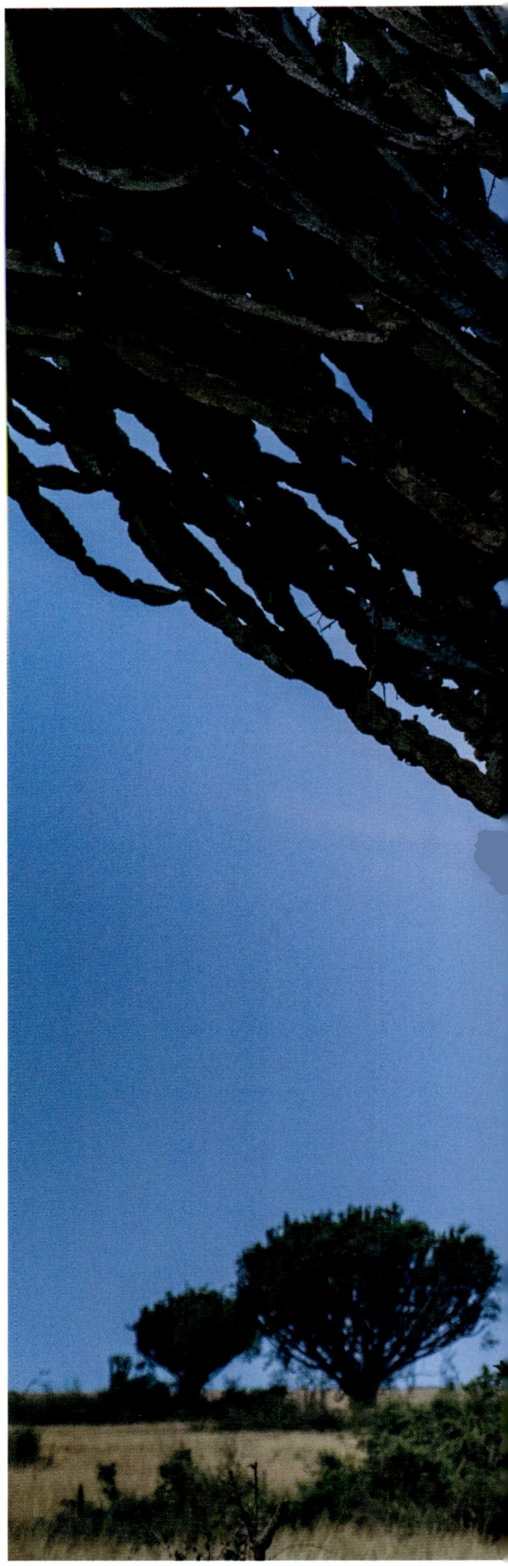

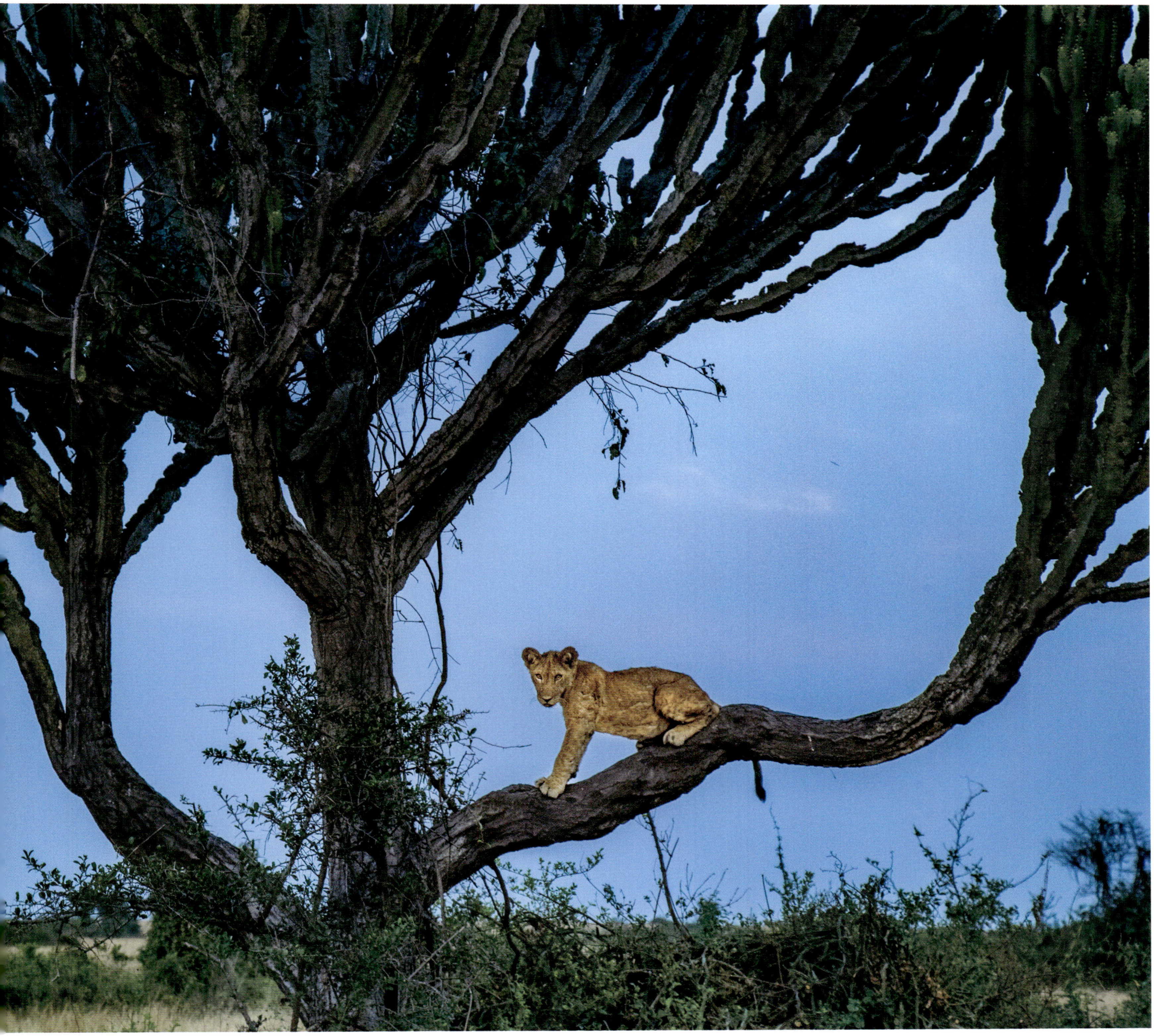

Lions, like this female, climb trees in Uganda's Queen Elizabeth National Park to escape biting flies and cool off.

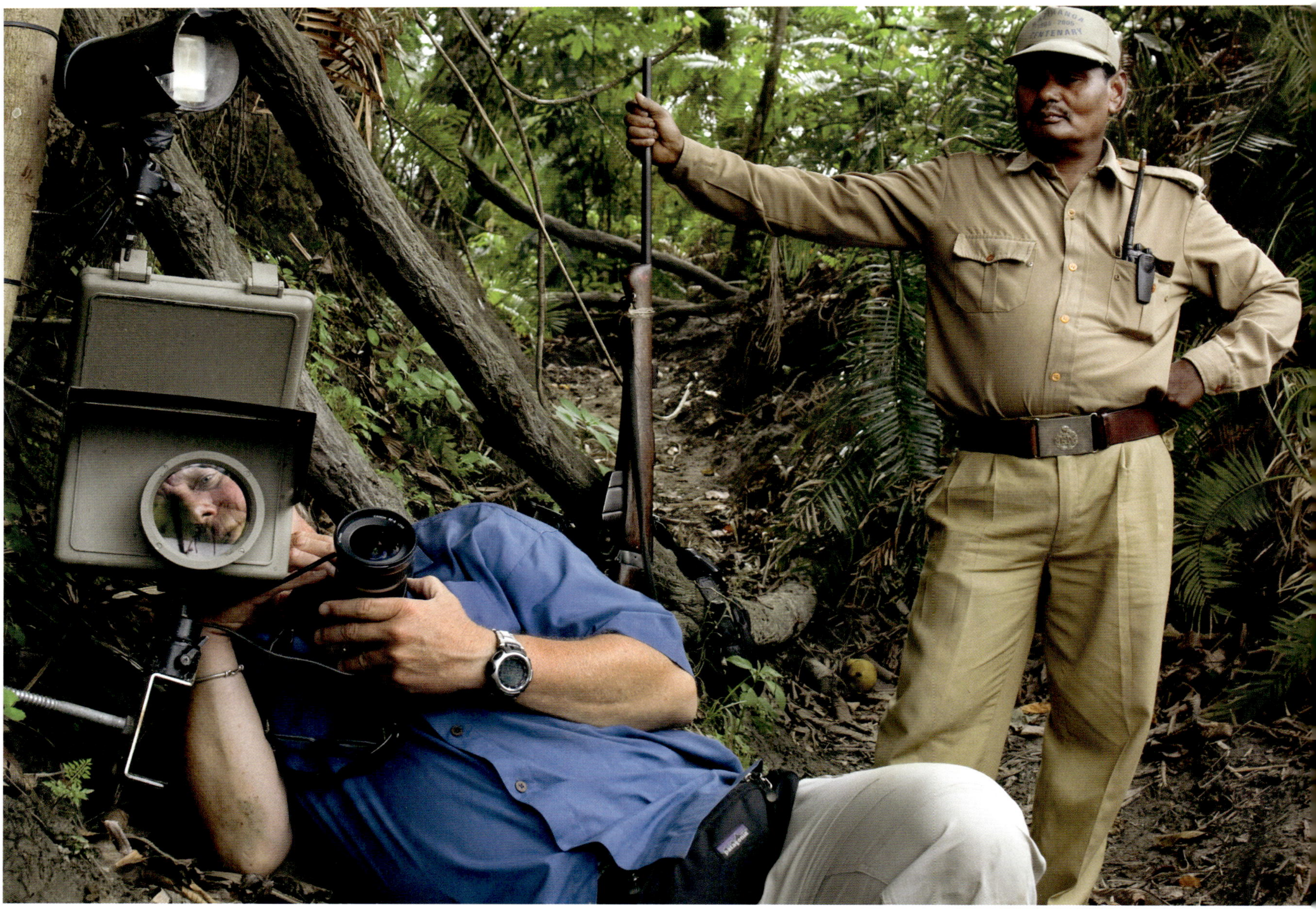

Under the watchful eye
of a guard in India's
Kaziranga National Park,
I set up camera traps to
capture images of Bengal
tigers, Asian elephants,
Indian one-horned rhinos,
and other animals.

It all began with a bird.

I was on top of a mountain, in the mist-draped Guatemalan cloud forest, working on my first feature for *National Geographic* magazine. I spent days in a hide, trying to photograph the resplendent quetzal: a bird with a crimson breast, shimmering iridescent green plumage, and an impossibly long tail, so majestic it was sacred to the ancient Maya and Aztec people.

The two men who helped conserve this forest had been shot by illegal loggers, so I was up there alone in a cabin. One night as I lay in my bunk, the stairs creaked outside. I was afraid that the poachers were after me, but then I heard sniffing and scratching at the door. I grabbed my walkie-talkie and radioed my contact, Juan Carlos. I told him what happened, and he laughed. "It's OK, Steve. It's just a *pantera negra*—a black panther."

"Why didn't you tell me about it?" I asked. I'd already been there for six weeks.

"I didn't want you to worry," he said.

The jaguar's presence explained a primal unease I sometimes felt walking through that towering forest, why the hair on the back of my neck stood up and I felt an inexplicable adrenaline rush.

A few days later, I was in my blind, photographing quetzals in a nest. The black jaguar crashed down through the branches of a nearby tree and disappeared into the fog, like an apparition.

And with that, cats chose me. My next project was on jaguars, launching a lifetime of photographs, stories, and films on the world's big cats.

I grew up in Indiana. My father gave me a camera when I was seven years old, and I pored over the pages of *National Geographic,* dreaming of traveling the world as a National Geographic photographer.

In my 20s, Michael "Nick" Nichols introduced me to the magazine world. I was his photo assistant for five years. But I didn't start out as a wildlife photographer. Early on, I covered news, wars, medicine, business, and science for a photo agency, Black Star. I didn't photograph an animal until I was 34.

At the time, most animal coverage centered on natural history. While capturing the beauty, intelligence, and unique behaviors of my wild subjects is crucial, I'm a photojournalist. I'm driven to tell stories. That means photographing the animals and humans they live beside, the lands they inhabit, the threats they face, the scientists who study them, and the conservationists trying to save them.

The world's wild cats are in trouble, their numbers dwindling as human populations skyrocket. I can't imagine a world without cougars, jaguars, leopards, snow leopards, tigers, or lions.

The words of George Schaller, the world's foremost field biologist and one of my heroes, continue to resonate: "Future generations would be truly saddened that this century had so little foresight, so little compassion, so lacked generosity of spirit for the future that we would eliminate [some of] the most beautiful and dramatic animals that the world has ever seen," he said. I want our granddaughter, Winter Rose, to live her life on a planet where these magnificent animals still roam.

The work collected here has had impact, bringing international attention to the plight of big cats, helping people coexist with them, protecting habitats, and passing laws. My cougar story sparked construction of the world's largest wildlife bridge. My wife, journalist Sharon Guynup, collaborated on some of these stories; together we shut down a tiger trafficking venue, the Thai Tiger Temple.

Big cats are umbrella species. Preserving the vast lands these wide-ranging predators need to survive saves the full spectrum of life they share the landscape with—entire ecosystems, living biomes that evolved in synchrony over millions of years. They also sequester carbon, fighting climate change. They generate the oxygen we breathe, source the water we drink, and maintain natural buffers against flooding.

I hope these images inspire you to think differently. Documenting the destruction of wild places has made me acutely aware of the need to prioritize protection, preservation, and health over the relentless quest for development. Only then can we save big cats and the entire array of life on this magnificent planet—and save ourselves.

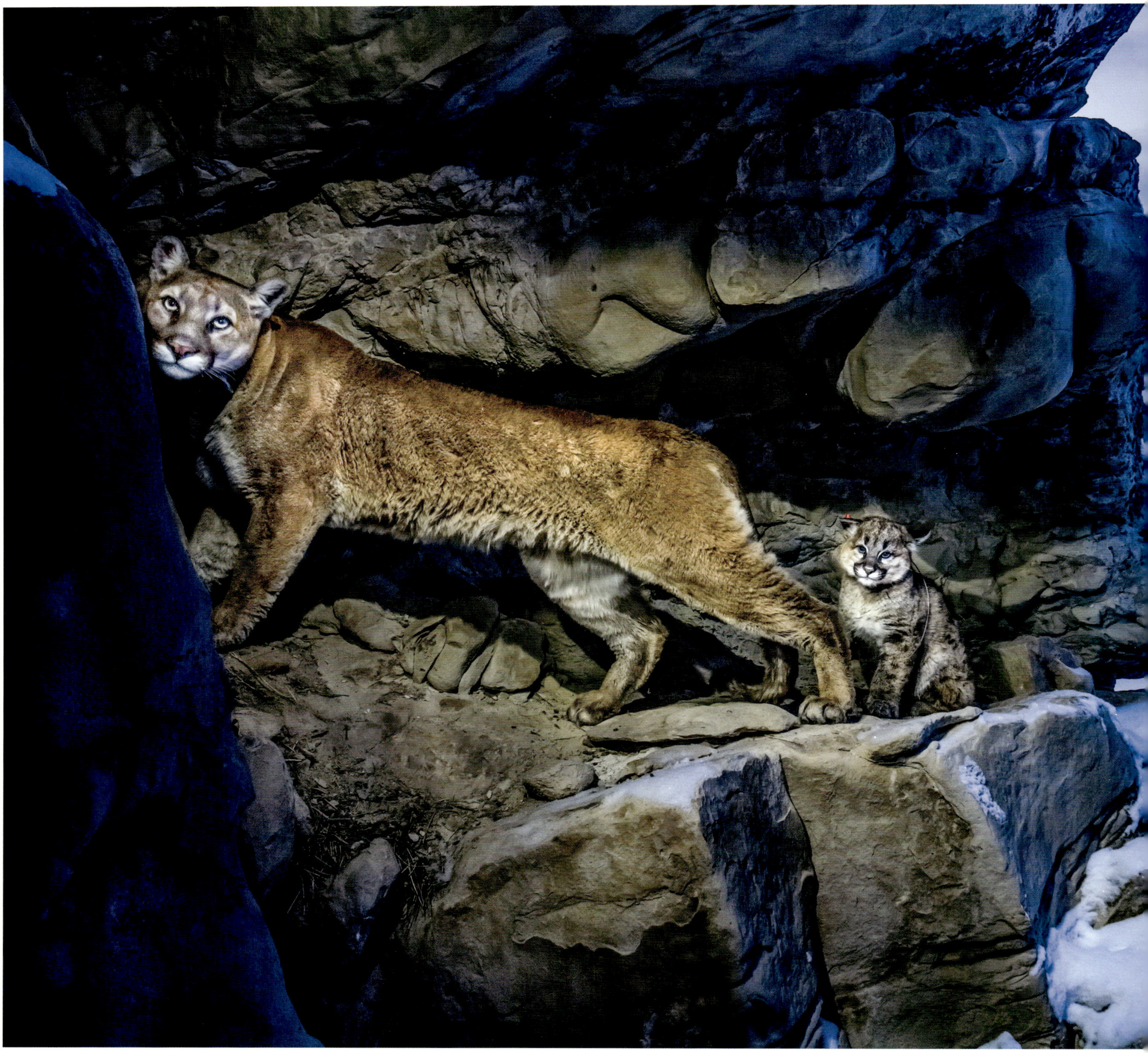

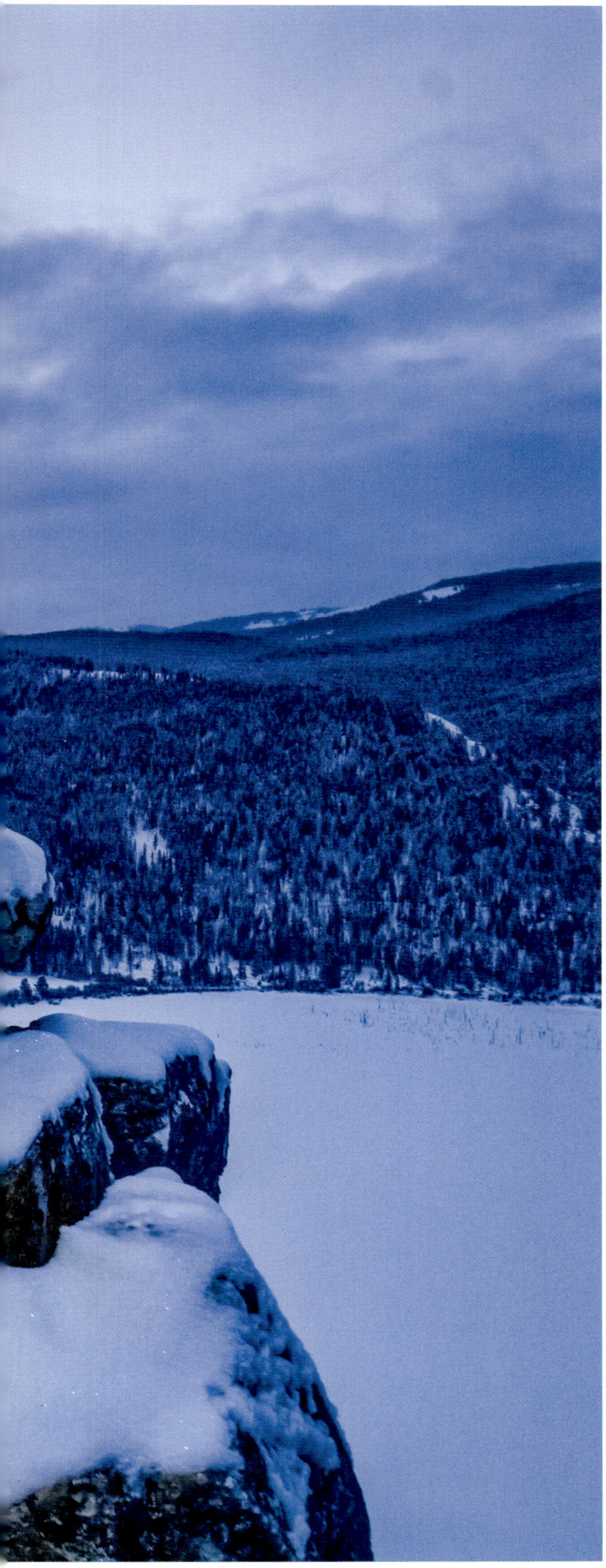

CHAPTER ONE

Cougars Ghost Cats and the Celebrity

Most cougars are hunted across their range, but one—P-22—became a star, inspiring Los Angeles to build the world's largest wildlife overpass.

A remote camera captures a GPS-collared mother named P-5 (Puma-5) and her kitten outside their den in Wyoming's Gros Ventre Mountains. Researchers tracked this female for a decade as part of the Teton Cougar Project.

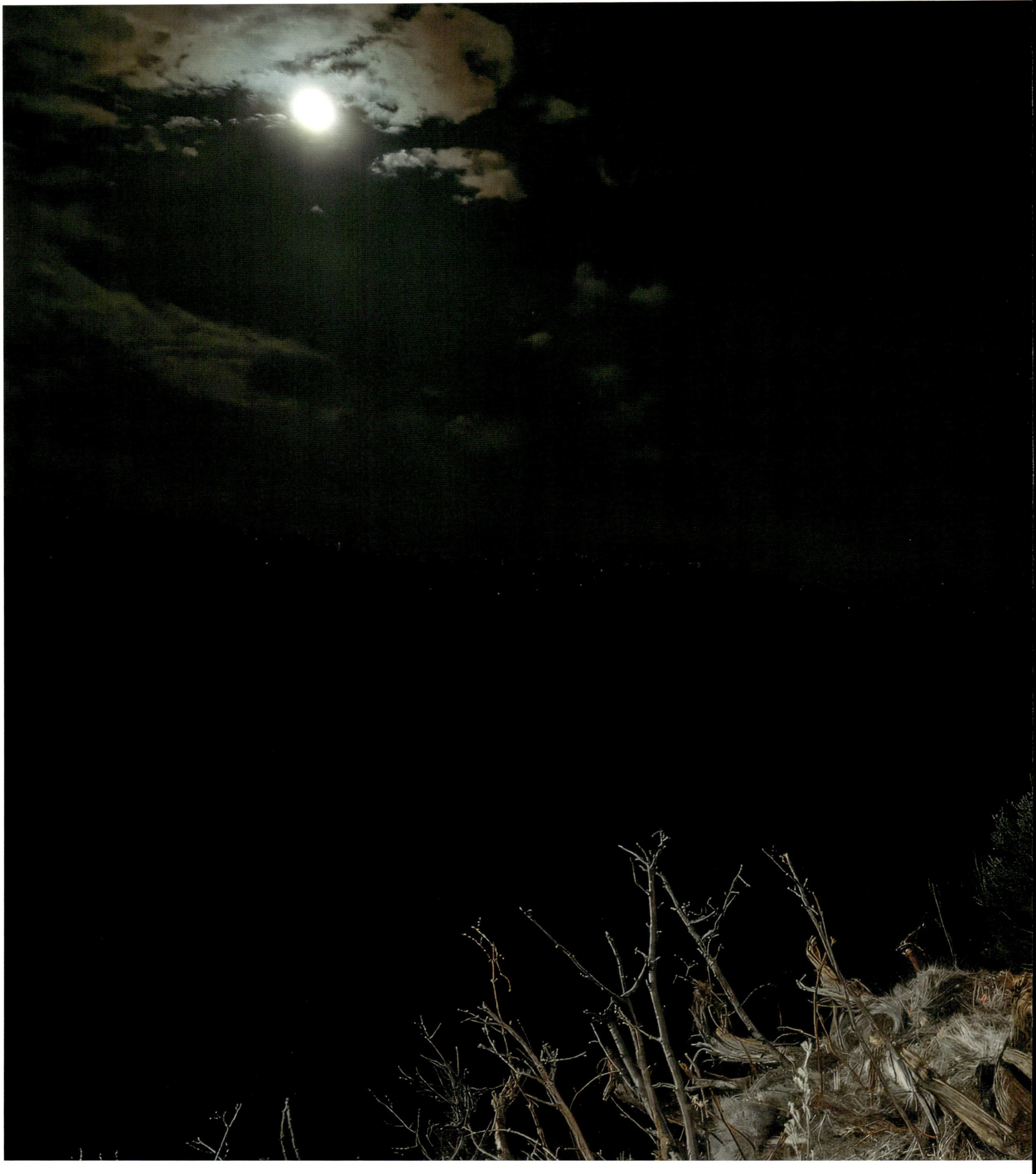

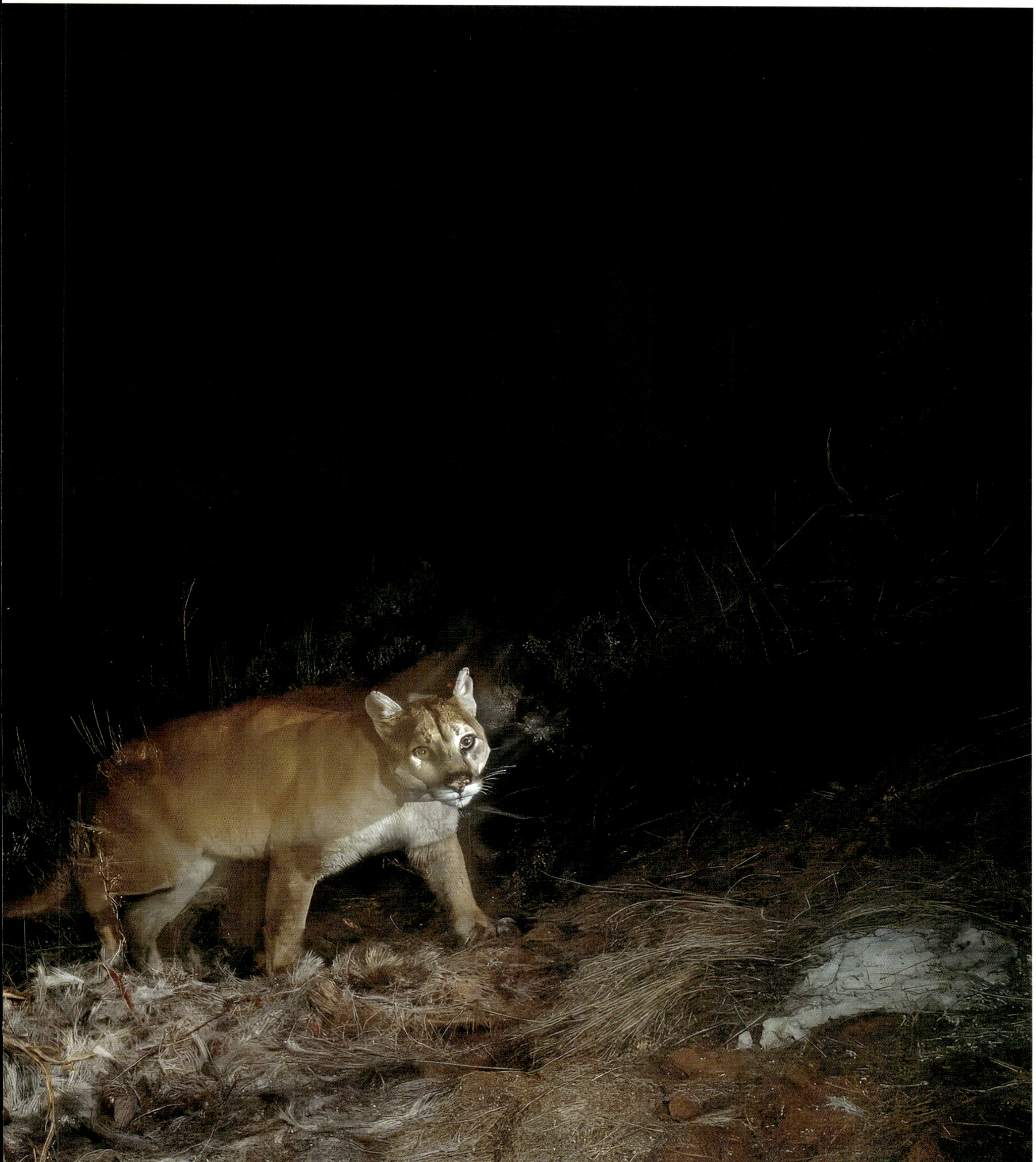

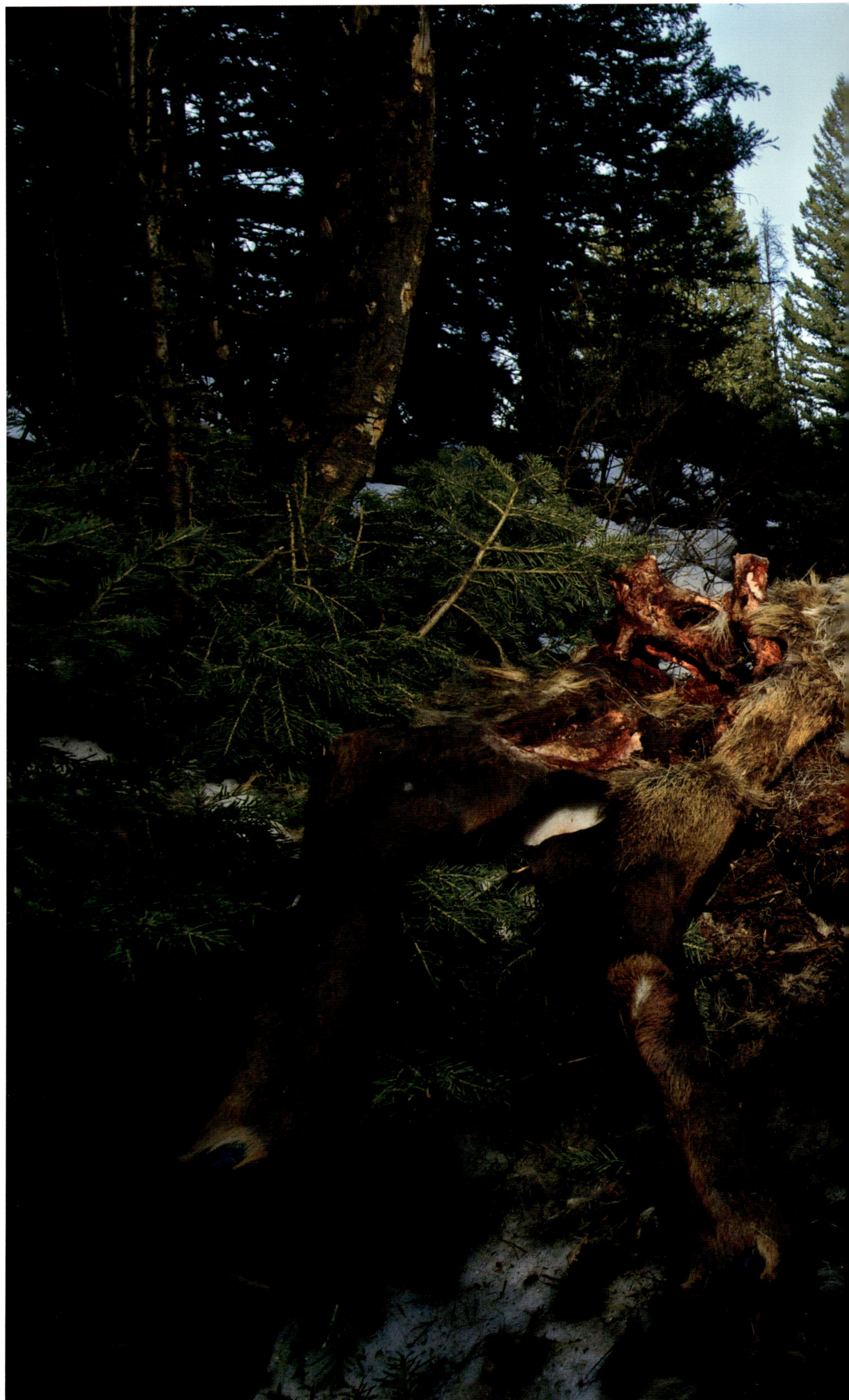

A four-month-old cougar kitten perches atop an elk her mother killed. The kitten survived a wolf attack that killed her two littermates. Teton Cougar Project researchers named her Lucky.

PAGES 22–23: To capture images of "ghost cats," I set up a remote camera on an elk carcass. Here, a GPS-collared cougar feeds under a full moon in Wyoming's Bridger-Teton National Forest.

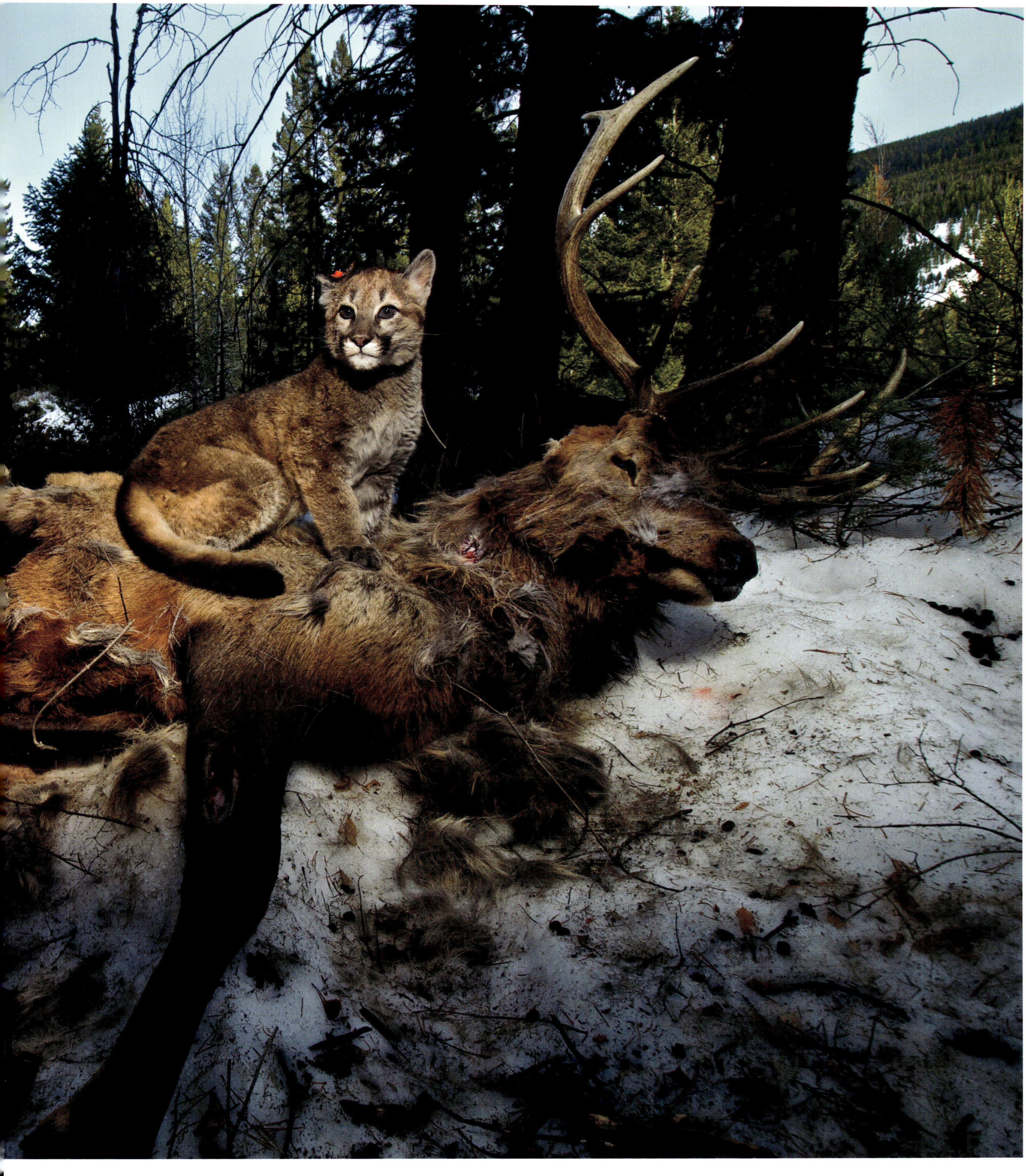

I'd been working on big cat stories

stories for more than a decade when I ran into my friend Howard Quigley at a conservation meeting in New York in 2011. He asked me why I'd never done a wildlife story in the U.S. "No reason," I said.

I knew Howard from his jaguar research in Latin America. He told me he was working with the Teton Cougar Project in Wyoming and I should come west to cover it. Cougars were behaving in ways biologists had never seen before. He attributed it to the reintroduction of wolves to Yellowstone in 1995—after 70 years of local extinction—which was reshuffling the entire ecosystem.

I pitched the story to my editor, Kathy Moran. It was quickly approved, and I jumped on a plane to Montana to attend a conference on cougars. One of the first people I met was Jeff Sikich, who tracked cougars in Southern California for the National Park Service. I thought his work with "urban" cats would add an important dimension to the story.

Of the 16 states that cougars call home, California is the only place they're not hunted. They have "specially protected mammal" status, though these stealthy cats, including one Jeff was tracking near an estate owned by Cher in Los Angeles, roam heavily populated areas.

I started asking questions. How could I show cougars living in the sprawl of L.A.? "Not sure," Jeff said. "They tend to avoid people." Do they walk on certain trails? "They definitely use game trails and hiking trails," he told me. That was hopeful. Is there a place where we might see the lights of the city behind a cat? "Possibly." He'd found a couple of places cougars moved along ridgelines in the Santa Monica Mountains—from there, you could see the 405 freeway and the city in the distance.

Finally, I described my dream shot. "You know what could *really* show this?" I asked. "A picture of a cougar under the Hollywood sign." Jeff looked at me like I was crazy. There were no cats in Griffith Park, where the sign is, he said. But I remembered that there used to be, back when I was a kid. My grandparents lived nearby, and they talked about cougars when I visited them during my summer vacations.

Though cougars rove across most of the New World—from the Canadian Yukon to Argentina's Andes Mountains—I decided to focus my story strictly in the United States, starting in Wyoming, where the Teton Cougar Project was nearing the end of a 12-year study. Then I'd go chase my dream picture in California, no matter how delusional it sounded.

This is an extremely adaptable feline that thrives anywhere: in steamy jungles, frigid peaks, mucky swamplands, and arid deserts.

Depending on where you encounter *Puma concolor,* it could be called a cougar, catamount, mountain lion, mountain demon, puma, panther, painter, or any of the 40-some names it goes by. It's also called "the American lion," but cheetahs and jaguarundis are its closest living relatives. These members of the Felidae family evolved from the same common ancestor some five to nine million years ago.

Perhaps the most accurate name is "ghost cat": The cougar is secretive and rarely seen everywhere it prowls; it poses little danger to humans. Over 156 years, from 1868 to 2024, there were 29 fatal attacks by cougars in North America. You're over 100 times more likely to be killed by a lightning strike.

But cougars have always been in the crosshairs in the U.S., hunted out of fear, shot as trophies, or killed in retaliation for livestock they occasionally prey on. During the 19th and 20th centuries, bounty hunters helped exterminate them from the eastern two-thirds of the country, and they were pushed to near extinction elsewhere. The survivors hid in desolate Southwestern deserts and the high mountains of the West. They still hang on in 16 states.

My assistant, Drew Rush, and I flew to Jackson, Wyoming, in November 2011. The jagged Rocky Mountain peaks and glacier-gashed valleys were cloaked in white, making it prime time for cougar research. With snow on the ground, researchers could follow paw prints, find the cats, and outfit them with GPS satellite collars to track their movements.

That first day on the ground in the Gros Ventre Range I saw what we were in for. Big cat tracker Boone Smith and his dogs led the way. We trekked up megasteep ridges, some towering to 11,000 feet. I live at sea level, and we'd had no time to adjust to the altitude. I could barely breathe. The temperatures hovered at or below zero.

We had beginner's luck. That afternoon, the pack treed a cat. The team's vet quickly shouldered her rifle and shot it with a tranquilizer dart. The next minutes were tense: Occasionally an animal bolts and hides before the drugs take effect.

Boone shinnied up the tree. He managed to loop a rope around a branch and around the cat's back leg without getting mauled. On the ground, five people lowered the sedated cougar, headfirst, then gently laid it in the snow.

Biologists used dogs to chase this cat into a tree, where they tranquilized the cat and collared it to monitor its movements with GPS.

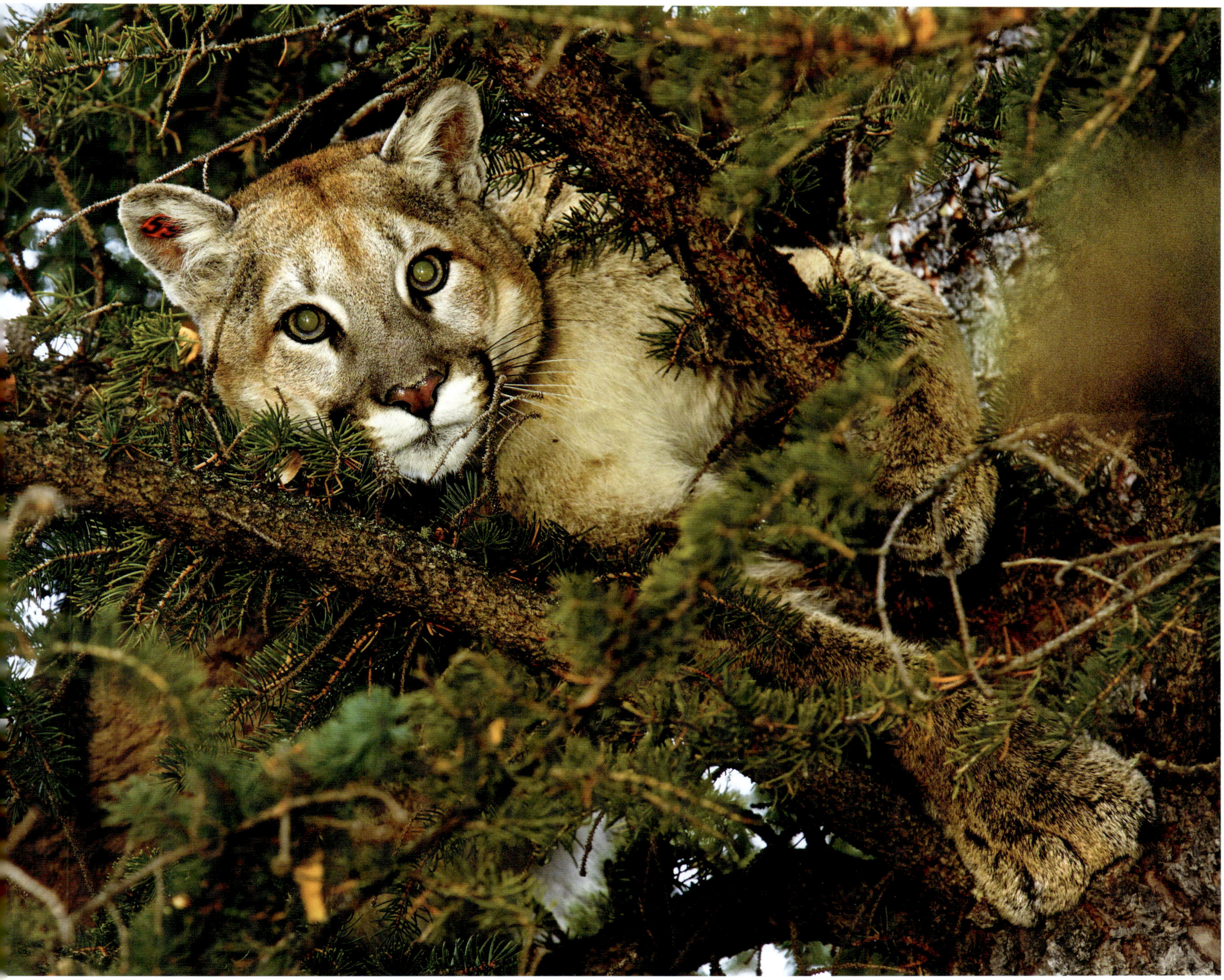

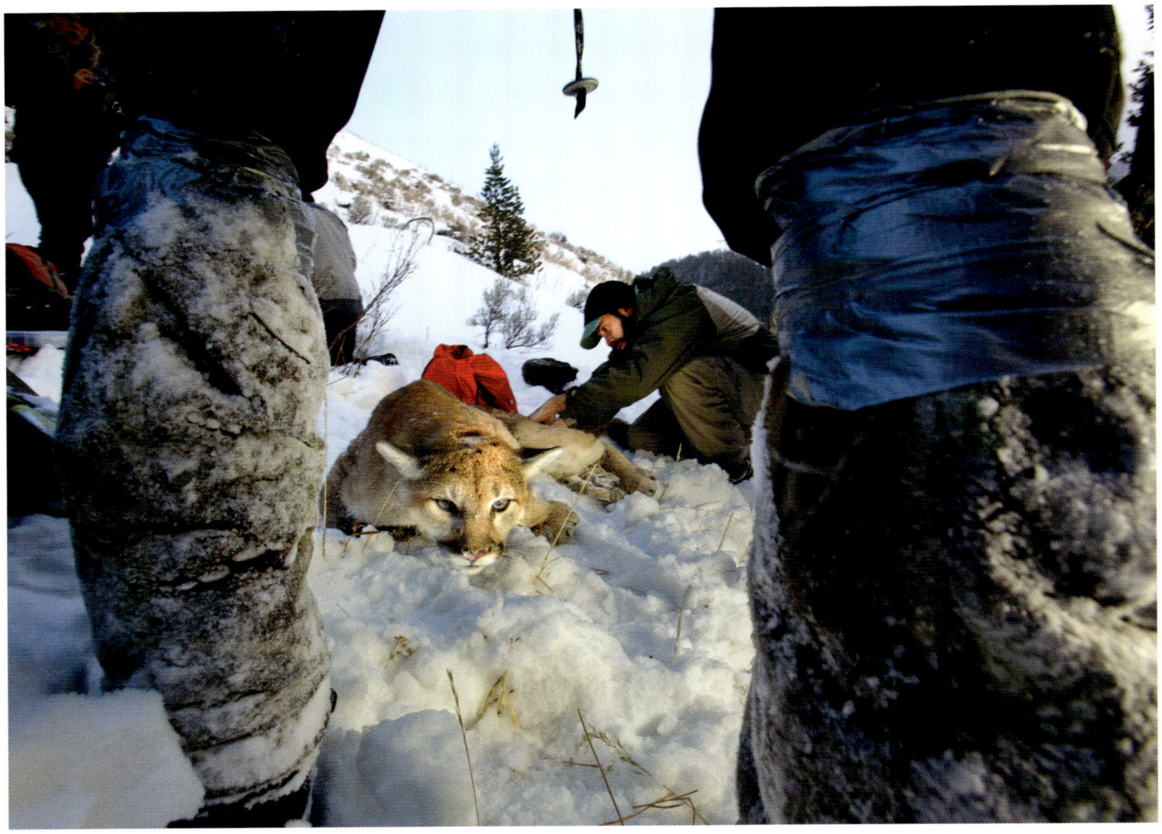

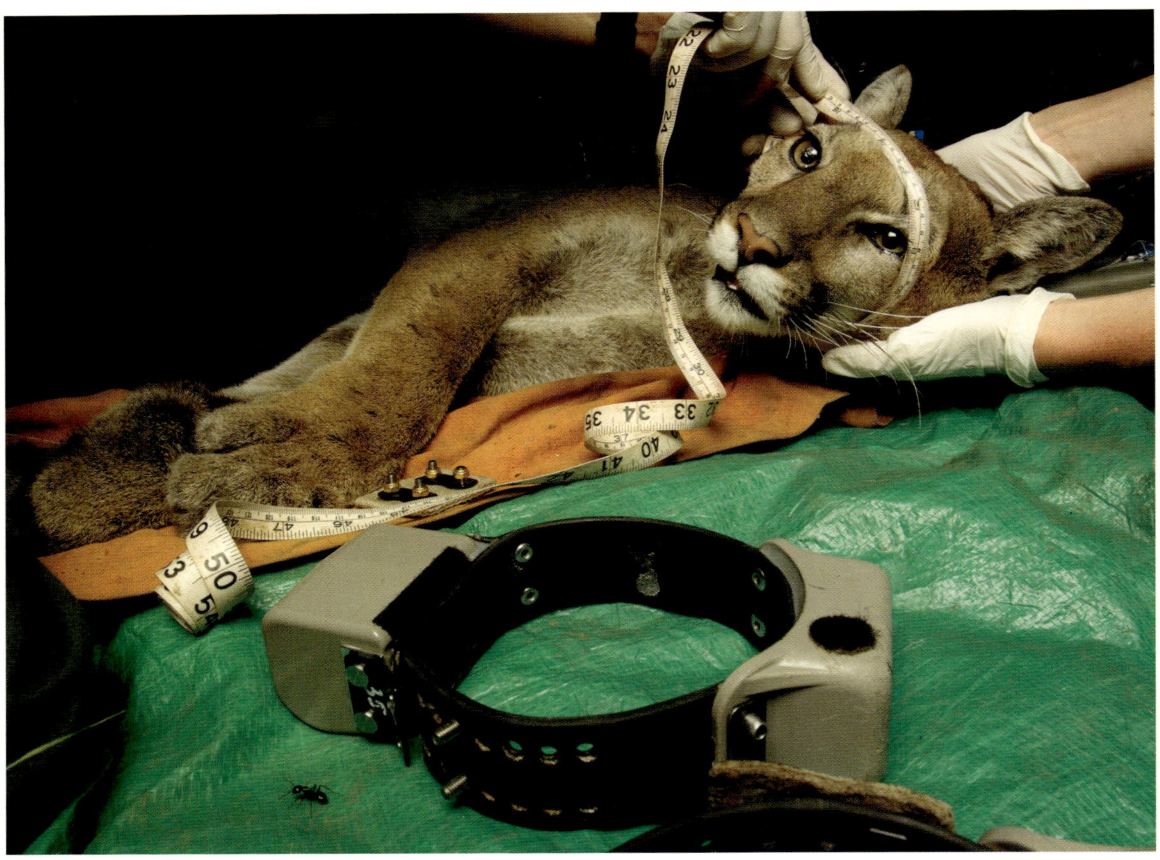

ABOVE: Members of the Teton Cougar Project measure, weigh, and run quick medical tests on a sedated cougar before fitting it with a GPS collar. Understanding the cats' range and habits provides critical data for future conservation decisions.

LEFT: Biologists measure this pregnant female before replacing the battery on her tracking collar and releasing her in the hills of Southern California. First captured as a kitten, she has provided data to scientists nearly all her life.

It was a young male, small but powerfully built. The team donned latex gloves and worked with the speed of a pit crew. The cat stared open-eyed: The drug affects an animal's blink reflex, so the vet applied salve and shielded his eyes with a bandana. Then came a quick health check. They collected blood, urine, and fur samples and hefted the cat into a sling to check his weight: 100 pounds. Then they fastened a GPS collar around his neck.

If all goes well, these collars transmit multiple daily locations for two years, showing where the cats travel, how much space they need, what they eat, how cubs fare, and other information that informs conservation efforts. The team captured other cougars, including two cubs they fitted with expandable collars. I photographed it all.

Each morning felt a little like Christmas as the researchers downloaded overnight collar data, tracking cougar movements in near real time from their field office. A cluster of map points meant a cat was on a kill, and we'd trudge up there—it always seemed to be up. Maybe the cats heard us coming, but we never encountered one face-to-face. We set up motion-triggered camera traps and flashes in waterproof housings next to elk carcasses and one at the mouth of a cave frequented by a mom and cub. Sometimes the only way to photograph notoriously shy animals is to simply not be there.

During four months of camera trapping in Wyoming, we documented new cougar behaviors. This is an evolving ecosystem, changes that began when 14 gray wolves were reintroduced to their former range in Yellowstone National Park some 16 years before. These wolves hunted an exploding elk population back into balance, and the herds that remained grew wary, no longer grazing in exposed valleys or along riverbanks. Forests regrew. Rivers changed course. Beavers, foxes, and hawks returned. Small mammals flourished.

This domino effect altered cougar habits in ways scientists were just learning about. When cougars made a kill, wolf packs often moved in, taking over their hard-won dinner. To fight them off, female cougars were priding up, like lions, with both relatives and unrelated neighbors. I documented multiple females sharing kills. Cougars were considered solitary animals, together only when mating or raising young. Was this new social structure a survival tactic, or a return to the way cougars had behaved before the wolves disappeared?

I was dreading the next part of the story—hunting—as I boarded a flight to South Dakota. Cougars are legally hunted for sport in 13 U.S. states, classified as game animals (or "varmints" in Texas) and the old Western frontier mentality reigns: Predators are often considered an enemy. Many ranchers can't fathom the concept of coexistence. Hunters hate cougars too, (wrongly) blaming them for wiping out prized game like bighorn sheep and pronghorns. With some 300 cougars in the state, South Dakota issued hunting licenses for 100 that year.

I joined a bowhunter in the Black Hills. Spring comes late there, and I fell hard walking across a frozen river. I limped through the rest of the day as we followed a cat, sometimes glimpsing it at a distance. The hunter never took a shot.

But he'd had recent success. He invited me to come with him to a taxidermist to pick up his stuffed trophy. I shot candids of him with the trophy. Then the owner opened a freezer and asked if I wanted to photograph a cougar that hung inside, upside down. Sometimes this job is brutal. Both images ran in *National Geographic* magazine.

Back at home a month later, I went to the doctor. My knee hurt. "Whatever you've been doing is working," he said. "The break is healing quite nicely." It seemed I'd fractured my tibia in South Dakota.

Wildlife photography is an extremely physical job. You haul heavy gear, work in harsh terrain for absurdly long hours, seven days a week. You deal with bugs, mud, leeches, heat, cold, and you tough it out. It's part of the gig. The pressure is intense: No matter what, you have to get pictures, images that have never been seen before.

Everyone asks me if I've been attacked by dangerous animals, and yes, there have been some close calls, but it's the microscopic ones that'll kill you. Karen Barry, National Geographic's nurse, recently showed me my three-inch-thick medical file. I'd had larvae that hatched under my skin, moving in a circle around my belly button, and a botfly larva that grew in my head—it felt like I was getting stabbed when it dug its way out. Sandwiched between pages about amoebas, giardia, Indian tick fever, and malaria was a letter from the Centers for Disease Control and Prevention warning that I mustn't get dengue fever again—it could become hemorrhagic and has no cure. I've now taken so many antibiotics, I'm allergic to most of them.

I'd been home a couple of weeks when a text from Jeff popped up on my phone. "CALL ME NOW!!!!!!!!!!"—which of course I did. "You're not going to believe this," he said. "Someone just got a trail cam picture of a cougar in Griffith Park."

It was a young male, and somehow, he'd navigated two of the nation's busiest freeways, the 405 and the 101, to reach the park, likely following a biological imperative to find his own territory. Jeff searched for tracks, and eight days later, he trapped and collared the cat. He was the 22nd puma captured in a National Park Service study. Jeff named him P-22.

The Park Service budget was tight, so I reached out to Rebecca Martin (who ran National Geographic Society's grant program) and asked for funding to buy three new, state-of-the-art satellite collars. Jeff soon had them. He re-collared P-22, and I flew to California.

This was a new challenge. I'd never set up camera traps in a city before. P-22's new home, 6.7-square-mile Griffith Park in the middle of Los Angeles, had 10 million visitors a year: hikers, bikers, golfers, picnickers, and those visiting the zoo. How do you set up cameras in an urban park without getting them stolen? I hired a welder, who made 70-pound camera boxes from quarter-inch steel and metal casings for the flashes. We attached everything to metal poles that we drove into the ground with a sledgehammer. Jeff pointed out possible locations for my five traps. The challenge was showing that P-22 lived in a city. It's completely against National Geographic's (and my own) ethical standards to bait or lure an animal to get the shot. I expected him to cruise through in the dark, when cougars are most active. I set the cameras for long exposures: I didn't want a brightly lit cat on a black backdrop.

P-22 soon made a fleeting appearance—a single bad frame. I had to keep moving the cameras because he kept changing trails. This was unnatural habitat: A male cougar's territory traditionally encompasses up to 500 square miles, and he was cramped within a tiny park.

I was getting hundreds of noncougar pictures. Racoons. People walking their dogs. A bobcat. People posing for photos they would never see. Deer. One guy who mooned the camera.

Then two of my heavily armored traps got stolen and someone made an attempt on a third. They were insured, but losing the memory cards was tough. I'll never know if I lost a great photo—or 20.

It took a year to capture an image of P-22 gazing into the lens with the lights of L.A. behind him. Sometimes there's crazy synchronicity in the timing of events, and this was one of those moments. An exhibition commemorating National Geographic's 125th anniversary was hanging at L.A.'s Annenberg Space for Photography. My editors broke protocol and gave the *Los Angeles Times* permission to use the P-22 photo, though it wouldn't run in the magazine for months. The *Times* put it on the front page, and Angelenos met P-22. The National Park Service office was flooded with calls. Some were from concerned citizens, but most people were excited. Awed. Fascinated.

I needed more pictures. Looking at points on Jeff's computer, we saw that P-22 had begun using an animal trail not open to the public. This was perfect, because no one would ransack my traps.

P-22 walked by, triggered the camera—and the lighting wasn't good. I flew back to California and added another flash.

Then, at the 15-month mark, I got my dream picture: P-22, walking at night, with the Hollywood sign on the hill behind him. The issue containing my cougar story was at the printer, and the editors literally stopped the presses to include the photo. It made P-22 a celebrity in the land of celebrities.

L.A. fell in love with their cougar. Murals appeared across the city. Occasional P-22 sightings made national news. The Natural History Museum of Los Angeles County installed a P-22 exhibit. Local schools adopted urban wildlife into their curricula. Every year in October, L.A. celebrates P-22 Day with a festival. In 28 years working with big cats, I've never seen a wild animal so profoundly transform a community.

But P-22's urban diet poisoned him. Rodenticide is used everywhere in an ongoing war with rats, even in California's most glamorous neighborhoods, and it bioaccumulates up the food chain. Jeff had to capture and treat P-22 for mange.

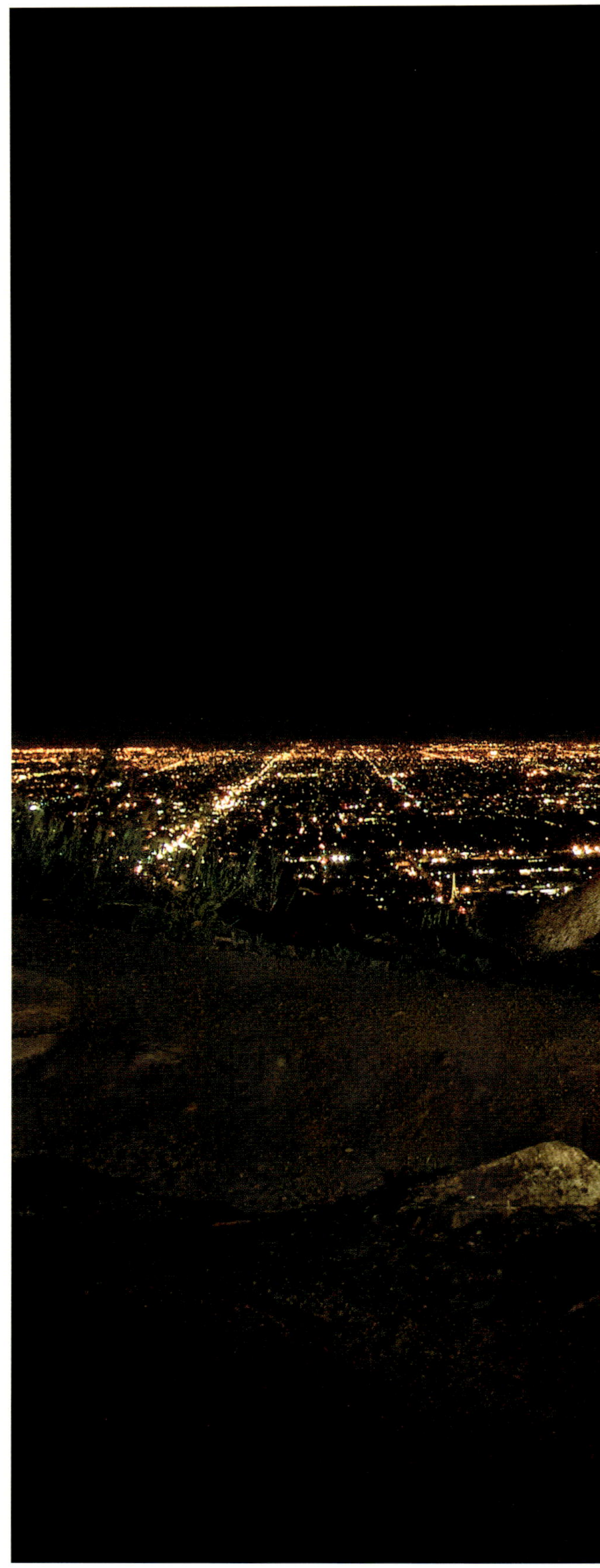

P-22 (Puma-22), a male cougar who crossed California's busy 101 and 405 freeways searching for his own territory, settled in Los Angeles's Griffith Park. He's photographed here with the lights of the city glowing behind him.

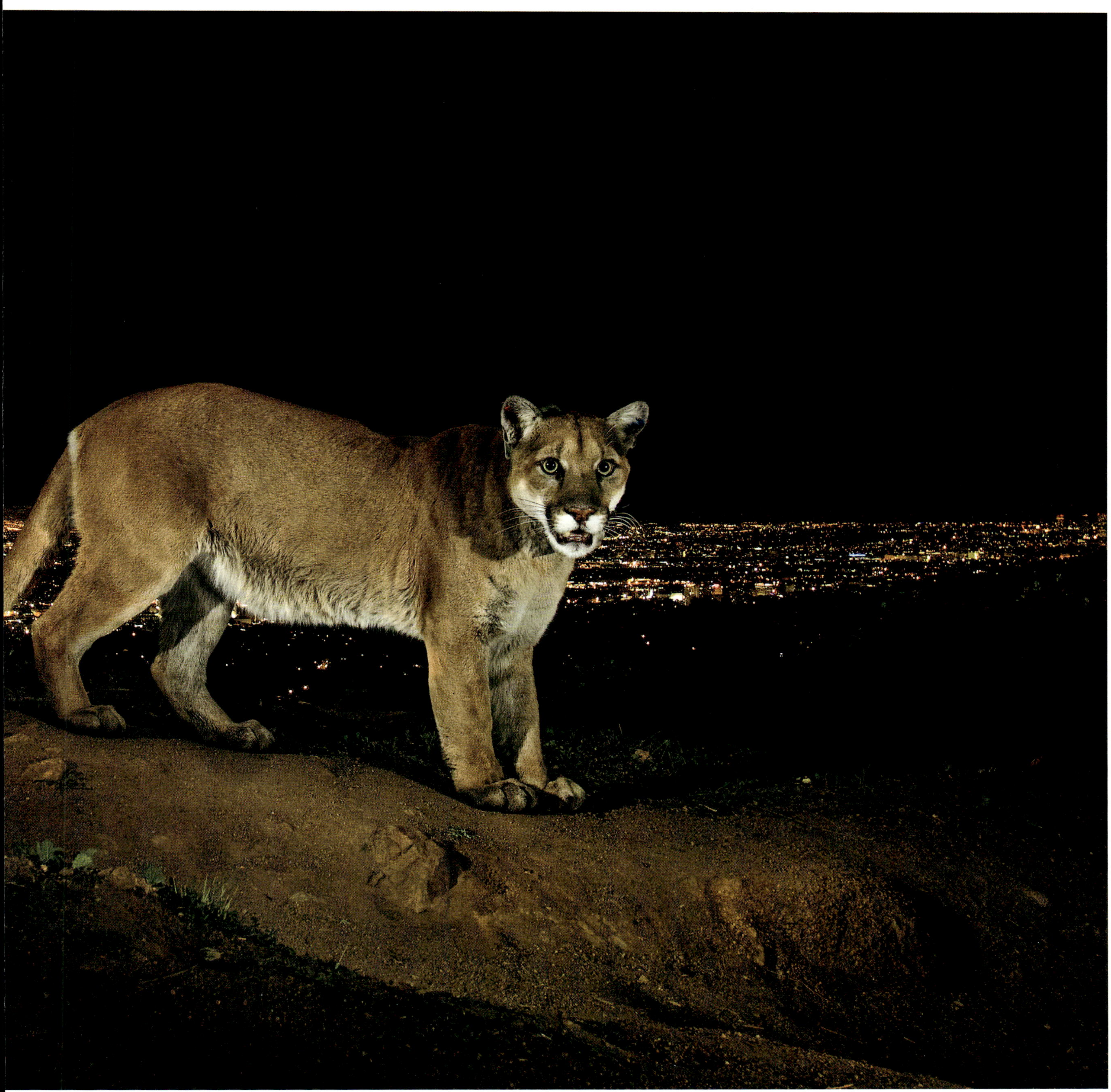

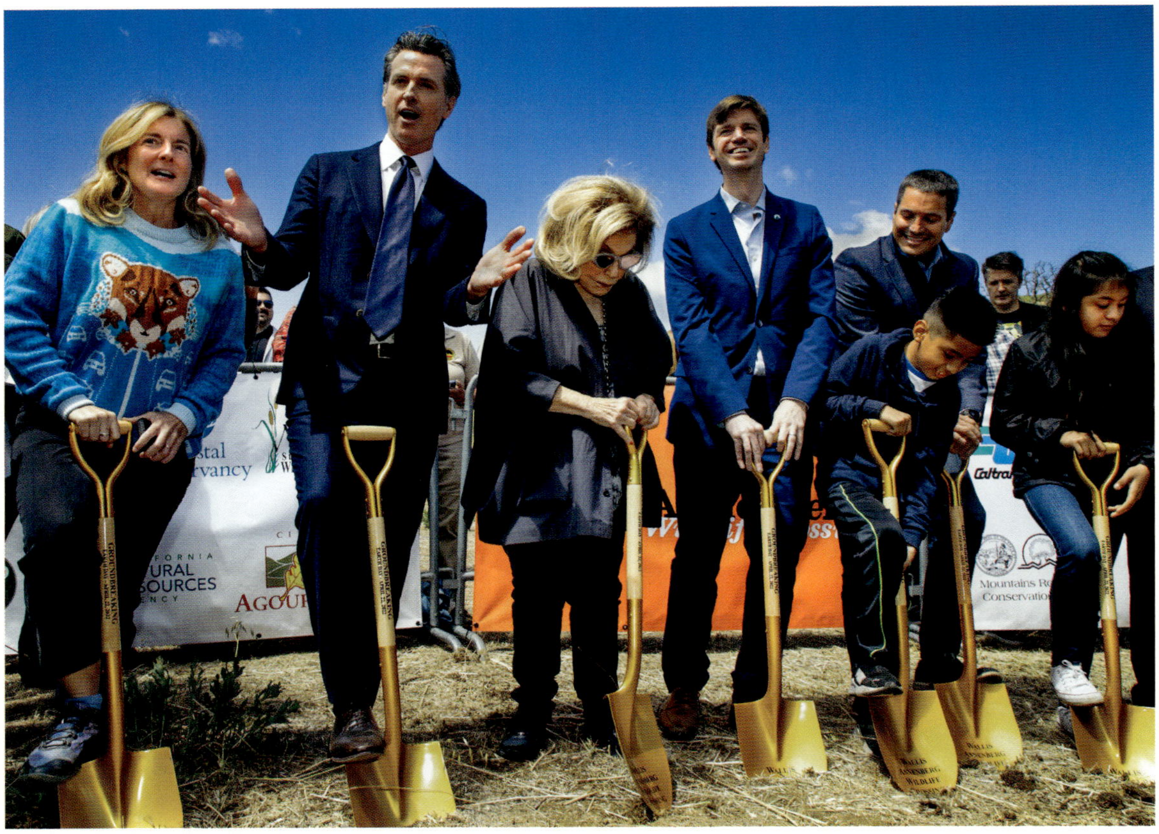

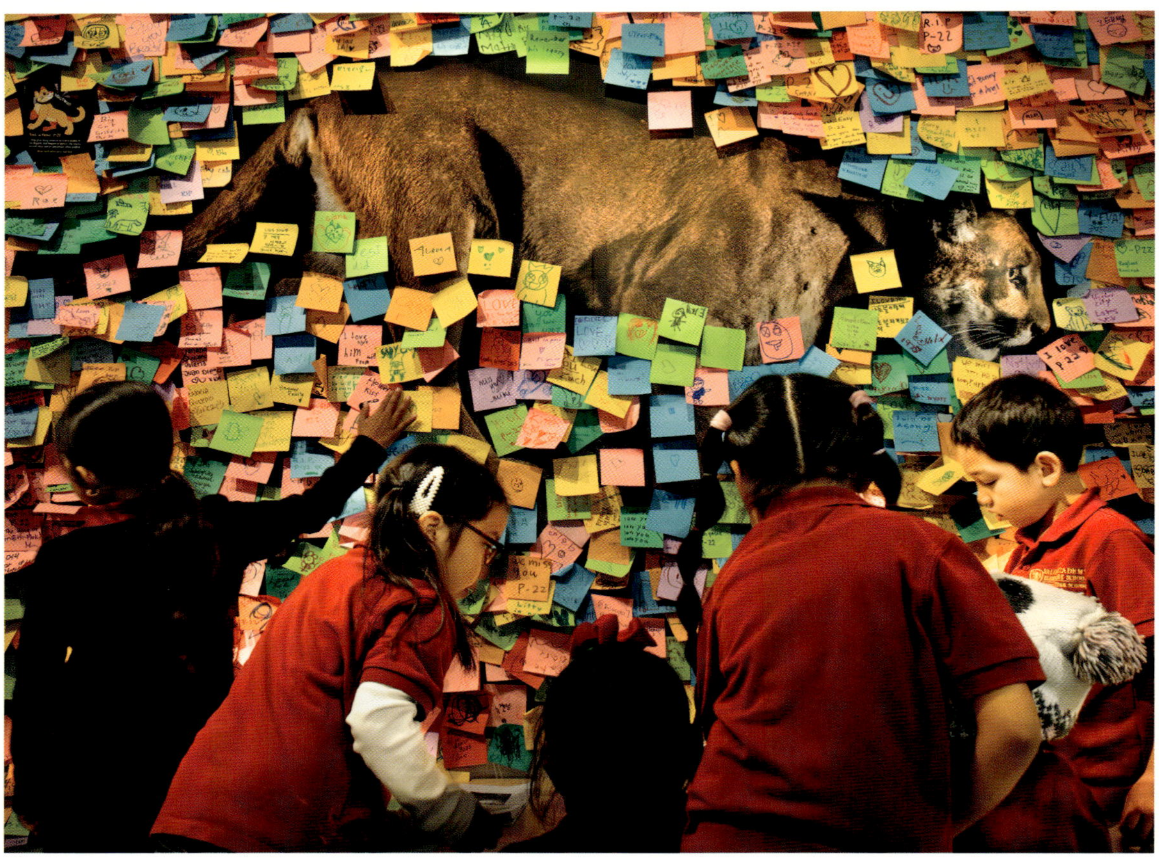

ABOVE: From left: Conservationist Beth Pratt, California Governor Gavin Newsom, and philanthropist Wallis Annenberg break ground on a wildlife overpass to reconnect Southern California habitats bisected by the 10-lane 101 freeway.

LEFT: Following P-22's death in December 2022, hordes of visitors to the Natural History Museum of Los Angeles County stuck notes on my photo of the cat on exhibit there – most were love notes.

And he lived alone, unable to mate. He was dubbed L.A.'s loneliest bachelor. Meanwhile, there was growing genetic evidence of inbreeding among Southern California's cougars, which doesn't bode well for long-term survival. Cubs were being born with birth defects and males had abnormal sperm.

Jeff was brainstorming with Beth Pratt (who heads California programs for the National Wildlife Federation) about how to create safe passage for SoCal's cougars through the sprawling chaos of L.A. These cats are trapped within fractured lands sliced and diced by housing developments, farms, strip malls, industry, highways, and a mass of humanity. Statewide, around 70 cats were dying on roadways each year.

A stretch of the 101 freeway just outside L.A. severed an important wildlife corridor between the Simi Hills and the Santa Monica Mountains, the Pacific Ocean, and wildlands to the north. It's a 10-lane highway, used by 300,000 vehicles a day. A wildlife bridge would allow cougars (and other animals) to safely roam in search of new mates, food, and water—and to escape deadly wildfires or flooding.

I donated my P-22 pictures and video to Beth for her "Save L.A. Cougars" campaign. She printed a life-size P-22 cutout that she carried around for nine years as she met with thousands of people: Governor Gavin Newsom. Conservation organizations. Executives at Caltrans, the state's transportation agency. Biologists. Engineers. Philanthropists. Indigenous Chumash leaders—the crossing would be on their ancestral lands—and residents. State and federal legislators. Hollywood actors. State wildlife and natural resources officials. She forged a massive coalition.

Beth launched a P-22 Facebook page and gave my photos and video clips to media outlets. They ran thousands of times around the world. P-22 became an international star.

Though raising money wasn't her forte, Beth fundraised a $25 million challenge grant from the Annenberg Foundation. It turbocharged support from the state and other investors, spurring a public-private miracle that raised $92 million.

The crossing broke ground on Earth Day 2022, attended by dignitaries including the governor and state and federal representatives. As everyone gathered, "Born to Be Wild" played over the PA system. The plan: The first animals would walk, hop, flutter, and crawl across this bridge in 2026.

Thought leaders ranked the Wallis Annenberg Wildlife Crossing as number 15 on the list of 50 Most Influential Projects of 2022, alongside the James Webb Space Telescope and the Human Genome Project.

P-22's story is a miracle, straight out of a Hollywood movie. Millions of people still love him, years after his passing.

In late 2022, Jeff noticed changes in P22's behavior that he hadn't seen in 10 years of tracking. He'd just gotten the green light to capture him for a health evaluation when he learned that P-22 may have been hit by a car.

On December 12, Jeff found him hidden in thick brush. He tranquilized him and sent him to a wild animal care facility. P-22 was now 12 or 13 years old, a senior citizen, and his medical exam showed he was not well. He was skinny. His skin was riddled with mange and ringworm. His skull was fractured, and his liver and kidneys were failing.

On December 17, 2022, P-22 was humanely euthanized. When I spoke to Jeff the next day, he told me that even before the vehicle strike, he thought the cat was suffering, and as he closed the door to his transport cage, he knew it was their final meeting.

This morning, he said, was the first time in more than a decade that he didn't start his day by checking P-22's coordinates. I cried.

I needed to say my own goodbye, so I went to the Natural History Museum of Los Angeles County. I was expecting quiet reflection at P-22's exhibit, but it was teeming with adults and schoolkids. Post-its and pens were on a table, and my "Hollywood cougar" photo and the walls around it were plastered with love notes, floor to ceiling.

Beth organized a memorial at the Greek Theatre in Griffith Park. It sold out within hours and crashed Ticketmaster. P-22 was given a four-hour tribute, including an unforgettable performance of "The Lion Sleeps Tonight." I spoke, reminding the crowd that the *Los Angeles Times* called him the most famous animal since Lassie—but there were nine Lassies and only one P-22. He'd also been called "the Brad Pitt of cougars."

P-22's story is a miracle, straight out of a Hollywood movie. Millions of people still love him, years after his passing. He was an ambassador for urban wildlife and the catalyst for construction of the world's largest wildlife bridge. He started a movement, and his legacy lives on: Dozens of wildlife crossings are being planned and built in California.

At P-22's memorial, Beth promised to honor his legacy, and true to her word, she launched the Wildlife Crossings Fund. Now Sharon and I are documenting this innovative effort to reconnect U.S. wildlands and provide safe passage for everything from turtles and salamanders to bobcats, pronghorns, salmon, ocelots, pelicans—and cougars.

Long live P-22!

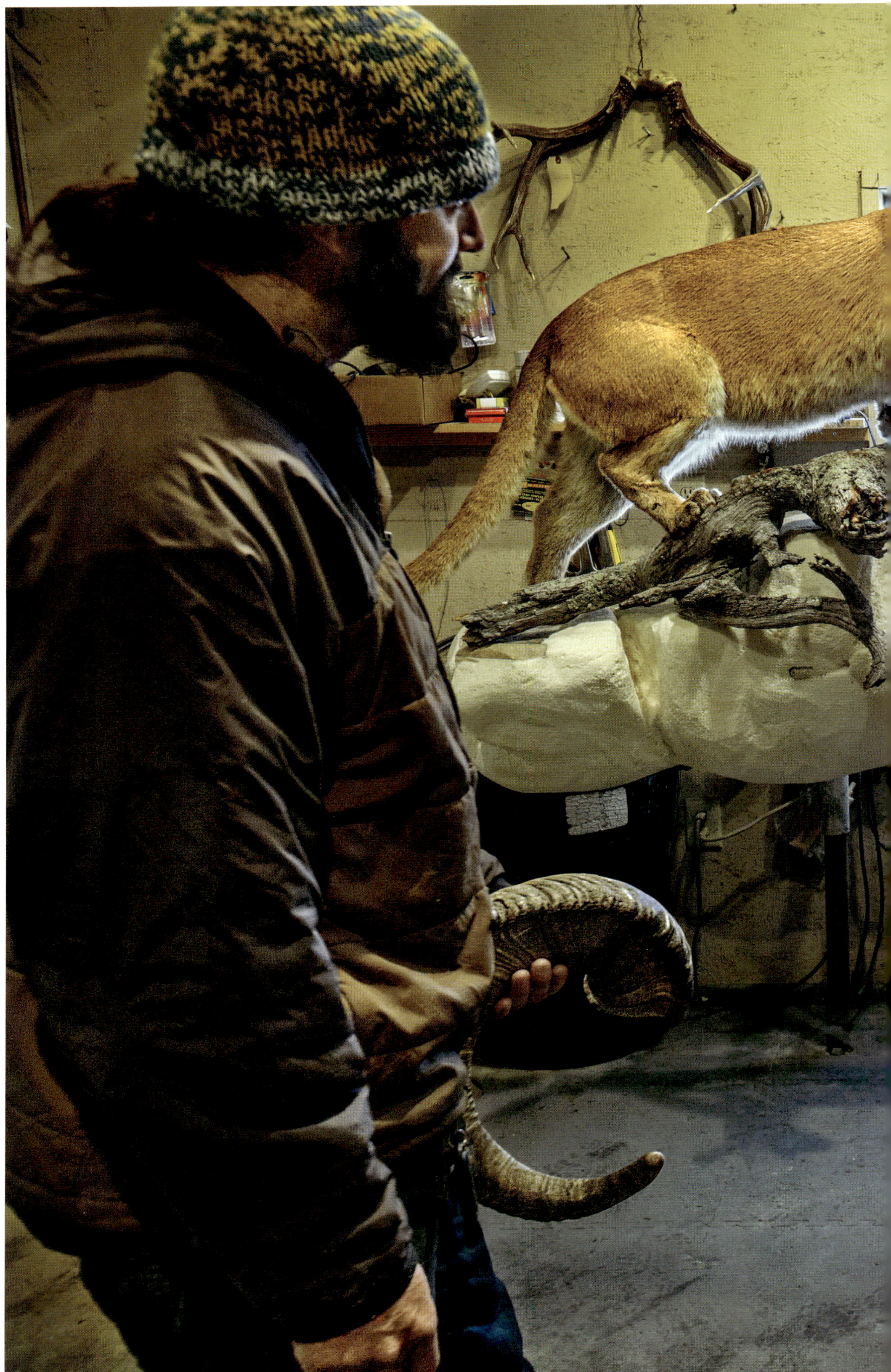

Cougars are legally hunted across their territory in the western United States, with the exception of California. Here, hunters admire the work of a local taxidermist in South Dakota's Black Hills.

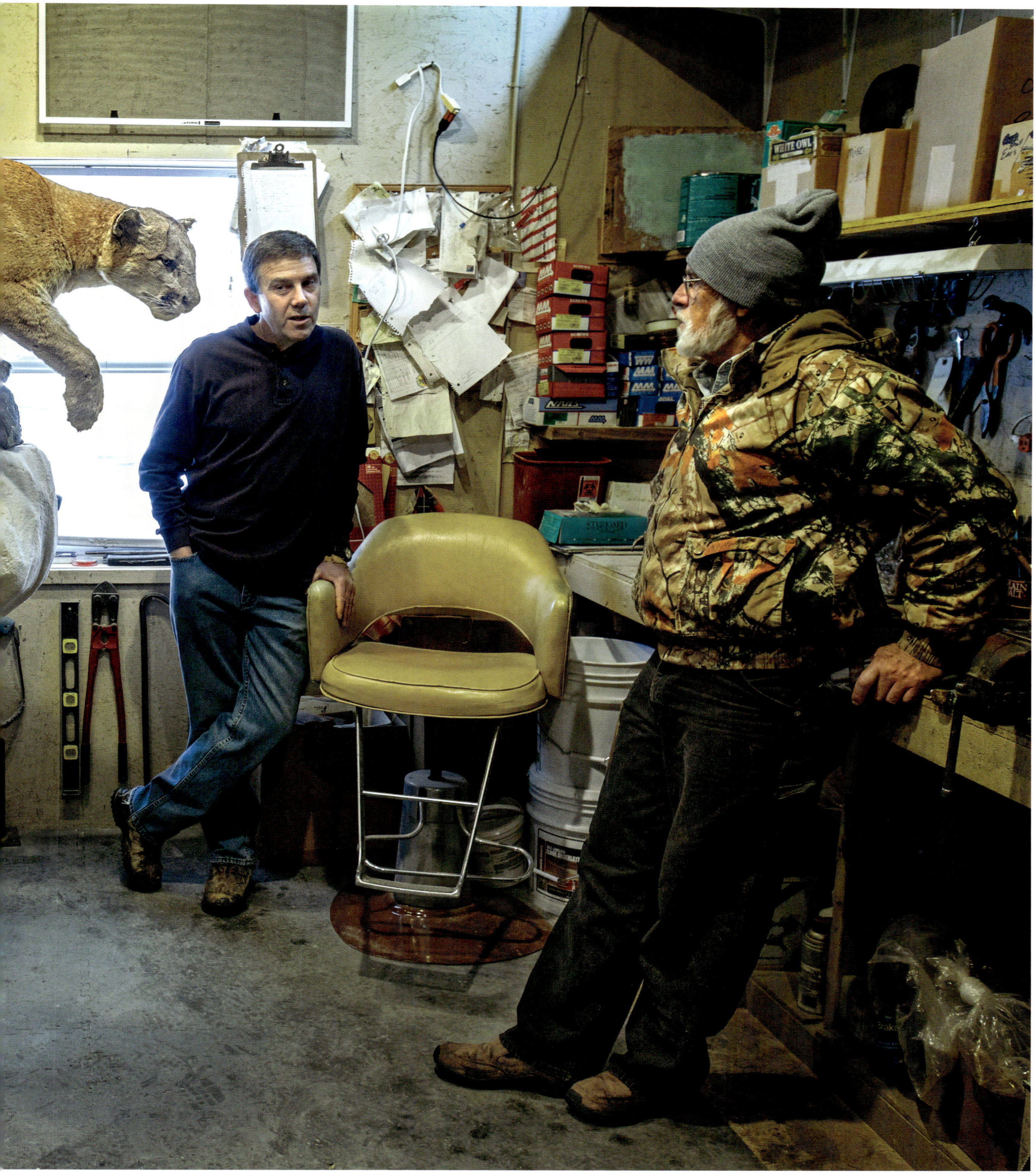

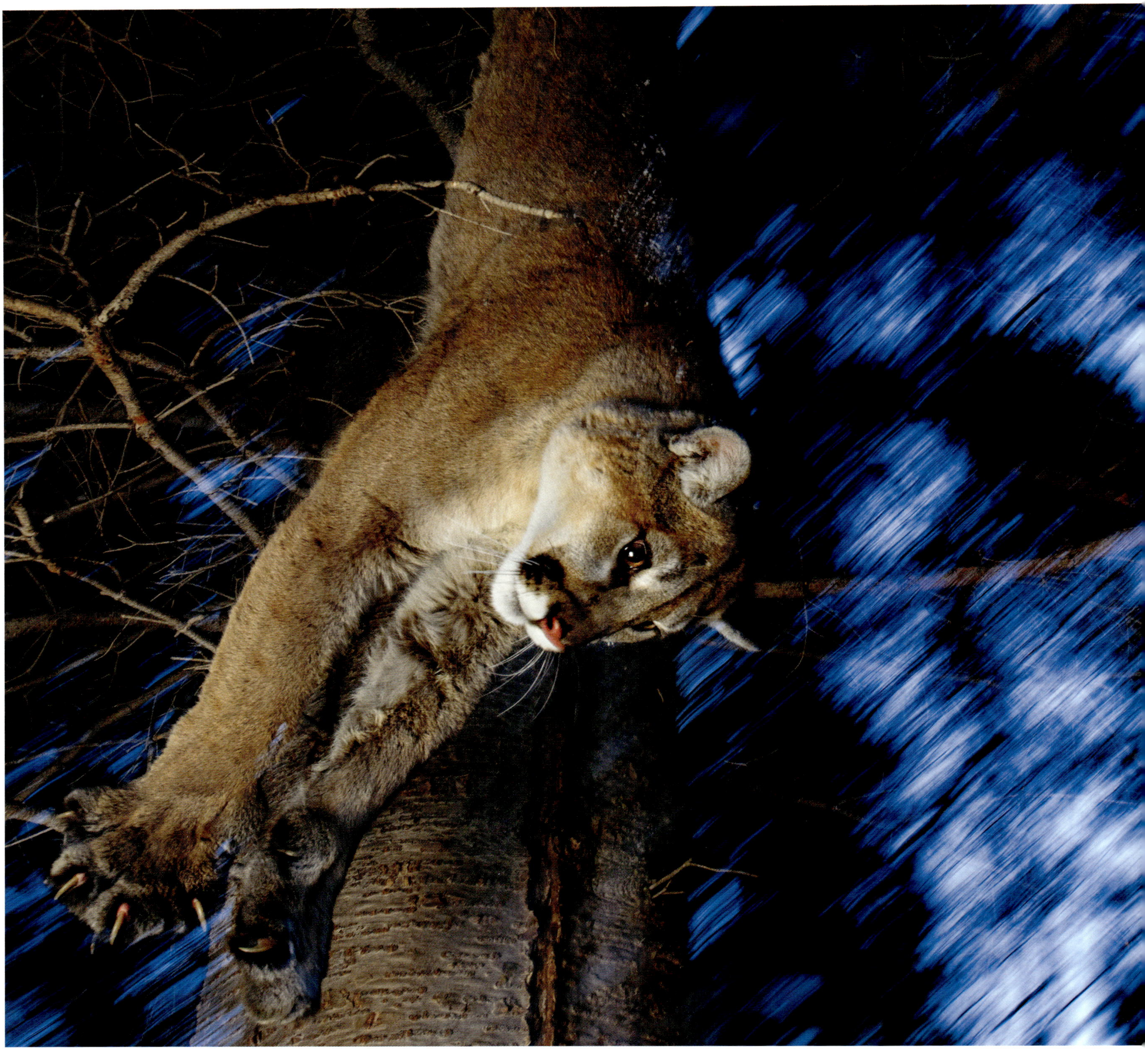

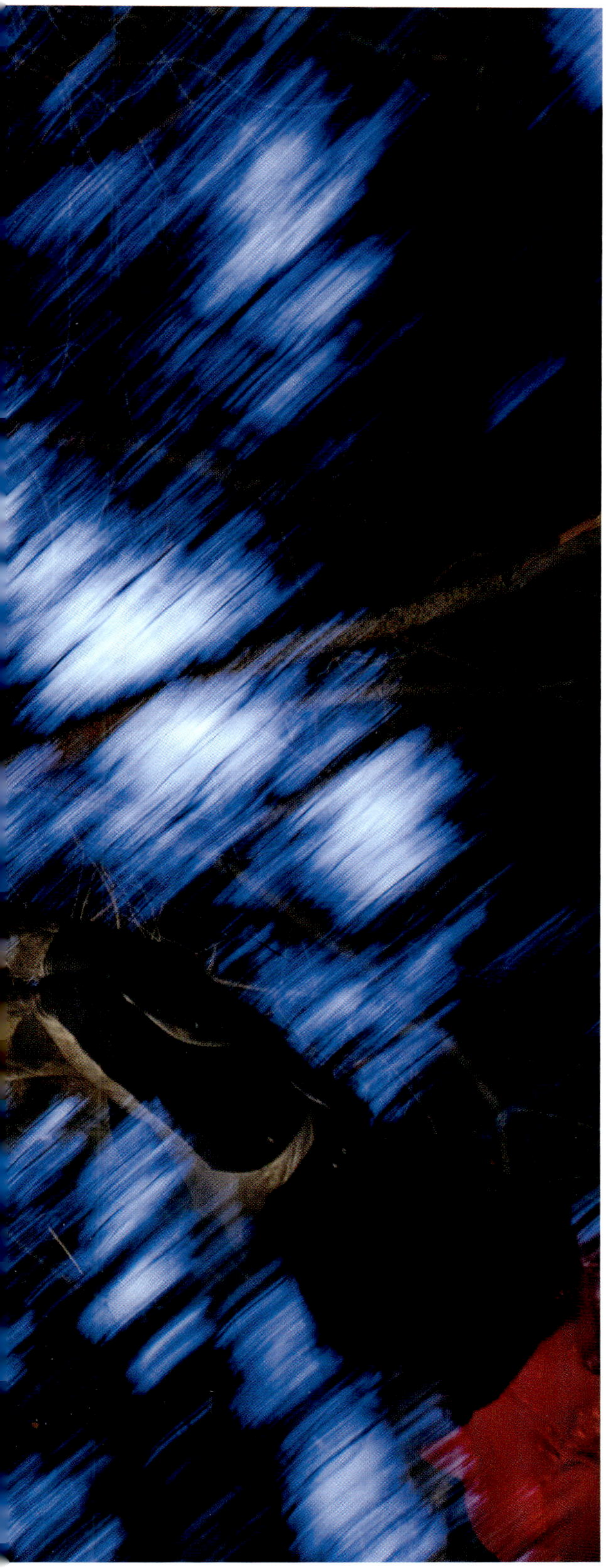

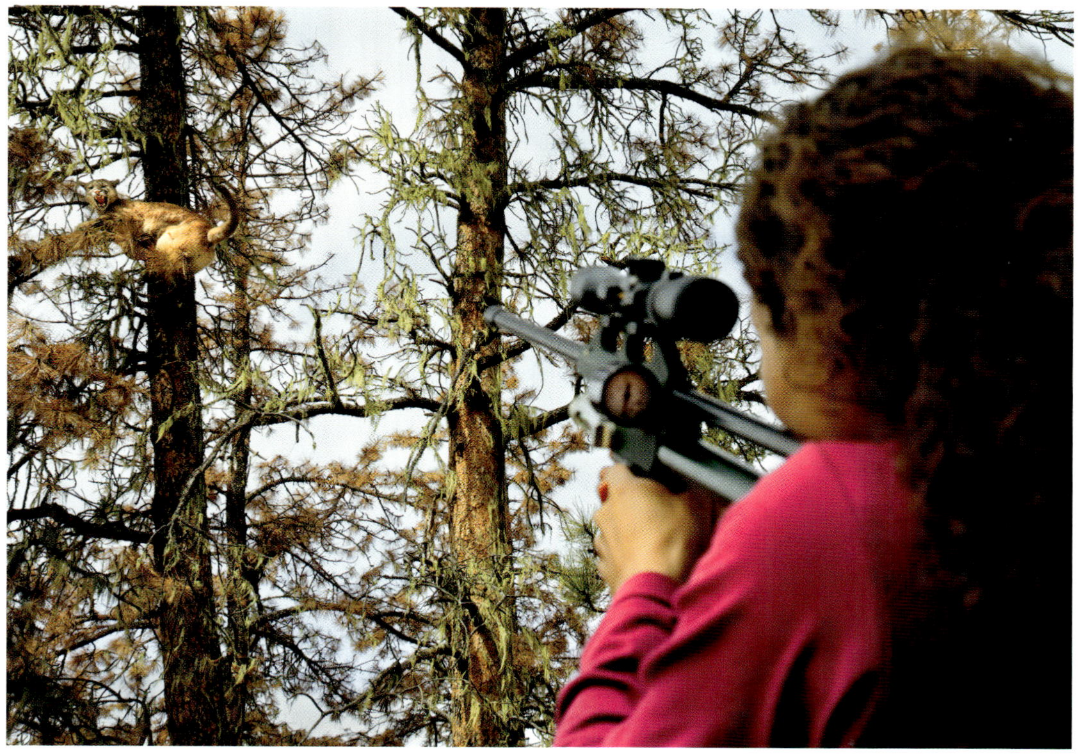

ABOVE: The only way to capture and collar a big cat is to anesthetize it. Here, a biologist aims a dart gun to sedate a cougar that climbed a tree.

LEFT: Treed by hounds and tranquilized, a semi-conscious two-year-old male cougar is carefully lowered to the ground in northwest Wyoming. He became a collared study animal in a 12-year research project.

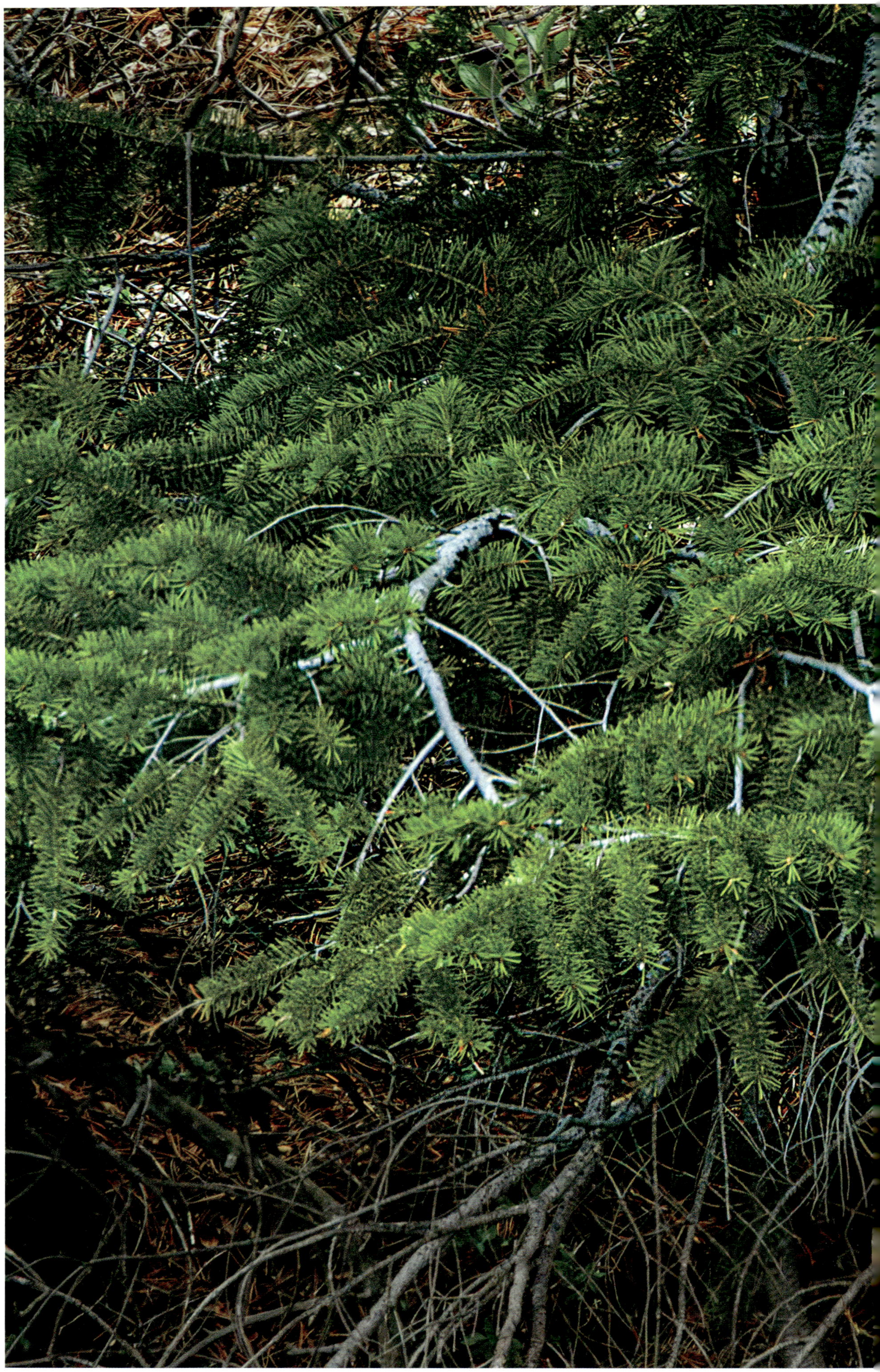

Young cougars rest under a pine tree in Bridger-Teton National Forest. Researchers are observing how these cats' behaviors have evolved since wolves were reintroduced to the area in 1995.

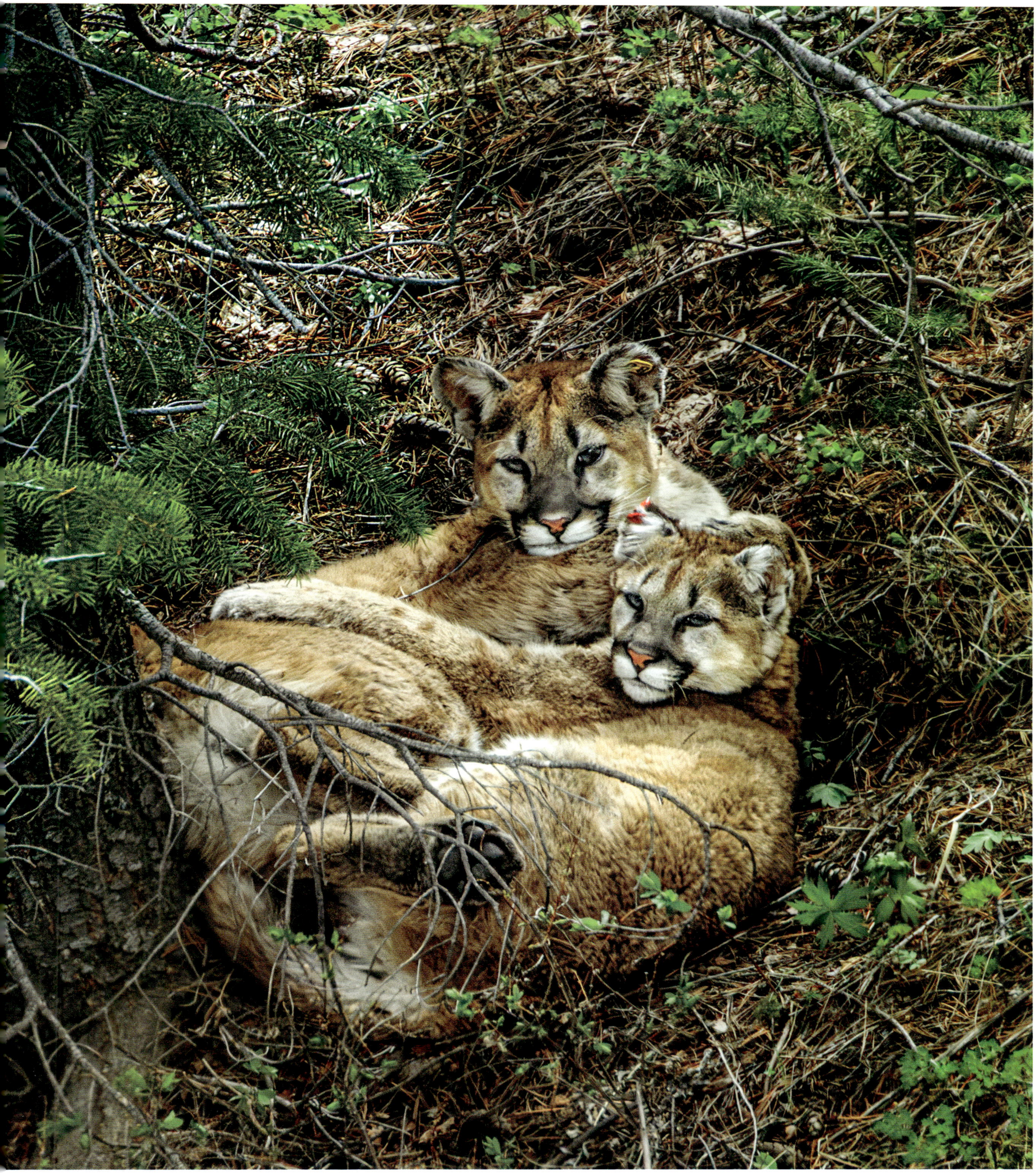

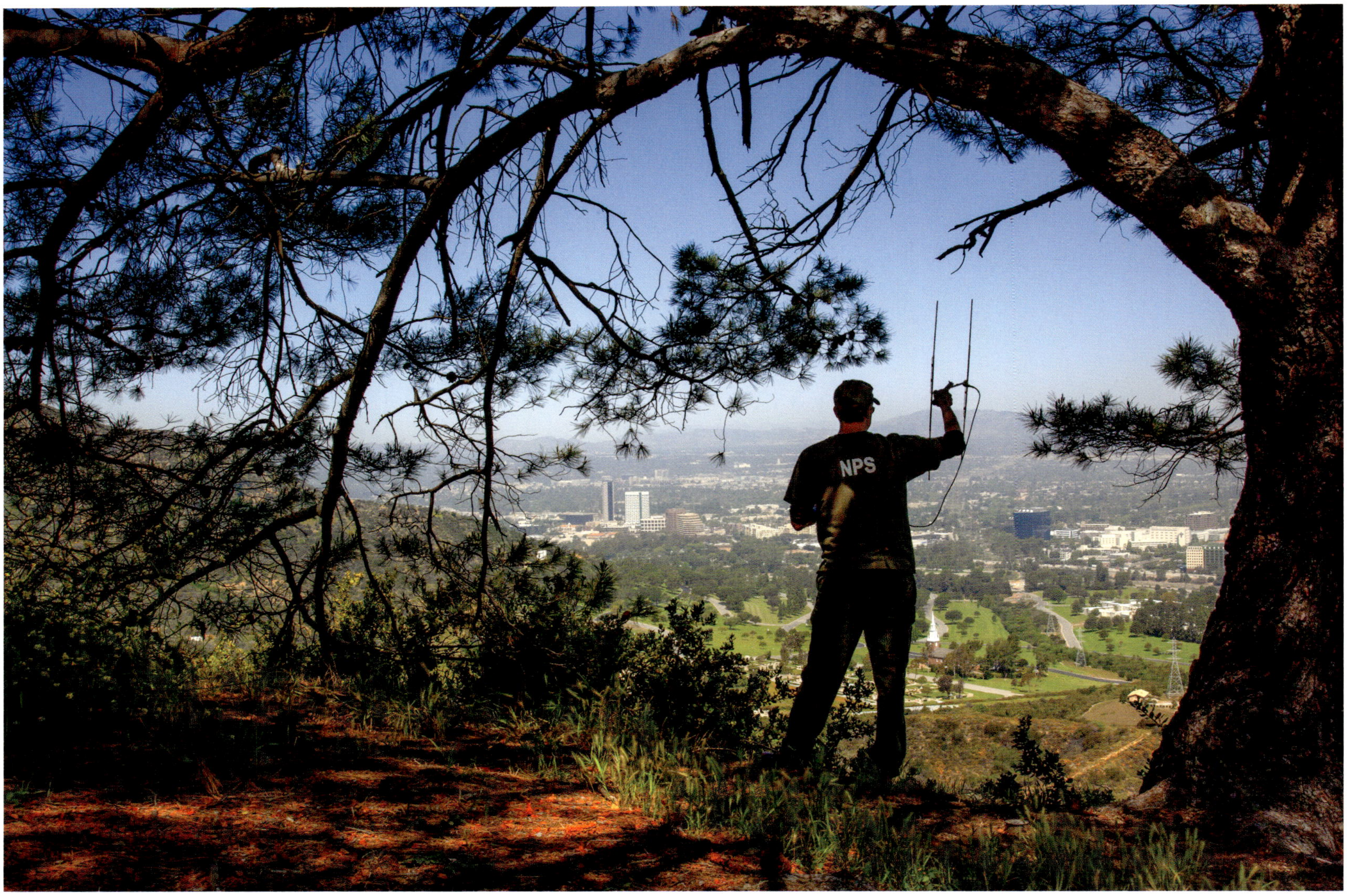

Jeff Sikich, a field biologist with the National Park Service, uses telemetry to track P-22 in Griffith Park, a 6.7-square-mile park in Los Angeles where the cat lived for a decade.

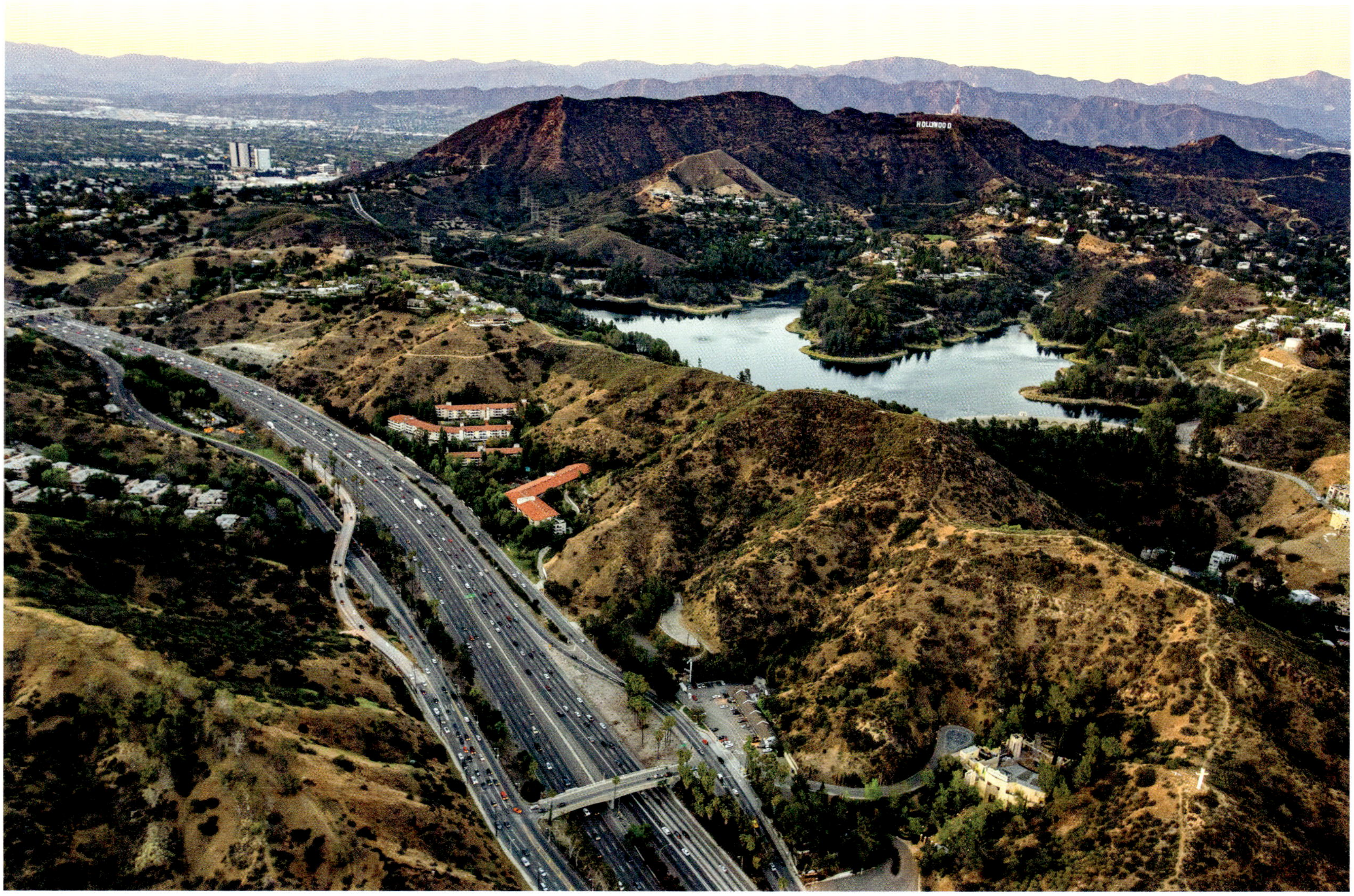

An aerial view of the landscape that young P-22 navigated when he left his family in search of his own territory. Somehow, he safely crossed two of the nation's busiest freeways, the 405 (pictured here) and the 101, to settle in Griffith Park.

PAGES 42–43: Hollywood's most reclusive star, P-22, was first seen in Griffith Park in Los Angeles in 2011. It took me 15 months to capture this image of him. After the photo was published, P-22 became an icon: Angelenos fell in love with him, and he became a poster child for urban wildlife conservation.

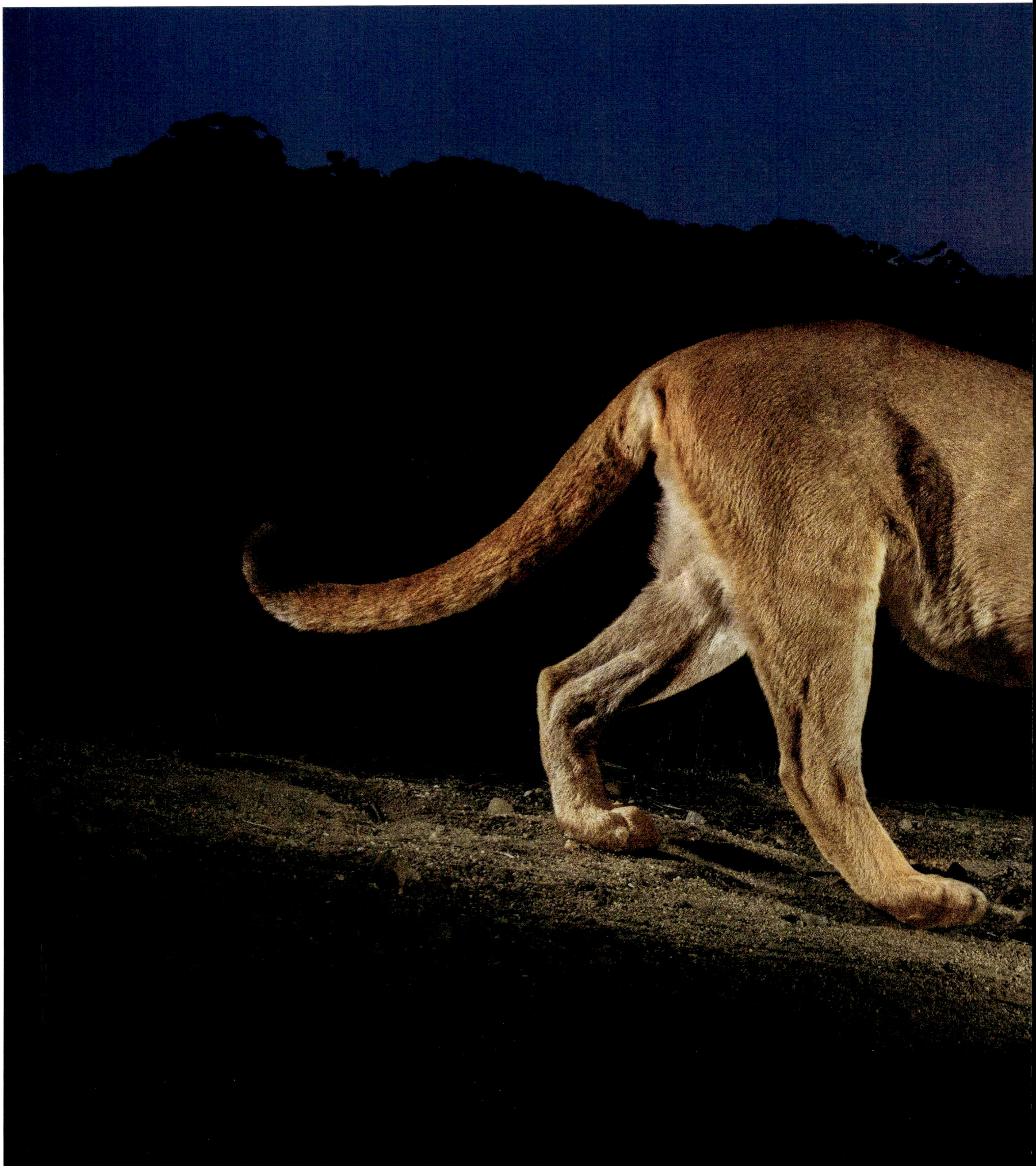

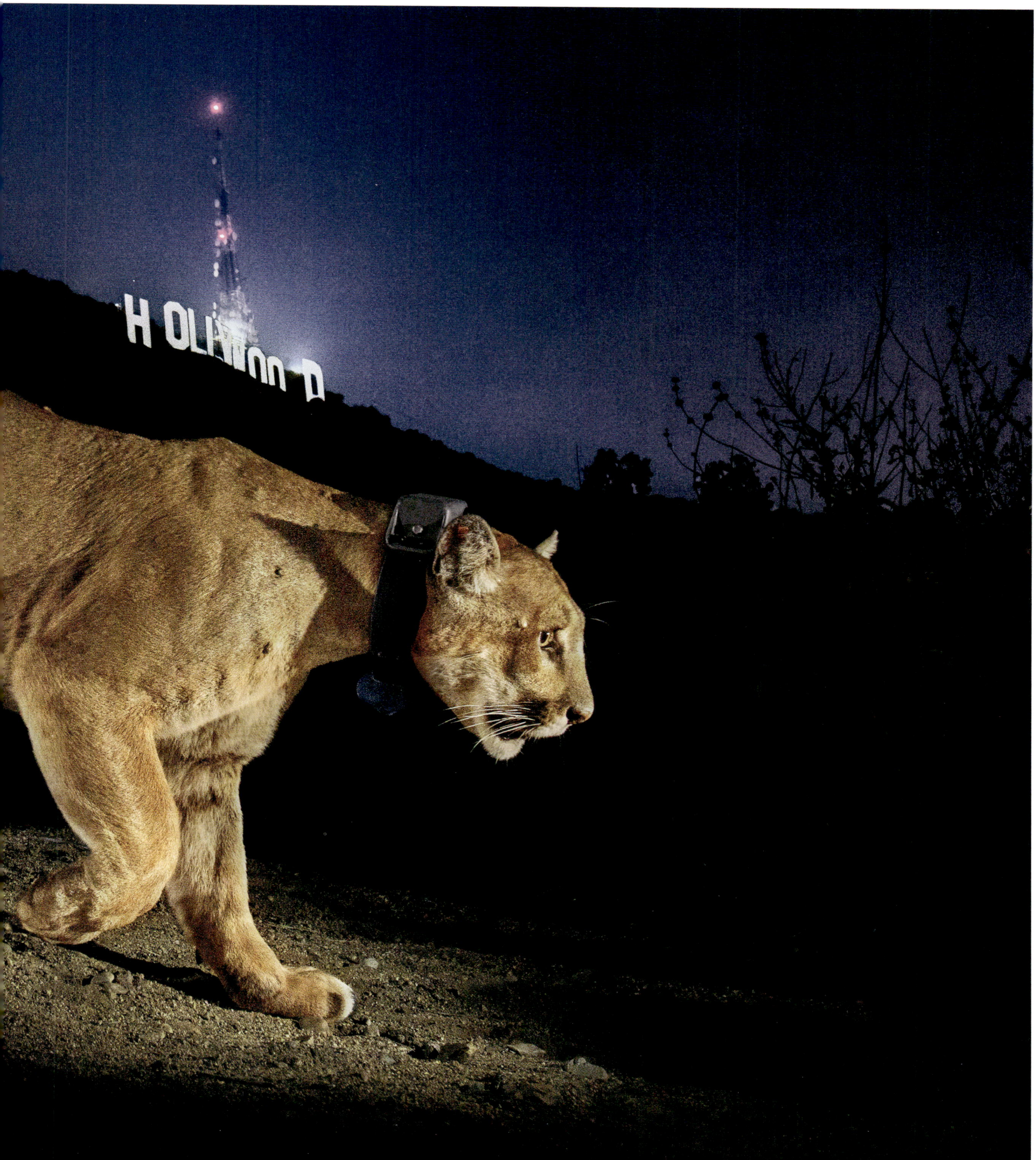

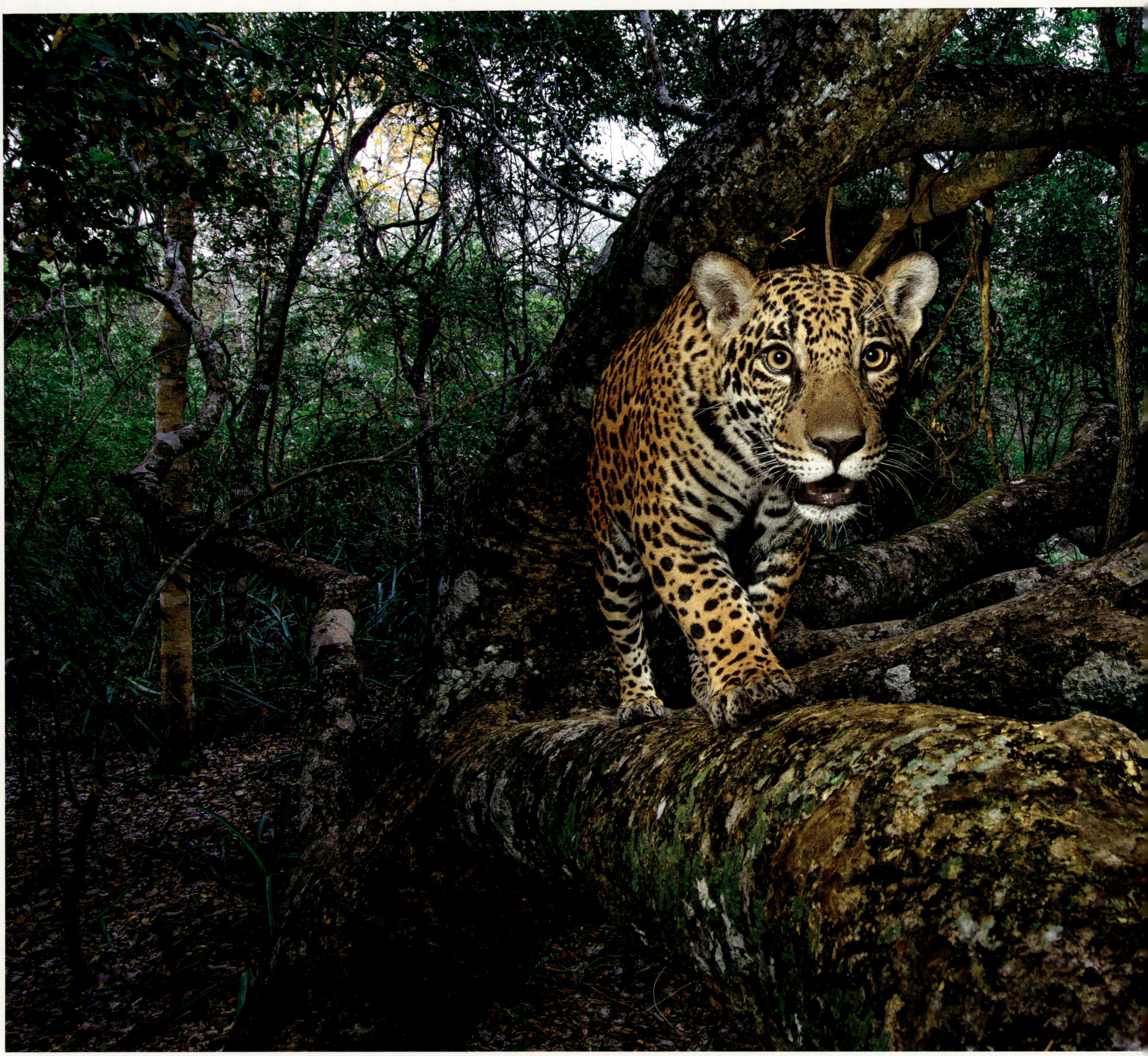

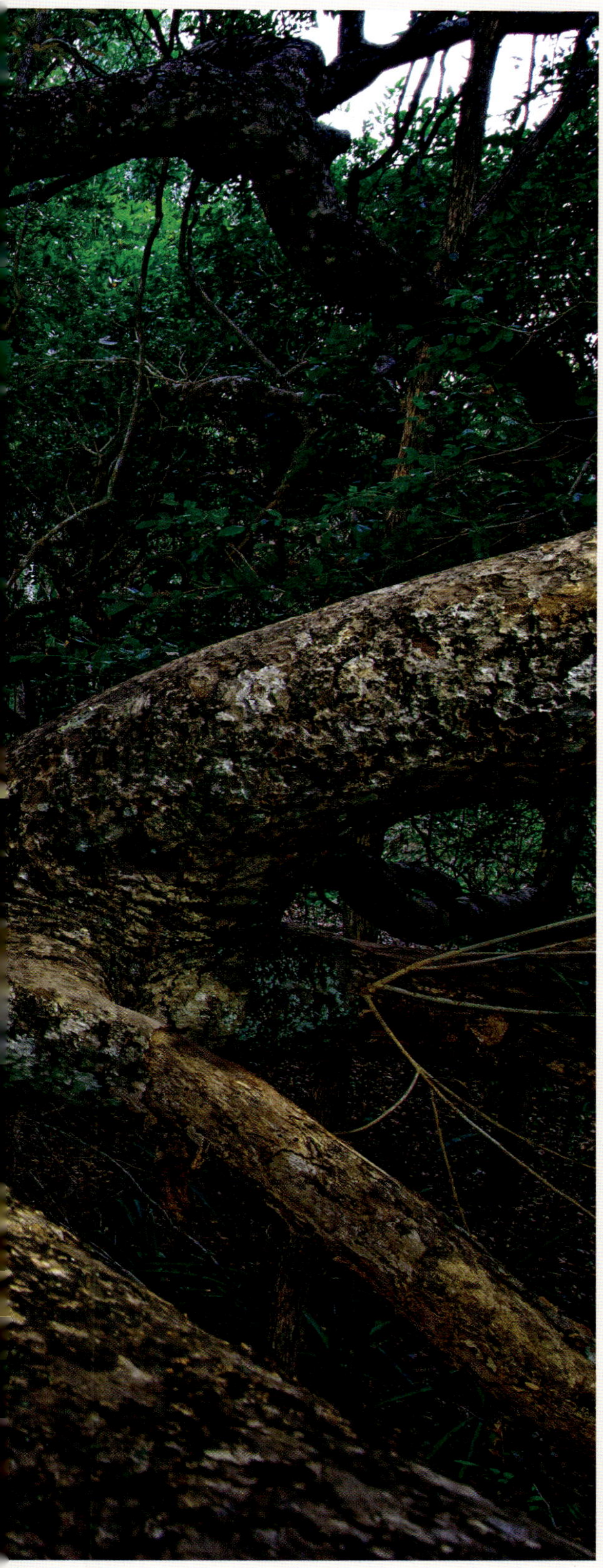

CHAPTER TWO

Jaguars The Indomitable Beasts

Jaguars in Brazil's Pantanal were once heavily targeted by ranchers. But now, thanks to ecotourism income and coexistence strategies, the cats are bouncing back.

A 10-month-old jaguar cub stalks along a tree trunk in the Brazilian Pantanal. These cats range from Mexico's Sonoran Desert to Argentina. Though there are no subspecies, jaguars in the Pantanal are the largest.

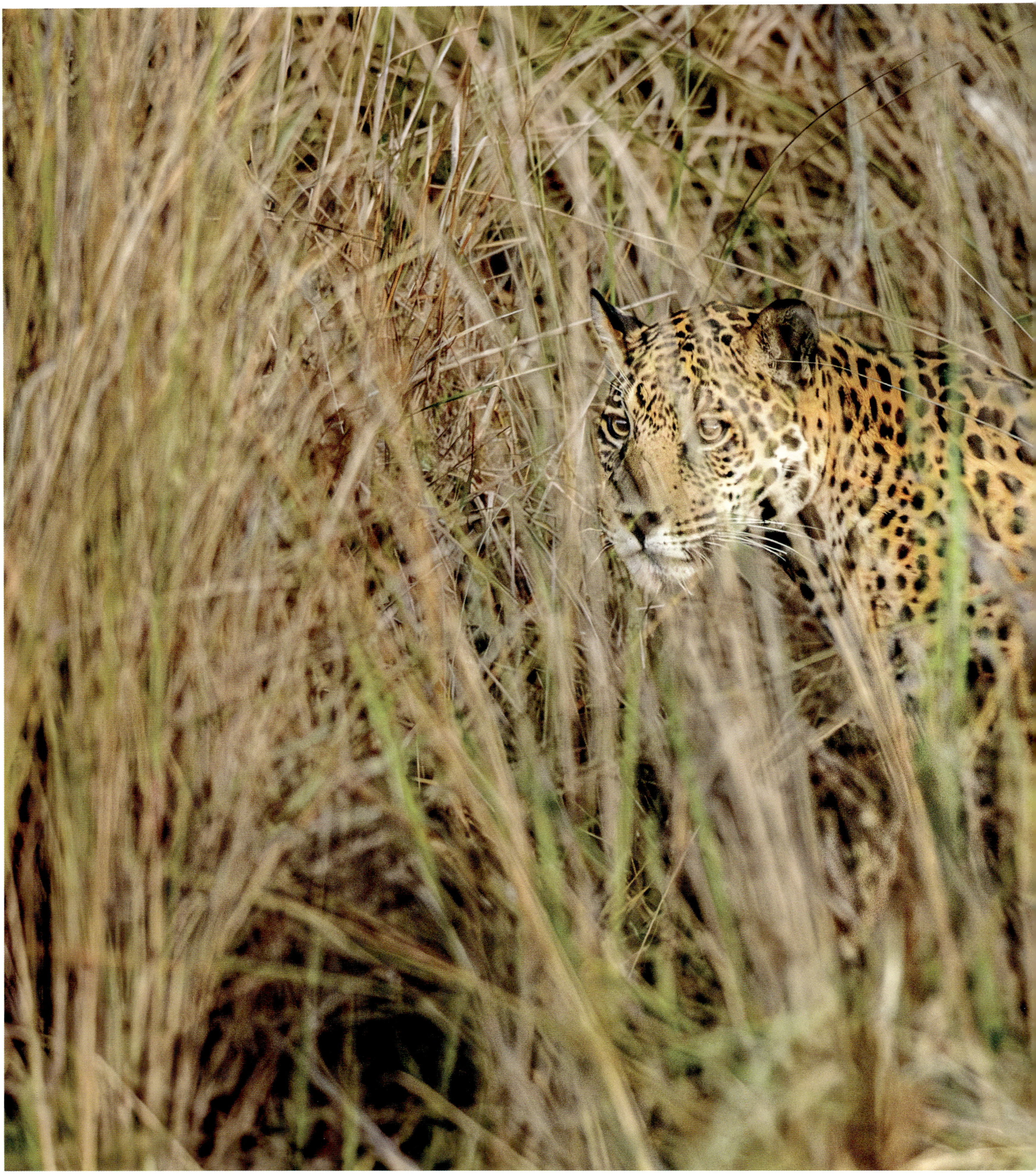

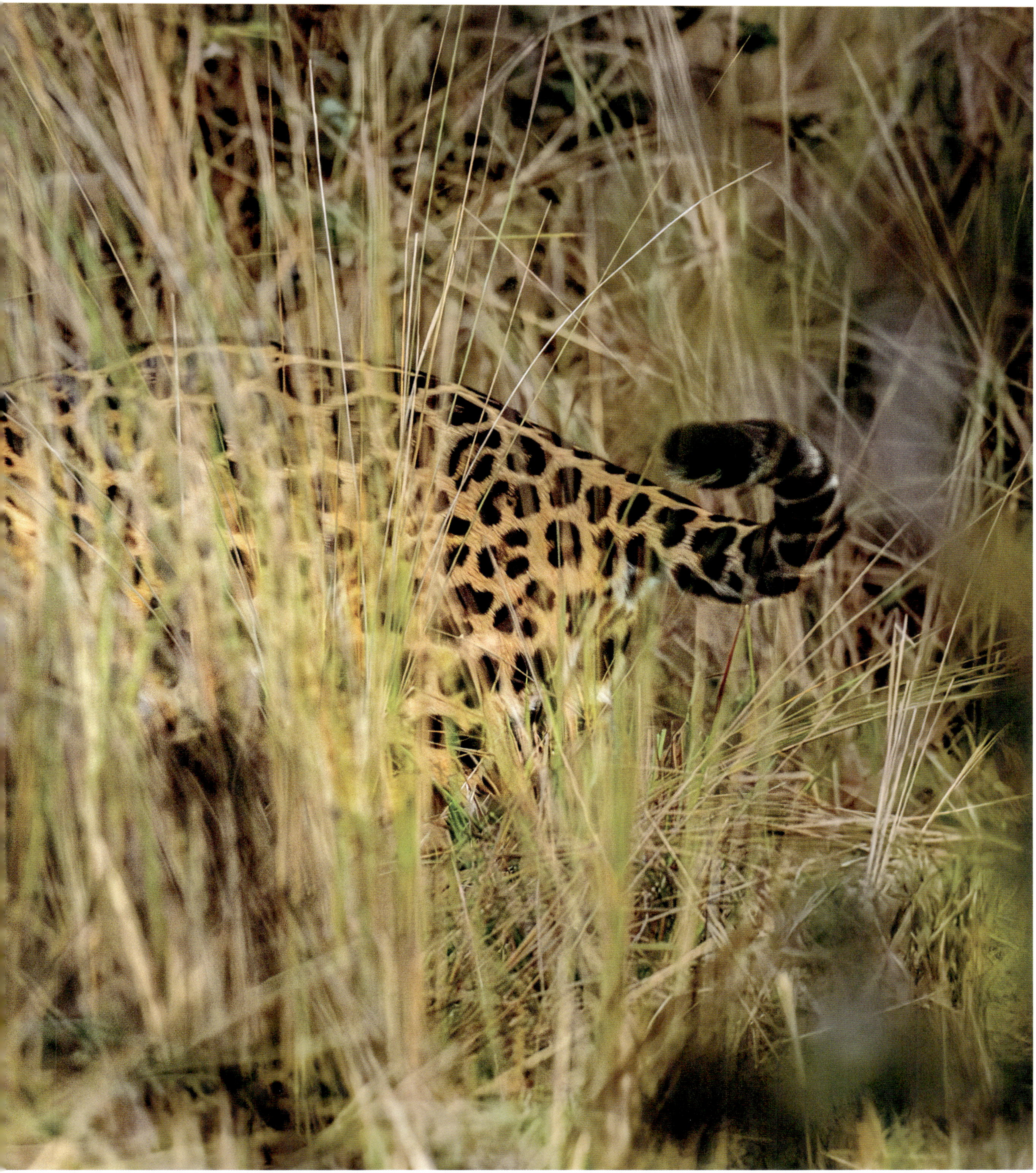

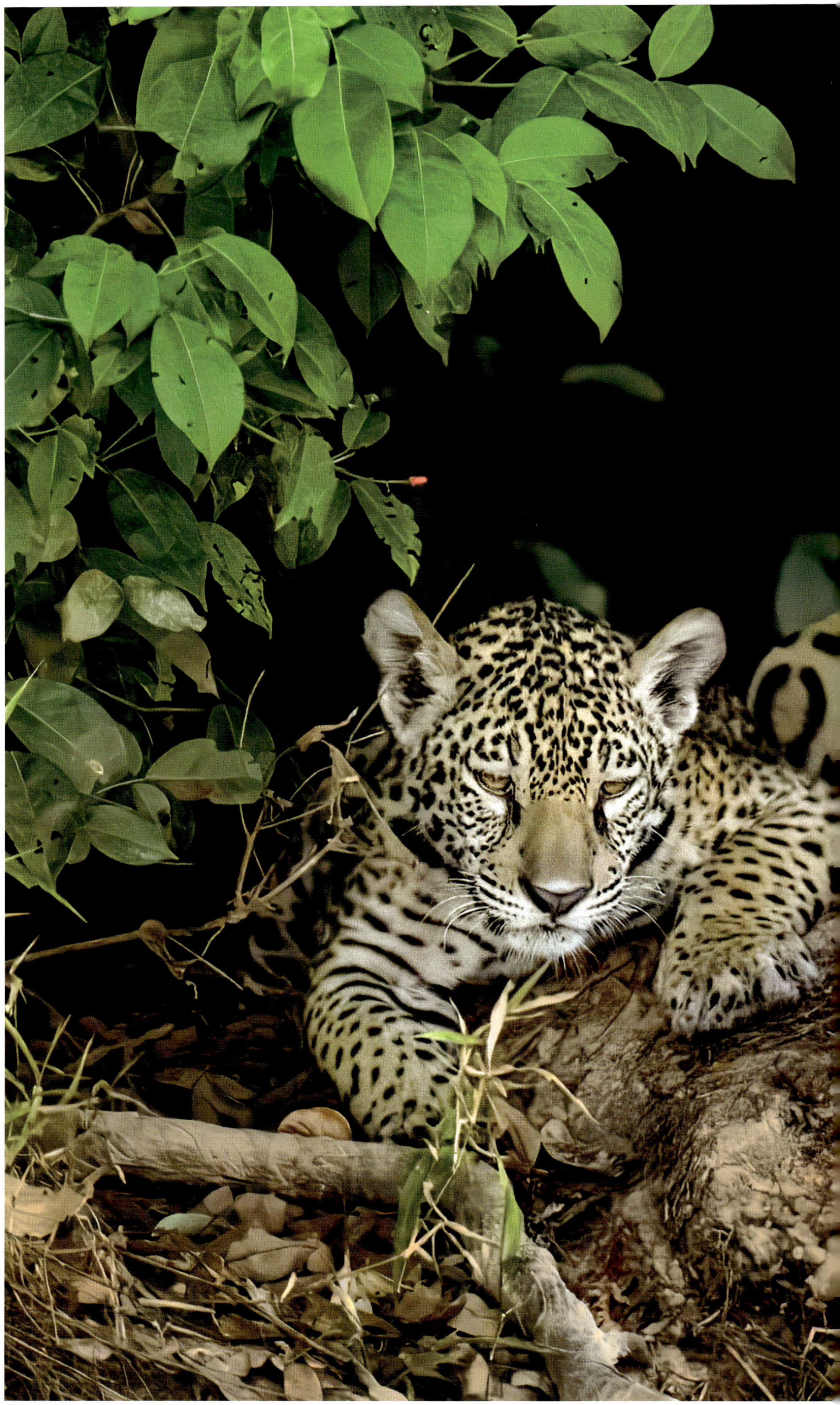

A mother grooms her cubs on the banks of the Três Irmãos River in the Brazilian Pantanal. Jaguars mate at any time of year – the only time these otherwise solitary cats interact – and bear up to four cubs.

PAGES 46–47: A jaguar scans for prey from the cover of tall, tawny grasses. The cat's name may have come from the Indigenous Tupi-Guarani *yaguarete,* which translates to "beast that kills its prey with a single bound."

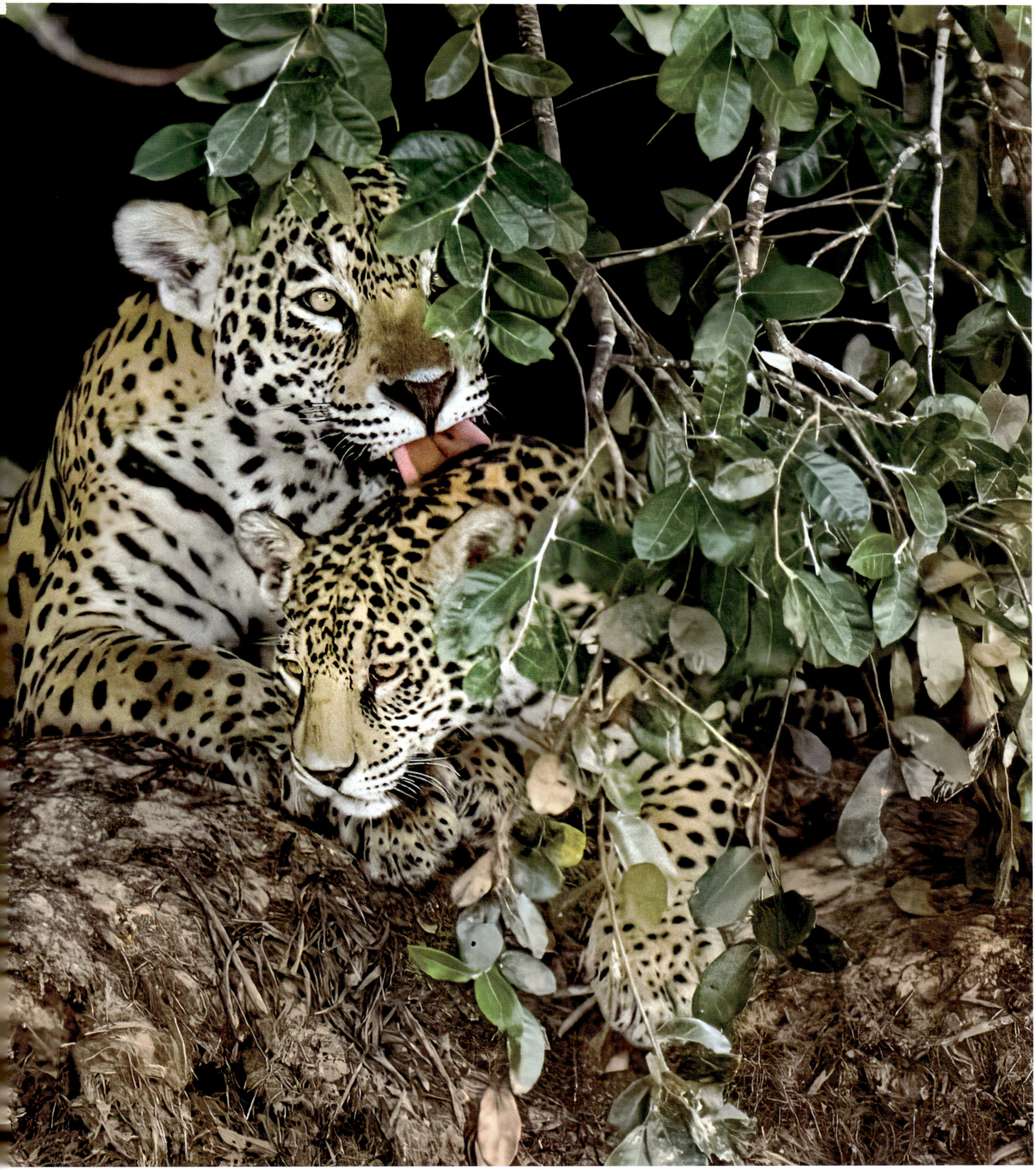

I never would have imagined, when I was photographing a coup in Haiti or shooting pictures of world leaders at a press conference, that I would choose to spend months in sweltering jungles seeking out deadly predators for the rest of my career. But something happened when I encountered my first big cat, that black jaguar, in Guatemala.

I still don't understand why this animal affected me so deeply, but it pawed its way into my subconscious. Perhaps it was primordial fascination or old instinct embedded deep within the primitive recesses of my brain. Jaguars have been both feared and revered for millennia by cultures across Central and South America. *Panthera onca* is the New World's largest and the world's third biggest cat.

The Olmec, who rose to power in Mesoamerica around 1200 B.C.E., believed they were born from a mystical union between a jaguar and a woman, imbuing them with animal powers. The Aztec emperor Montezuma donned jaguar skins in battle. The ancient Maya thought jaguars were the gatekeepers of the underworld, facilitating communication with the dead. These cats have superpowered night vision and silently appear out of nowhere, as I learned firsthand in that Guatemalan cloud forest. They hunt with ghostlike stealth.

Back at home after my quetzal story, I had jaguars on the brain when I saw a *New York Times* story about a cowboy, Warner Glenn, who'd taken a snapshot of one in Arizona. They were thought to be long gone, trapped and hunted to extinction, mostly by ranchers. The last known female was shot near the Grand Canyon in 1963. But it seemed that a Sonoran cat had roamed north. That was before the border wall cut off any possible movement into their former U.S. territory.

Alan Rabinowitz, a legendary wildlife biologist and jaguar expert, was investigating the sighting. After years of field research in Belize, he'd persuaded the government there to protect a substantial chunk of the cat's forest home, and the Cockscomb Basin Wildlife Sanctuary became the world's first jaguar preserve.

I was nervous when I cold-called Alan to find out more about this mysterious jaguar, as I have no background in science. I met him at the Bronx Zoo, where his organization, the Wildlife Conservation Society, has their offices. He resembled the animal he studied, built like a prizefighter. I would learn over many years working with him that he was a relentless warrior for big cats.

After that experience in Guatemala, I was compelled to do a story on jaguars, though it wasn't going to be easy. Just finding *"el tigre"* would be tough, let alone photographing the cat. Jaguars are solitary phantoms that avoid humans and are active at night or in the murky light of dawn and dusk. They're constantly on the move and hide in thick vegetation. This might have been why *National Geographic* had never produced a jaguar story in its then-109-year history. That fact was a selling point to my editor, Kathy Moran.

I was shocked to learn that at the time, in 1997, the jaguar was the least studied big cat. There was minimal understanding of its daily habits. Just three biologists had studied them: George Schaller, Howard Quigley, and Alan Rabinowitz. I flew to Rabinowitz's jaguar haven in Belize, joined by biologist Doug Chadwick, the writer on the story, and Howard Quigley.

Photographing in the Cockscomb jungle was tough. The tall canopy acts like a big blackout curtain, enveloping the forest in semi-darkness. This was before digital cameras, and I was shooting color slide film, Velvia, that I pushed to 400 ISO, so different from today when you can shoot in the dark on an iPhone.

At the time, camera traps—essentially professional trail cams with better equipment and flashes that were triggered by motion—were a relatively new technology. Since few trails cut through the dense rainforest, tracking animals or predicting where to find them was challenging at best. I needed local knowledge to figure out where to install the cameras.

So I hired the best hunters from a nearby Maya village: This forest was their backyard, and they knew the wildlife. We spent sweaty 12-hour days out in the field, sometimes in sucking mud up to our ankles, trudging through tropical downpours, searching for paw prints—and setting up remote cameras.

A big male walked in front of my traps a few times, but then it disappeared. We found out it was killed by loggers when it wandered near their camp. We'd have to find another cat.

One day, as the sun lowered, we hurried to check a trap before nightfall. You *did not* stay in this forest after dark. That was the moment I knew I was failing: There wasn't a single frame on that camera. No pictures. After three months in the jungle, I had only a handful of jaguar images.

This was only my second story for *National Geographic,* and I was afraid it would be my last. But I don't dwell on setbacks; I always focus on what I'm going to do tomorrow to advance the story. There

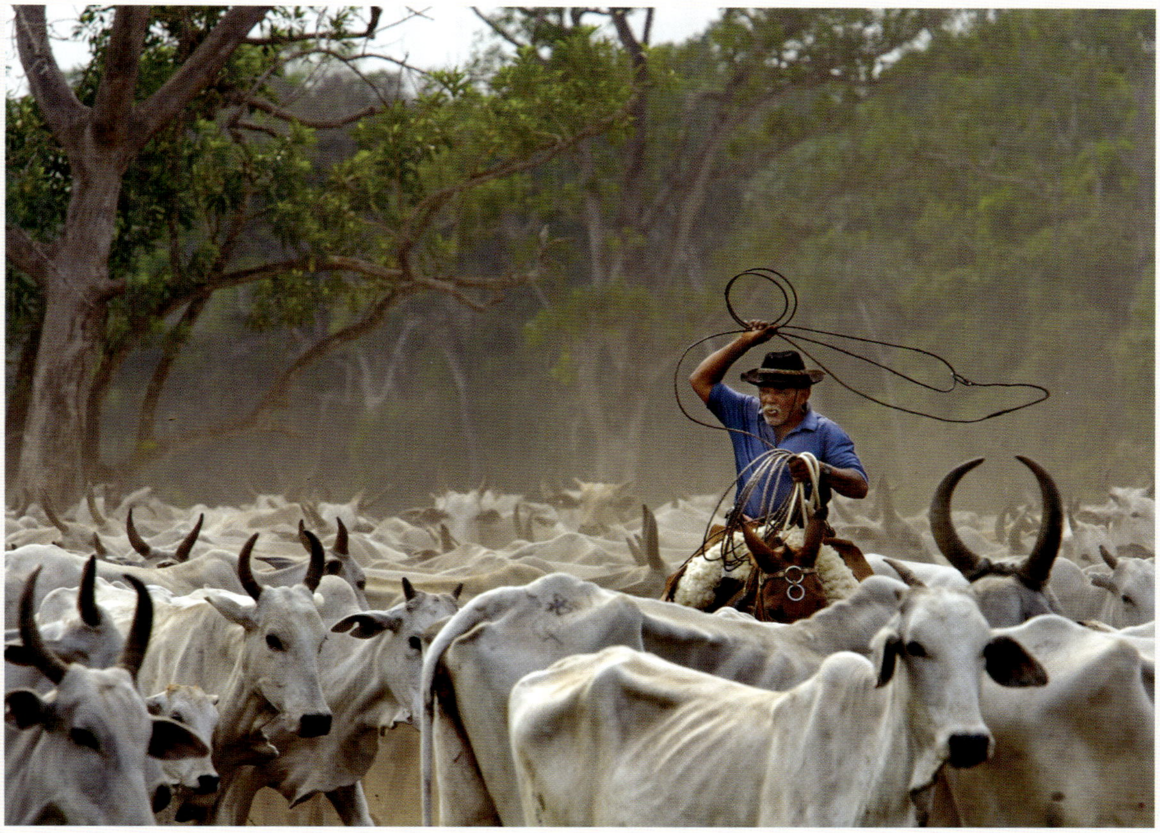

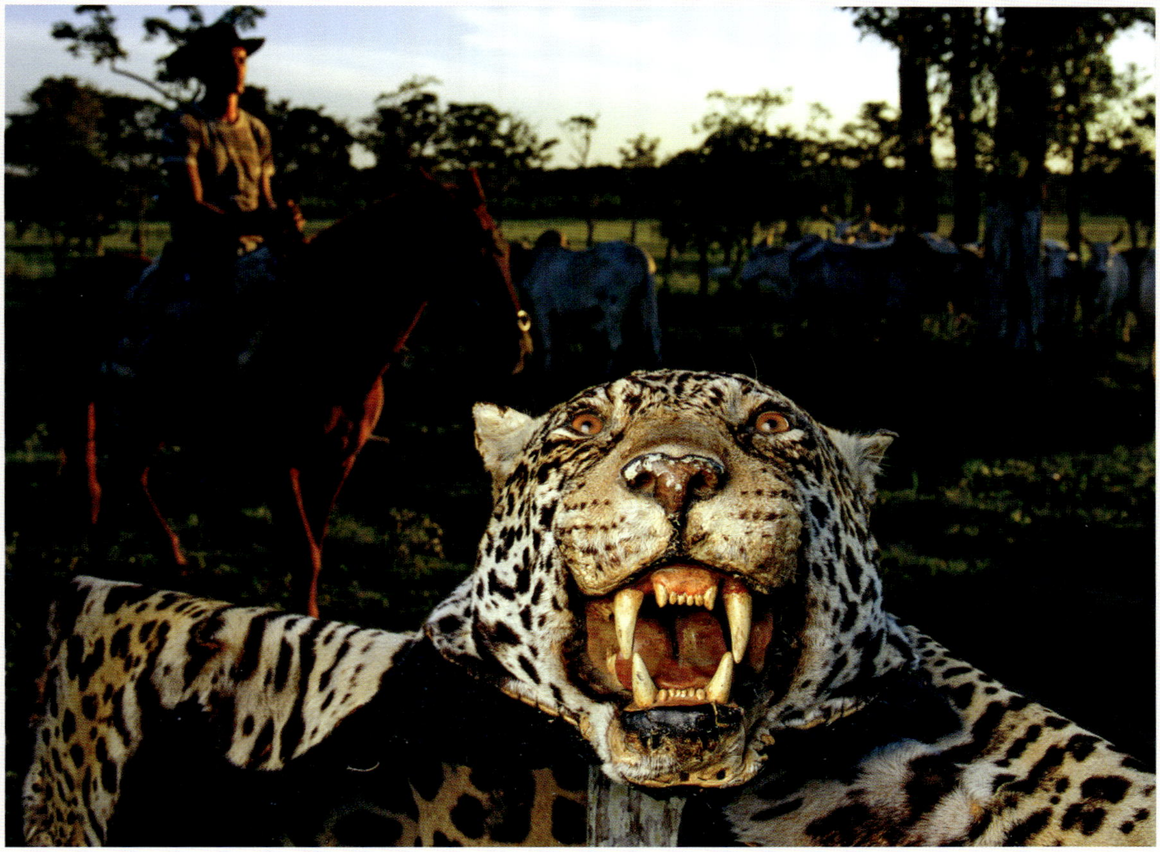

ABOVE: More than 95 percent of Brazil's Pantanal is privately owned, mostly by ranchers. Until recently, *pantaneiros* (cowboys) heavily hunted jaguars: All were assumed to be cattle killers.

RIGHT: Research shows that less than 5 percent of cattle deaths can be attributed to jaguars. This revelation, along with growing income from jaguar ecotourism, has lessened jaguar killings in the Pantanal.

had to be somewhere I could actually see and photograph these cats—and someone who could help me.

That place was Brazil's Pantanal, which means "great swamp" in Portuguese. At some 75,000 square miles, it's the world's largest inland wetland. It's also a jaguar stronghold—and it's remote. It took three days to get there from New Jersey. The last leg was eight hours on a gut-jolting dirt road that had massive holes in its hundred-plus wooden bridges.

The Transpantaneira Highway ended at the Cuiabá River, where Doug Chadwick and I boarded a small skiff with biologist Ricardo Boulhosa. He spent a few days giving us the lay of the land, but headed off for graduate study, leaving us in the hands of a former hunter and excellent guide—who spoke only Portuguese.

It was a biodiverse wonderland. Hyacinth macaws flew overhead. Giant river otters performed gymnastics in the water beside legions of capybara. Rivers and swamps teemed with piranha and caiman. It was brutally hot. We lived on rehydration salts.

We cruised the murky, coffee-colored shallows of the Corixo Negro ("black channel"), and there, camouflaged under sunlight-dappled bushes, I glimpsed rosette-spotted fur. A large male regarded us with golden eyes. My first jaguar. That week, I saw four more. There, jaguars had completely different habits. They were active any time of day or night. It was the dry season, water was scarce, and they hung out near the rivers. We met "Bob" a few times—a male jaguar that sometimes sauntered across a levy at dusk.

We heard about another cat that had wandered around a lodge nearby where Theodore Roosevelt once hunted. And in fact, near dusk, I saw a jaguar way down the road. He was massive. As I walked toward him, he disappeared. A half hour passed and I thought I'd lost him, but then I saw his eyes flash in the grass some 12 feet away. The cat crouched, ears flat, ready to pounce. I squeezed off a few frames, literally shaking, and slowly backed away. He stood, stretched, and wandered back down the road, his compact, muscled heft radiating power. One of those photos became the story opener.

There was just one rustic fishing lodge on the river, so we stayed in a ranch bunkhouse. Some 95 percent of the Pantanal is privately owned, mostly cattle outfits, and our bunkmates were not exactly conservationists. Jaguars had no cultural or religious significance here. They were demonized. The cowboys hated them.

Ranchers blamed jaguars for every dead cow, and the only good jaguar was a dead jaguar. Armando Klabin, a paper company magnate who owned the ranch we were staying on, claimed that jaguars killed 120 cows a year from his herd of 7,000.

That made retaliation killings the cats' greatest threat. Though Brazil had banned jaguar hunting decades before, in 1967, the law was unenforced. Ranch hands and *onçeiros*—hired jaguar hunters—tracked them with packs of hounds and shot them. To illustrate this,

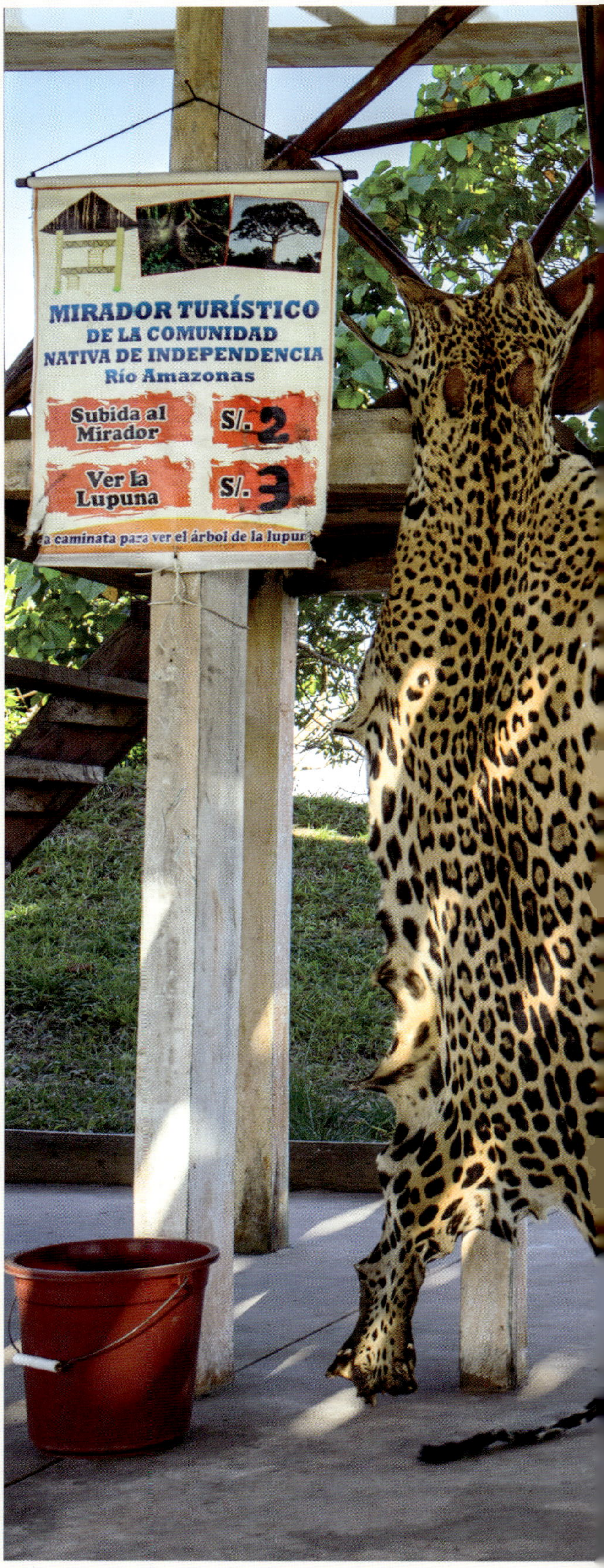

A vendor sells jaguar skins at a tourist destination outside Iquitos, Peru. Jaguars are an internationally protected species, but demand for their skins and canine teeth has increased in recent years, putting the cats back in the crosshairs.

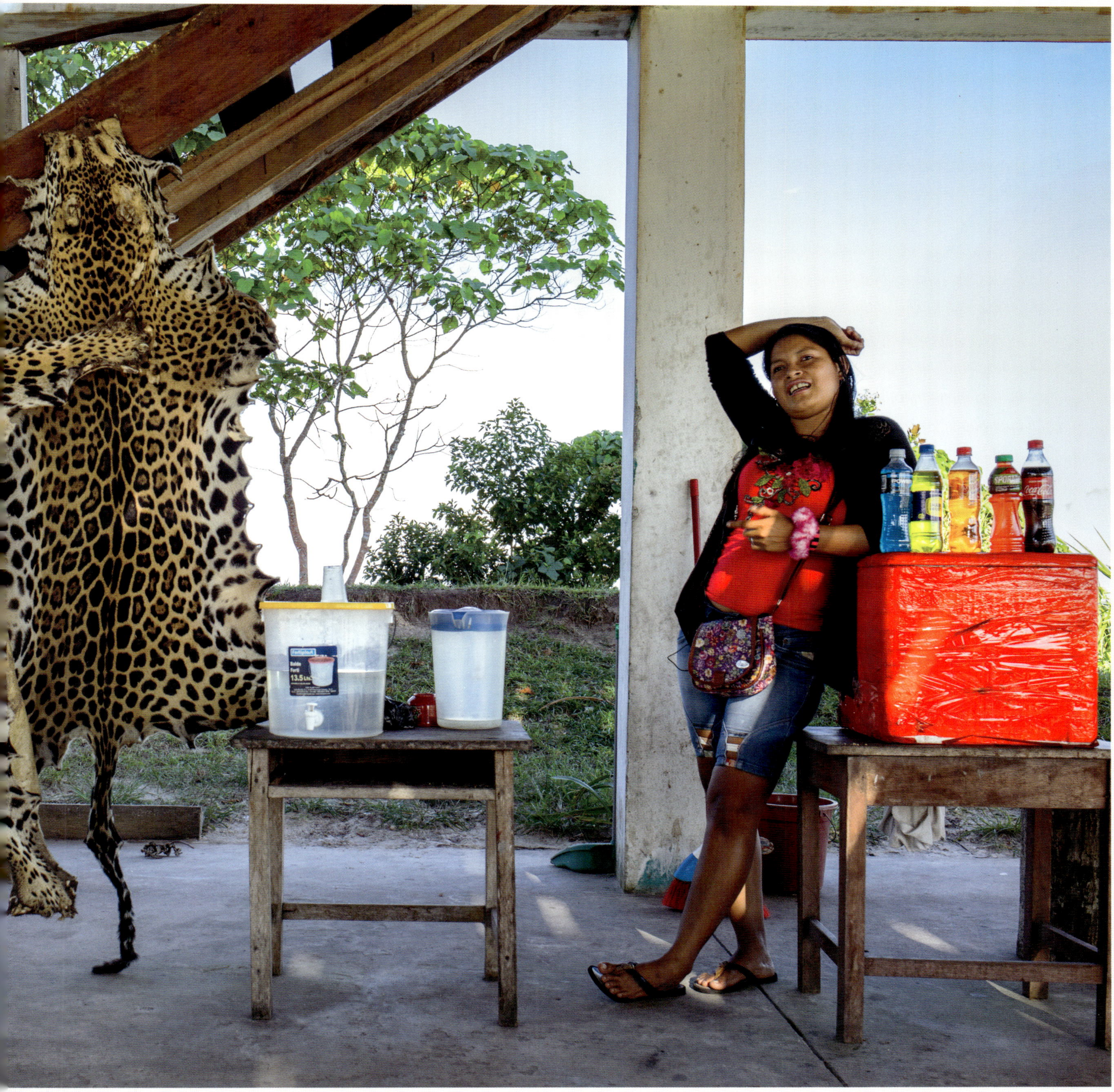

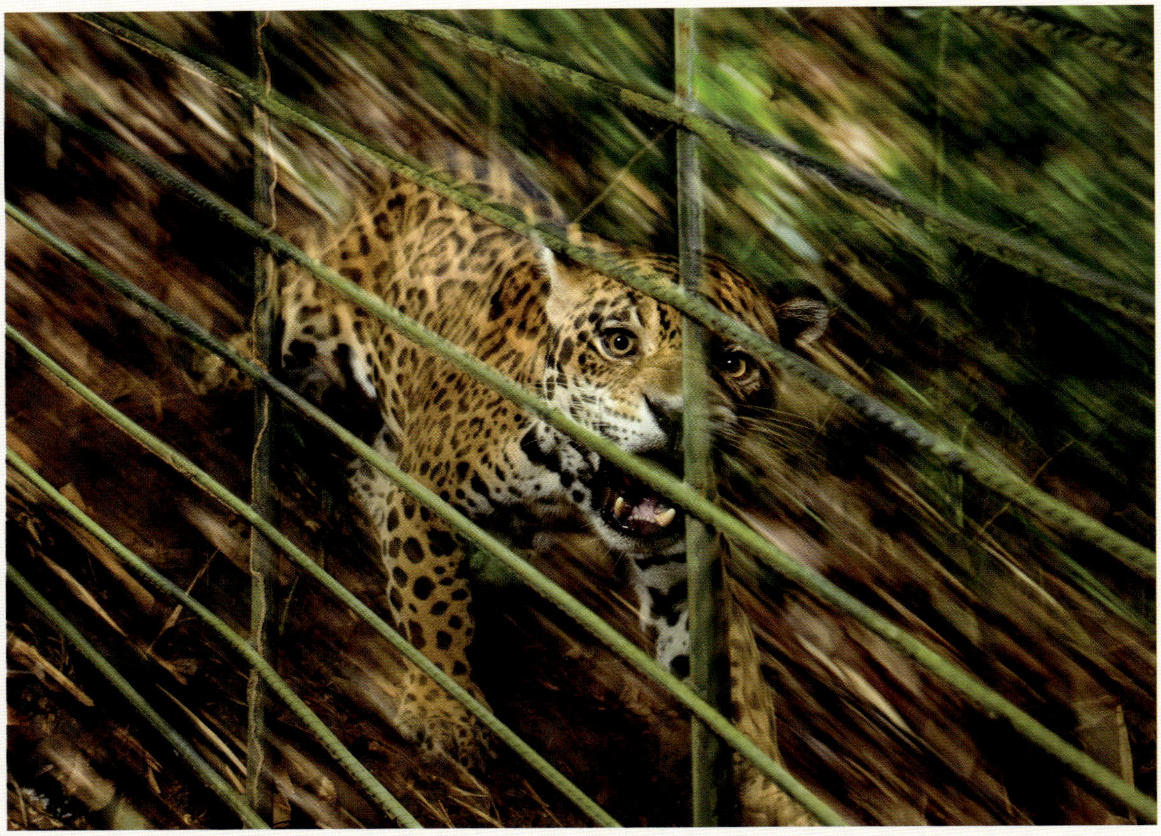

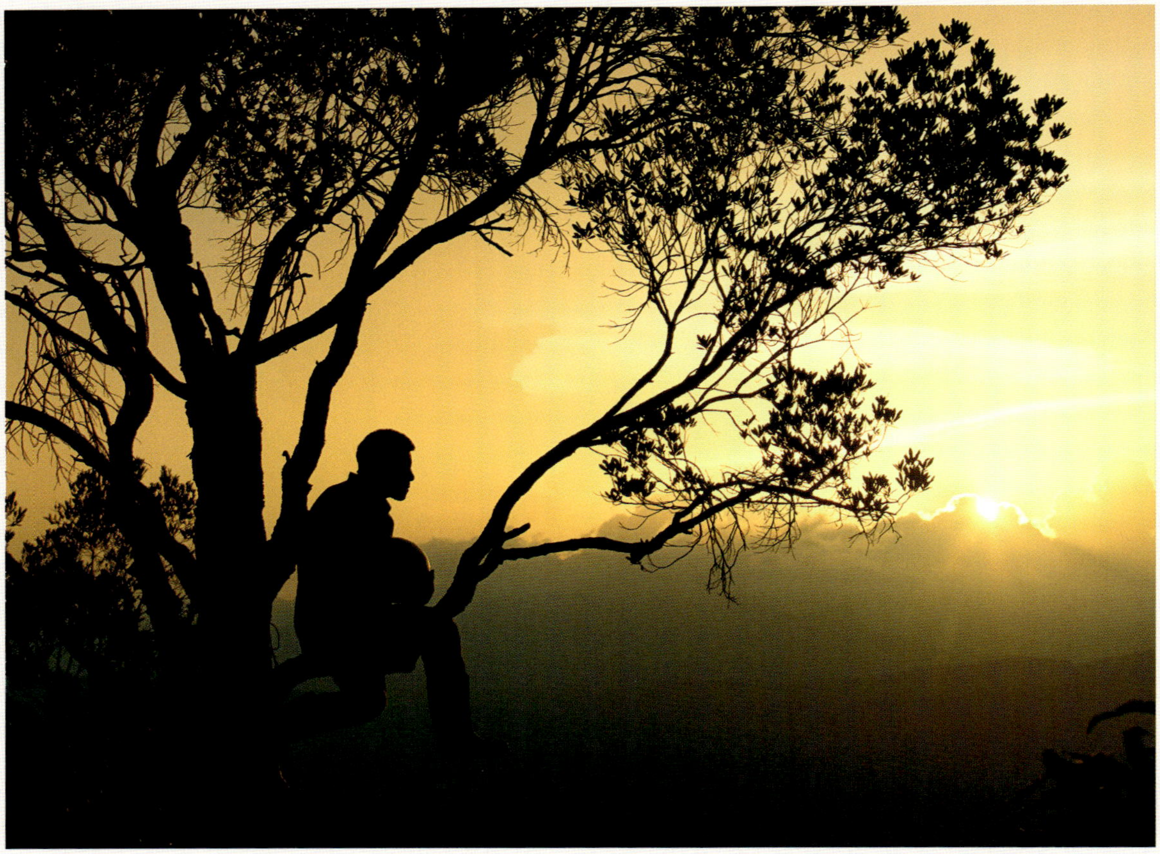

ABOVE: This jaguar was rescued as a cub from the pet trade in southern Colombia. He will spend his life in captivity at this facility in Sabana de Torres, as he was raised by and has habituated to humans.

LEFT: A Maya jaguar hunter in Belize now uses his skills for the tourism industry, attracting the big cats by mimicking their mating call and luring them with a cassava gourd and banana peel soaked in honey.

I documented a jaguar pelt drying on a fence with cattle grazing behind. Then I was introduced to a man who'd killed a dozen jaguars during the previous year. I arranged the skulls in the crooks of a tree and photographed them using soft, fashion-style lighting.

The cats were demonized partially because there was no information on livestock predation, making it easy to blame jaguars, and hunting them was a macho Sunday afternoon tradition. Biologist Sandra Cavalcanti was working to remedy that. She'd outfitted a few cats with old-fashioned radio collars and tracked them on horseback. Doug and I mounted up and joined her on forays. She tore across the landscape after her dogs, lofting an antenna in the air—and I held on as best I could with two cameras around my neck. I am no horseman. They nicknamed me *Mau* Cowboy: bad cowboy.

It was an impossible landscape. Sometimes Sandra lost the cats in thick vegetation or couldn't follow them through a mucky marsh or across a river. I called Alan Rabinowitz and convinced him to pay for 10 GPS collars; they would ping the collared cats' locations to Sandra's laptop every two hours. I went with her on the first collaring attempt, and it was an adrenaline moment. A jaguar leaped out of a tree, landed maybe a foot away from me—and then took off.

A few weeks later, Sandra got a call from Armando. He'd allowed her to track collared cats on his land, and he was angry. "One of your jaguars killed one of my cows last night," he said. Sandra opened her computer and pulled up maps that pinpointed the cats' movements. The night before, the accused was about seven miles away from Armando's property. The jaguar was innocent.

Armando was contrite. "This is the science we've been waiting for," he said. After 30 months of tracking, Sandra proved that less than a third of jaguar meals were beef. Snakes, puma attacks, disease, and parasites were far more deadly to cattle herds.

Sandra's work, alongside our 2001 *National Geographic* story, sparked a shift in attitudes about jaguars in the Pantanal. The story also revealed that Brazil's Pantanal was *the* place to see jaguars in the wild, and they became superstars in a booming ecotourism industry. A 2017 study revealed that a cow was worth $2,000 in its lifetime, but each year, tourism brought in between eight and nine million dollars, with each jaguar contributing $108,000 to the local economy. That made them far more valuable alive than dead, and jaguar populations continue to rise in the Pantanal. To save wildlife, people must benefit economically, especially if the animals are dangerous.

Jaguar numbers were hammered by the expanding human presence across the Americas, and then the post–World War II fashion craze for fur decimated big cat populations worldwide. Brazil alone exported some 30,000 jaguar skins each year during the 1960s. An international treaty put a stop to the carnage in 1973, banning international trade. Meanwhile, humans continued to level the jaguars' homelands, eradicating *el tigre* from nearly half its original range within a century.

Two scientific discoveries changed that trajectory. In the 1930s, scientists categorized jaguars into 16 subspecies, but by the end of the century, the revolution in genomic technology contradicted this theory. DNA samples revealed few genetic differences, making jaguars the world's only large, wide-ranging carnivore with no subspecies.

The implications for conservation were earth-shattering. From Mexico's deserts to the dry Pampas of northern Argentina, jaguars had been wandering great distances to breed with each other, even swimming across the Panama Canal along what was an ancient, possibly prehistoric wildlife corridor that was once unbroken wildland.

On another front, Alan had unleashed a small army of researchers to locate where jags were and where they went. They found dozens of hot spots with perhaps 50 jaguars each. Then, when he overlaid the GPS data points onto a map, a path appeared, composed of 82 corridors that spiderwebbed through jaguar territory.

Alan conceived a grand conservation strategy to connect 90 jaguar populations across the Americas. Protecting a wide-ranging mammal throughout its entire range had never been attempted before. His plan spanned 1.7 million square miles.

He envisioned an adolescent cat striking out to find unoccupied territory, moving along a mostly contiguous "jaguar freeway." Alan likened this corridor to the Underground Railroad. Farmland and forest patches linked protected areas or offered a place to rest for young cats searching for a new domain and mate, thus invigorating the gene pool. Gene flow is crucial for long-term survival. Trapped populations are susceptible to inbreeding and often march the inexorable road toward extinction. And now, with the growing climate emergency, the ability to move safely into new areas may determine life or death.

I traveled with Alan to Colombia and Costa Rica as he met with business owners, government officials, and landowners to convince them to protect the Jaguar Corridor—or to at least allow safe passage for traveling cats. This was a new 21st-century conservation paradigm. Saving these wide-ranging felines would save the entire realm, including the plants and animals living beside them. Since then, stitching together habitat and rewilding it with its former array of species has become a global movement.

I wanted to share the stealthy grace and beauty of this animal with a wider audience, so I returned to the Pantanal in 2016 with filmmaker Bertie Gregory. The cats had become far less skittish. Many had grown up around tourists and had never fled from hunters. They leaped off low cliffs to hunt caiman right in front of our boat. Their numbers had grown.

Alan frequently reminded me that this cat was a god, and it was the thread binding together cultures throughout the Americas.

Jaguars will always be my first big cat love. I can't fathom a world without them.

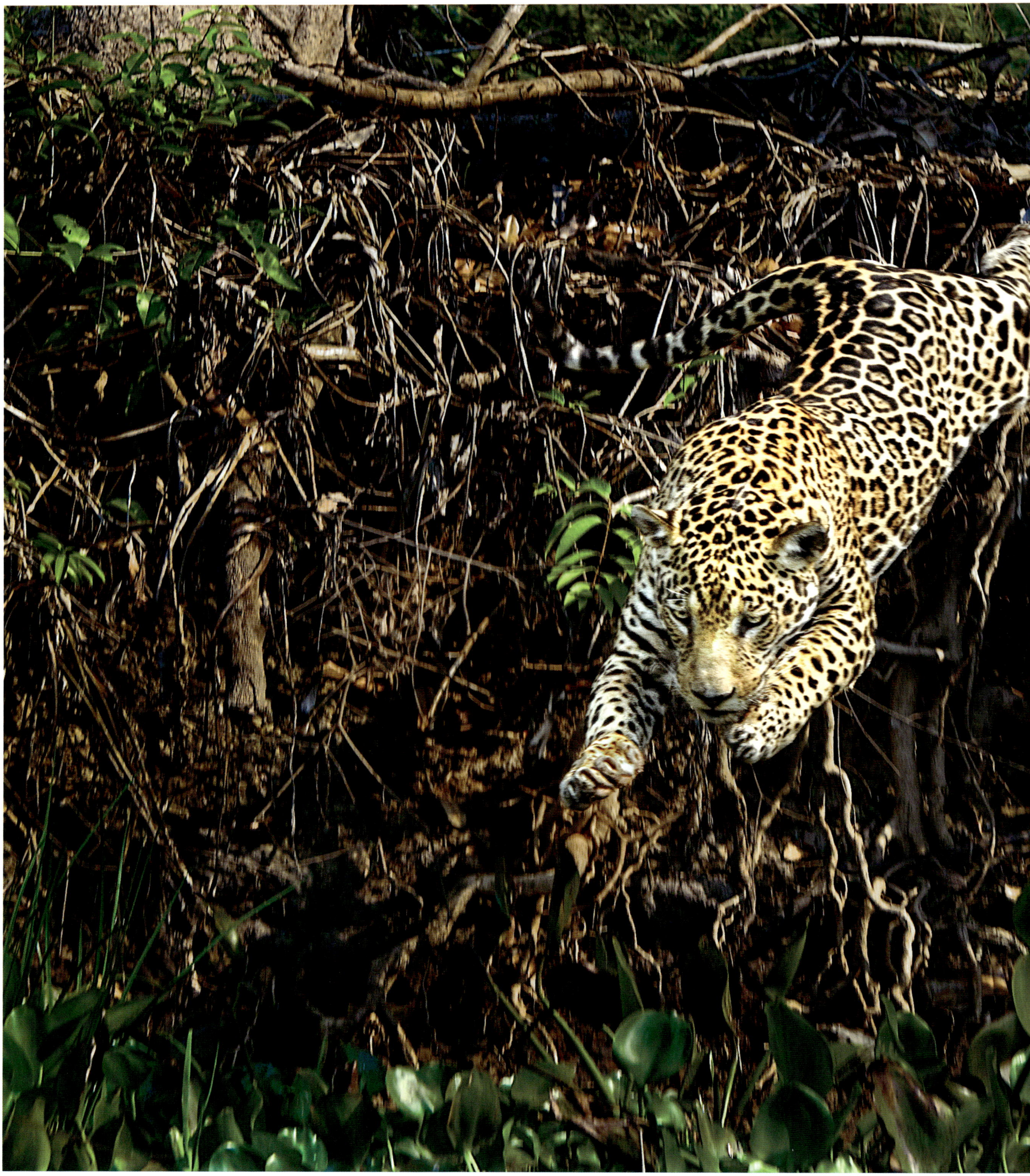

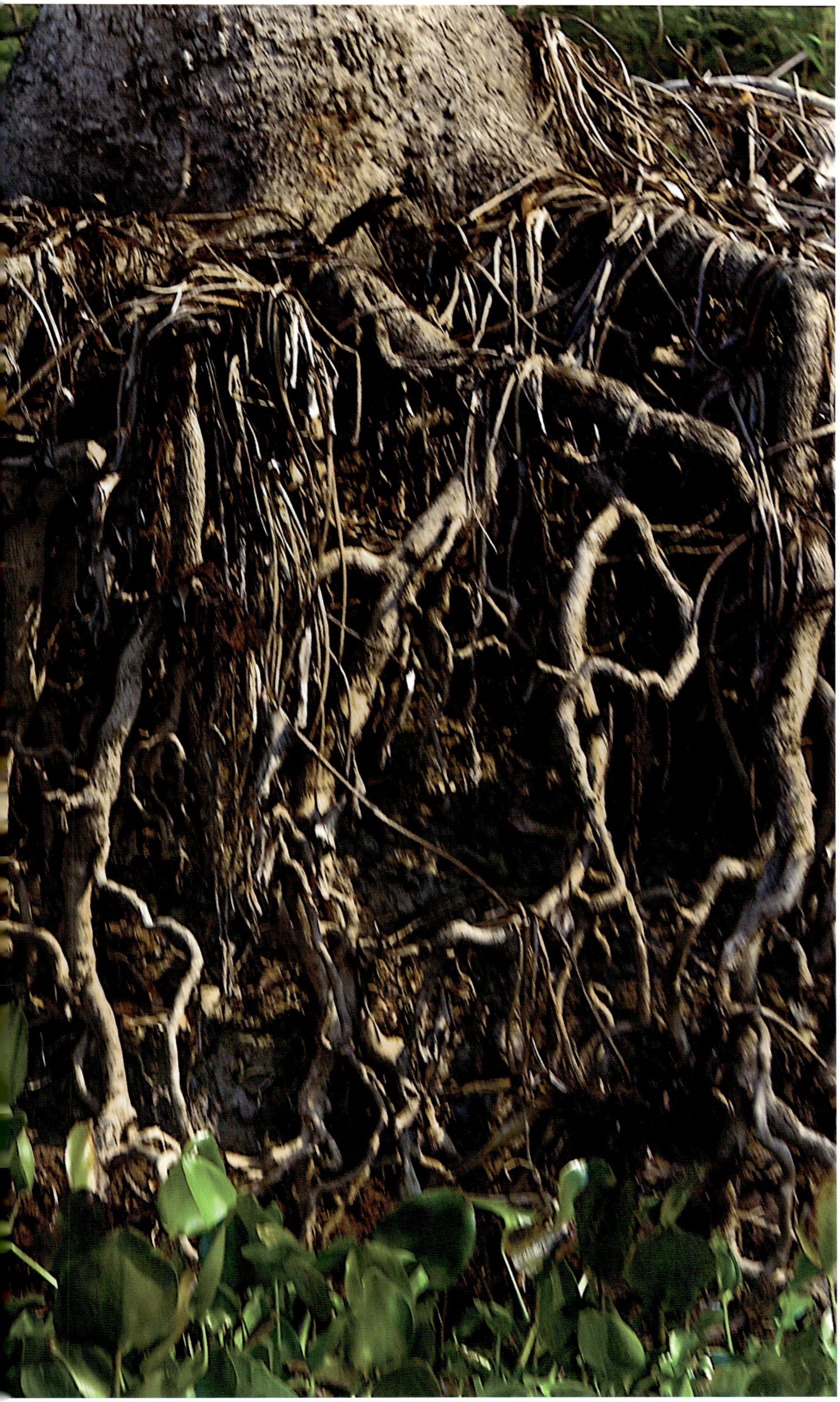

A jaguar leaps into Brazil's Cuiabá River hunting a caiman, its favorite food. Jaguars kill prey with a precision bite to the skull. They have the strongest bite force, relative to size, of any large cat.

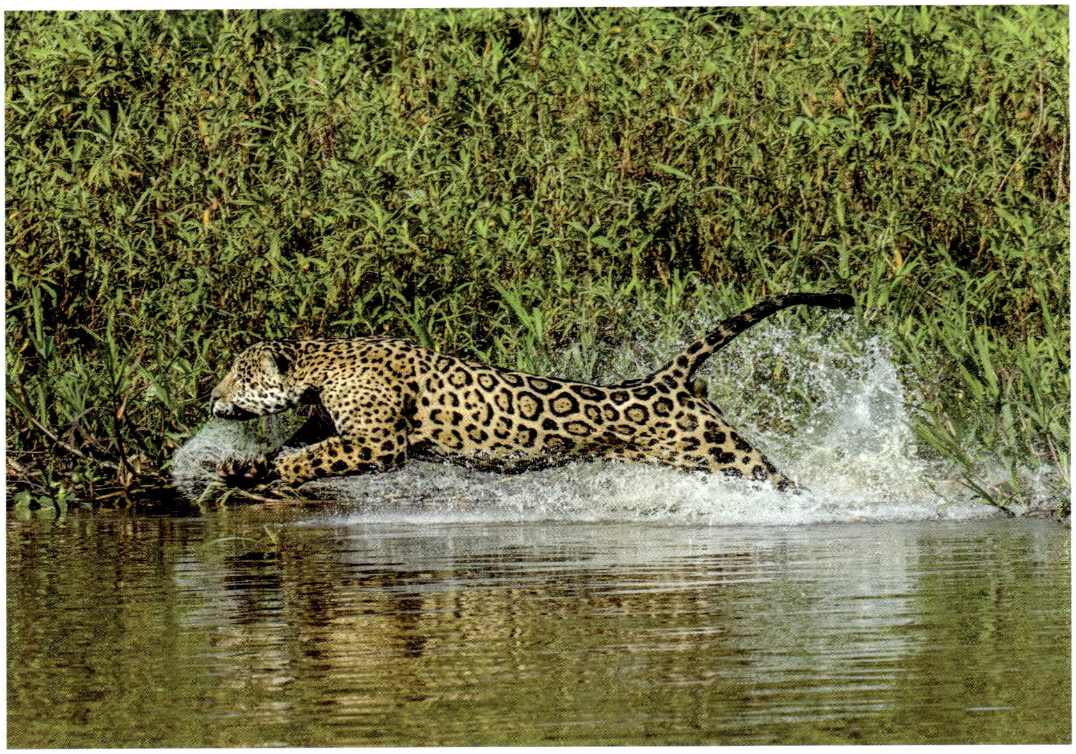

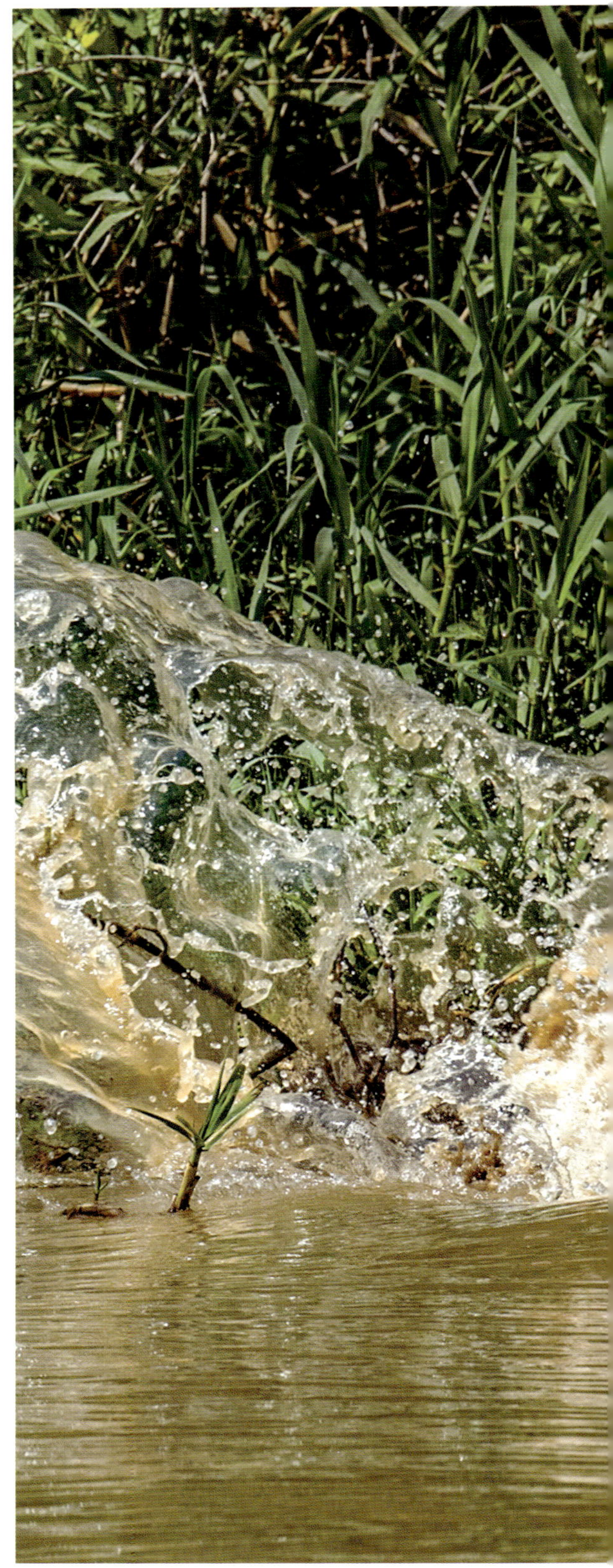

ABOVE: A jaguar lunges after a caiman in the Pantanal. Given the cats' extensive diet and their comfort in water, prowling a riverbank is like scanning a buffet. With their broad palate, they also hunt on land.

RIGHT: Jaguars succeed in their hunts approximately once every 10 attempts. Wrangling a caiman is particularly difficult.

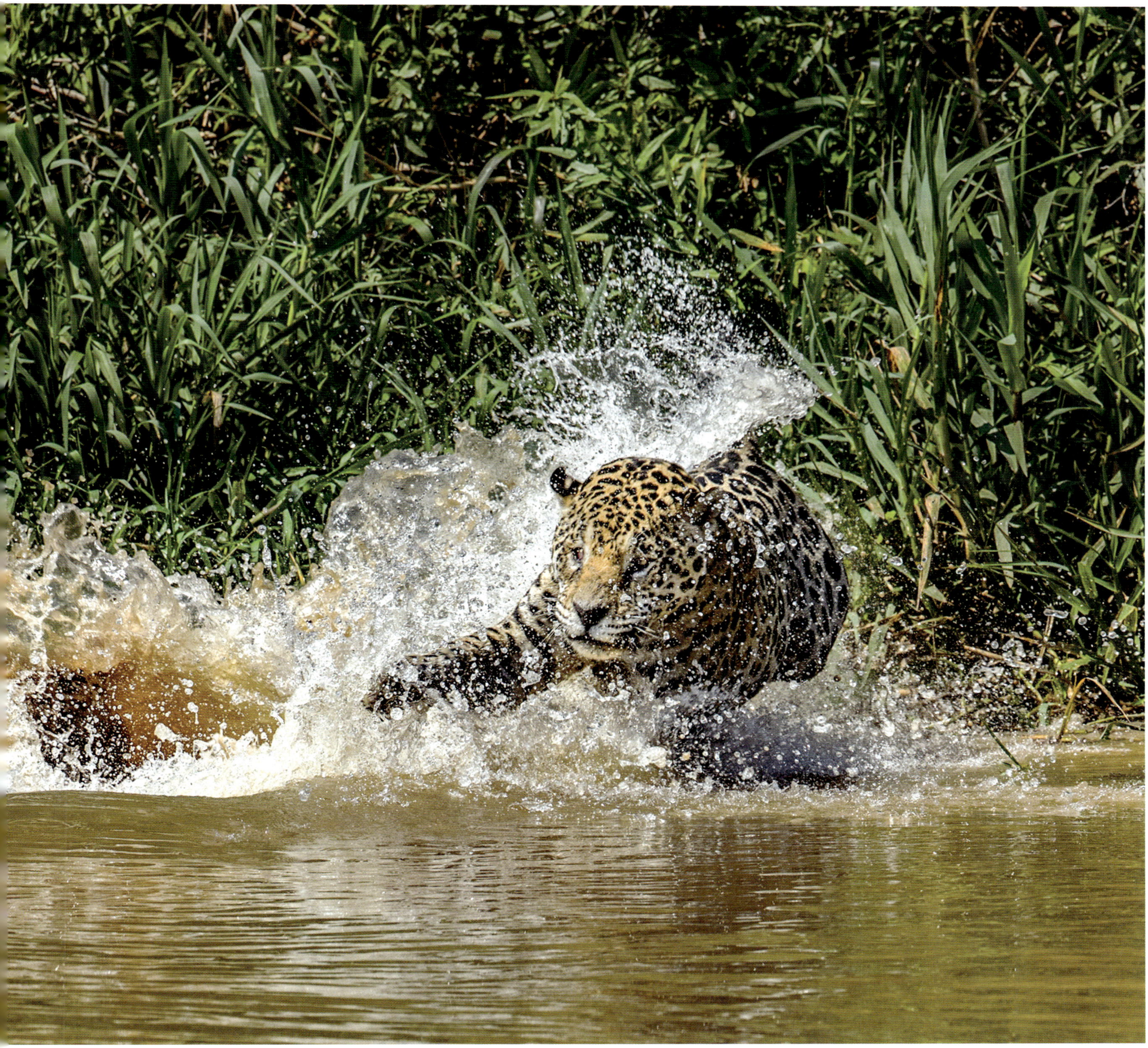

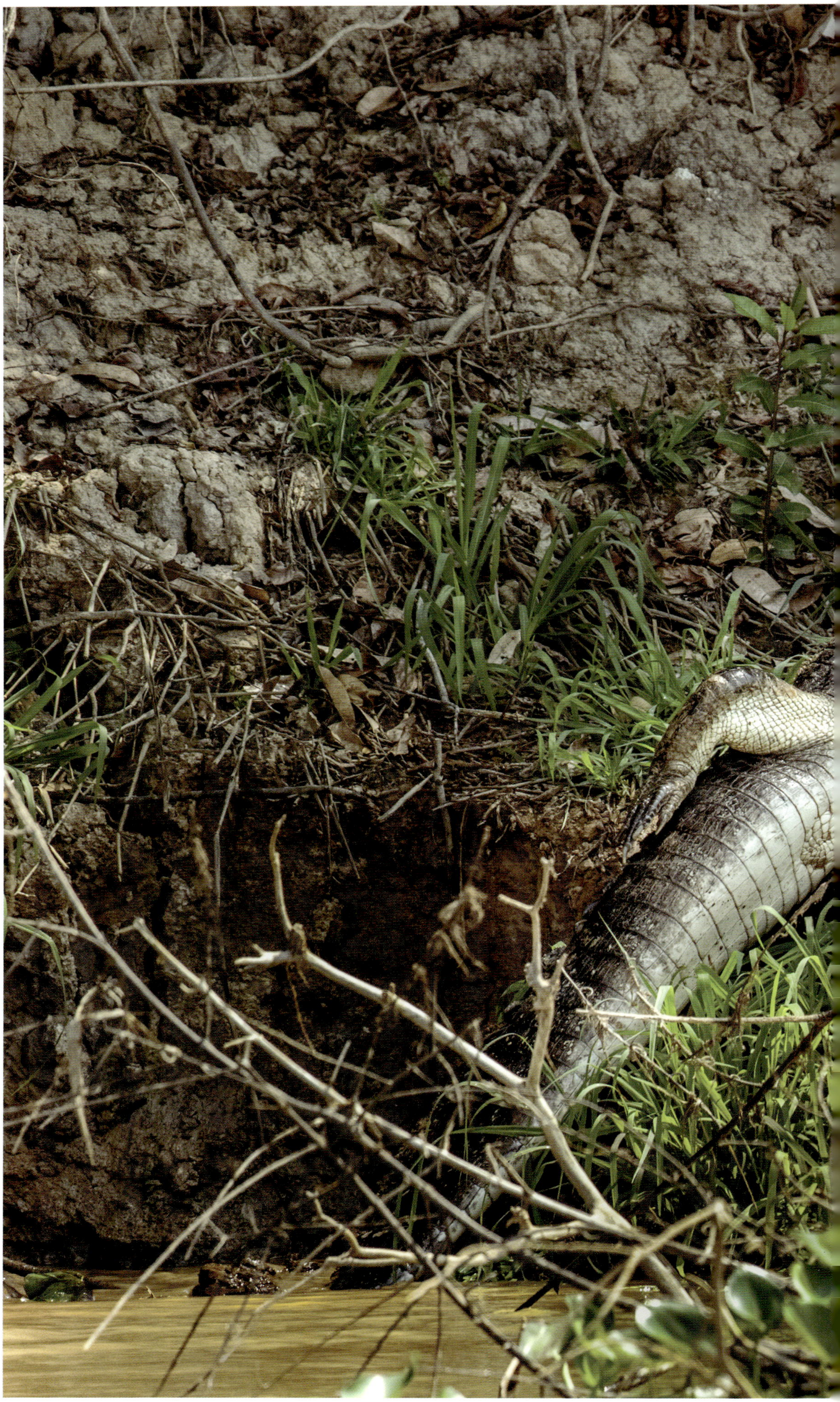

After a dramatic underwater fight, this caiman became a meal. Built like prizefighters, jaguars are muscled, compact, and strong enough to haul a 100-pound caiman up a steep riverbank.

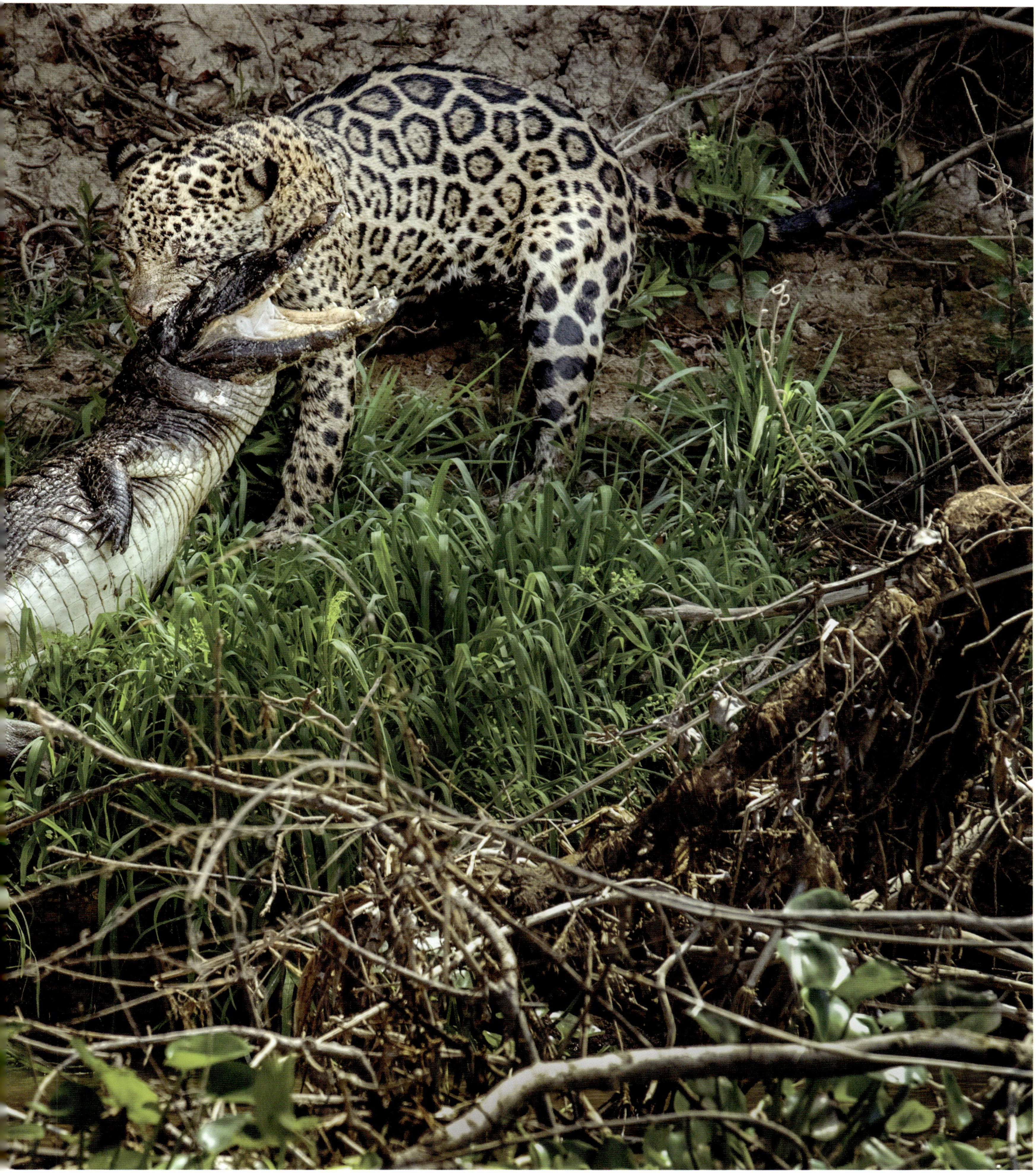

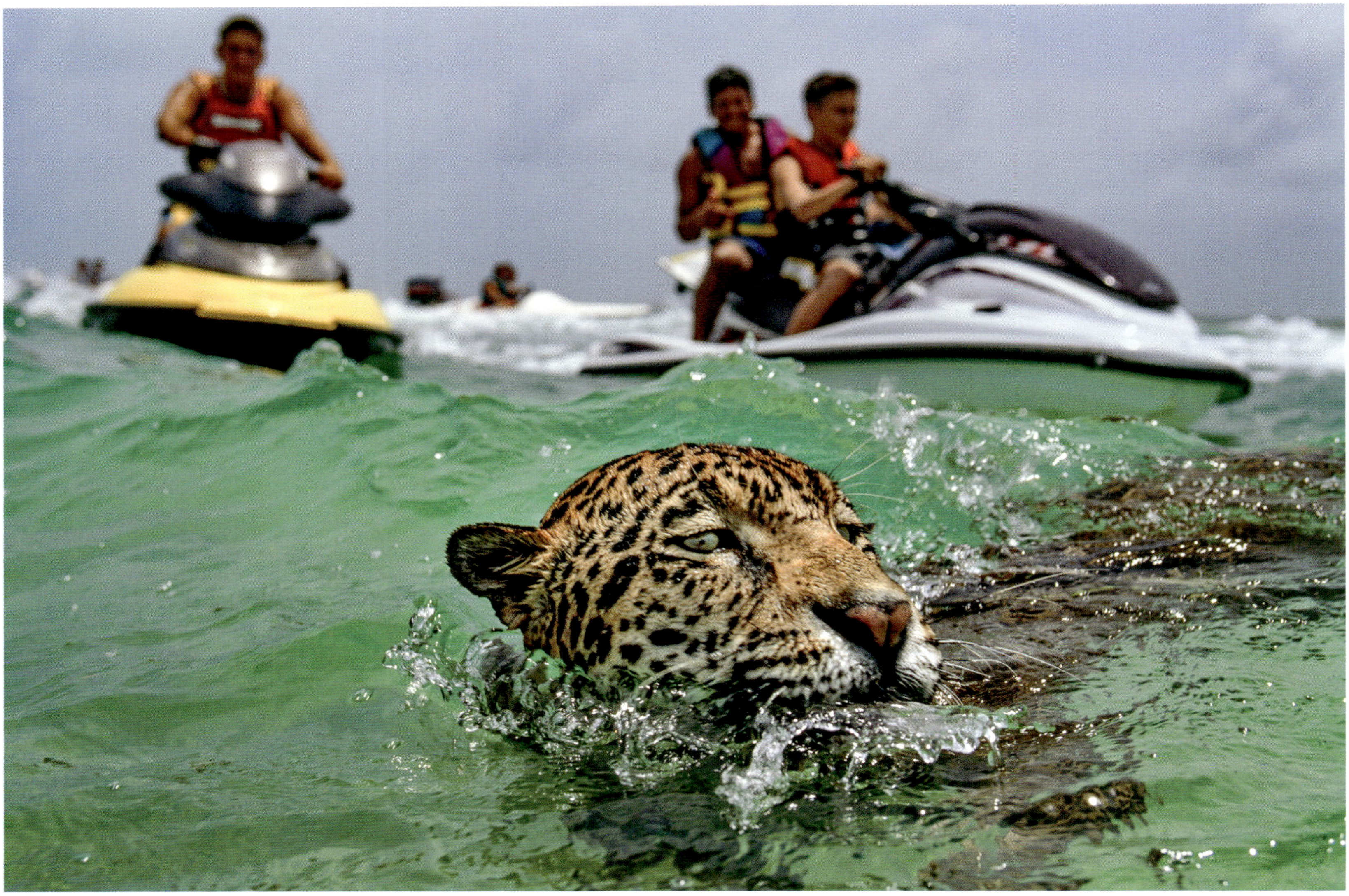

In Cancún, Mexico, a jaguar purportedly rescued from the illegal wildlife trade was trained to swim with tourists. More tourists watch from nearby.

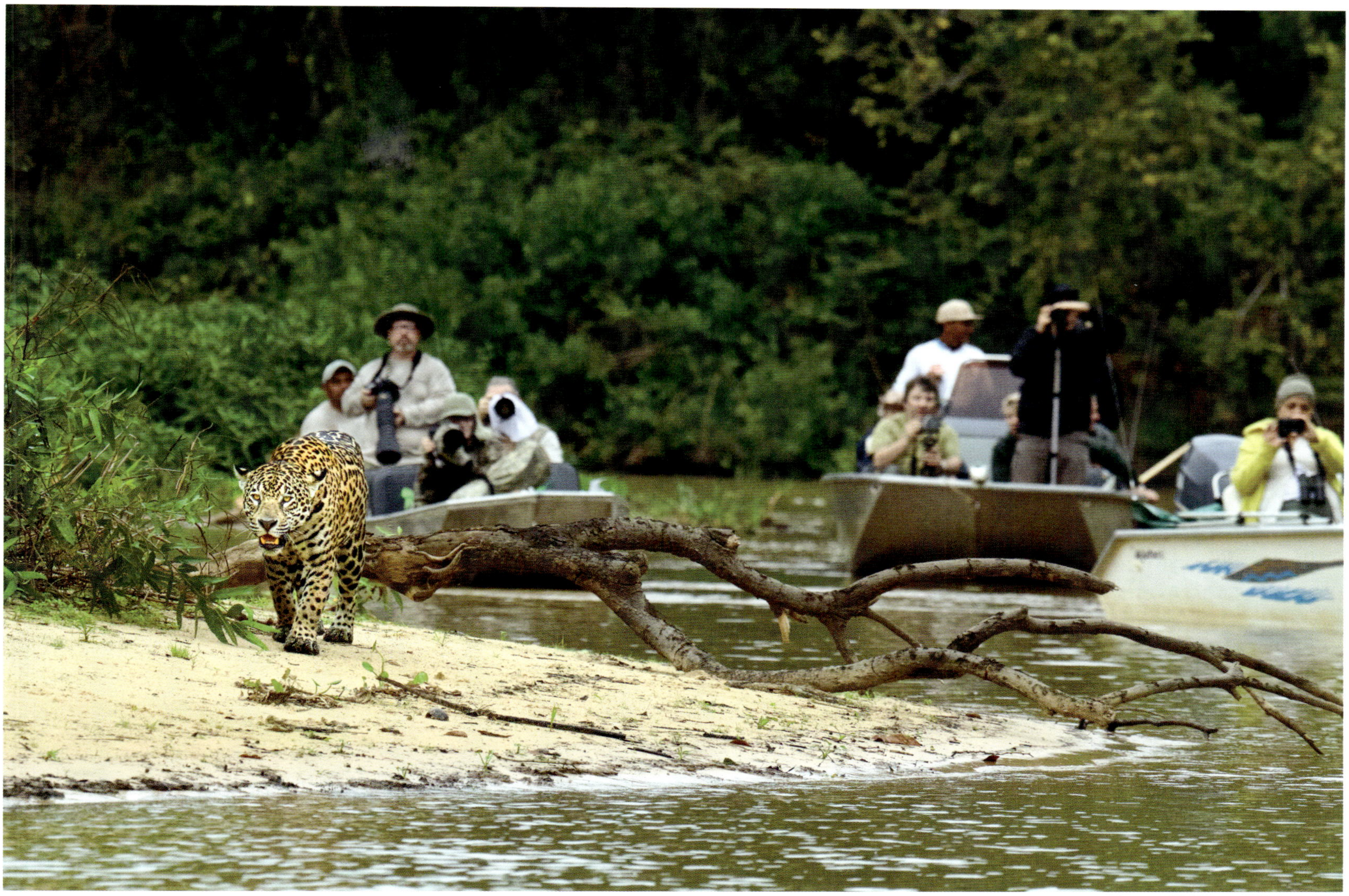

Substantial income from ecotourism in Brazil's Pantanal has made jaguar conservation a priority for locals. Though the species' territory stretches across 18 countries from Mexico to Argentina, this is the best place to see them.

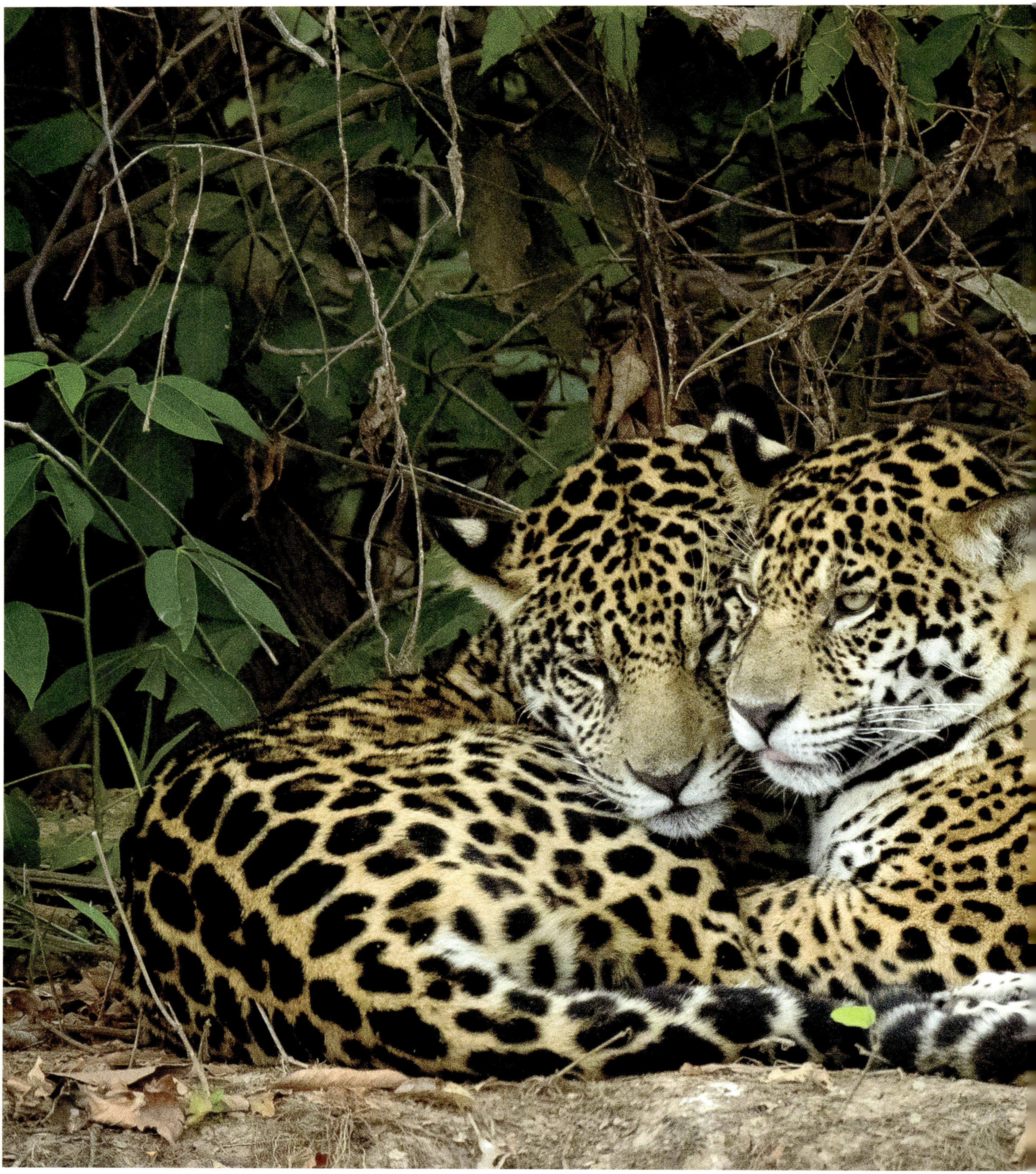

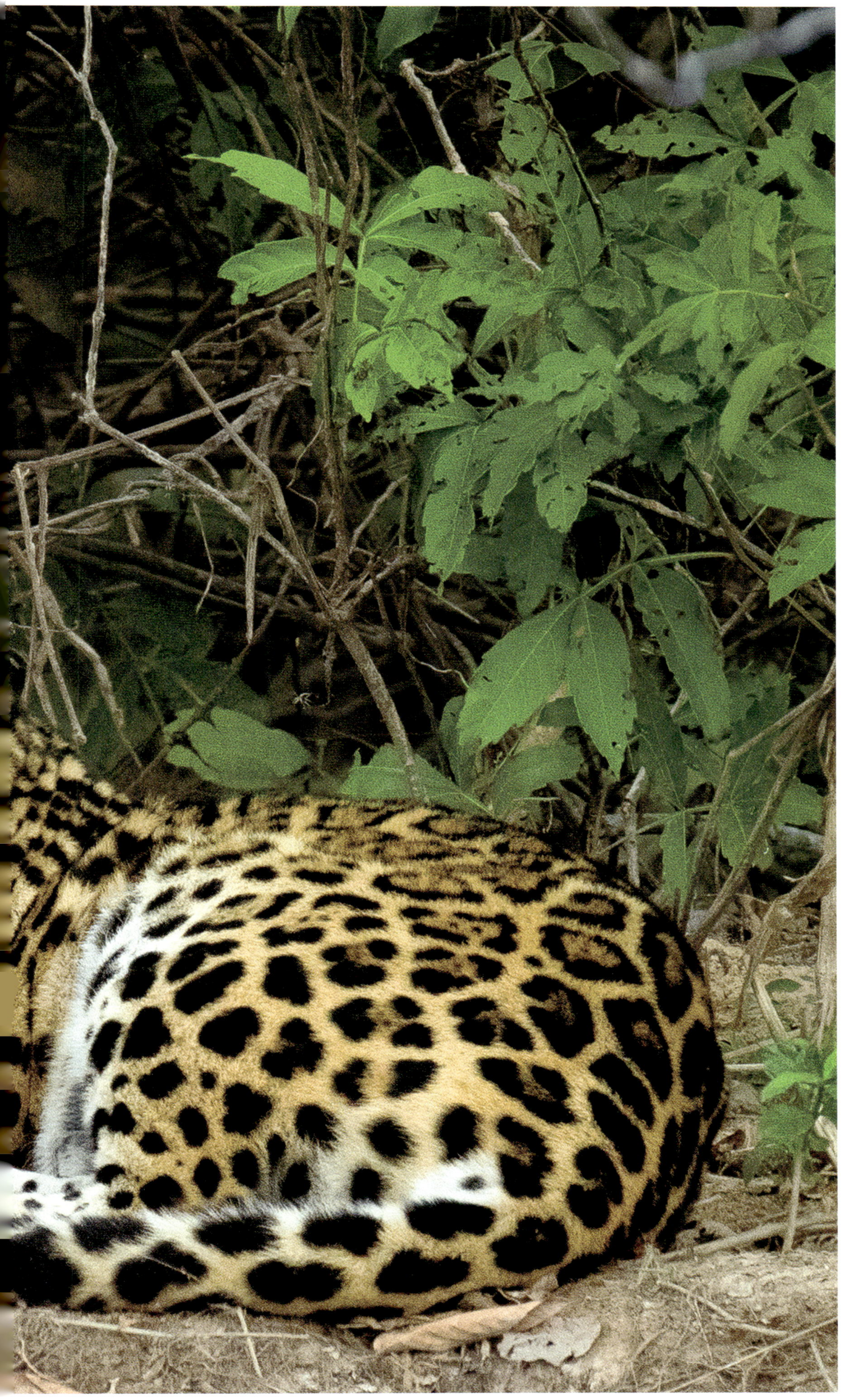

Now too old to live with their mother, these young jaguar brothers will stay together until they head off to find their own territories.

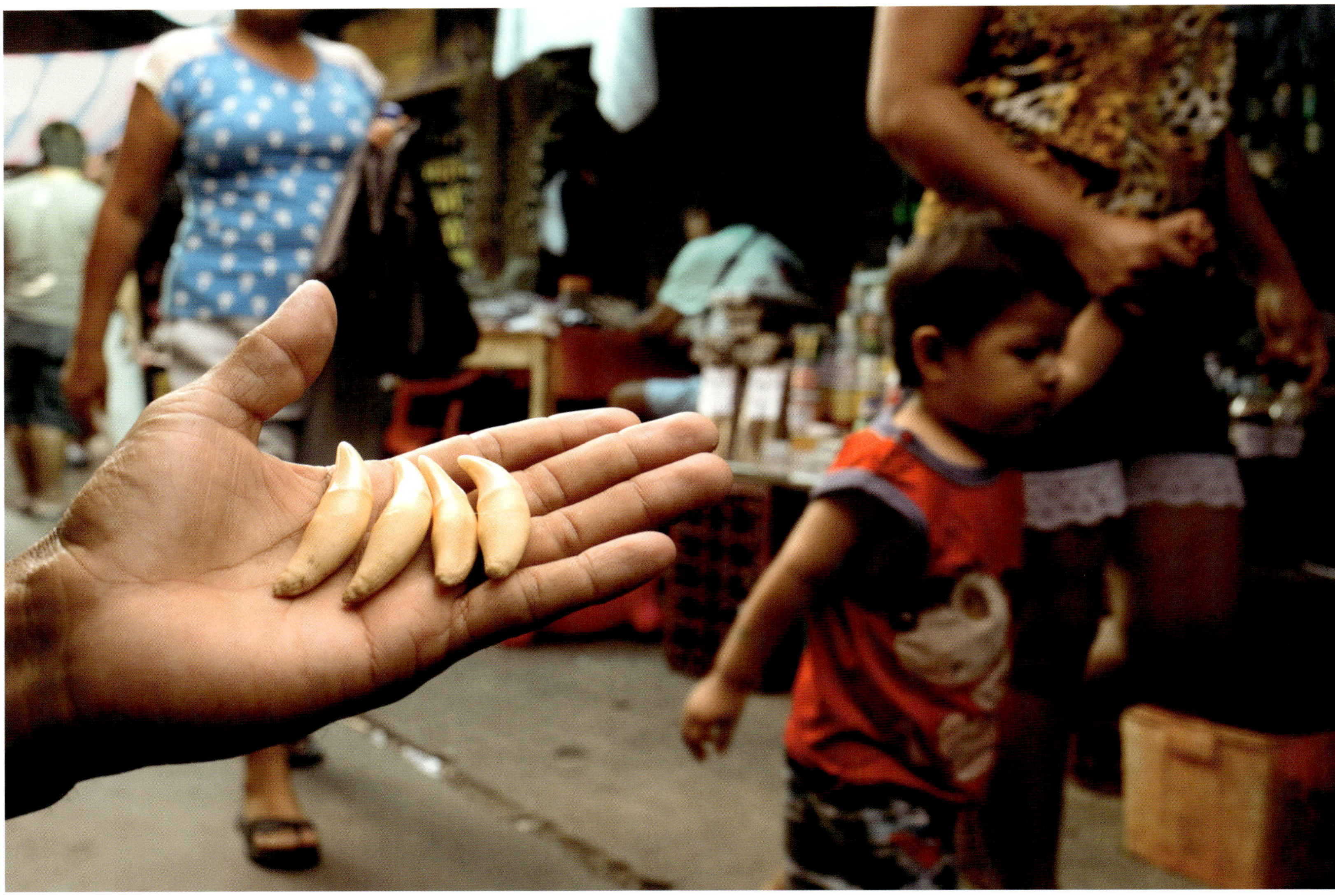

Jaguar skins, canine teeth, and claws are sold at the back of Belén market in the Amazonian city of Iquitos. As jaguars are a protected species, commercial trade of any kind is illegal, but Peru is a wildlife poaching and trafficking hot spot.

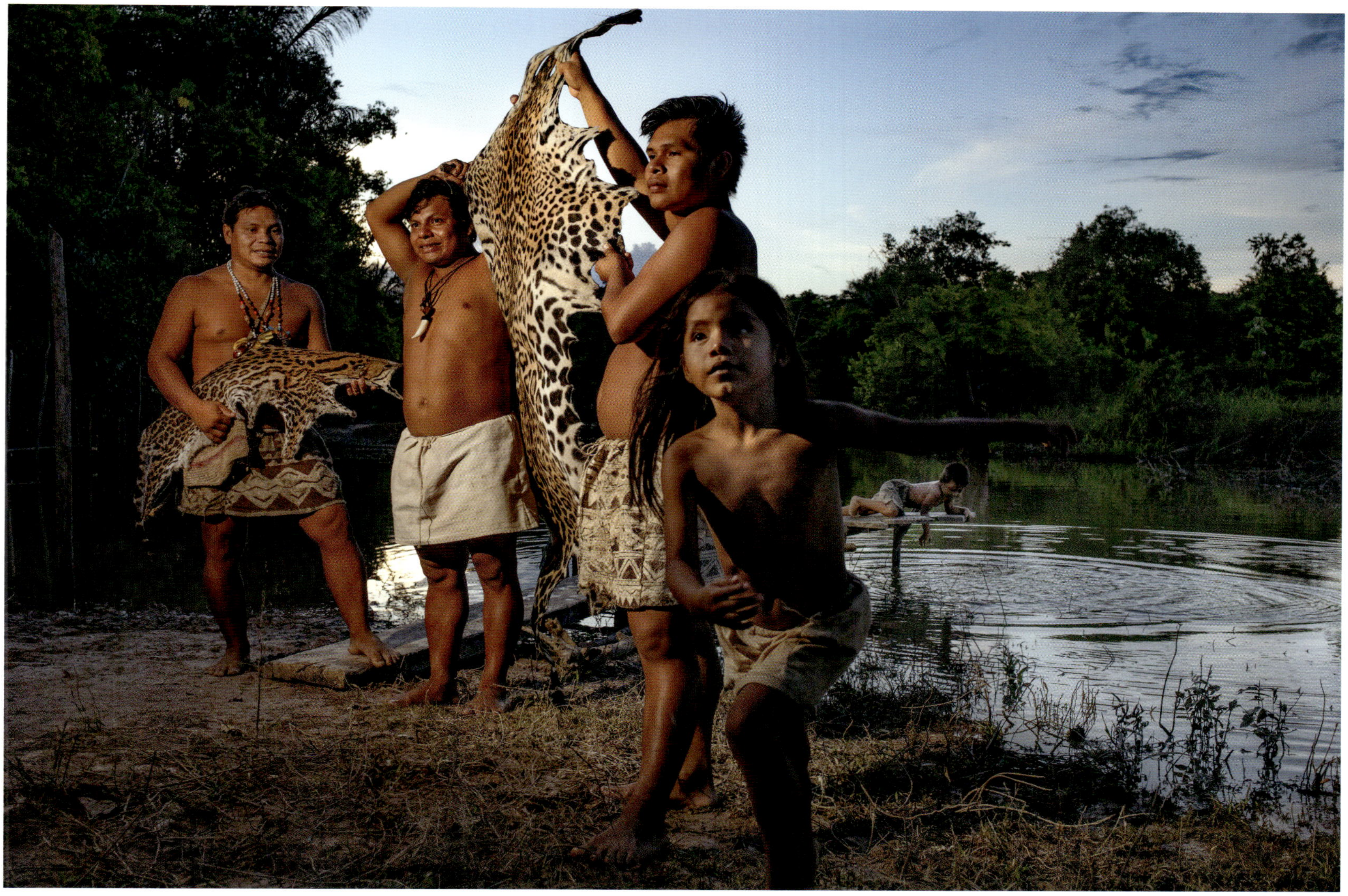

Members of the Boras Comunidad, a Peruvian Indigenous group, display a jaguar skin at a tourist destination. The elders told me that they sell skins to tourists and dealers, but Chinese buyers also commission them to hunt jaguars for their canine teeth.

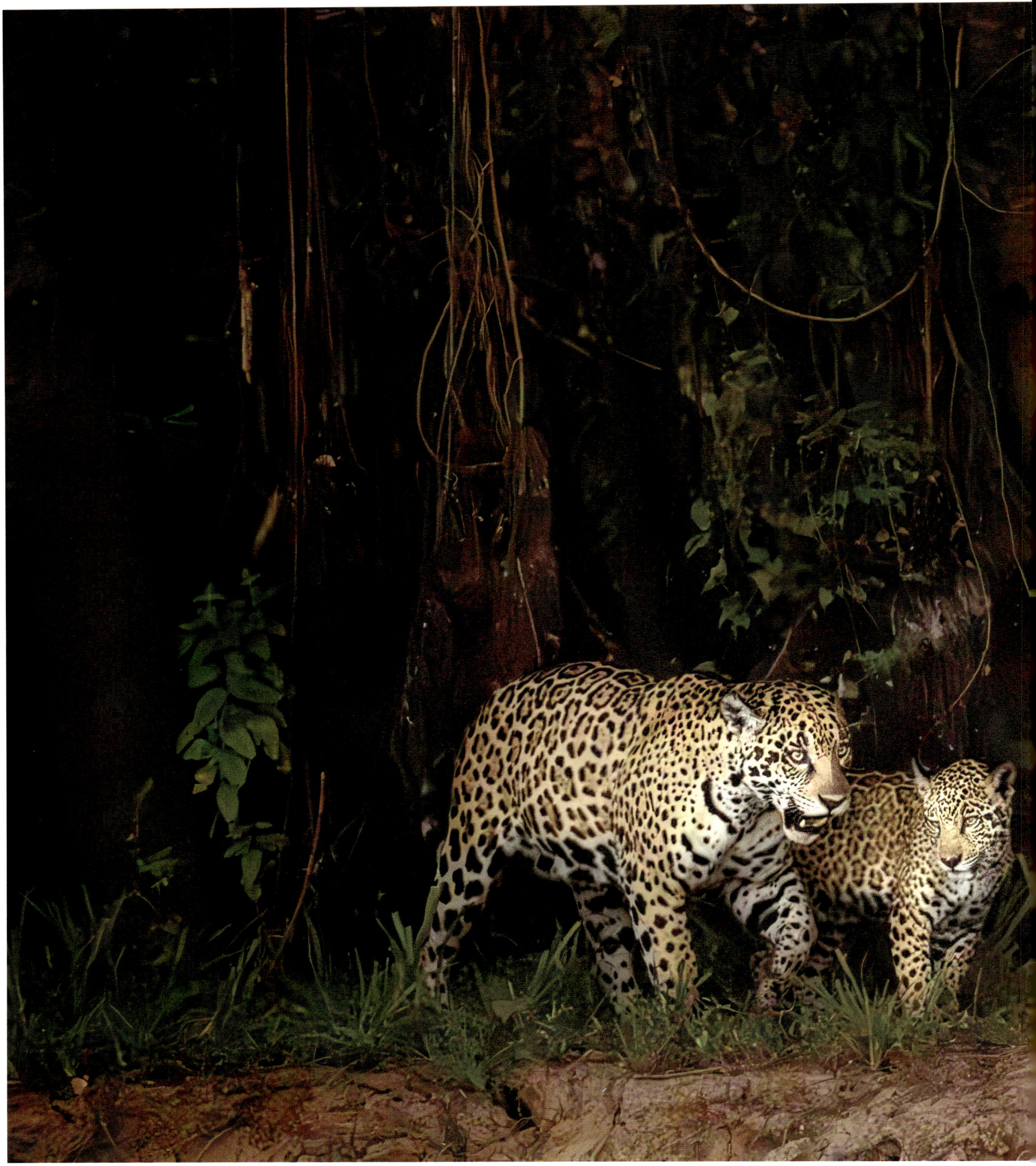

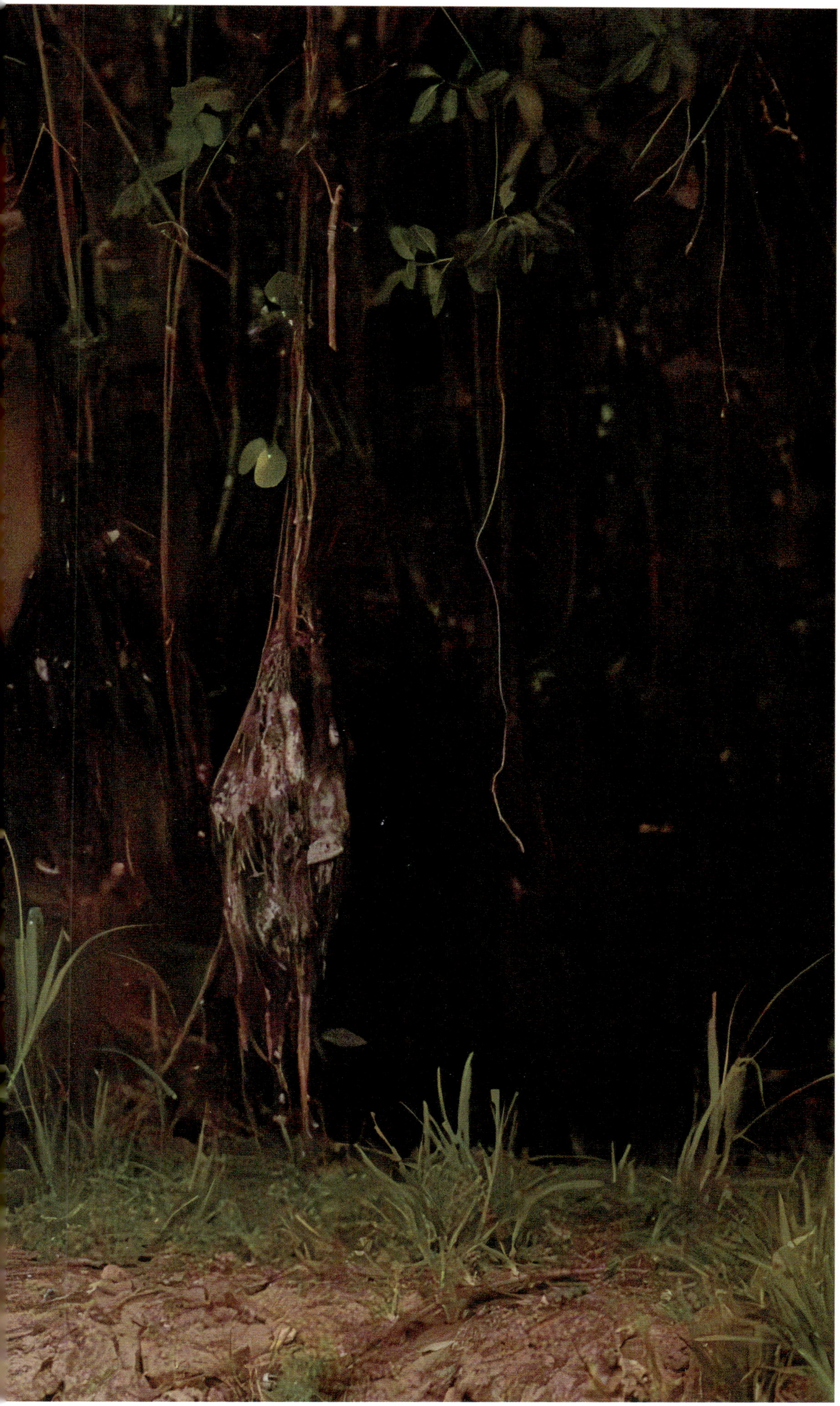

A mother and cub patrol the Cuiabá River just after dusk, prime time for hunting. The ancient Maya believed that jaguar spirits battled underworld forces at night to allow the sun to rise and life to continue.

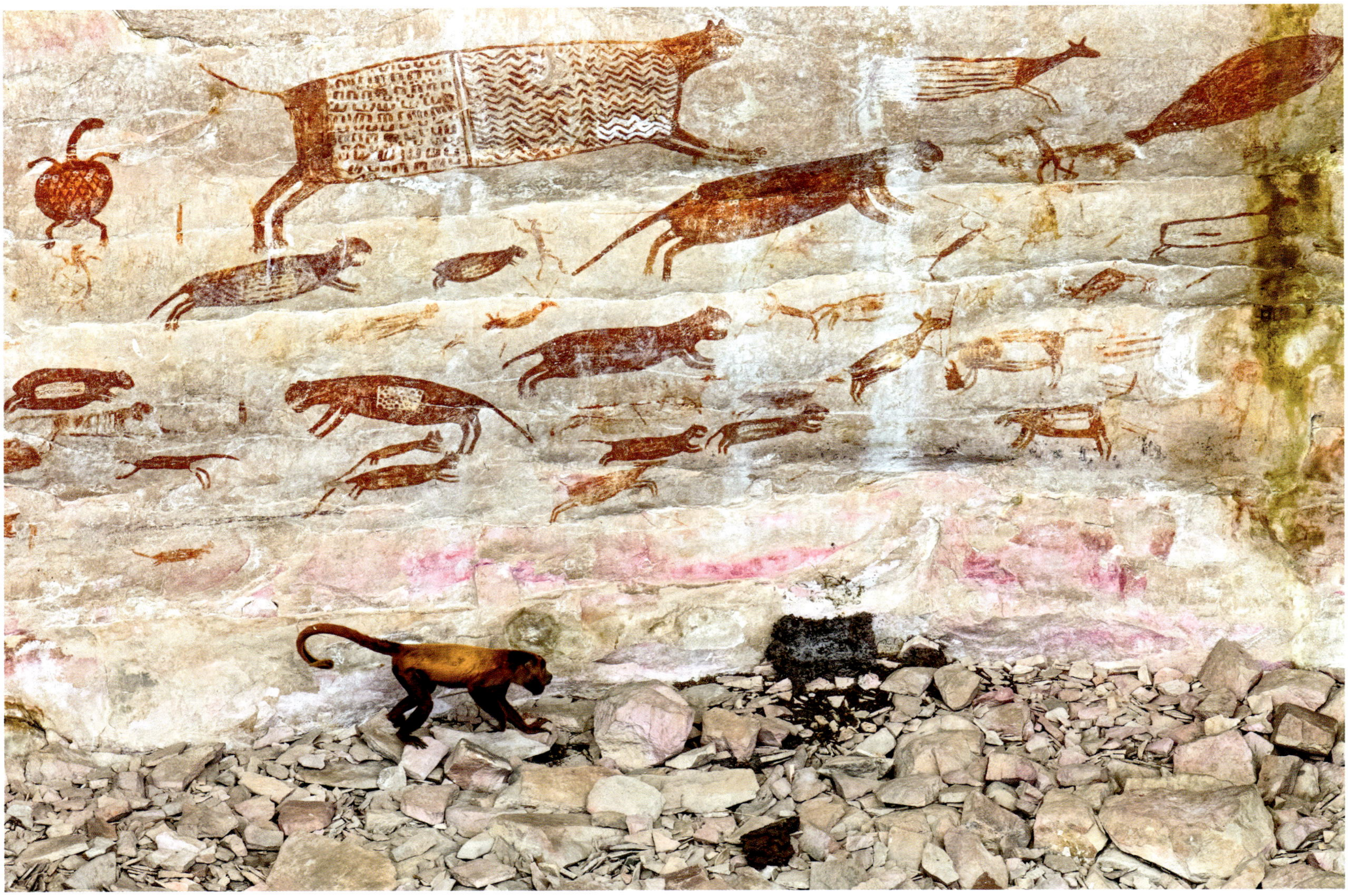

These ancient paintings of jaguars and other animals, high up on *tepuis* (tabletop mountains) in Colombia's Chiribiquete National Park, are thought to be some 20,000 years old.

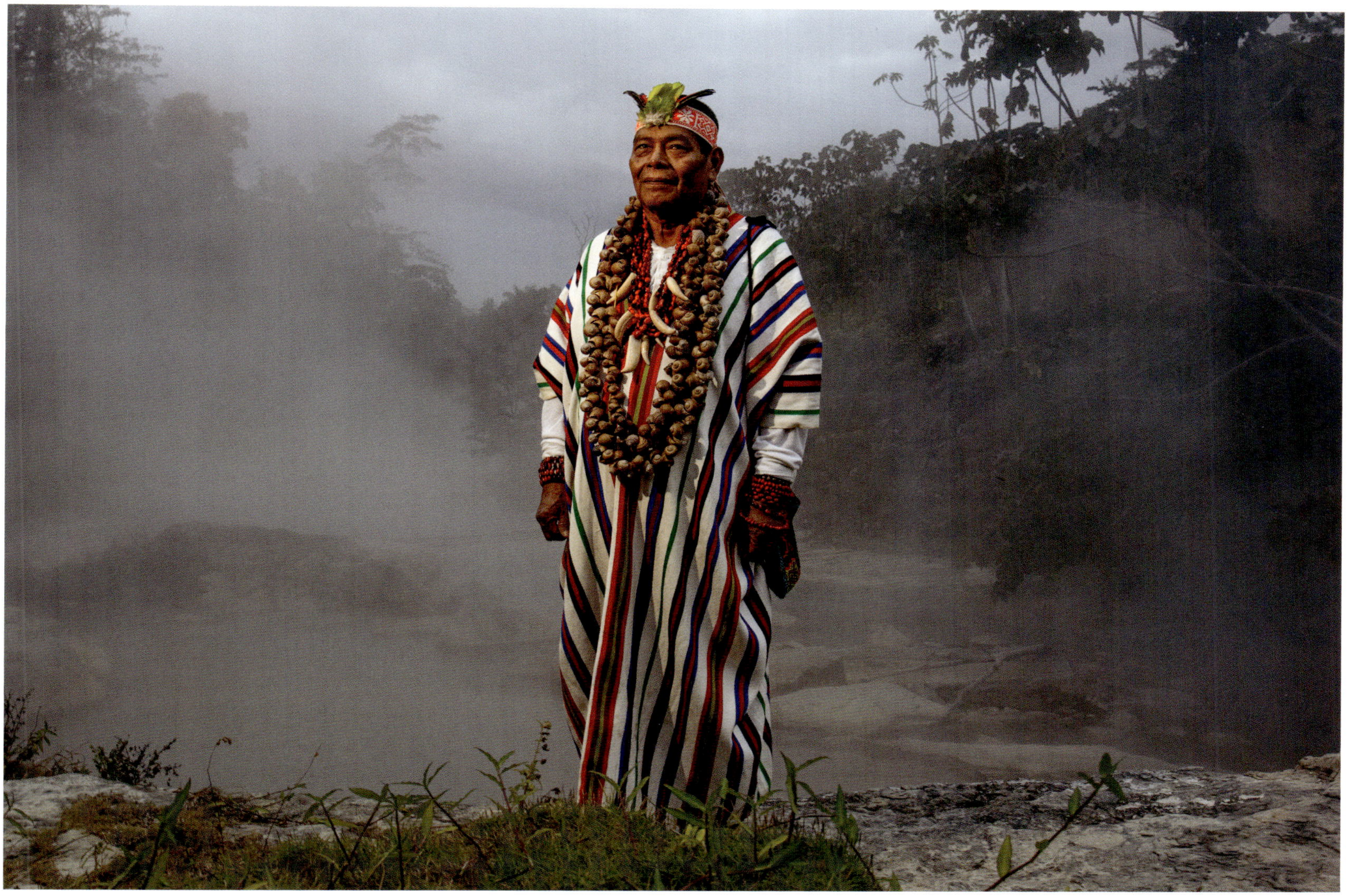

Peruvian shaman Maestro Juan Flores stands by the Boiling River, which was once avoided by local people because of jaguars. The site was also thought to be ruled by otherworldly forces.

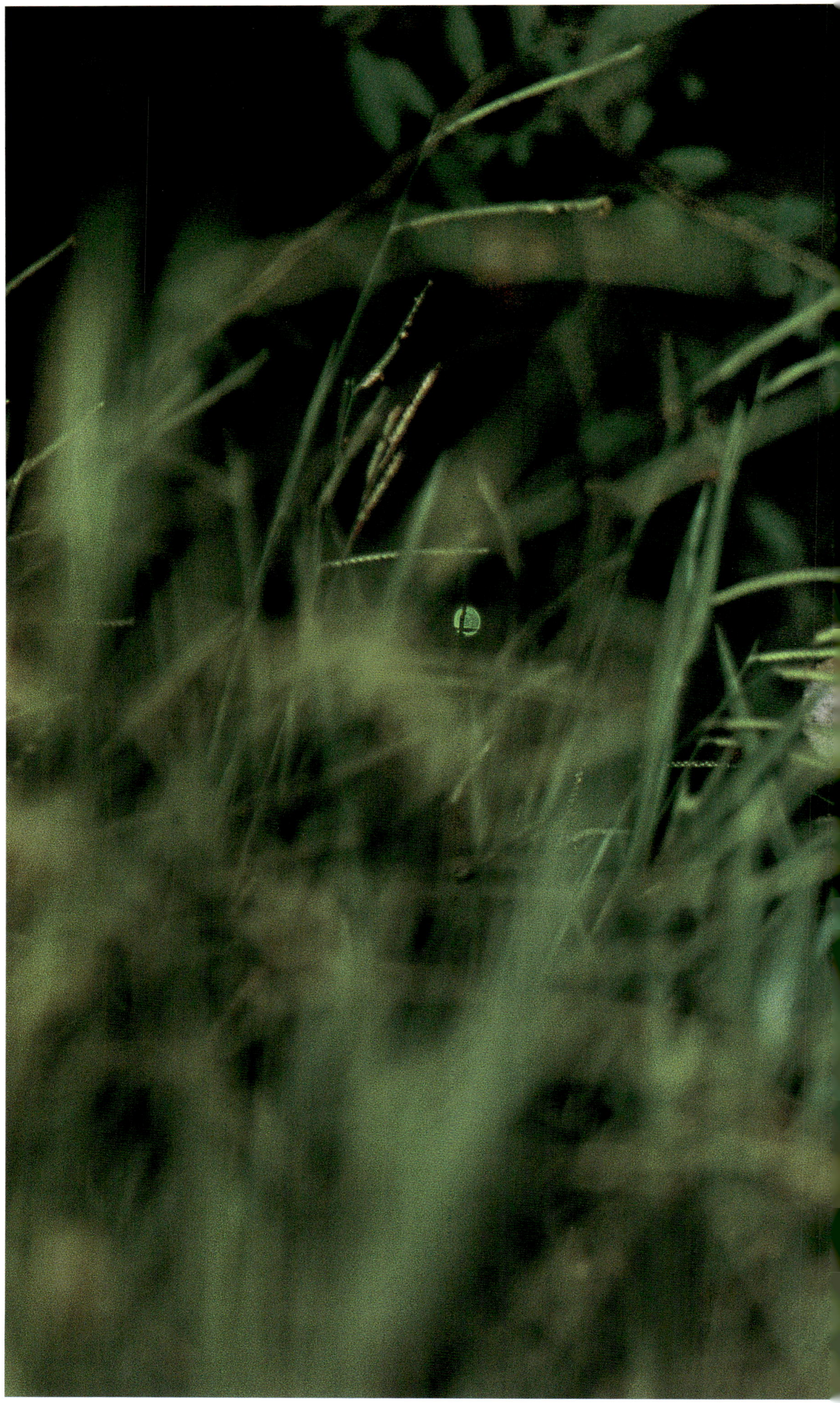

A huge male cat peers through the tall grass in Brazil's Pantanal—my first face-to-face encounter with a wild jaguar. The vast abundance of food may be why these jaguars are larger than those in the Amazon or anywhere else across their range.

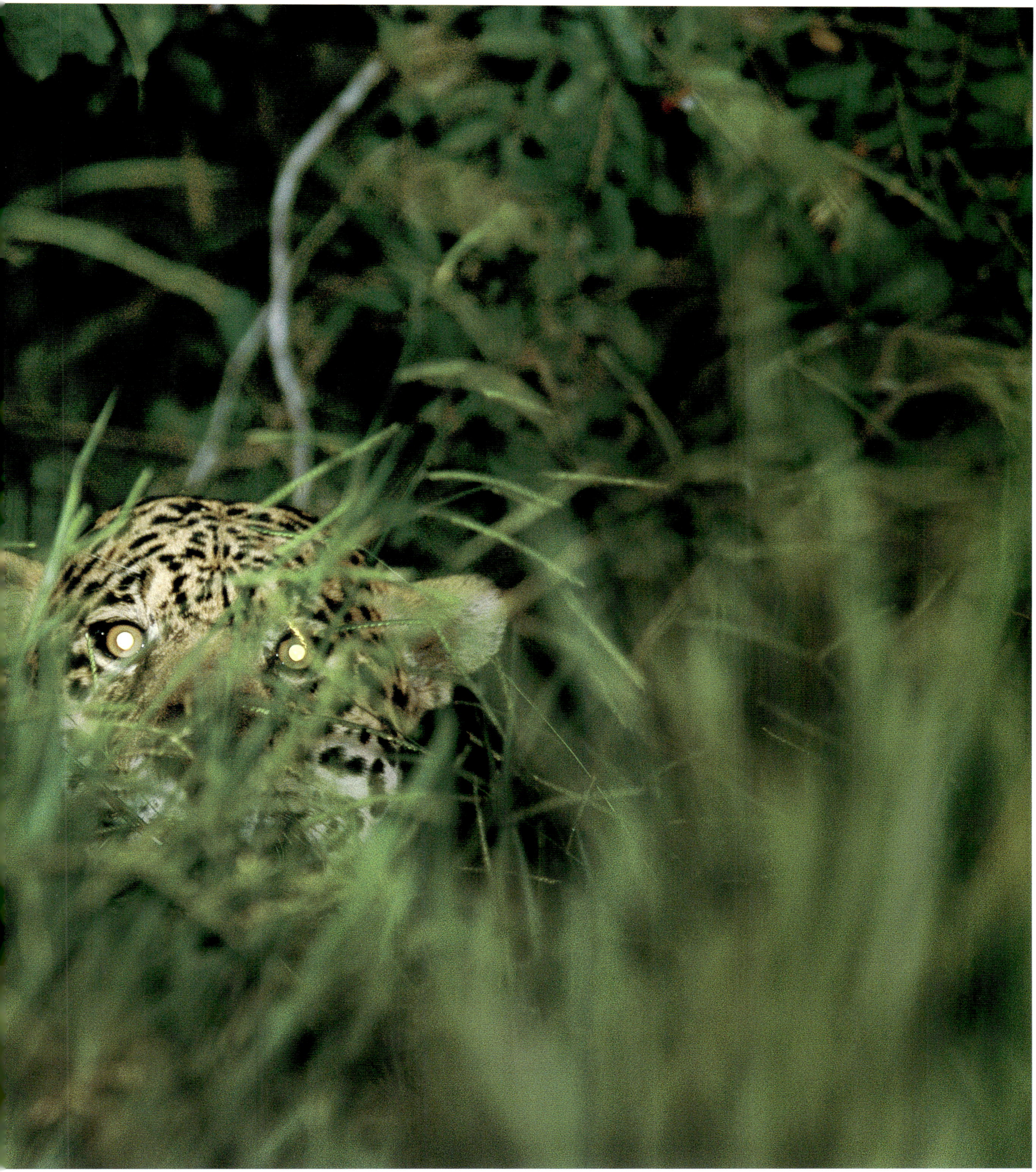

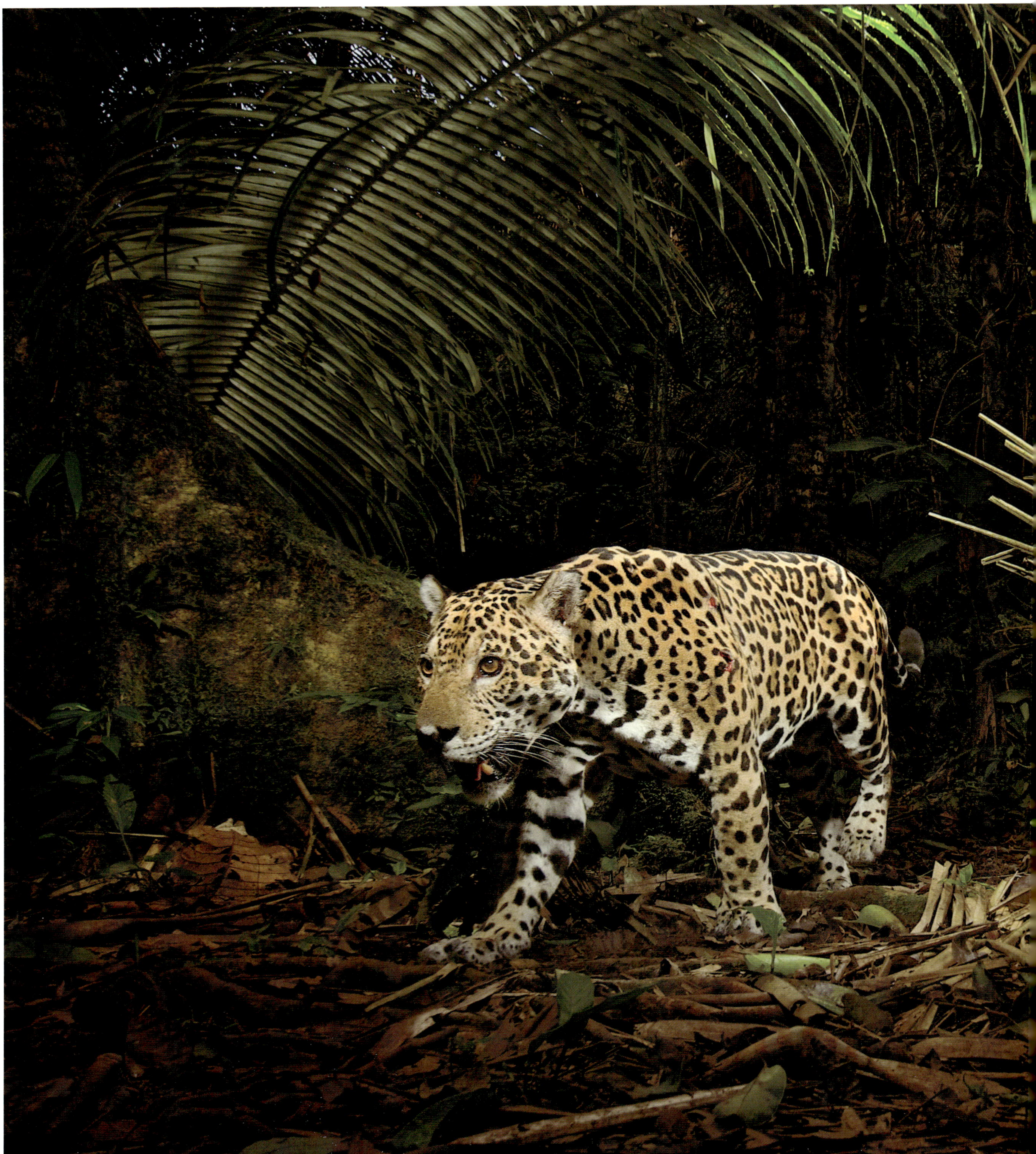

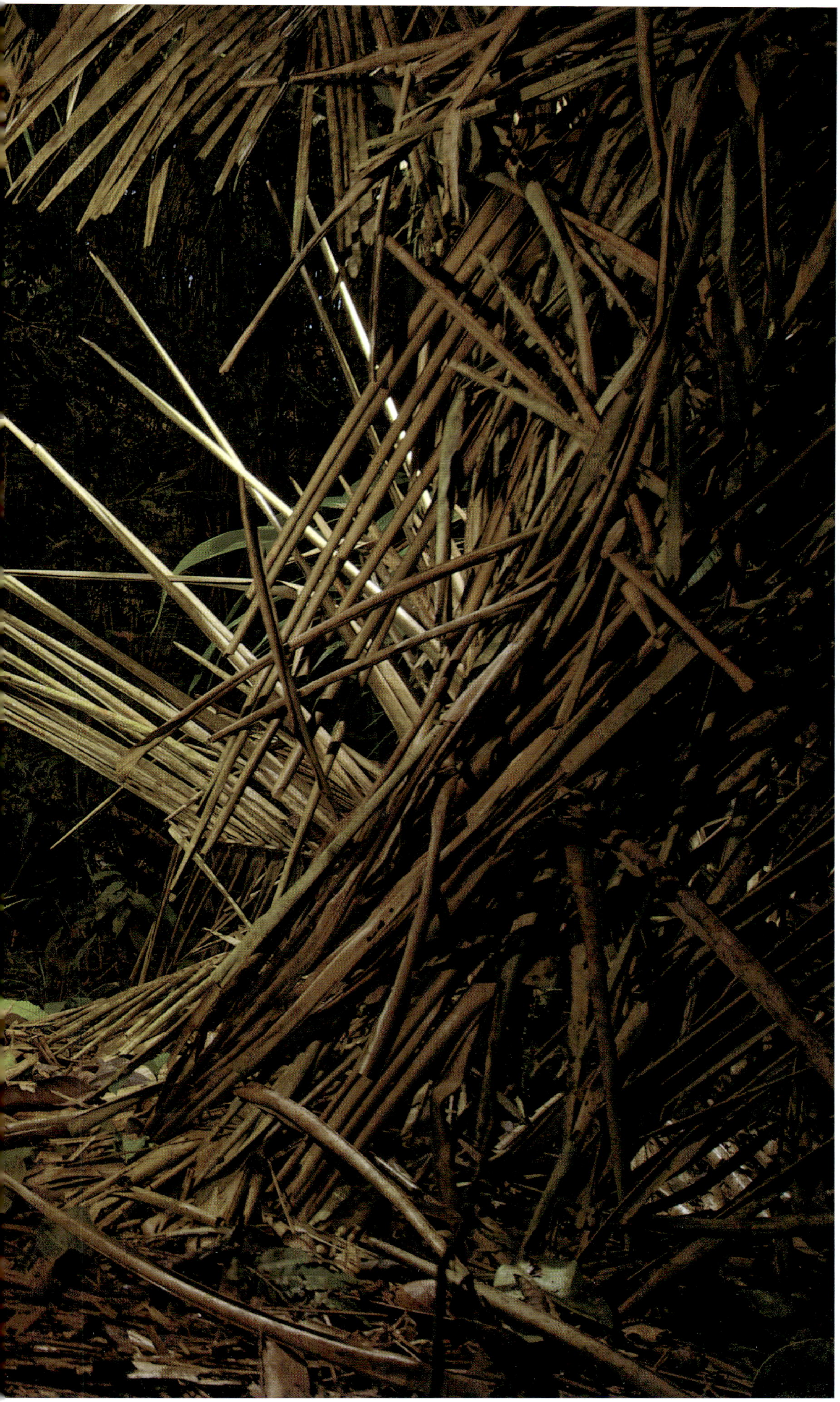

A jaguar stalks on a trail in Yasuní National Park, the last contiguous virgin rainforest in eastern Ecuador. At the intersection of the Amazon and the Andes, this is a biodiversity hot spot with a healthy jaguar population.

PAGES 76–77: A jaguar slips beneath a cattle fence on a ranch in Brazil. Jaguars, like all big cats, sometimes prey on livestock. Establishing economic incentives for locals to protect these cats is critical to their acceptance – and their ultimate survival.

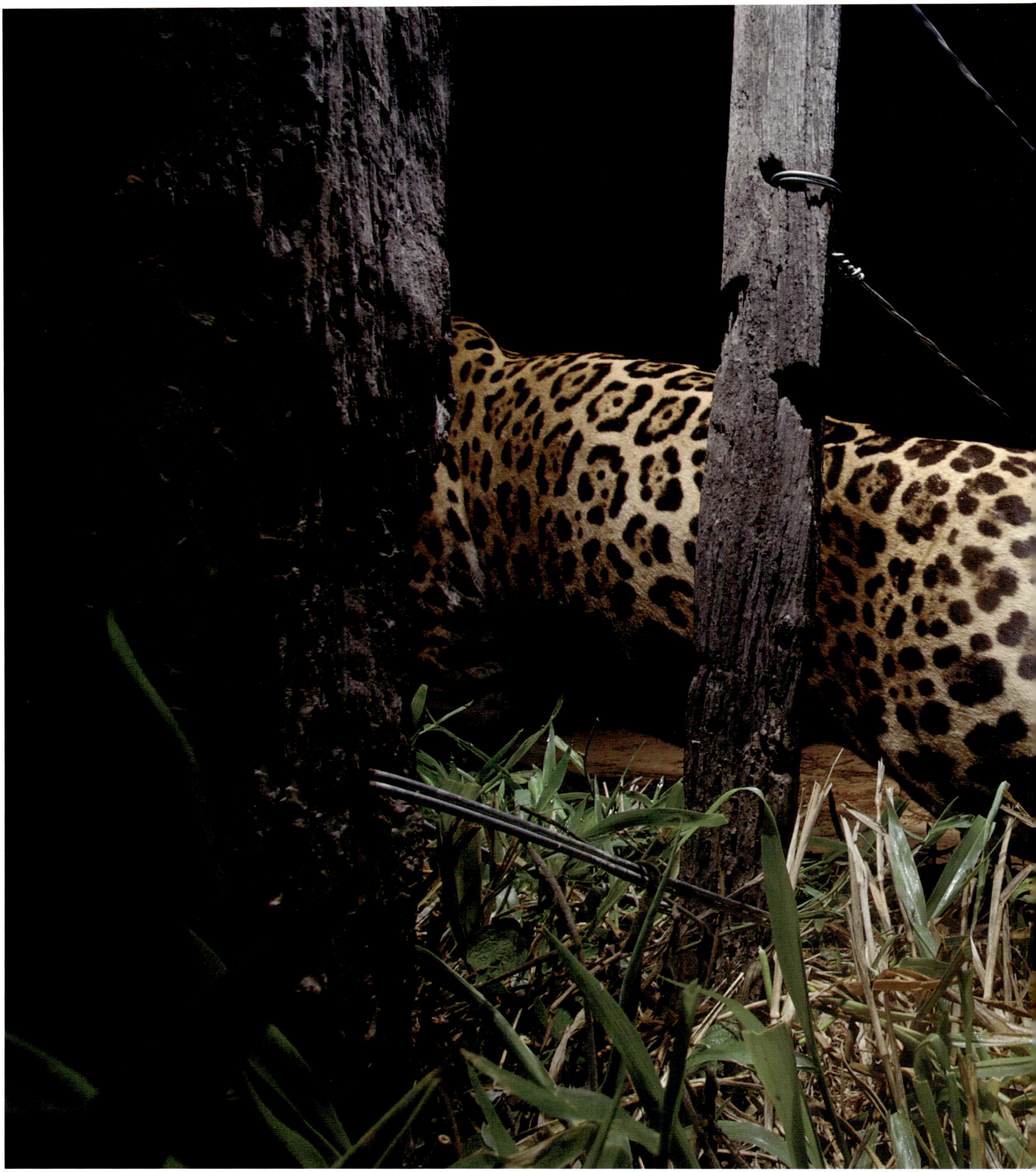

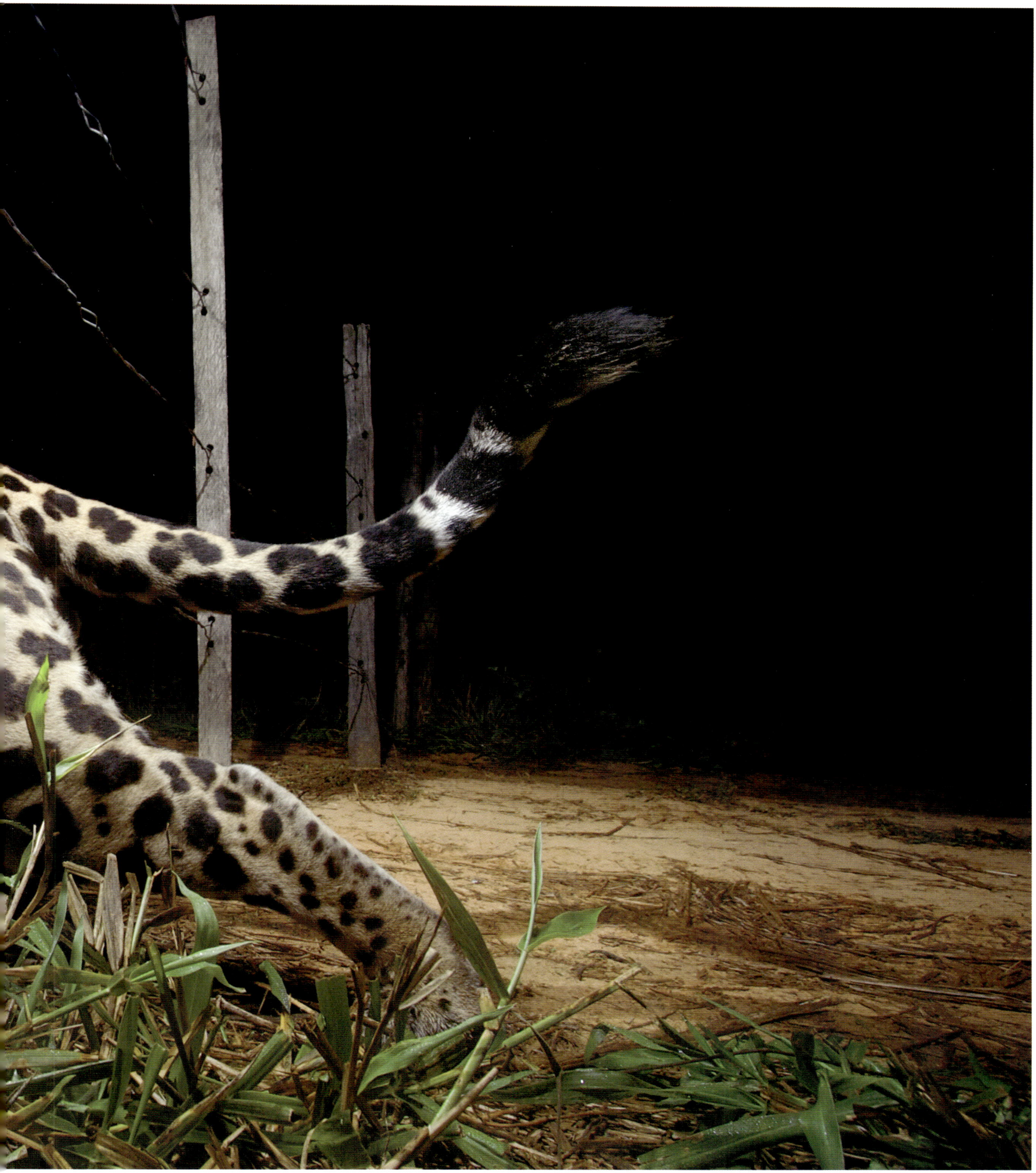

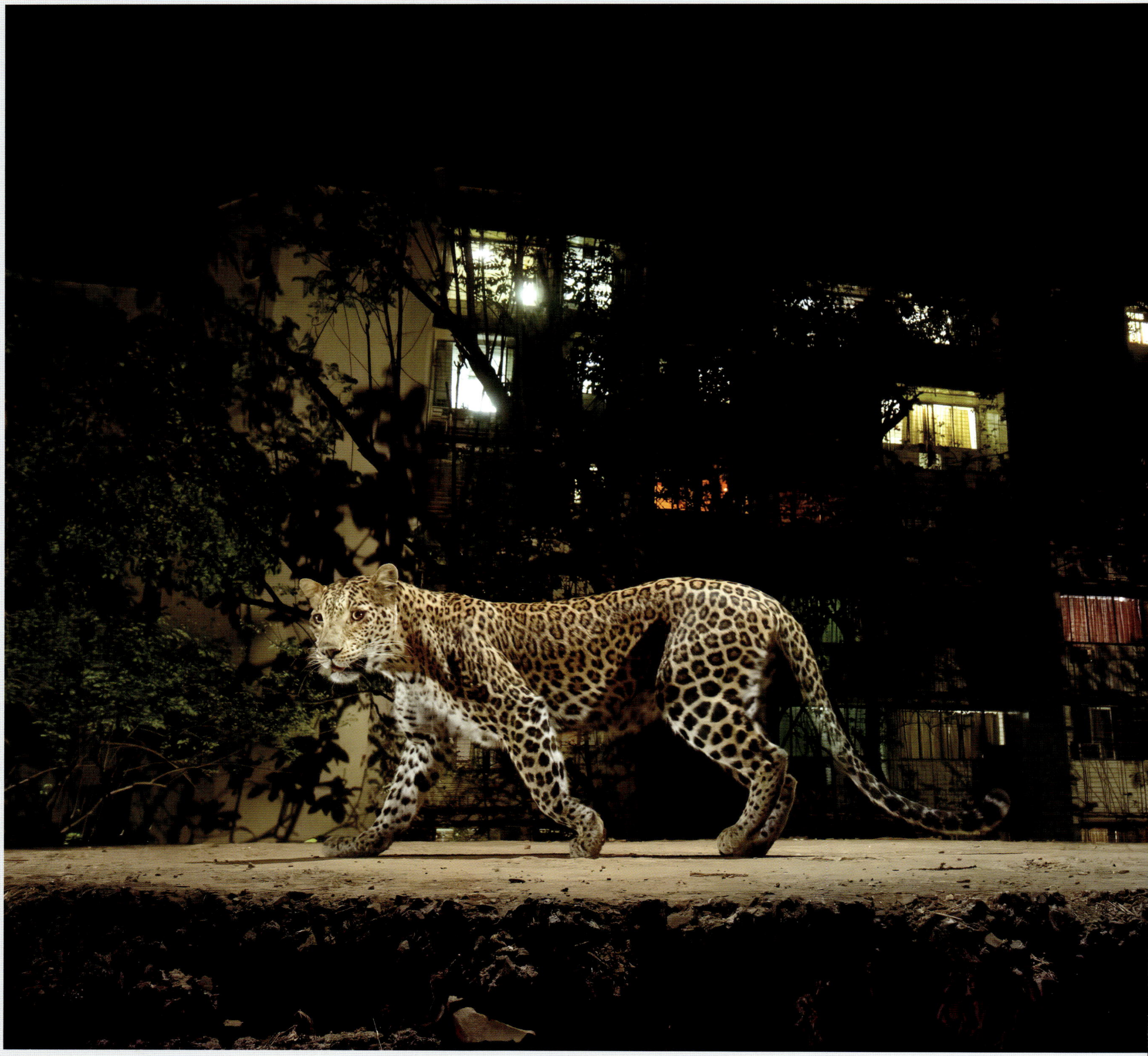

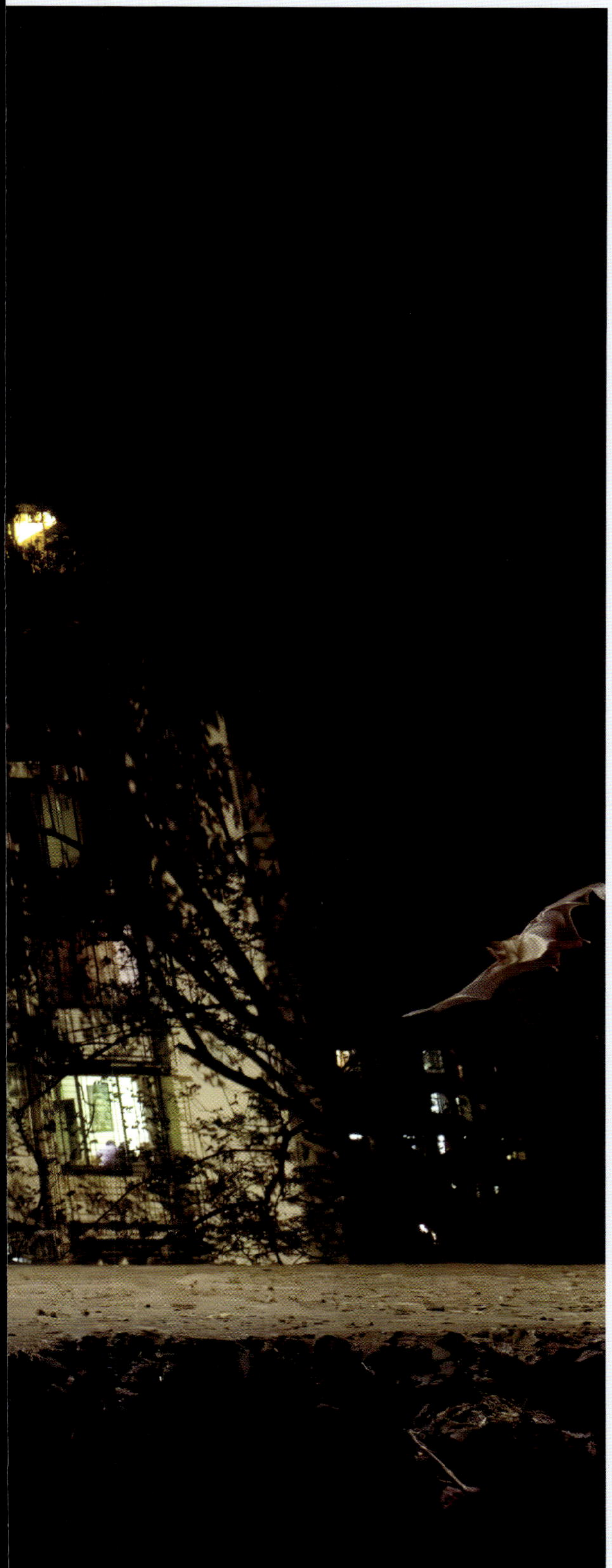

CHAPTER THREE

Leopards The Ultimate Survivors

Leopards can live nearly anywhere, feed on nearly anything, live close to people—or survive in remote, wild places.

Lit by a remote camera trap flash and the glow of urban Mumbai, a leopard prowls the edge of India's Sanjay Gandhi National Park in view of nearby apartment buildings.

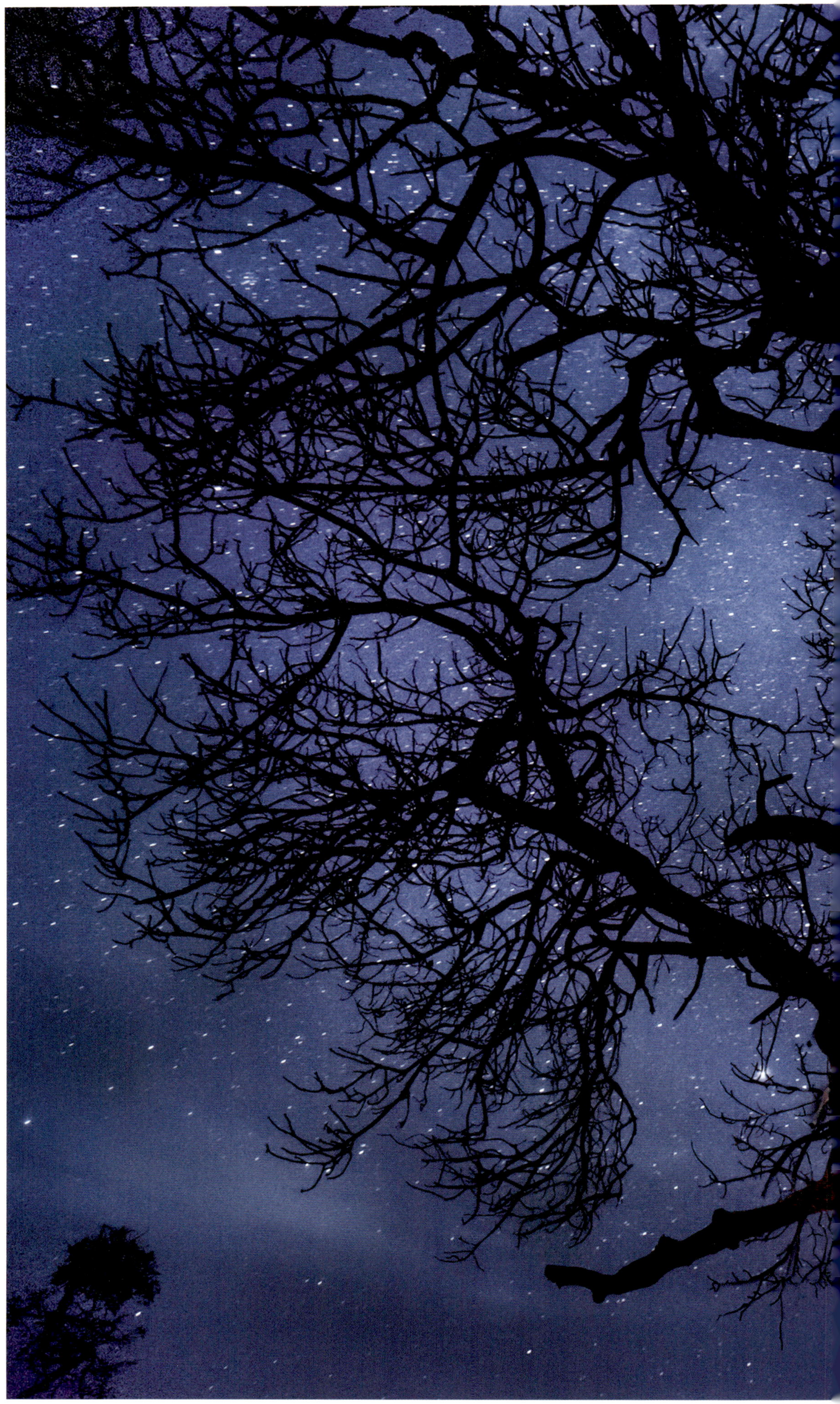

A leopard perches in a tree beneath star-studded skies in South Africa's Timbavati Nature Reserve.

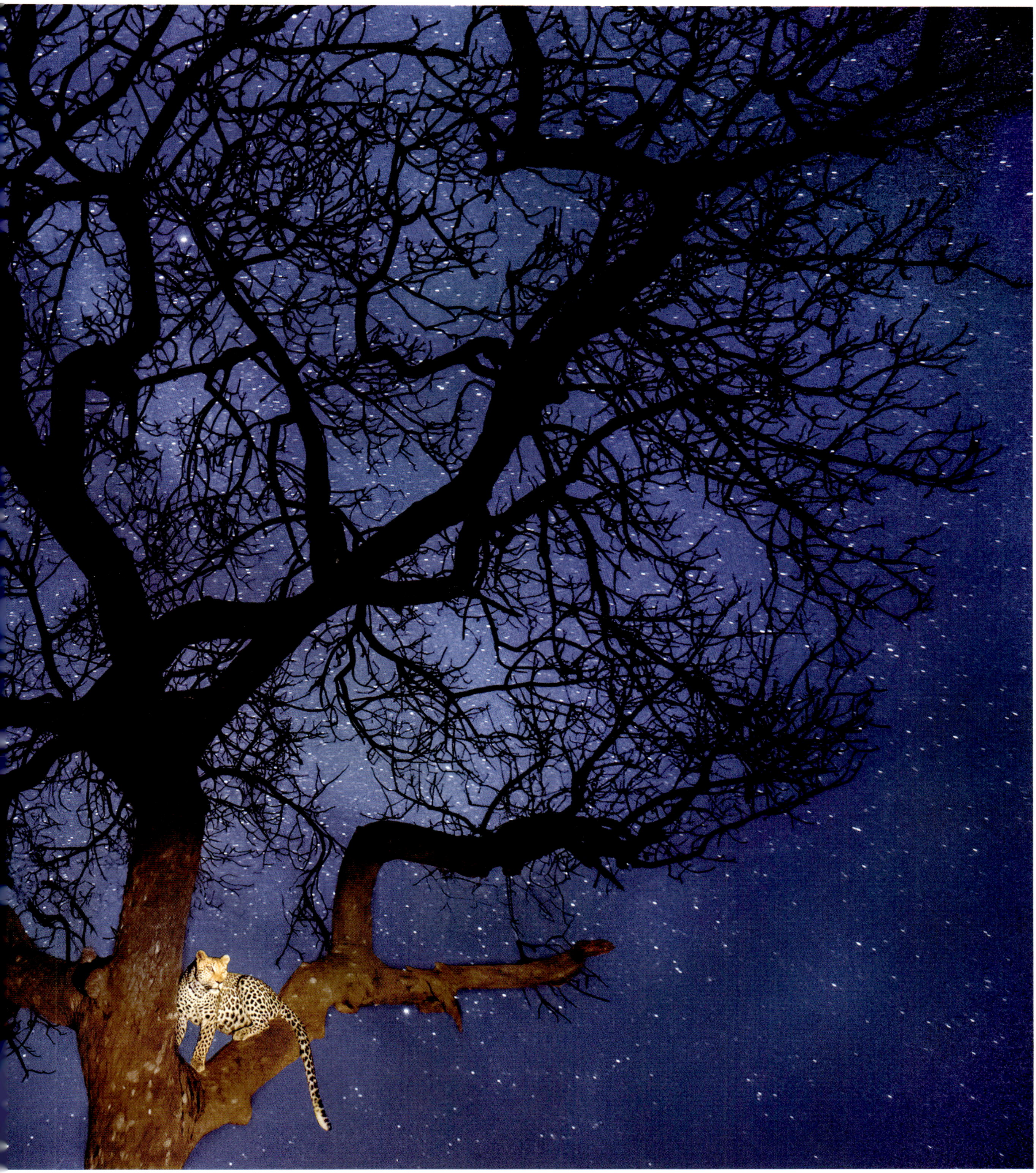

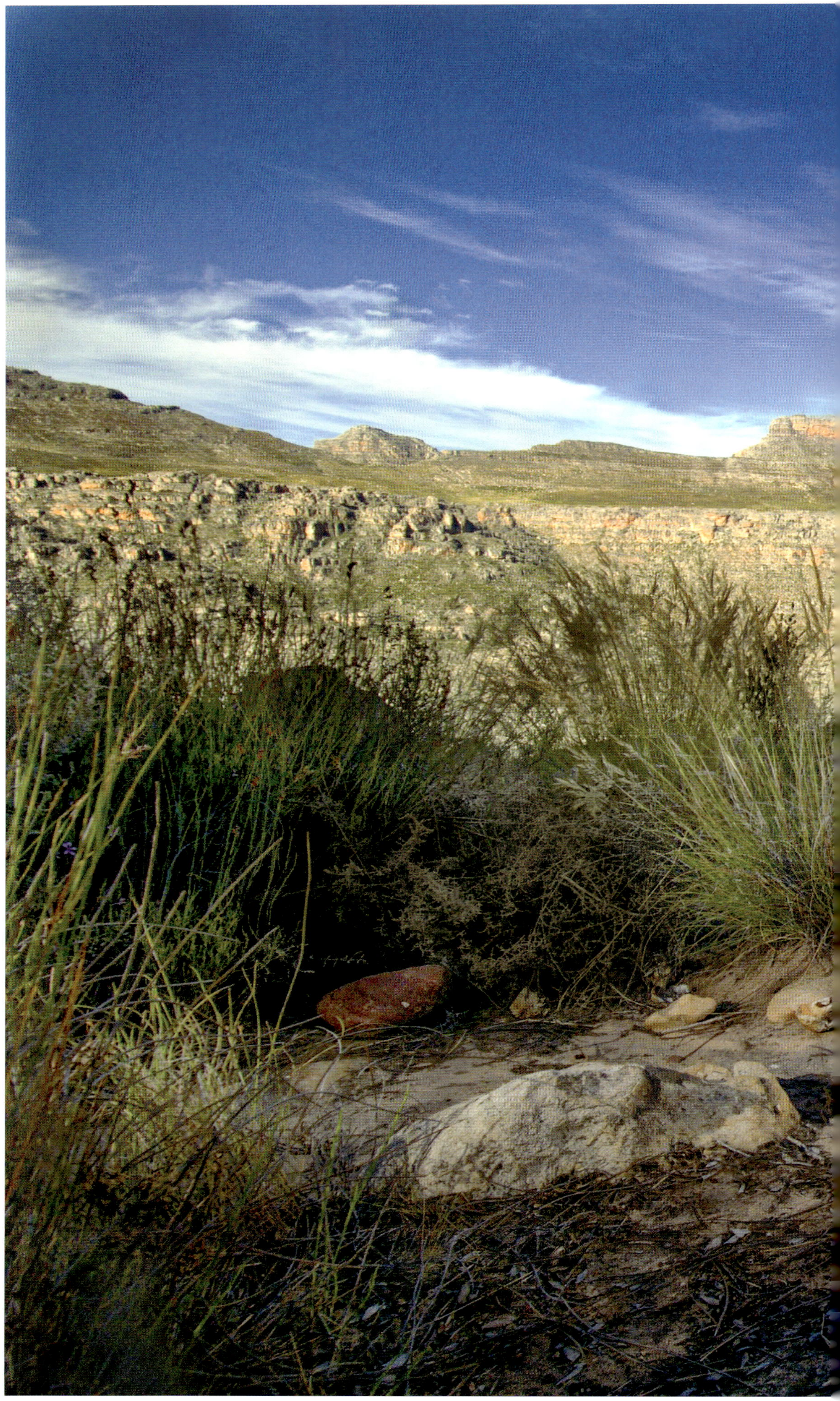

A Cape leopard cub stares straight into the lens of a camera trap set in South Africa's Cederberg Wilderness Area. Though not classified as a subspecies, these shy mountain cats are smaller than their brethren that roam the African savannas.

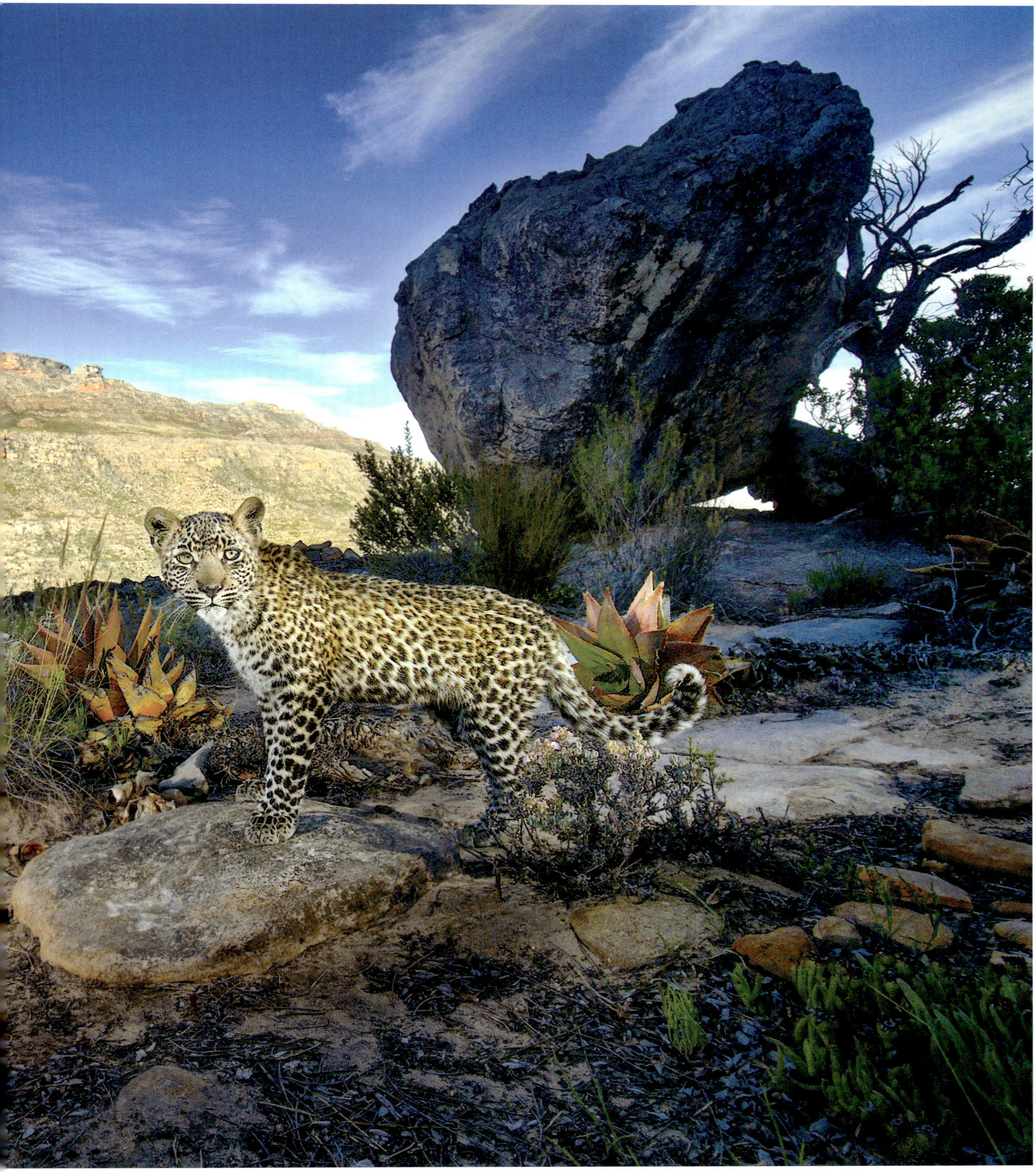

The road snarled into a chaos of traffic

and humanity as we neared Inanda in South Africa's eastern province of KwaZulu-Natal. Every Sunday in July, tens of thousands of Nazareth Baptist Church worshippers climb a sacred hill for a religious gathering, the Shembe Festival. Their faith melds Zulu customs and Christian beliefs. I was one of four non-Zulus at the gathering and the first foreign journalist who'd ever received permission to attend. I was welcomed because I was with a South African zoologist, Tristan Dickerson.

"You must walk up the mountain barefoot," a man told me. We took off our shoes and were pulled into a throng that danced, sang, blew horns, and drummed all the way to the top. They came from all walks of life: tradesmen, accountants, farmers, lawyers, government officials. But most wore traditional Zulu warrior regalia, including shields, brightly colored staffs, and leopard skin capes. These pelts were once worn by Zulu royalty, symbolizing power. Shembe devotees said that wearing leopard skins brought them closer to their ancestors—and to God.

This was in 2013. Tristan had learned about a rampant leopard skin trade a few years before, while he was studying the cats at South Africa's Phinda Game Reserve. Police raided a nearby village and confiscated 92 skins, some sewn into clothing for church members.

The Shembe gatherings have been called the "biggest display of illegal wildlife contraband on Earth." Fewer than 7,000 leopards were left in South Africa, and there were roughly five million Shembe worshippers, with membership growing. The skins lasted maybe five or six years until they fell apart and needed to be replaced. New ones were expensive: A cape made from the front of a leopard cost about $390; one fashioned from the back, $425. The average per capita income was less than $13,000.

Tristan brainstormed a solution that both saved leopards' lives and preserved cultural traditions: faux fur capes. Church leaders agreed to use them if they looked real. Tristan consulted designers and clothing companies. He visited factories in China to find lush fabrics that mimicked real skins. He was going for the "fake Rolex effect," he said. His employer, the nonprofit organization Panthera, gave the capes away for free.

At the time, I was documenting the cats and the threats that endangered them for *National Geographic.* As I looked out upon thousands of dancers, I saw thousands of leopards that were no longer walking in the bush or birthing cubs.

But 30 to 40 percent of dancers, men of all ages, were wearing the faux pelts. At one point, a group of young men posed, grinning, in front of my camera, all of them dressed in fake fur. Tristan translated what one of them told me: "It is our land, our South Africa, and we need to be sure that there are leopards in the bush for the future."

I donated use of those images to Panthera. It has since partnered with other organizations to expand the effort. As of 2025, 22,000 faux furs have been given away to Shembe worshippers.

I spent two years photographing leopards. All eight subspecies are threatened across Asia and Africa. Some have stabilized, but overall, they have declined by a third during the last 22 years (three leopard generations) from the same threats that face the lions and tigers they sometimes coexist with. But leopards are more dangerous to people than other cats: Attacks are relatively rare, but they ambush humans (particularly women and children) more frequently. This makes the leopard the most persecuted big cat.

Wildlife biologist Andrew Stein calls them "the ultimate survivors." They're highly intelligent, and they're elusive. They camouflage into the shadows so effectively that Indian leopard expert Vidya Athreya calls them "living ghosts." They can live nearly anywhere, feed on nearly anything (over 100 prey items are on their menu), and they can live close to people or in remote wild places. They are stunningly strong and agile. I've seen leopards drag an animal much bigger than they are up a tree. I've seen others leap 20 feet in the air.

Leopards call 62 nations home, from Senegal to Russia. They're found from sea level to 17,000 feet, in jungles, towering peaks, savannas, swamps, deserts, and even in an Indian metropolis. As I developed my story for *National Geographic,* I decided to include three subspecies: African leopards, Cape leopards, and Indian leopards.

In South Africa, I went to Sabi Sand Nature Reserve, adjoining Kruger National Park. It's a haven: Leopards are well guarded and abundant and have no reason to fear humans. I spent hours, sometimes days, with leopards: mating pairs, single males, and cubs that played, slept, nursed, or practiced hunting with small animals their mother brought them. If they were disturbed by my presence, I left. Conservation is my top priority. It's their home, not mine.

When we were out searching for cats, my guide reminded me to

(Continued on page 89)

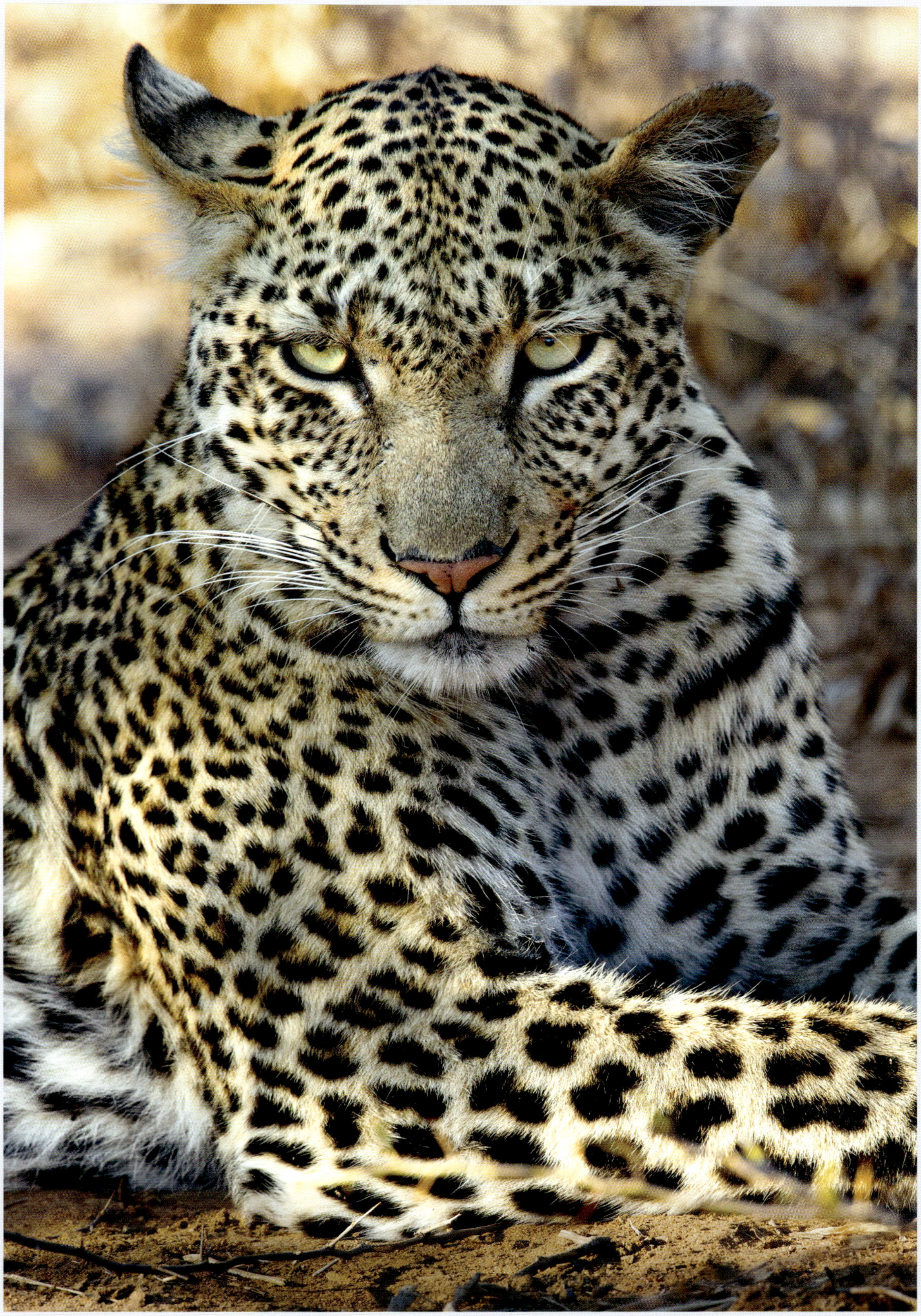

A leopard at rest in Timbavati Nature Reserve, South Africa. Leopards are the most persecuted of the big cats, likely because they are more aggressive than other species and lack the same iconic status of lions and tigers.

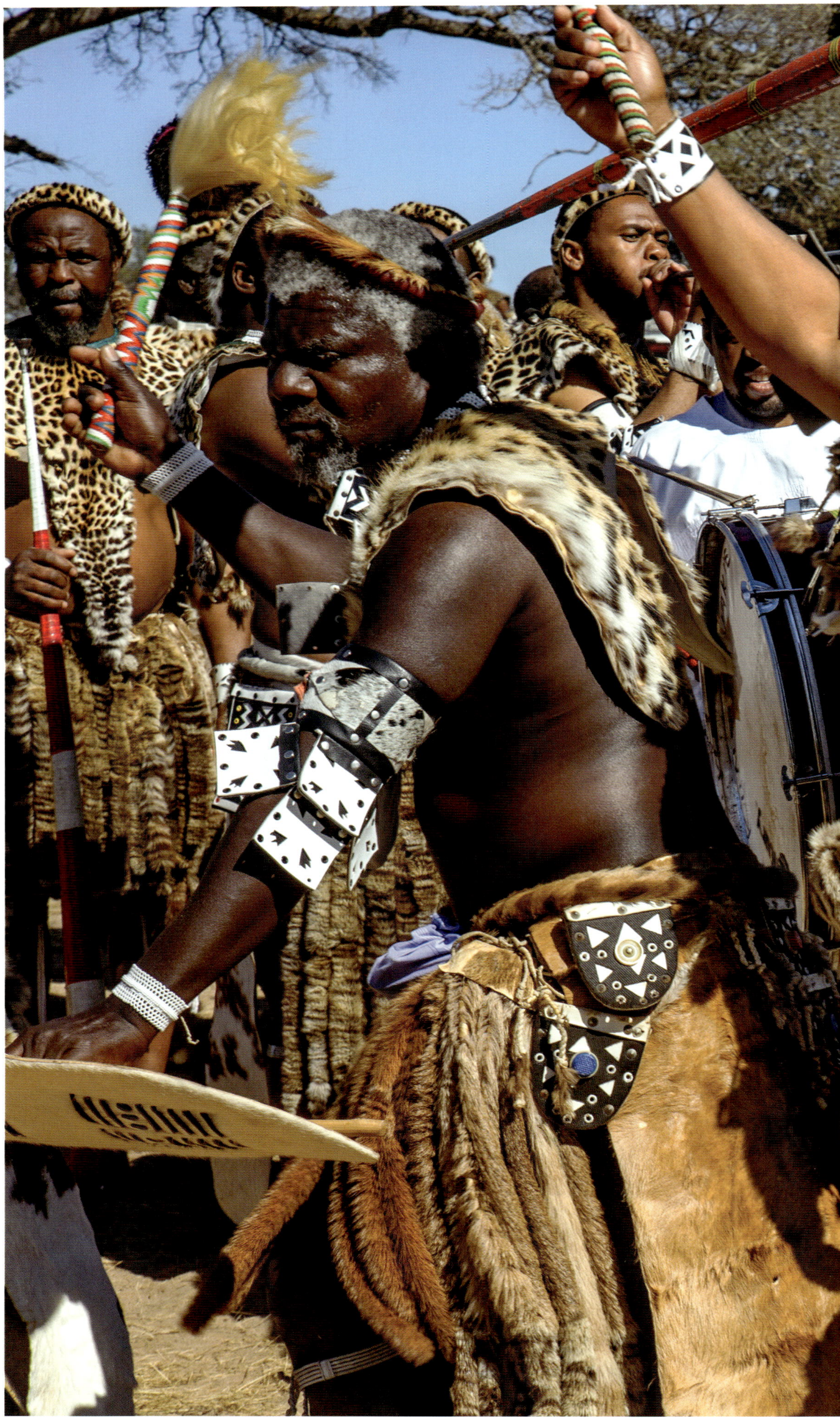

Demand for leopard skin shawls worn in Shembe religious festivals like this one near Durban, South Africa, drives poaching of these cats. As of 2025, 22,000 faux fur substitutes have been given to worshippers by Panthera, a conservation group, to help protect the wild cats.

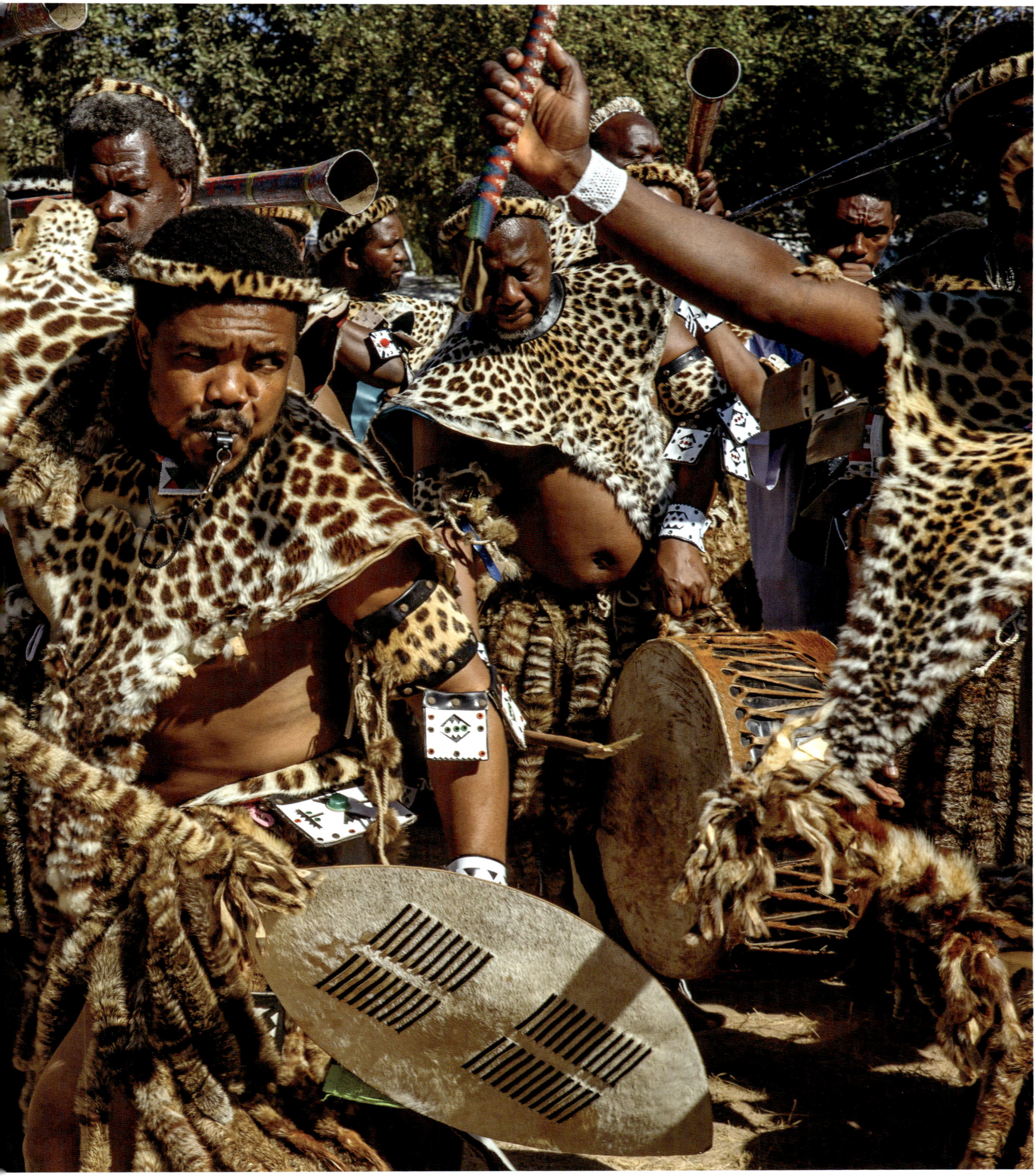

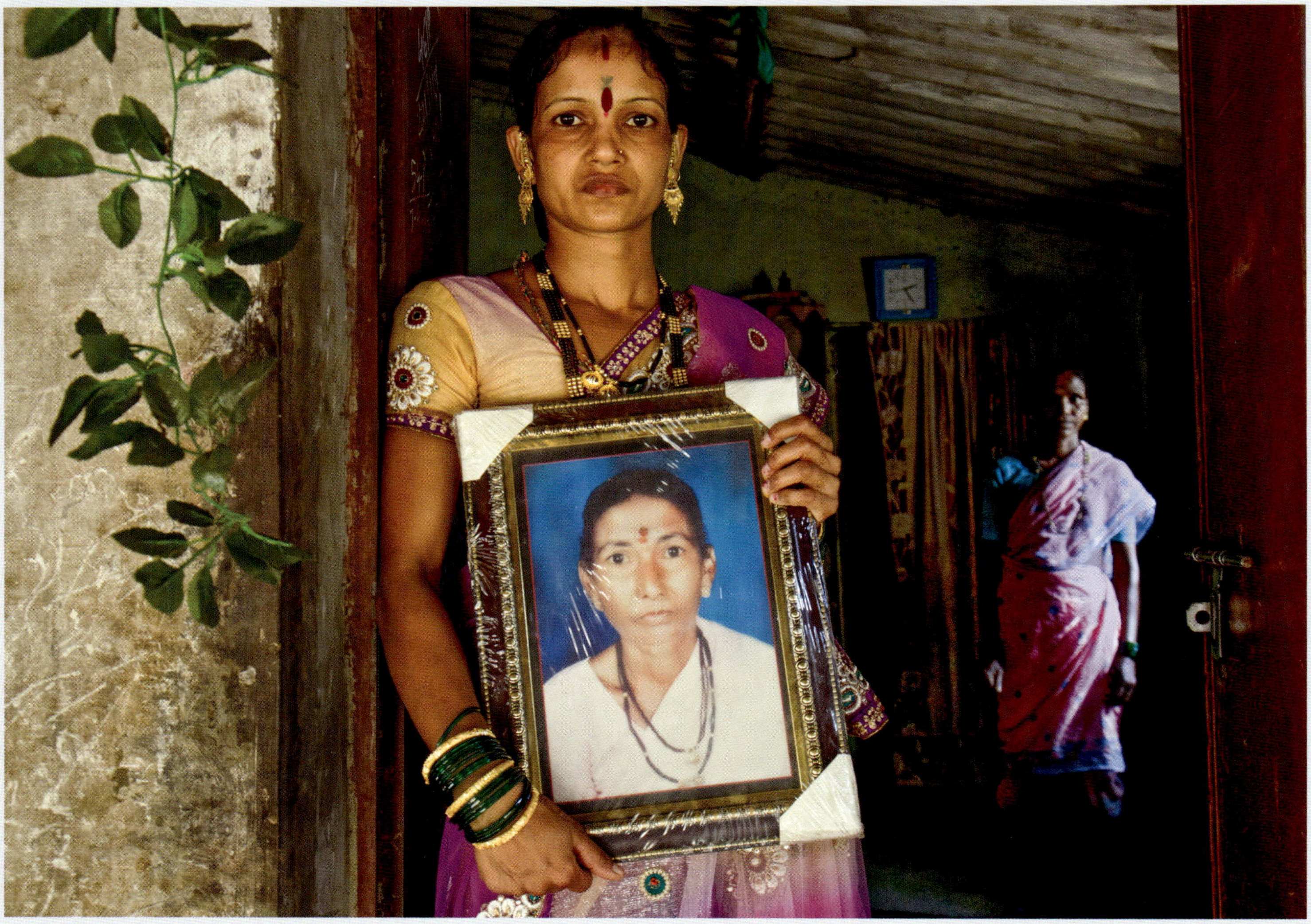

A woman holds a photo of her mother, who was killed by a leopard outside her home near Sanjay Gandhi National Park in Mumbai.

(Continued from page 84)
look up. That yielded some of my favorite images: a leopard eating its kill high on a bough just after sunset, and another sprawled on a branch under a canopy of stars.

Thanks to strong research and management, Indian leopards are doing well when compared with their critically endangered cousins, the Arabian and Amur leopards. About five times as many wandered the Indian subcontinent a century ago, frequently living outside protected areas—in shocking proximity to people. Anthropologist Sunetro Ghosal told us that there's "a history of sharing space" and even "mutual accommodation," with both leopards and humans avoiding confrontations. However, human populations continue to expand, and people and animals are fighting for the same territory.

This is what brought me to Mumbai's Sanjay Gandhi National Park with cinematographer Bertie Gregory. We wanted to film and photograph people and leopards living side by side. Some 40 leopards live in the park amid this metropolis of 22 million people.

In the late 1800s the park was designated as reserve forest to guard against the city's rapid growth. In 1950, it became a national park. Then, in 2002, authorities began trapping "problem leopards" that strayed into villages in other parts of Maharashtra state—and bringing them here.

People use the park during daylight hours: Kids play cricket, and couples and families stroll or walk their dogs, enjoying the green respite from the city. But at night, the leopards come out.

I wanted my photos to show that these were urban cats. The best location was a trail along the park's eastern flank, which crossed a bridge right beside a multistory apartment complex. It had forest on one side and skyscrapers looming beyond the trees on the other. We set up still and video camera traps at what would be eye level for a cat. I hoped to capture images of the area's huge, dominant male. I nicknamed him Big Daddy.

To install the camera traps, we erected 40-foot bamboo platforms. We used a receiver with an infrared beam: When something moves through and breaks the beam, it sends a signal to the transmitter, firing the camera and flashes for still images and a floodlight for film. We needed light.

Next, we had to show the human side of the story. When Bertie and I walked around Mumbai, we learned that many residents didn't know that leopards padded by right outside their homes—and I was surprised to hear that so many were excited they were there. One woman I spoke with called them her neighbors, grinning proudly.

A Hindu temple devoted to the leopard is near the park. Some consider the cat a guardian. Lord Shiva is sometimes depicted wearing a leopard skin, and he's believed to have granted the cat permission to eat whatever it wished.

Nationwide, some show empathy for leopards, risking their lives to rescue one if it falls into an abandoned well or is otherwise trapped or injured. But others fear and vilify this cat. Whole villages have descended on a leopard that ventured too close, and killed it.

Many of those who live in Mumbai's "informal settlements"—makeshift housing just inside the park or on its outskirts—are fearful. This 40-square-mile sanctuary, as big as 30 Central Parks, is chock-full of leopards, with the highest density in the world: one per square mile. The cats readily move 10 miles a day.

Only so many spotted deer and other prey lived in the core area, so leopards didn't stay in the park. Surveillance footage frequently showed them hunting stray dogs, which were everywhere—about 100,000 roamed the city at that time—and documented leopards wandering near homes and going after livestock.

When authorities began moving relocated leopards into the park in 2002, they thought it would provide them safe haven. But in trying to solve one problem, they created another. These displaced predators fought over territory and food in their new, unfamiliar surroundings. The newcomers settled on the park's fringes, near people.

It was disastrous. There were 84 attacks on humans during the first two years after these highly stressed leopards were released there. In June 2004 alone, they killed nine people. The relocation program ended in 2006. Attacks diminished but didn't stop, and settlements that lacked toilet facilities and electricity remained most at risk. Some 80 percent of ambushes happen when people answer nature's call in the dark. While we were working on the story, a cat killed a seven-year-old girl.

I set up a camera trap in a courtyard just beyond the park boundary, where a man slept outside in a cage to keep watch over his goats and chickens at night, and I captured images of a leopard walking through—twice. We met with people who'd lost family members in leopard attacks, including parents of young children.

Across India, officials capture killer cats and move them to a so-called rescue facility. I went to one. A male cat growled aggressively and hurled himself against the bars of his cage as I photographed him. Just two of the captives were man-eaters. Others included an injured female that was stoned by villagers. Another leopard lost two toes when he was caught in a poacher's snare. Both were recovering there, and both were locked up for life.

We were on deadline for both the story and the film, racing against the weather: Monsoon season was coming. We needed images and footage showing that these cats lived in a city. Bertie spent a night in a blind in the park, using remote-controlled cameras that were developed for reality TV. He got Big Daddy. And so did I.

After the story ran in *National Geographic,* nonprofit organizations came in and built toilets in settlements bordering the park, saving lives.

As writer Richard Conniff wrote in our story, "Leopards are entirely capable of living among humans. The question is whether humans can learn to live with leopards." Our ability to coexist will determine the future for leopards everywhere they live.

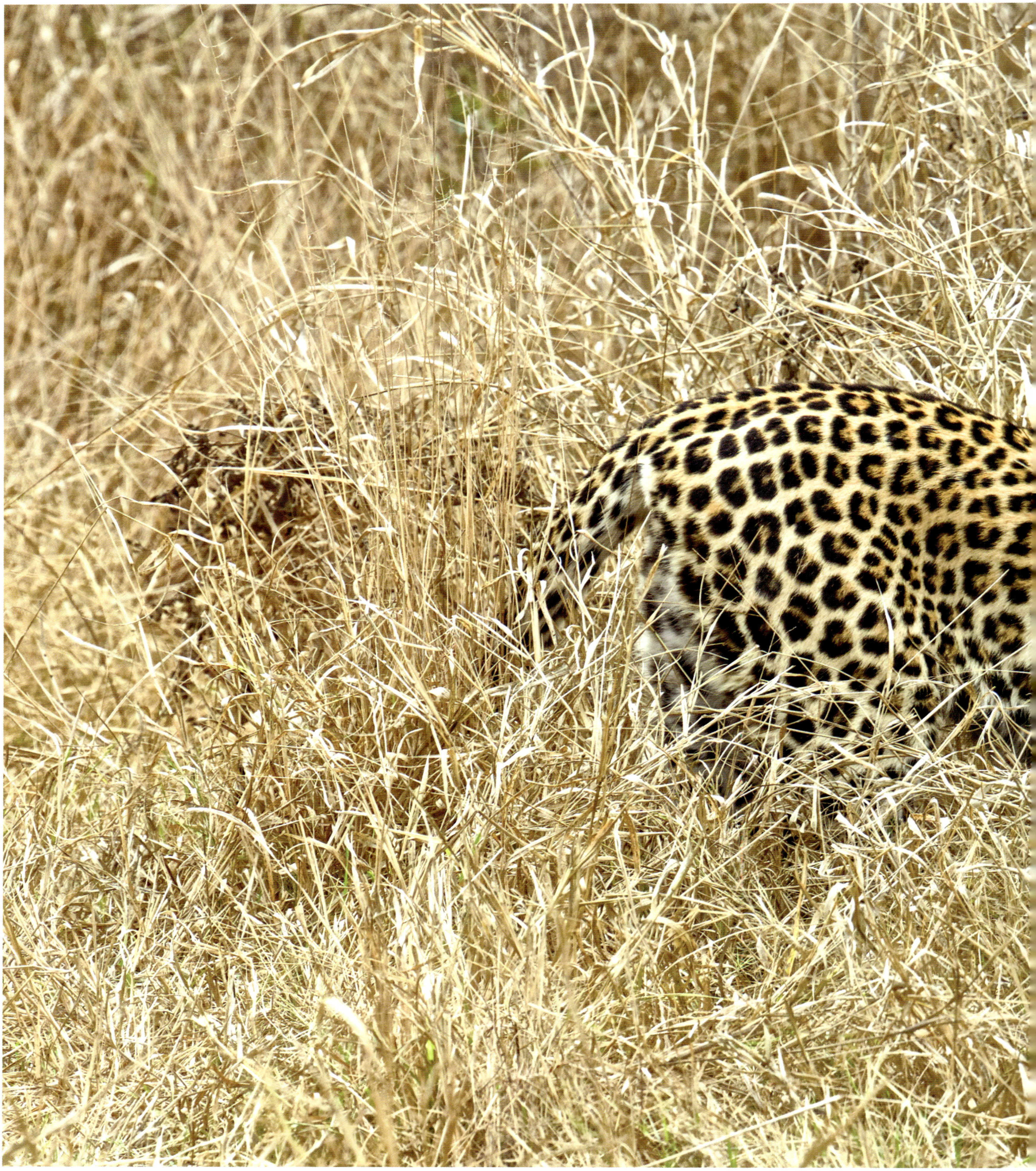

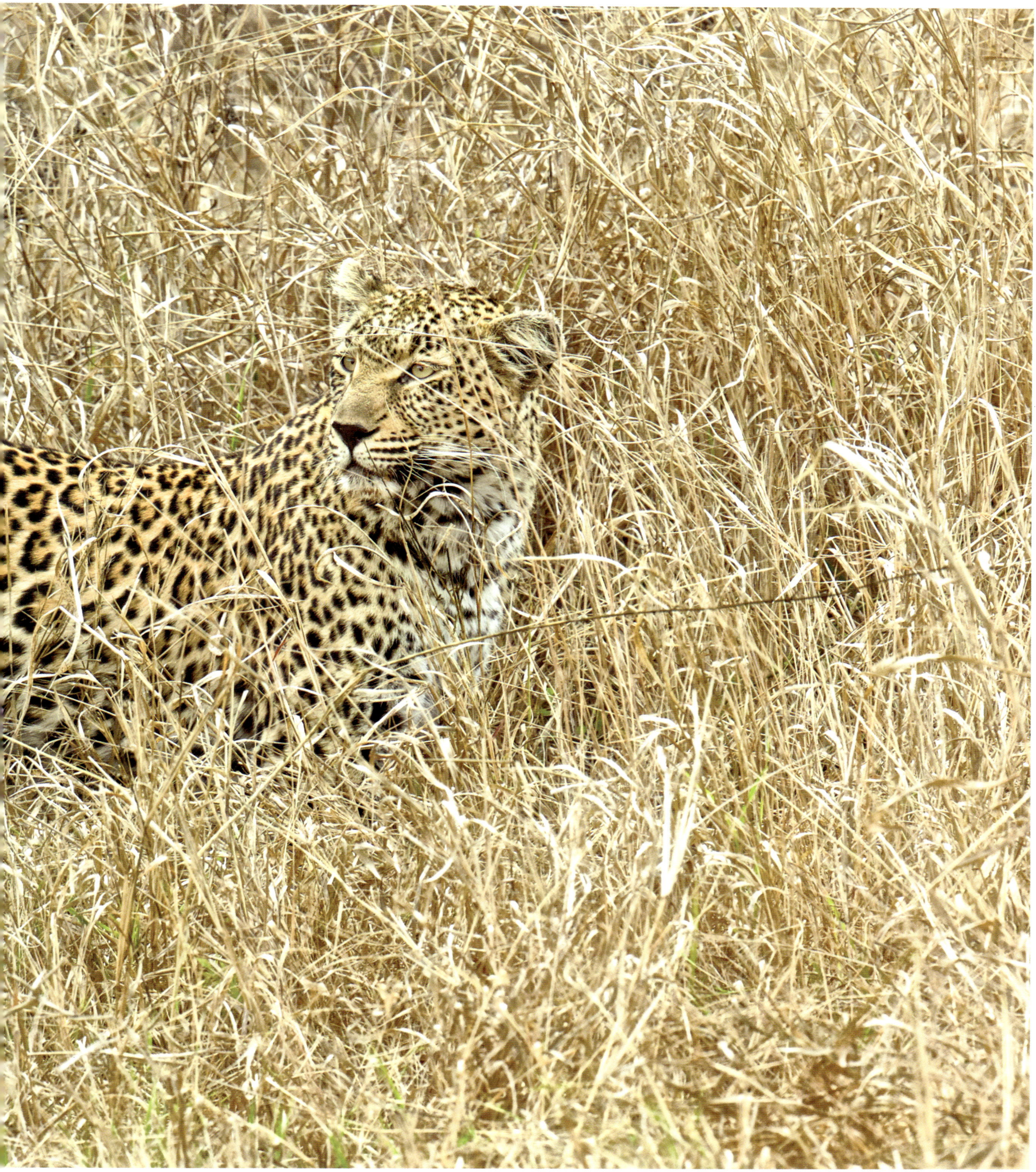

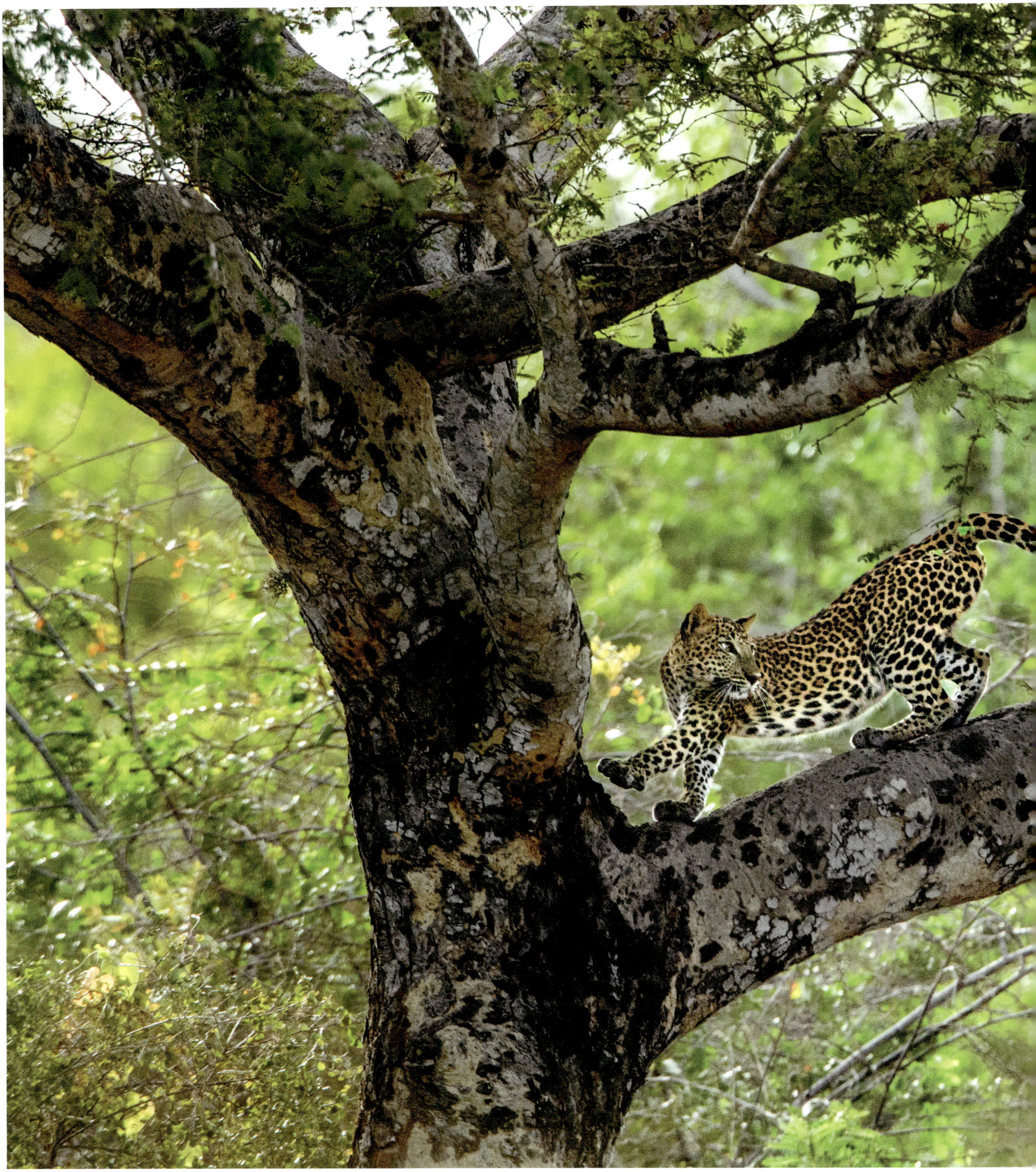

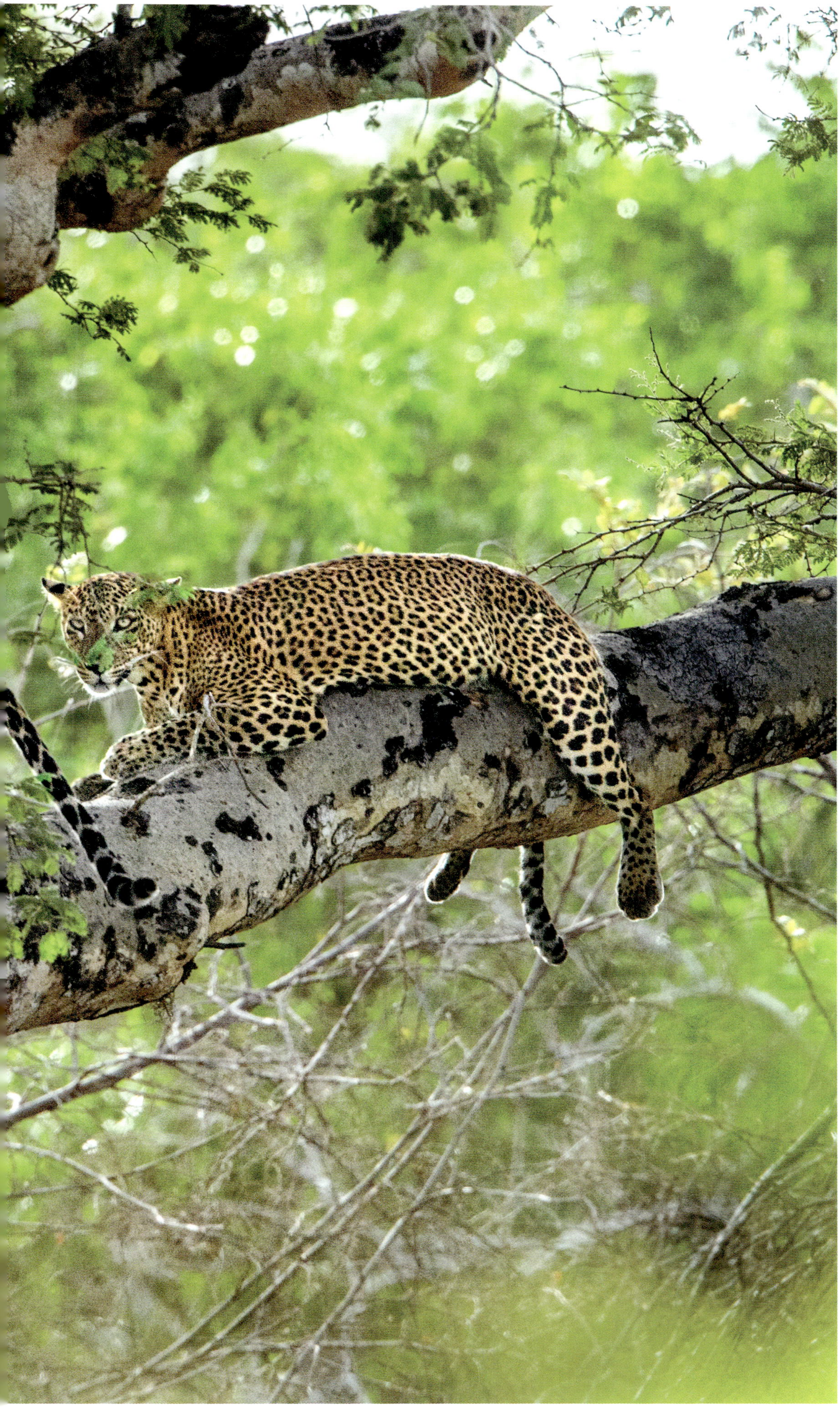

Leopards spend significant time in trees, often hauling their prey aloft to protect it from lions, hyenas, and other carnivores, but also just resting, like this female and cub in Yala National Park, Sri Lanka.

PAGES 90–91: The high grasses of Sabi Sand Nature Reserve, South Africa, provide perfect camouflage for a leopard.

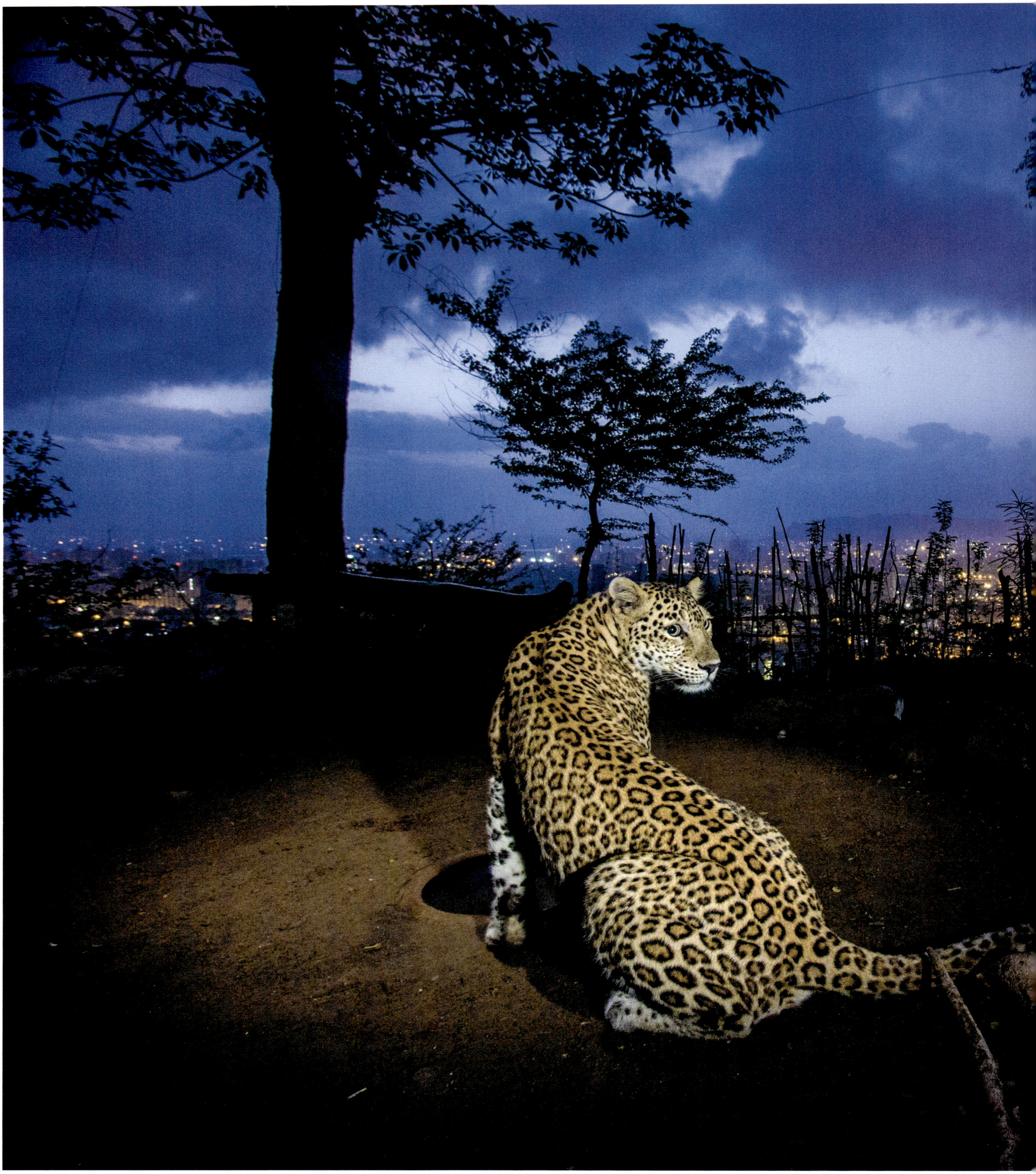

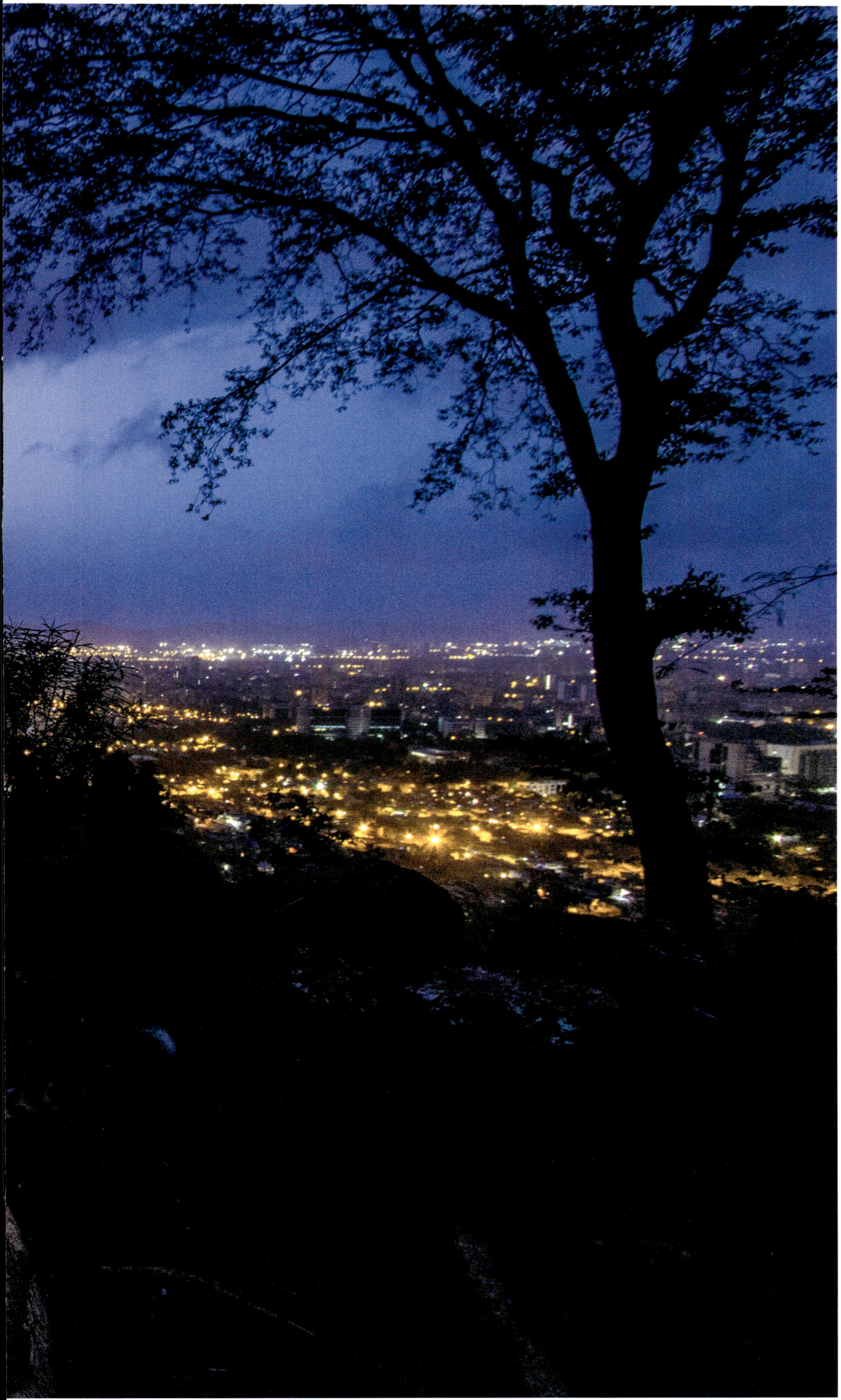

On a hill overlooking Mumbai, an artificial water hole attracts one of the 40 or more leopards that lived in and around Sanjay Gandhi National Park, which is home to the highest density of leopards in the world, once estimated at one per square mile.

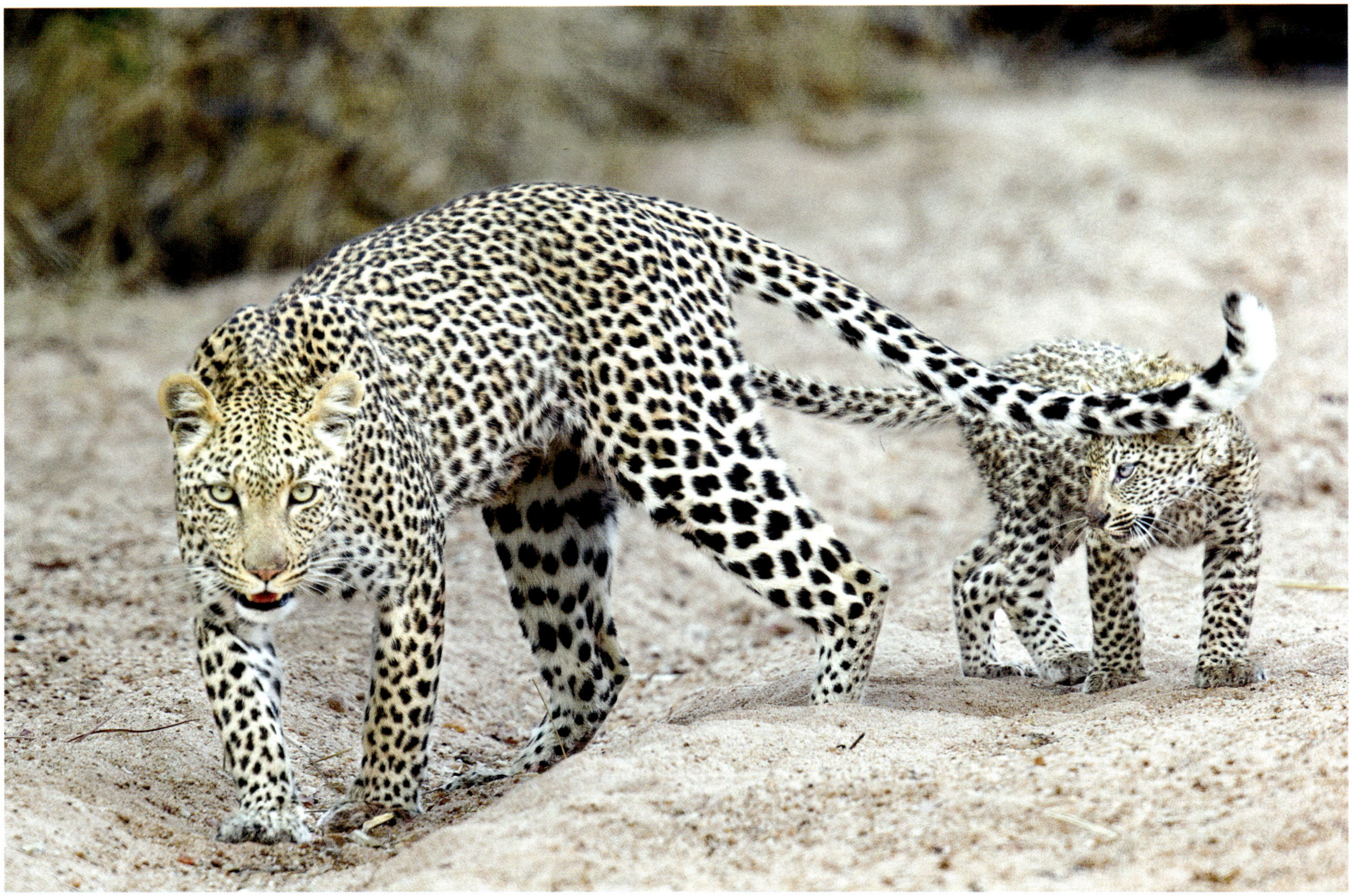

A female and her cub prowl South Africa's Timbavati Nature Reserve. Leopards are the most adaptable big cat: They can live near humans or in remote places, and they feed on at least 100 types of prey.

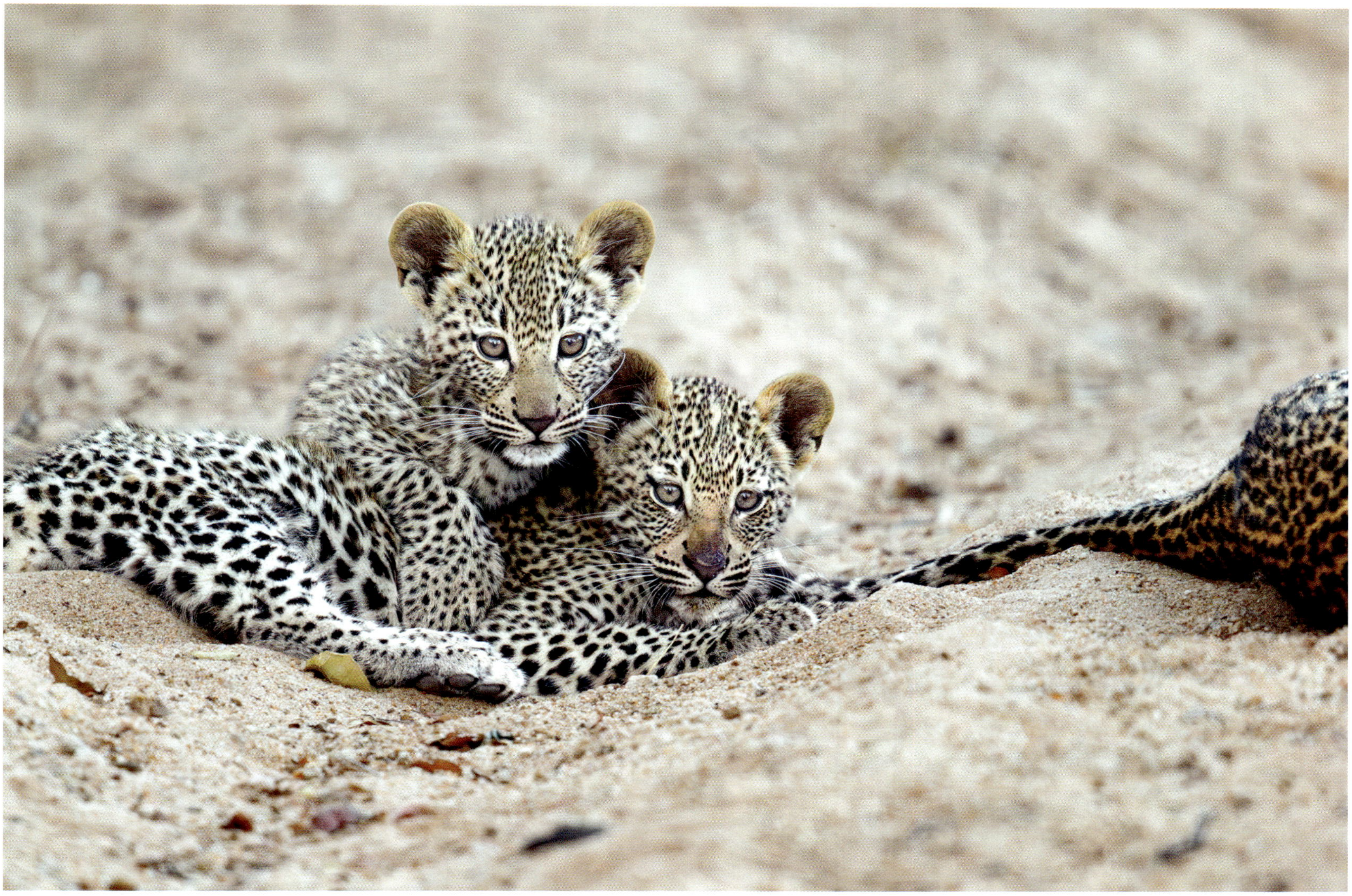

Leopard cubs snuggle in the sand in Timbavati. Mothers typically birth two to four cubs, who stay near for about two years before striking out in search of unoccupied territory.

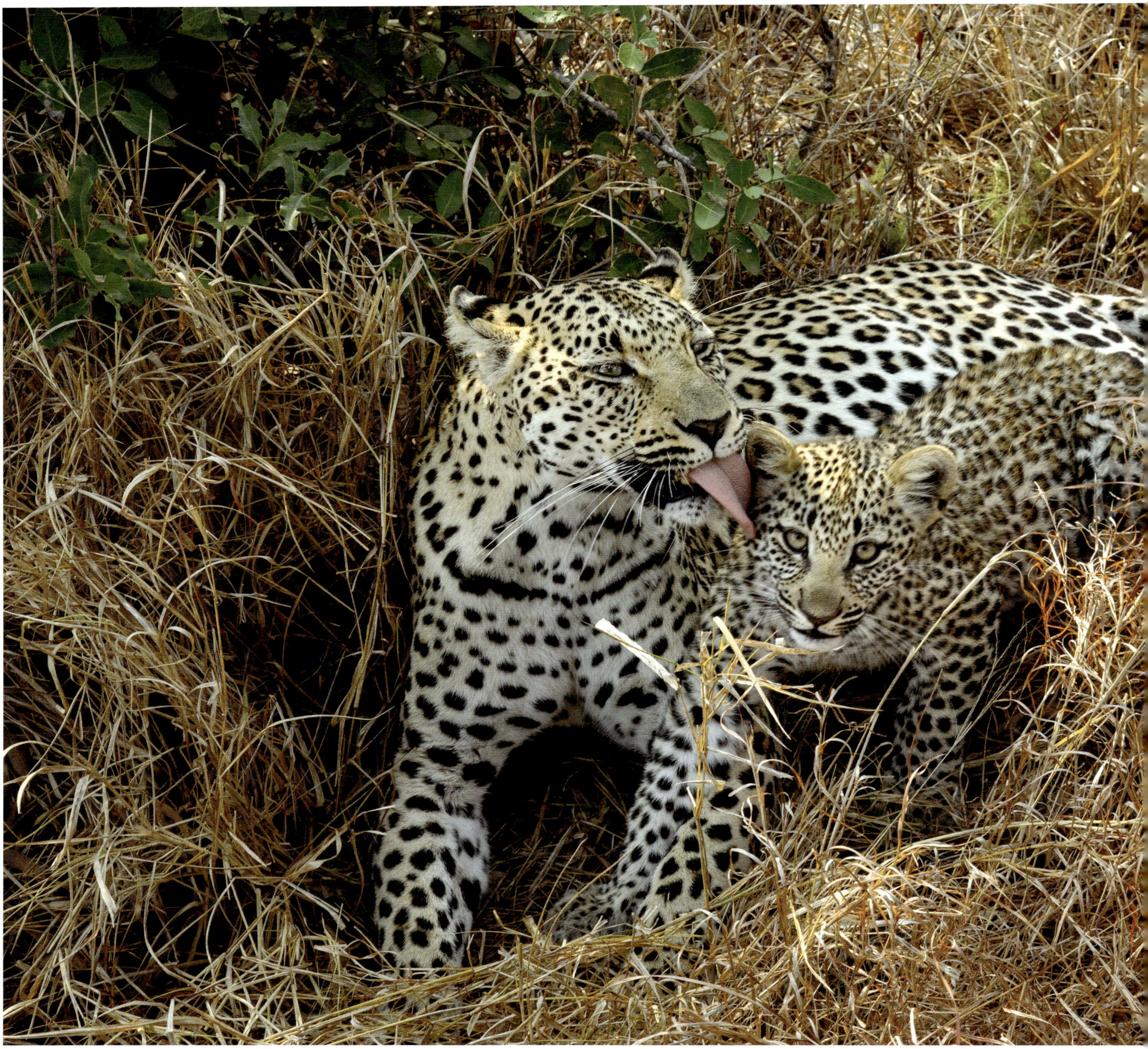

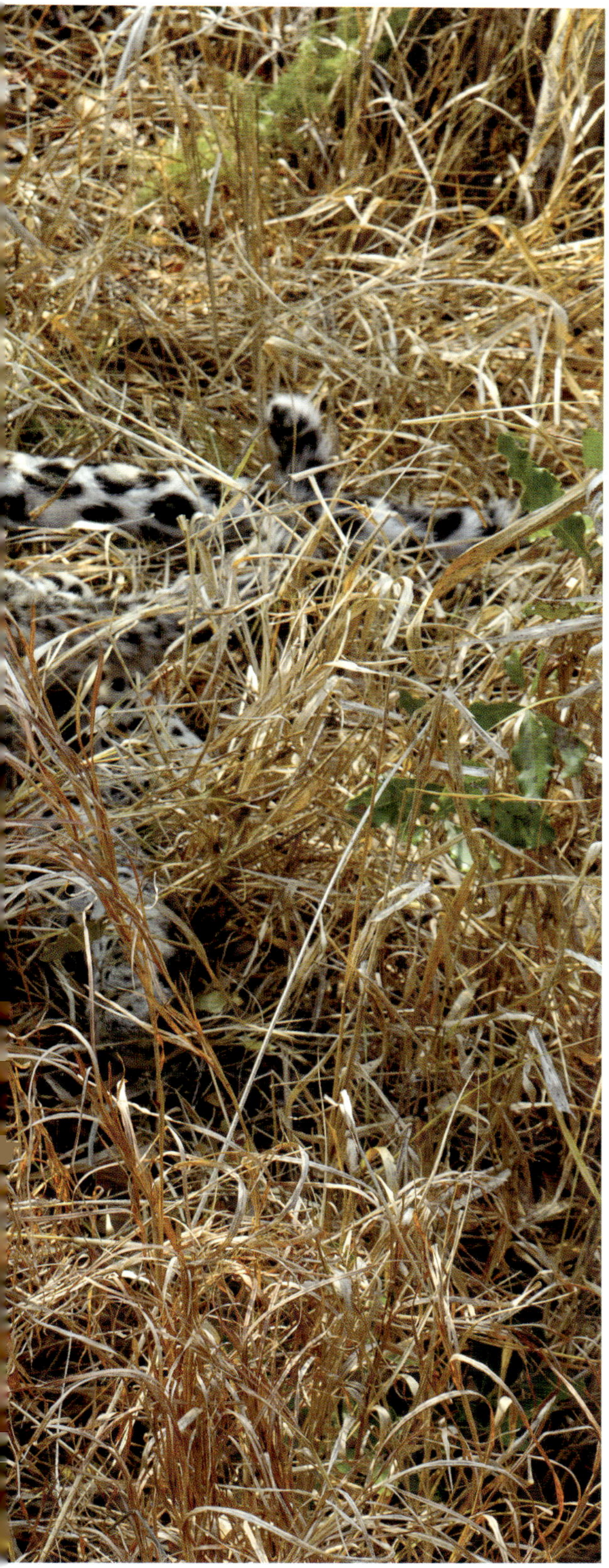

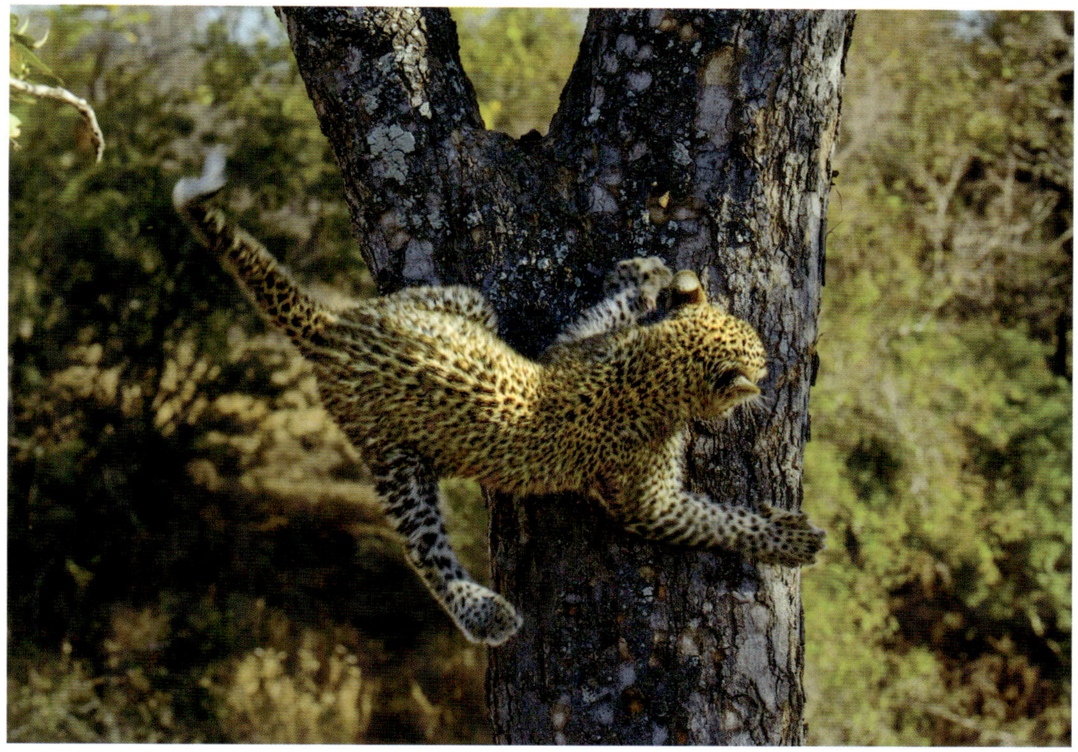

ABOVE: A leopard cub learns to climb down a tree where its mother has stashed an impala kill in Africa's Sabi Sand Nature Reserve.

LEFT: A female leopard grooms her cub in Sabi Sand, where wildlife is well protected. These animals showed no fear of humans.

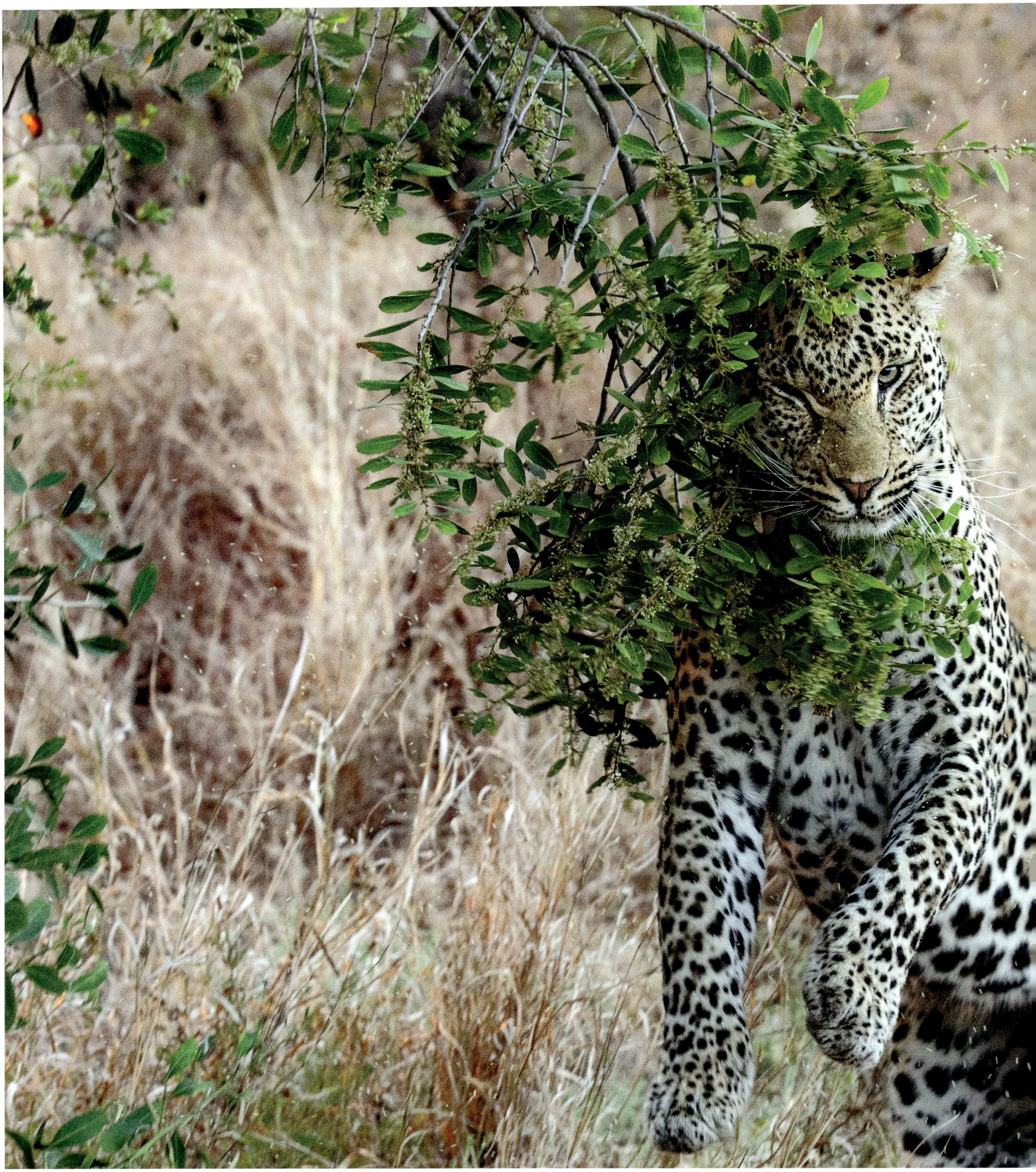

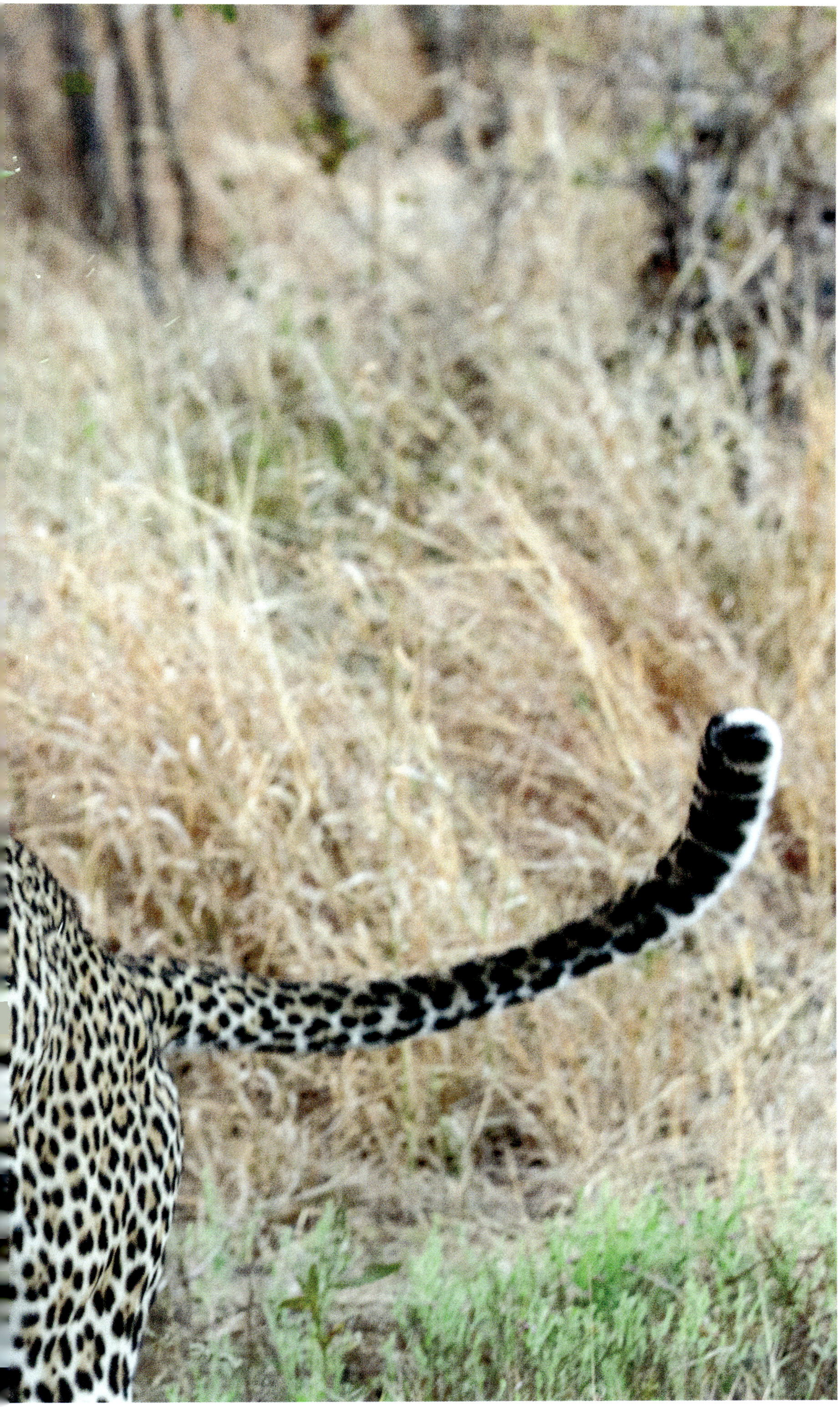

A leopard marks its territory by spraying on a tree and then rubbing on it to transfer the scent, leaving behind olfactory messages for other leopards.

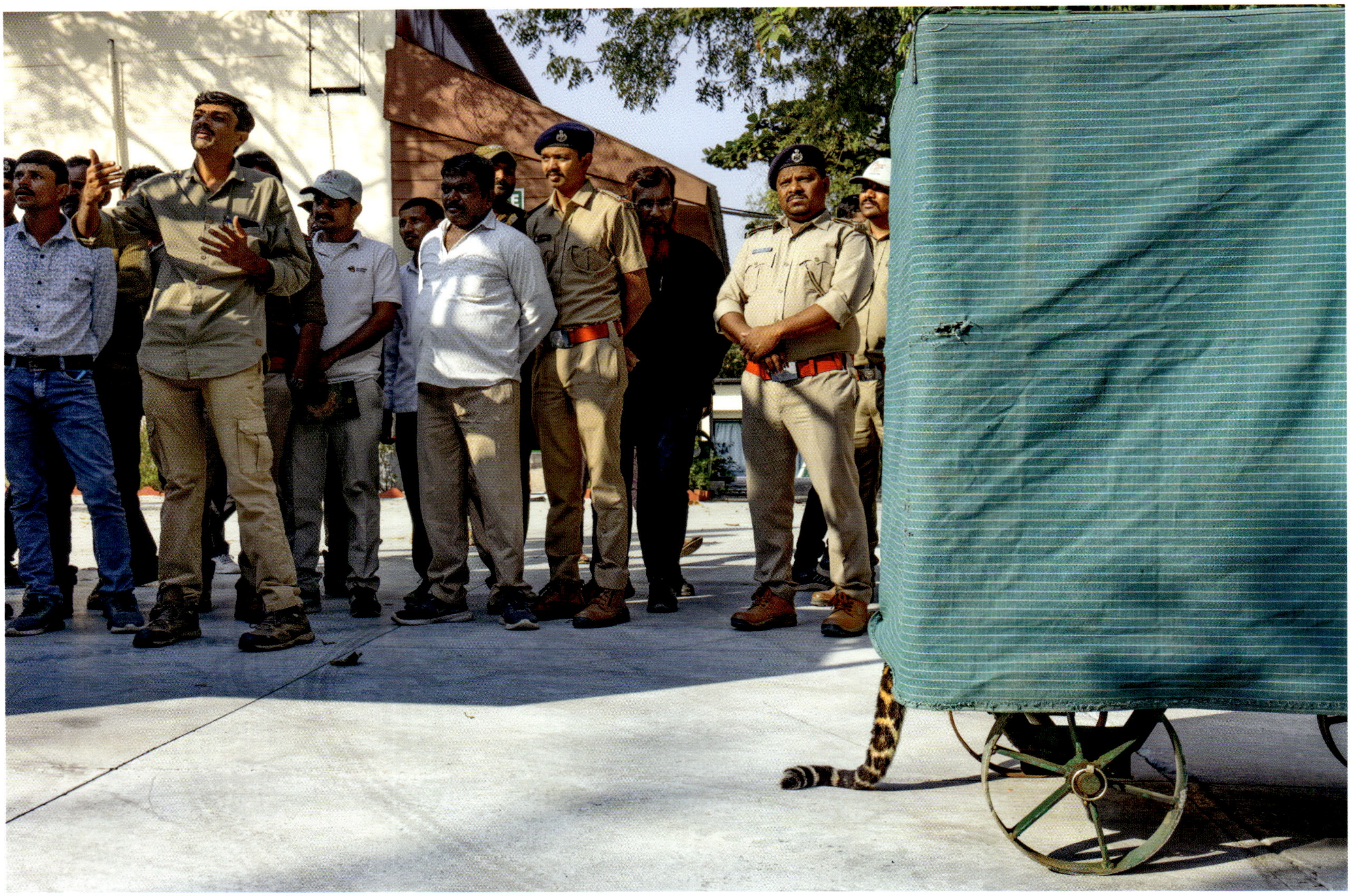

When a leopard strayed near a village outside India's Gir National Park and began hunting livestock, it was captured and brought to a wildlife hospital. Here, park guards and medical staff prepare to load the cat into a truck, headed for release back into the wild.

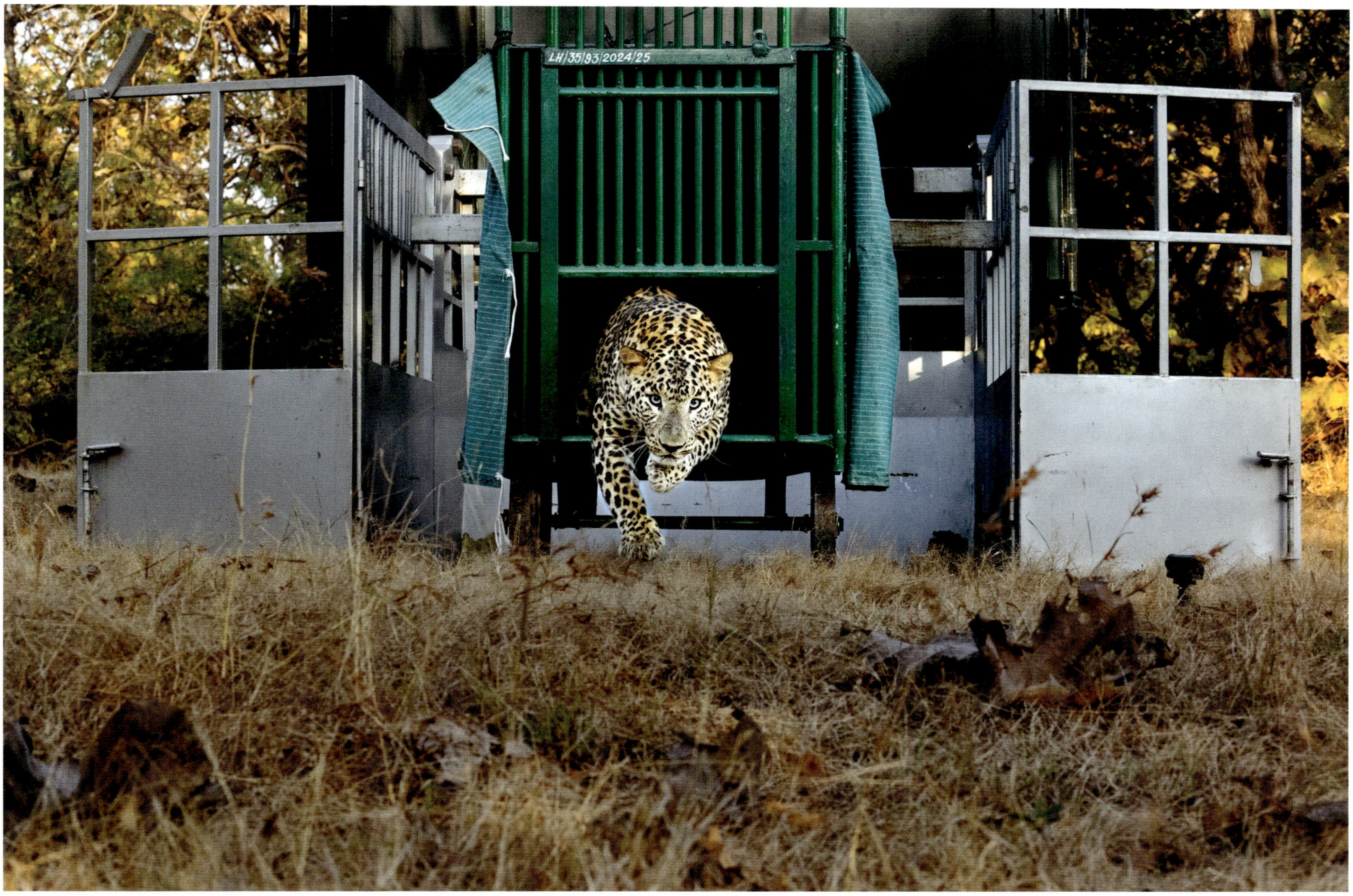

After treatment in a wildlife hospital, this leopard was released into a remote location in India's Gir National Park. Because of the park's high density of both Asiatic lions and Asiatic leopards, the Gujarat Forest Department actively manages the cats to try to keep both animals and people safe.

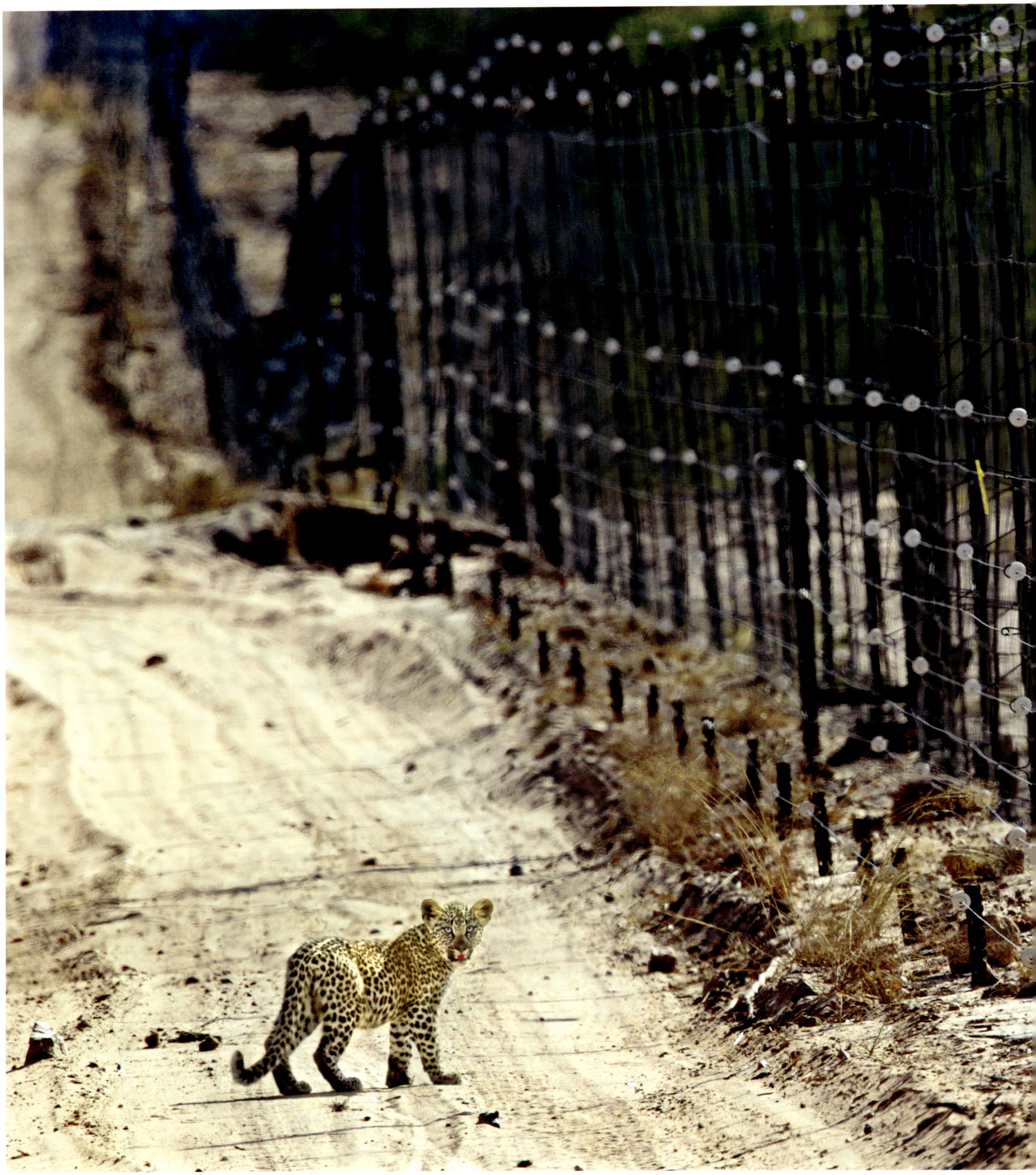

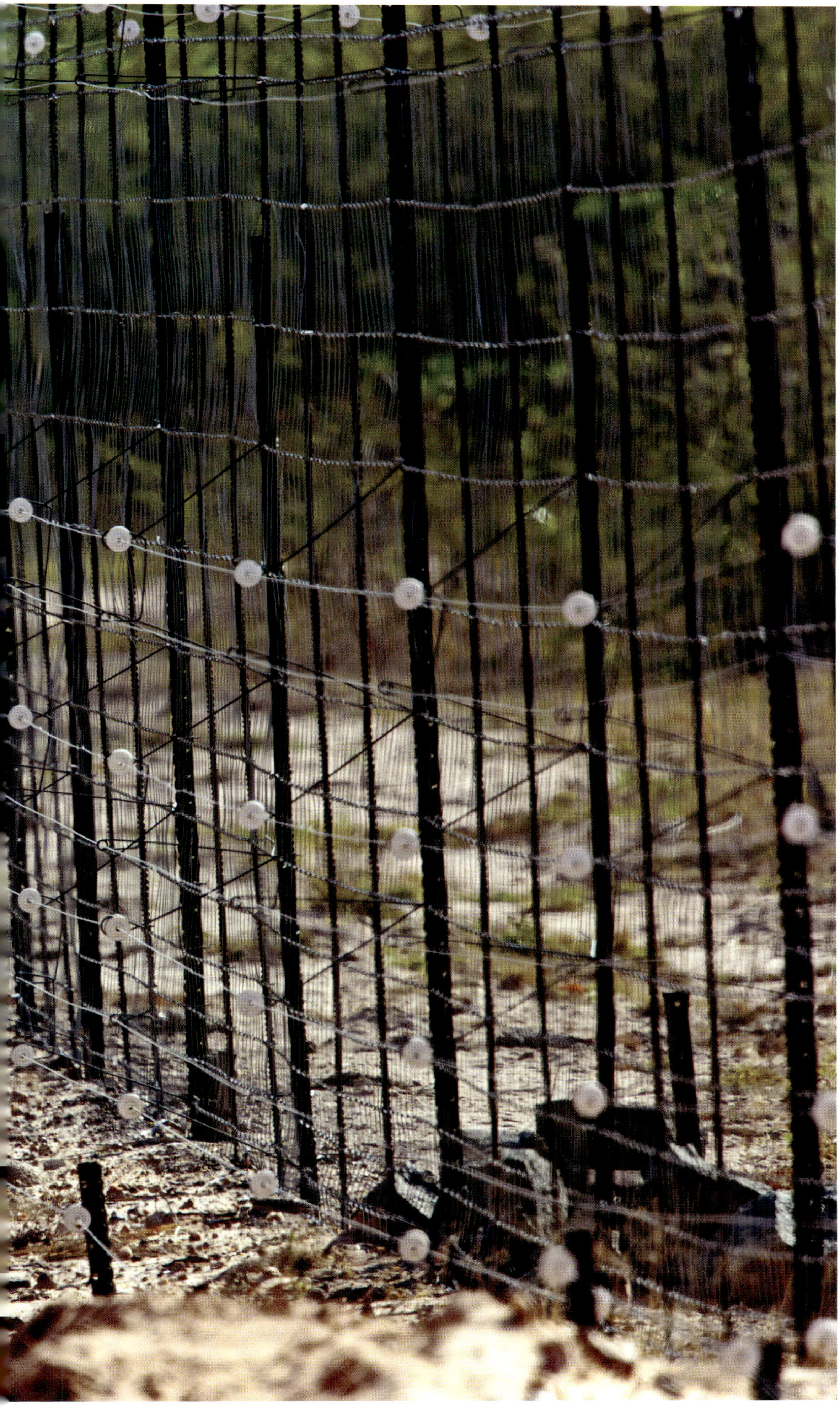

A six- to seven-month-old African leopard cub roams along a fence that keeps wildlife in South Africa's Sabi Sand Nature Reserve from straying into villages – and onto land where livestock grazes.

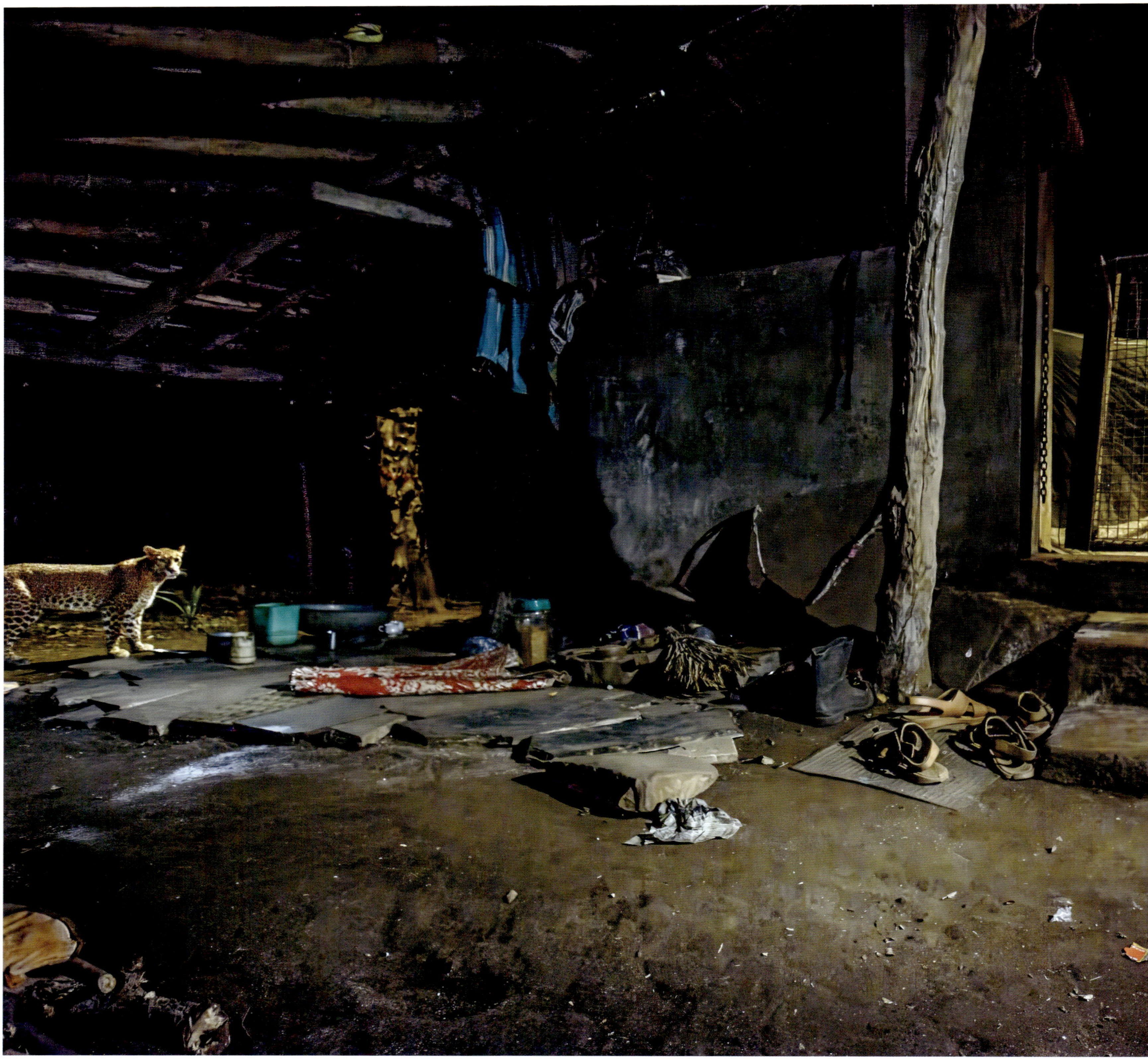

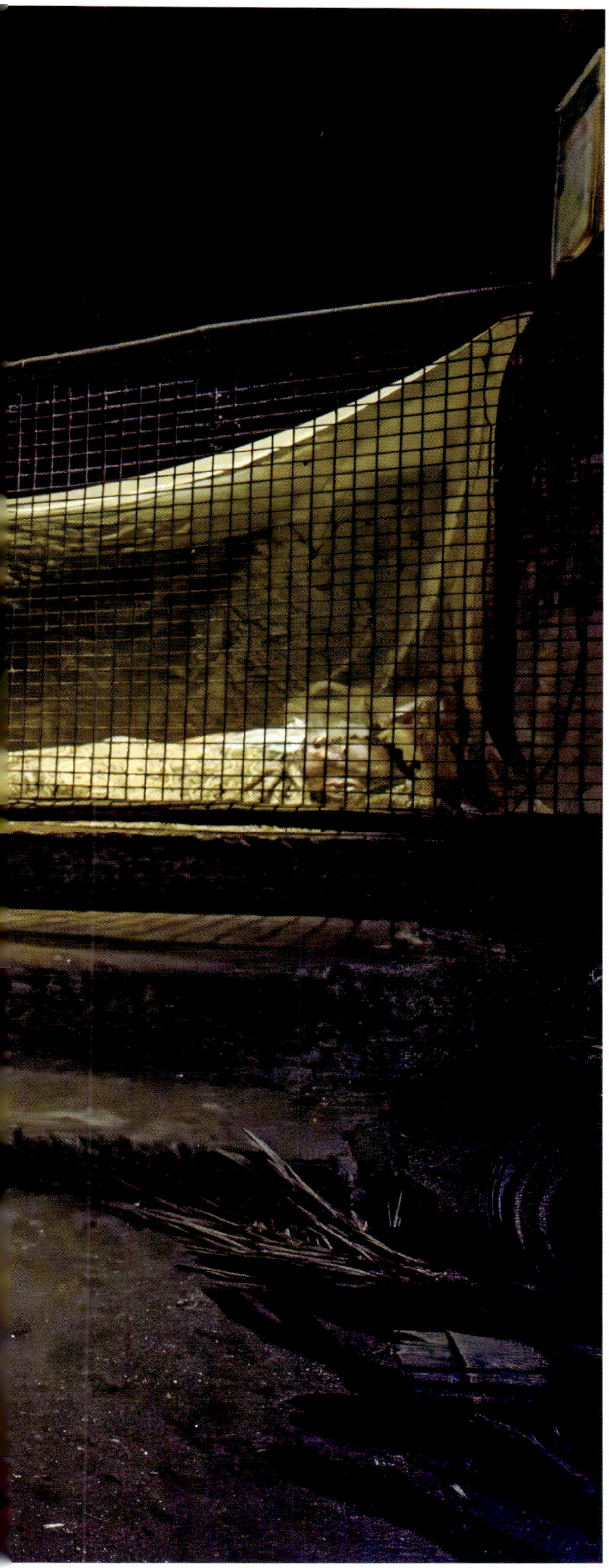

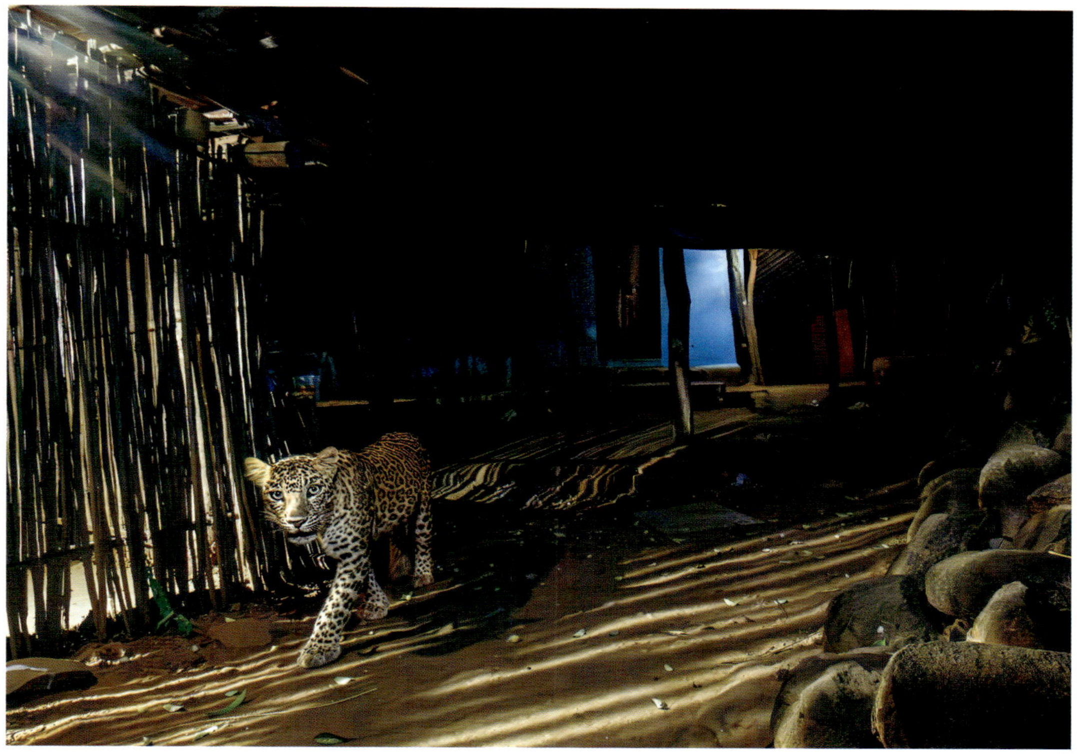

ABOVE: Leopards sometimes leave Sanjay Gandhi National Park to hunt chickens, goats, and packs of feral dogs. This cat strolls along a fence meant to protect this household.

LEFT: A remote camera captures a leopard walking near a sleeping man in Mumbai's Aarey Milk Colony neighborhood. At night, the leopards in Sanjay Gandhi National Park sometimes stray into the outskirts of Mumbai.

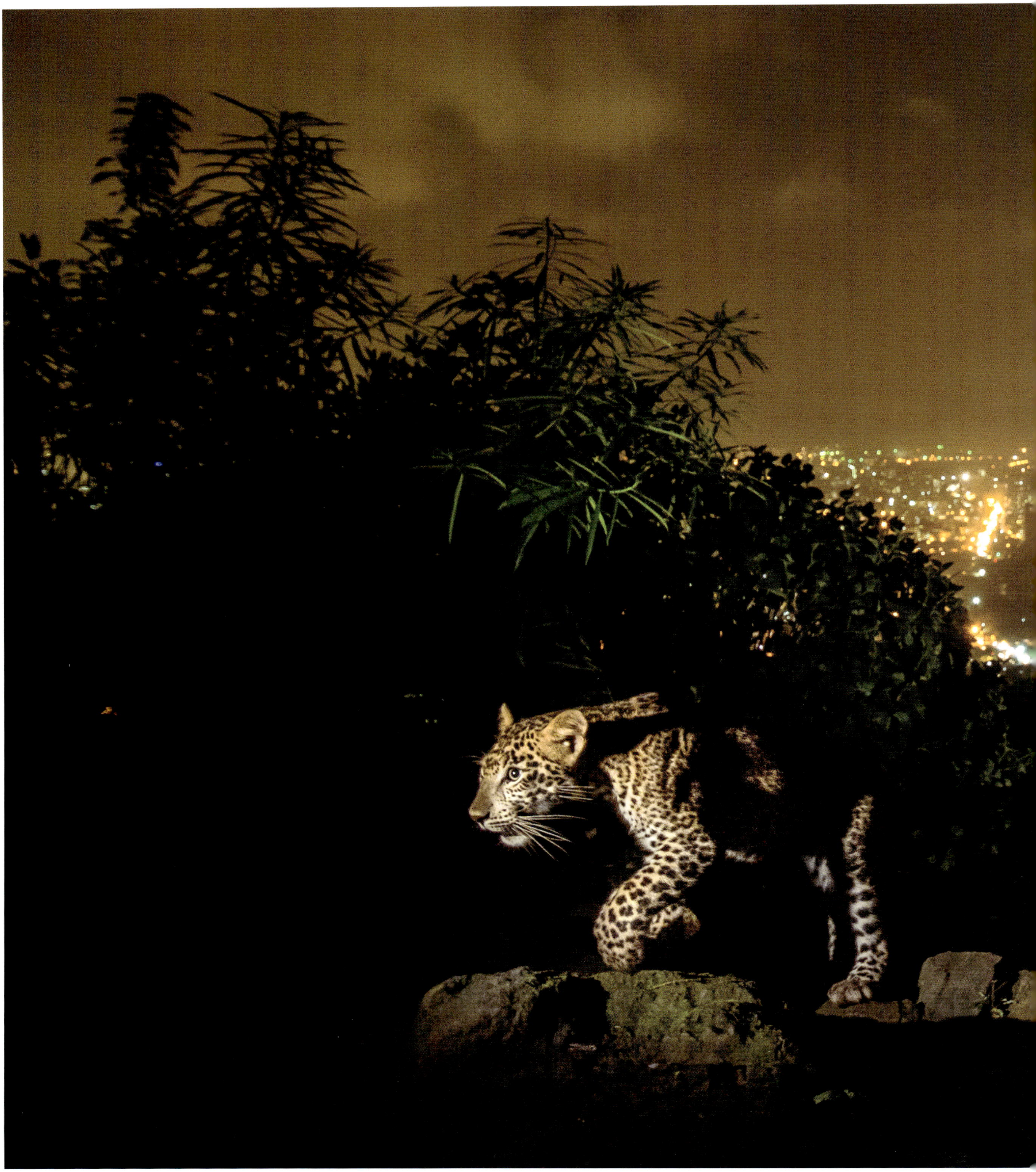

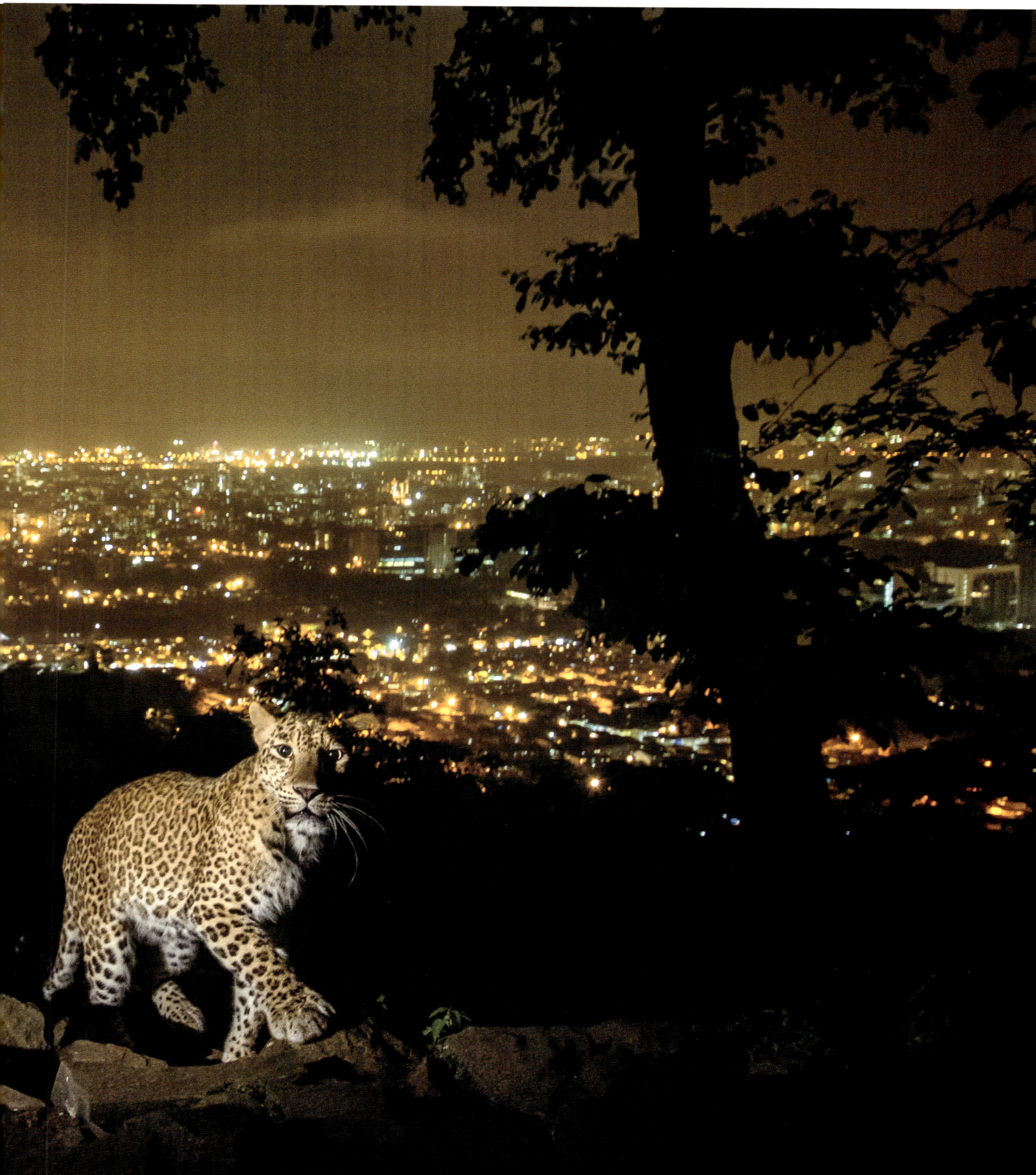

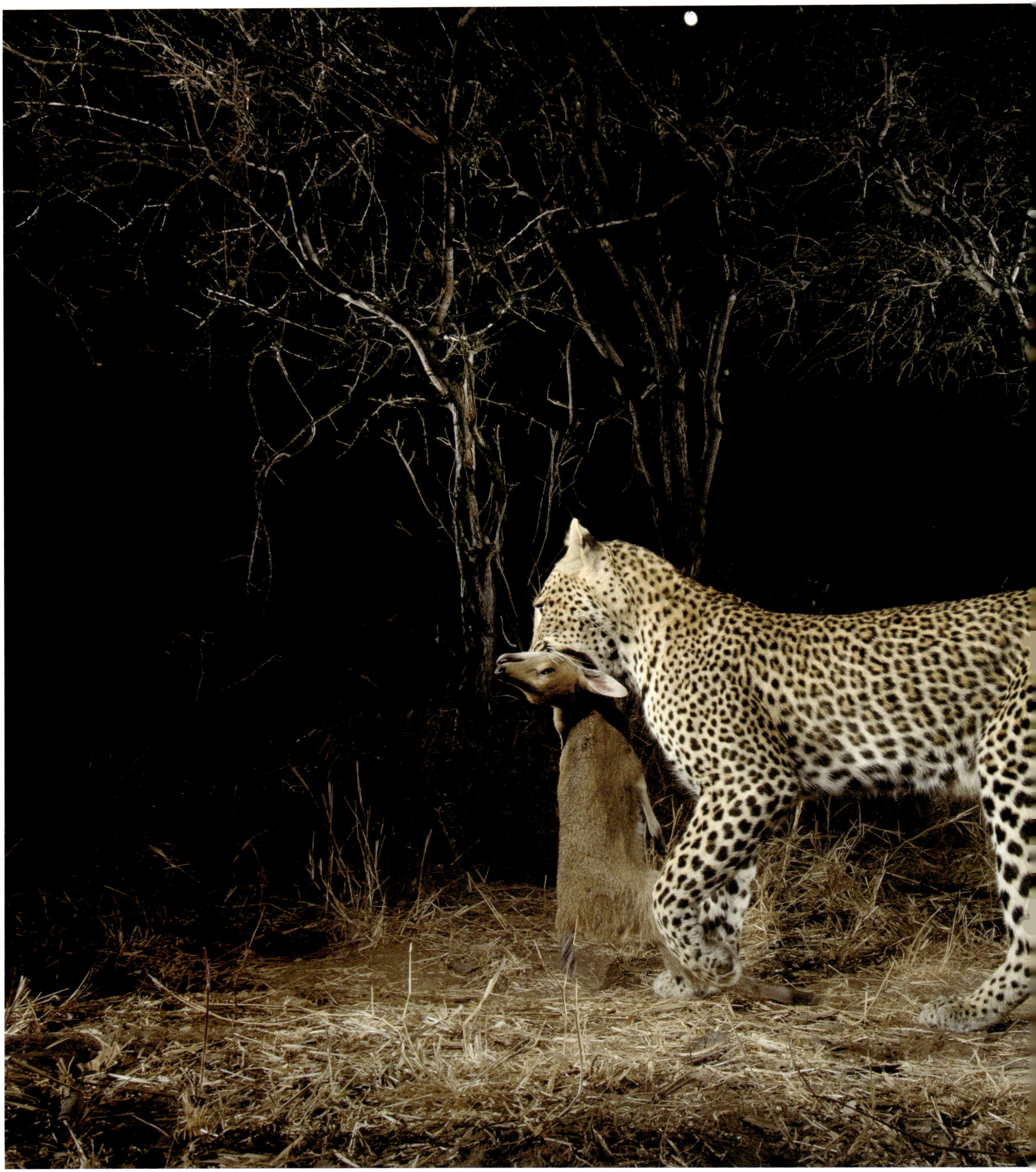

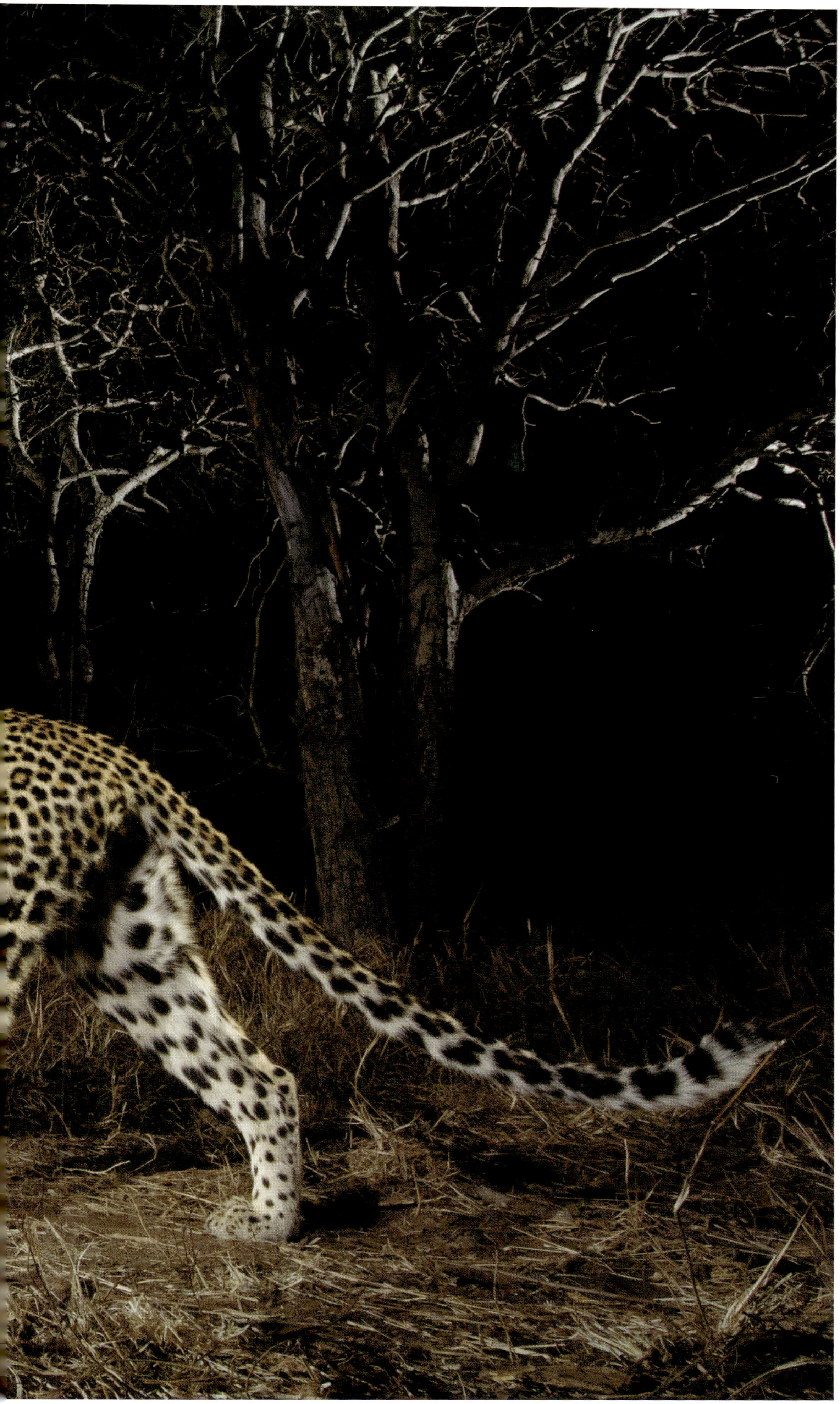

A female African leopard carries a steenbok – a small antelope – back to her two young cubs so they can learn how to eat prey. Soon, they'll be out hunting with her.

PAGES 108–109: A remote camera captures leopard cubs on a trail in Sanjay Gandhi National Park. This 40-square-mile sanctuary – 30 times bigger than New York City's Central Park – is in the center of Mumbai, a city of more than 22 million people.

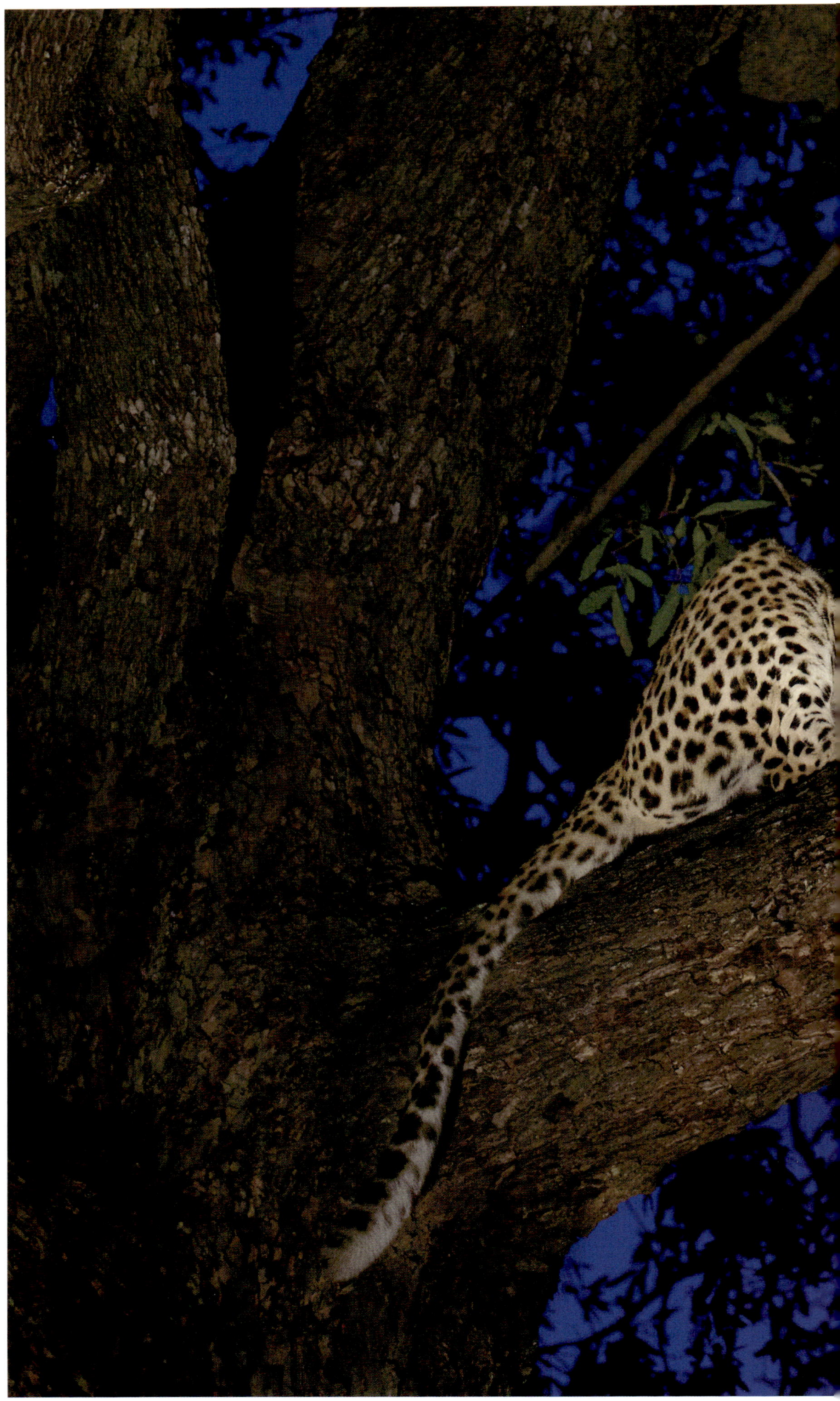

Resting high on a branch so as not to be disturbed, a leopard feeds on an impala. Leopards are incredibly athletic, capable of leaping 20 feet into the air and of dragging a large animal up a tree.

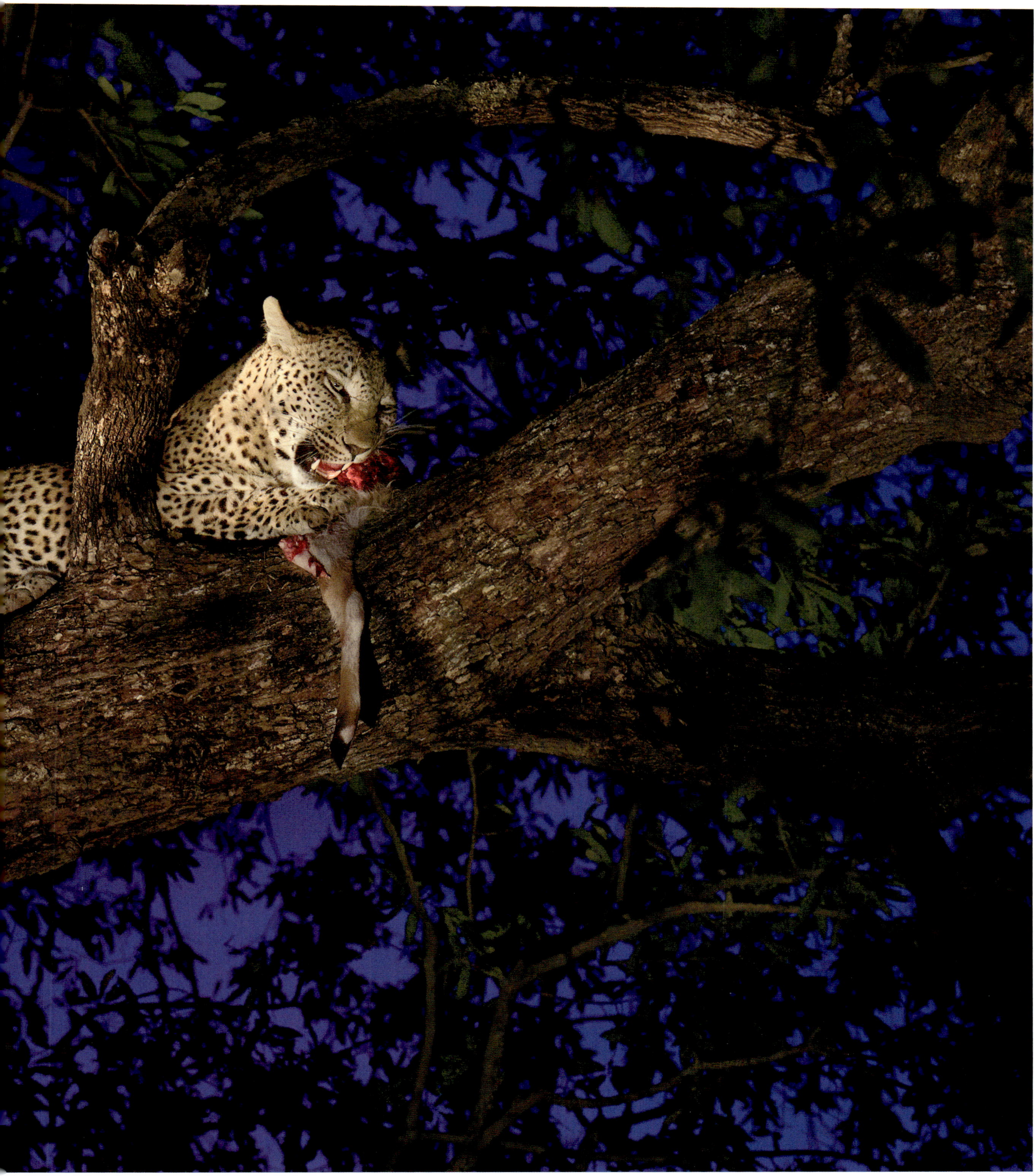

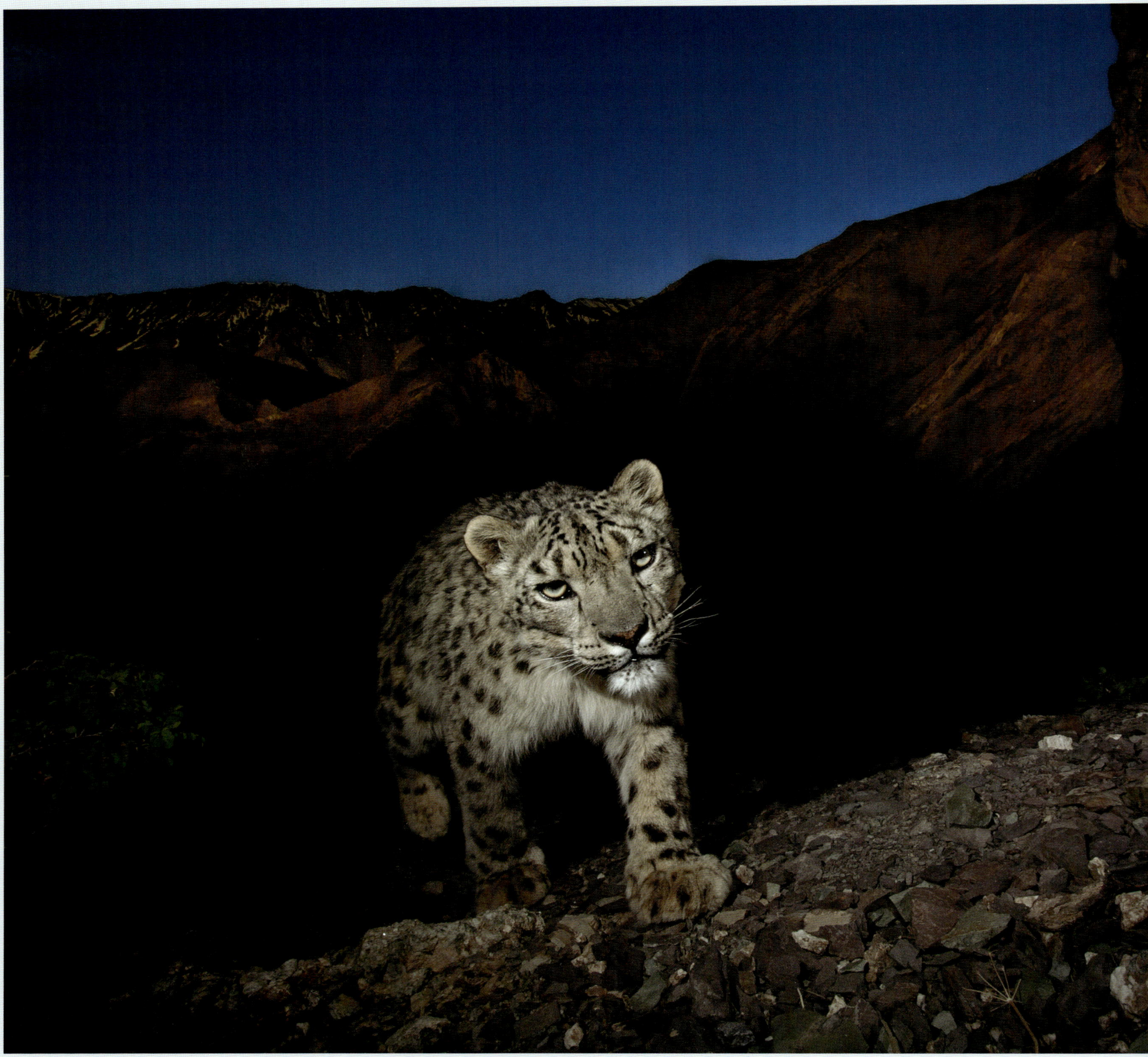

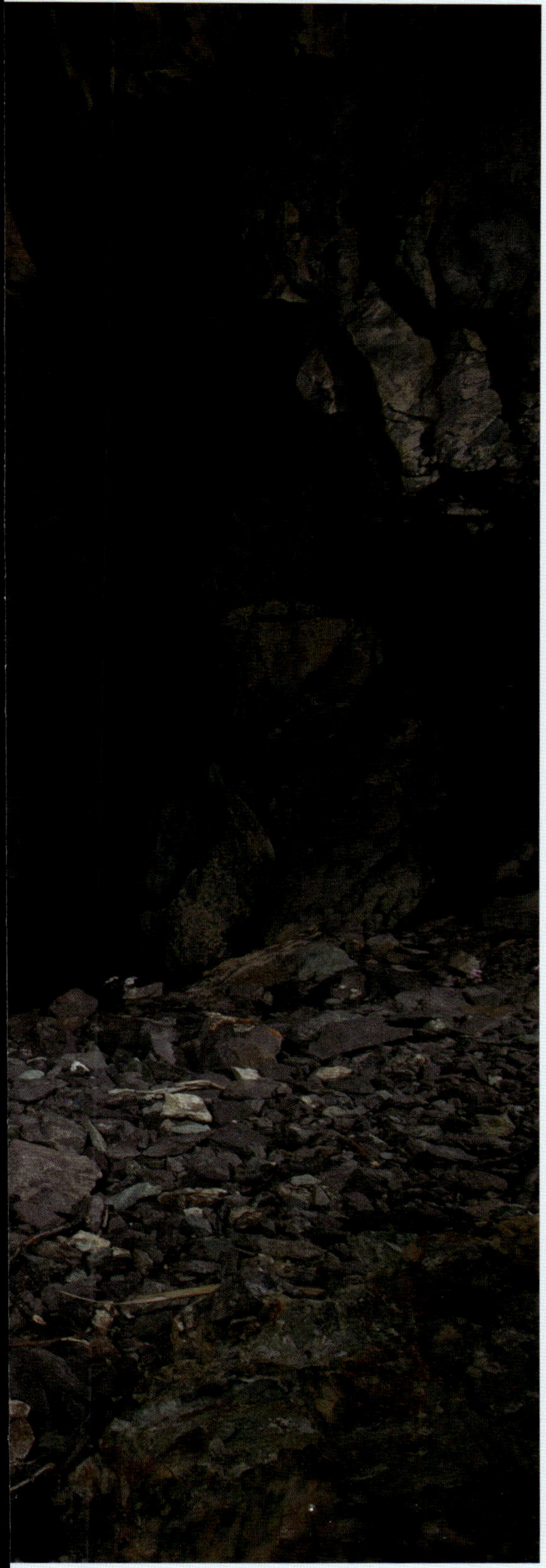

CHAPTER FOUR

Snow Leopards The Elusive Cats of the Himalaya

Conservation efforts and community programs are working to save these threatened cats, which roam Asia's vast high-altitude wilderness.

A camera trap captures an image of an elusive snow leopard on a high ridge trail in Hemis National Park, India.

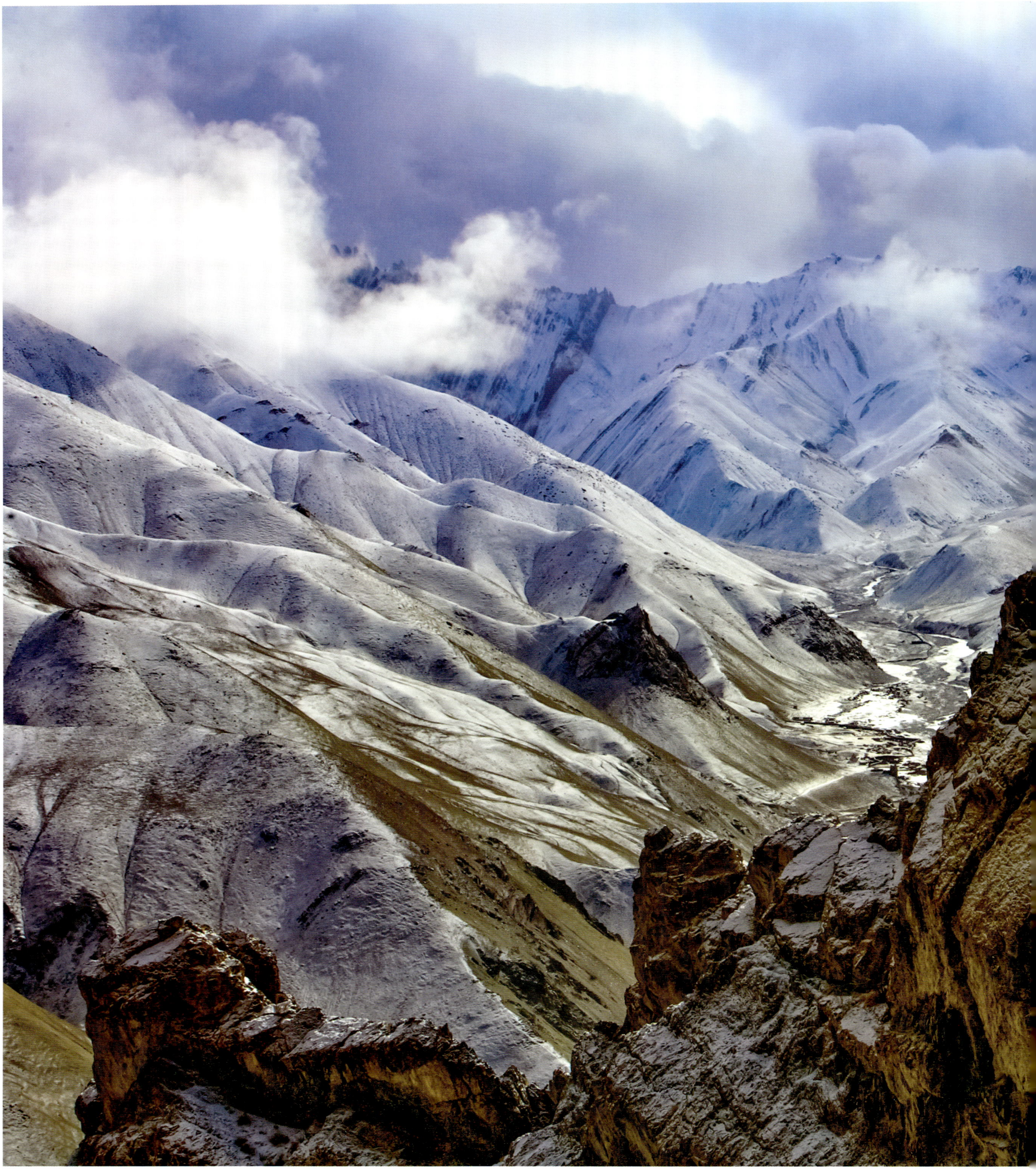

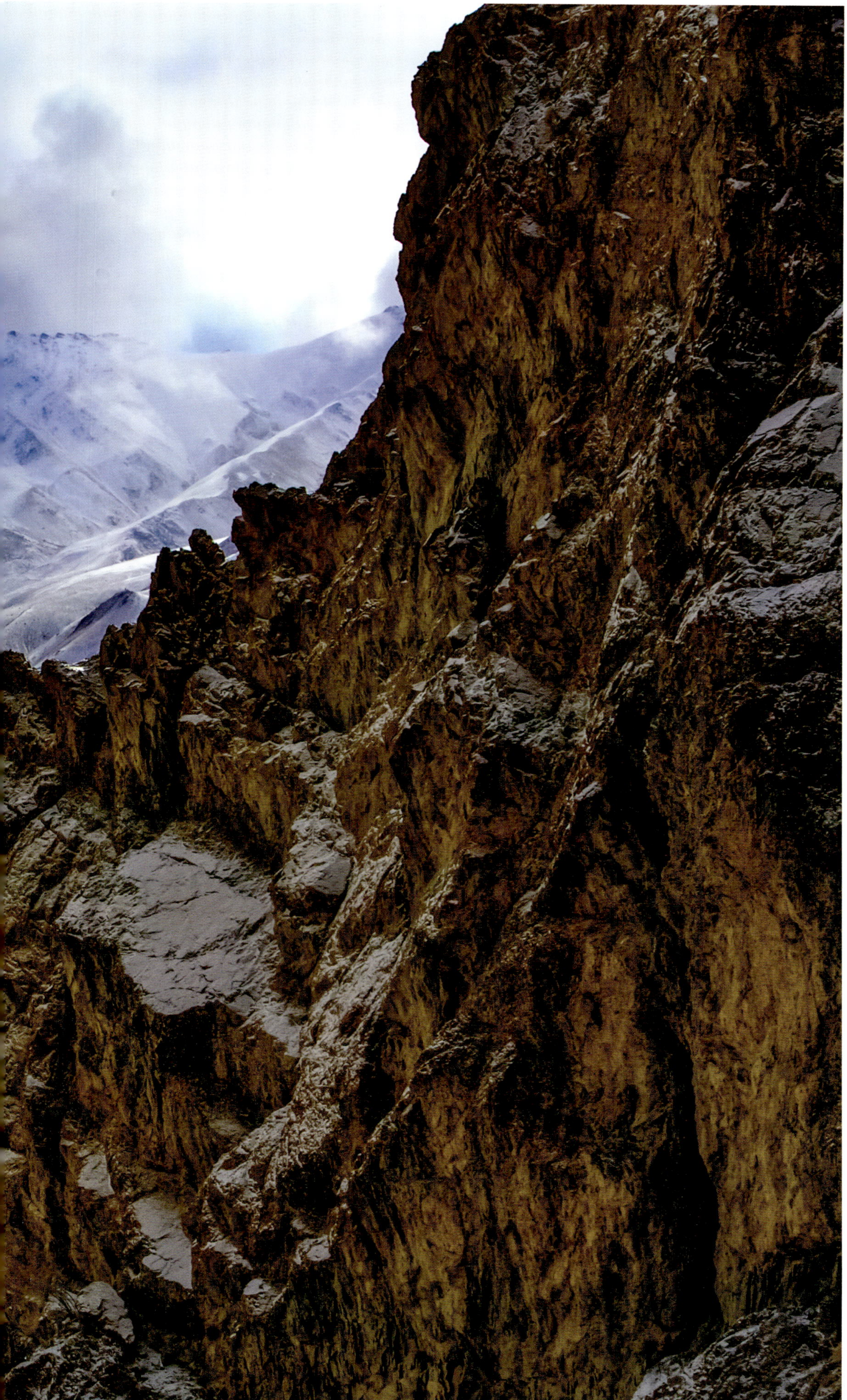

Snow leopards live in remote regions in 12 Asian countries. In India, home is the bare, rugged, high-altitude Himalaya. Long, muscular limbs and tails nearly as long as their bodies help the cats careen down rocky crags and maintain balance as they hunt blue sheep and other prey.

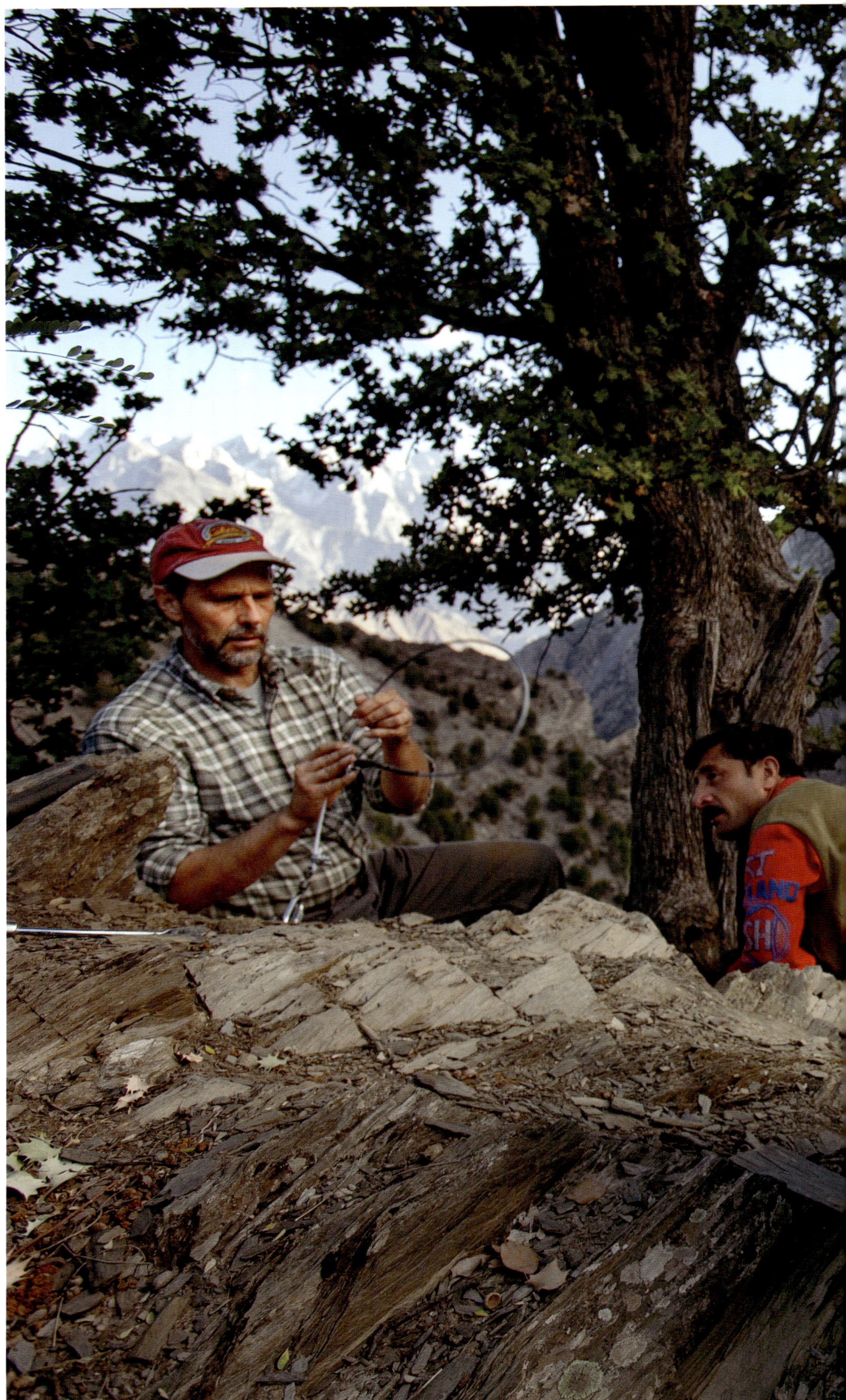

Tom McCarthy (left) and his team of local biologists scan for snow leopards in Pakistan's Chitral National Park. The team captured and outfitted the cats with GPS collars to map their territory.

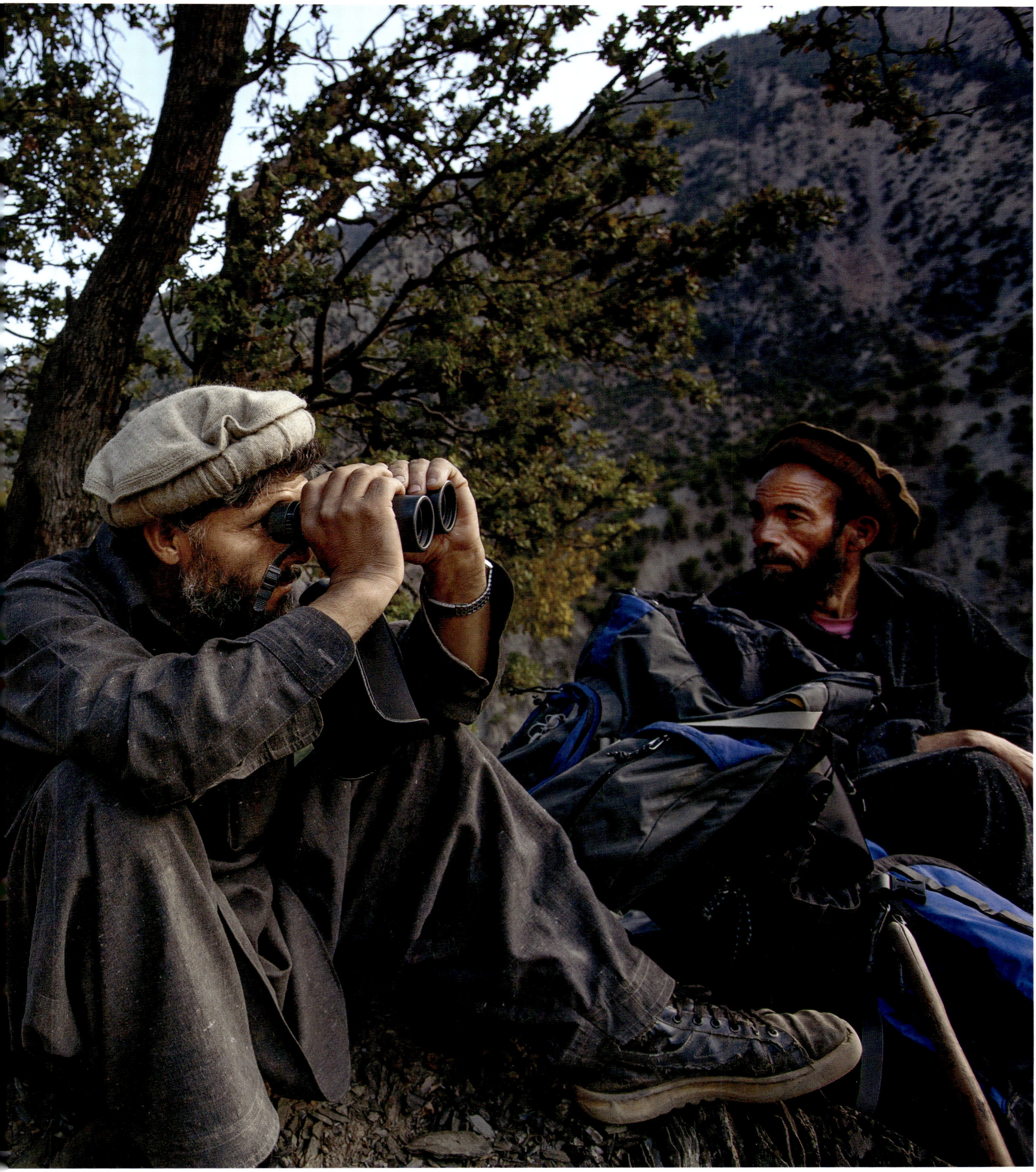

I was trying to figure out what to do

after my time in Brazil with jaguars when Kathy Moran, my editor at *National Geographic,* sent me an email. "What is your dream story?" she asked.

When I was in college, I'd been captivated by Peter Matthiessen's book *The Snow Leopard,* which chronicled his 1973 expedition into remote Nepalese mountains with George Schaller, the world's preeminent field biologist. At the time, George was studying Himalayan blue sheep and searching for snow leopards—which hadn't been seen by a Westerner in 25 years. I'd just done a story on a cat that had taken months to find, so I thought, Why not do a story on a cat I might never see?

Kathy gave the go-ahead, and I began my research on what became a 13-month quest. Known as the "snow cheetah" in Urdu, this solitary animal ranges through 12 countries and two million square miles. So the first step was to find a biologist to work with. That person was Tom McCarthy at the Snow Leopard Trust. He'd studied them for years, including research in Mongolia that tracked one cat across its 386-square-mile range. He was headed to Chitral Gol National Park in northern Pakistan to collar snow leopards.

I flew to Islamabad, boarded a Pakistani Air Force plane with soldiers, and flew to Peshawar, then on to Chitral. Chitral's mayor was on the plane, and "for my safety," he took my passport. He was surprised to learn that research was going on in the mountains there.

Tom was really upset when I told him. He'd tried to stay off the radar, but I didn't know that. He called his boss to discuss the situation, including kidnapping insurance. They figured something out, and I stayed for three weeks.

But we weren't seeing any live prey—or sign of snow leopards. They're cold-weather animals. It was extraordinarily hot for November, and the cats must have stayed up high, at cooler altitudes. My budget was tight, and I decided to head home.

Just then, terrorist bombings rocked the city of Peshawar, about eight hours south of us. Tom's colleagues, local biologists with the World Wildlife Fund, showed up and told me I had to leave immediately. They dressed me in traditional clothing—an Afghan hat, a knee-length kurta, loose-fitting pants, and sunglasses to obscure my blue eyes—and put me in the back seat of an SUV between two Pakistanis.

With the unrest, I was in danger. This was 2005, four years into the U.S. invasion of Afghanistan, and we weren't far from that border. Protests had erupted in Afghanistan and Pakistan after a *Newsweek* article alleged that U.S. interrogators in Guantanamo Bay desecrated a Quran. Thousands marched, chanted, and burned American flags. Some of these protests turned violent, including one in Peshawar. As an American, I'd be a valuable person to abduct—or punish.

As I got into that vehicle to drive to Islamabad, I called my partner, Sharon, on my satellite phone, and knowing there was surveillance everywhere, I used it as a platform. I said, "Whoever is listening, I'm Steve Winter. I'm with *National Geographic.* I'm getting in a Toyota Hi-Lux pickup with biologists from the World Wildlife Fund, and they're driving me to Islamabad. I'll be there by midnight. Call my home number if I don't arrive." Anyone who was listening could easily get my phone number.

First we had to retrieve my passport from the mayor in Chitral. Disguised or not, I'd have to show my passport at military checkpoints and at the airport. At the first stop, a soldier spoke English and was excited to meet me—he had family in the States. Other checkpoints were thankfully uneventful. We took a circuitous route to avoid Peshawar, and I was never so happy to arrive at a hotel as I was when I collapsed on my bed in Islamabad and called home.

Three weeks later, the weather in Chitral shifted, and a snow leopard came with the cold. Tom set a snare on the exact spot where George Schaller had taken a photo of a snow leopard decades before. He caught the cat there and put a tracking collar on it. He followed its movements deep into Afghanistan and then into far northern Pakistan. Animals don't understand wars. Or borders.

I decided to do the story in India, but I needed more recon to settle on a site. In January, I joined mountain climber David Breashears on an expedition to Ladakh, in northern India. We walked up the frozen Zanskar River with a team of Sherpas, scouring the area for signs of snow leopards.

David pushed us hard: fast pace, long days. Sometimes we waded through running water or edged across questionable ice. Late one afternoon, I fell through the ice up to my waist. I instinctively splayed my elbows to keep from going under altogether, and our guides quickly hauled me out.

We were above the tree line at 12,000 feet, so there was no wood for a fire. The only way to get warm was to set up the cook tent and light the kerosene stove. But we were in a rocky chasm with no

(Continued on page 125)

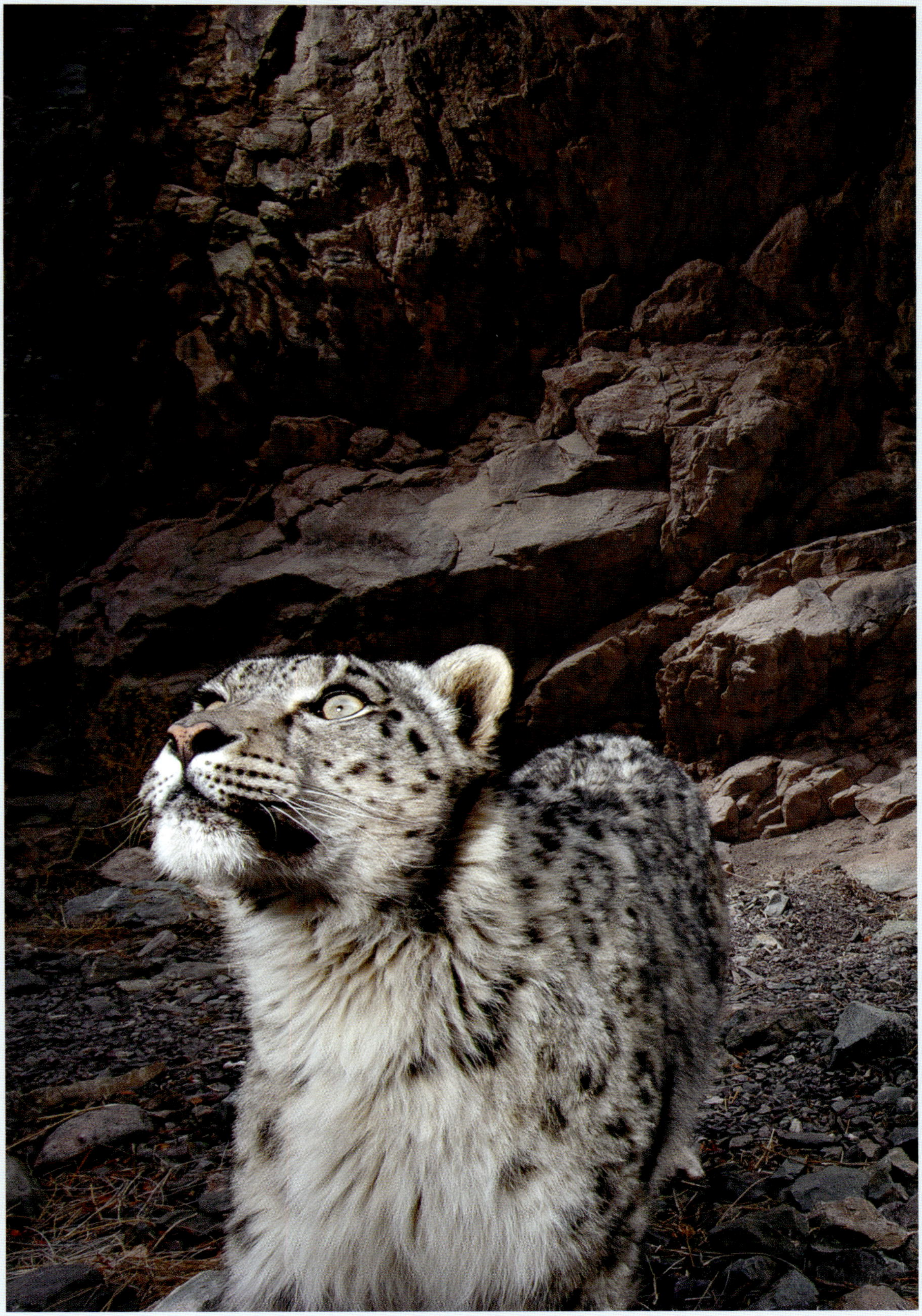

A snow leopard gazes at a camera trap in India's Hemis National Park. These cats, with their large eyes and exceptional low-light vision, can hunt in near total darkness.

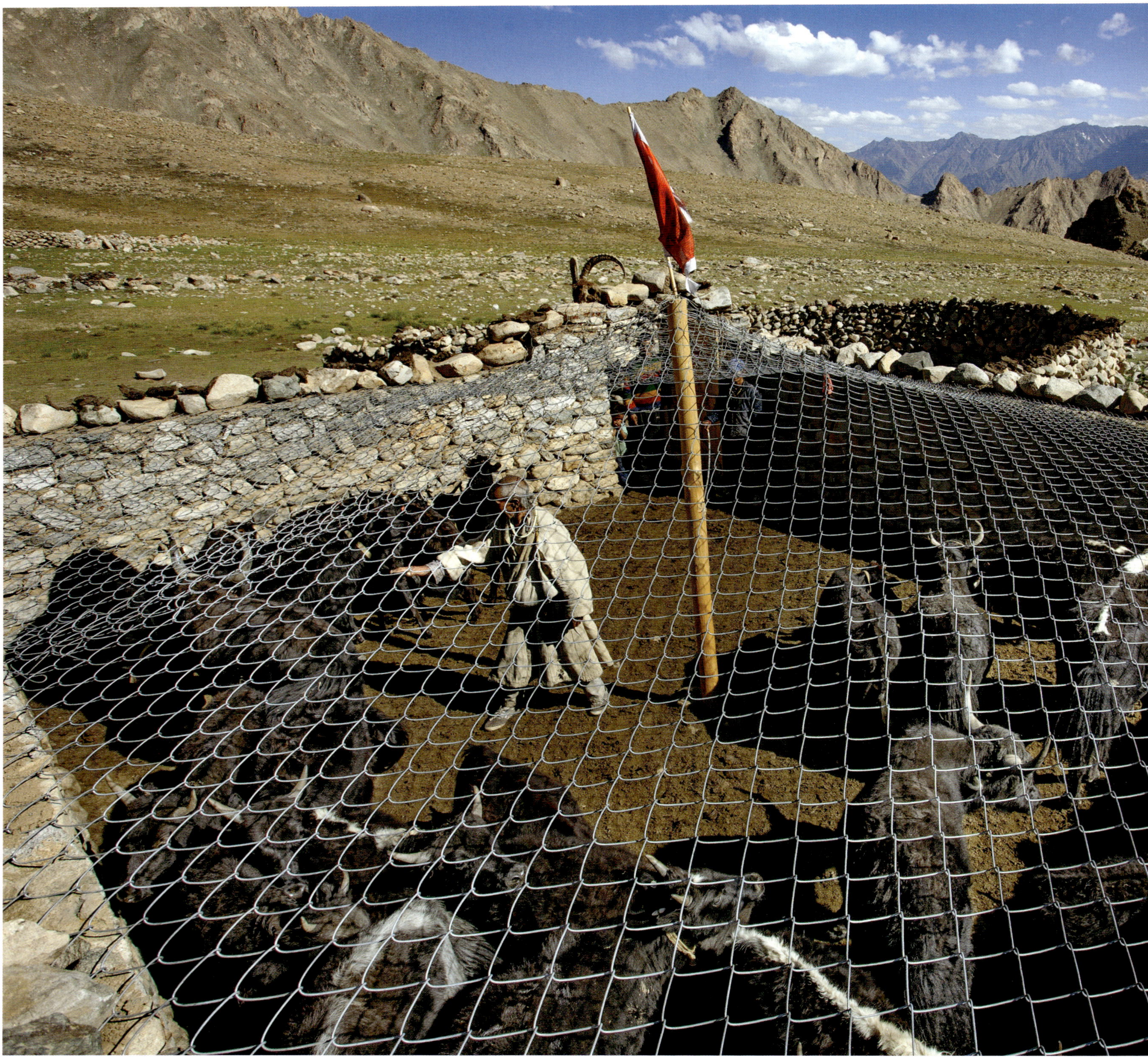

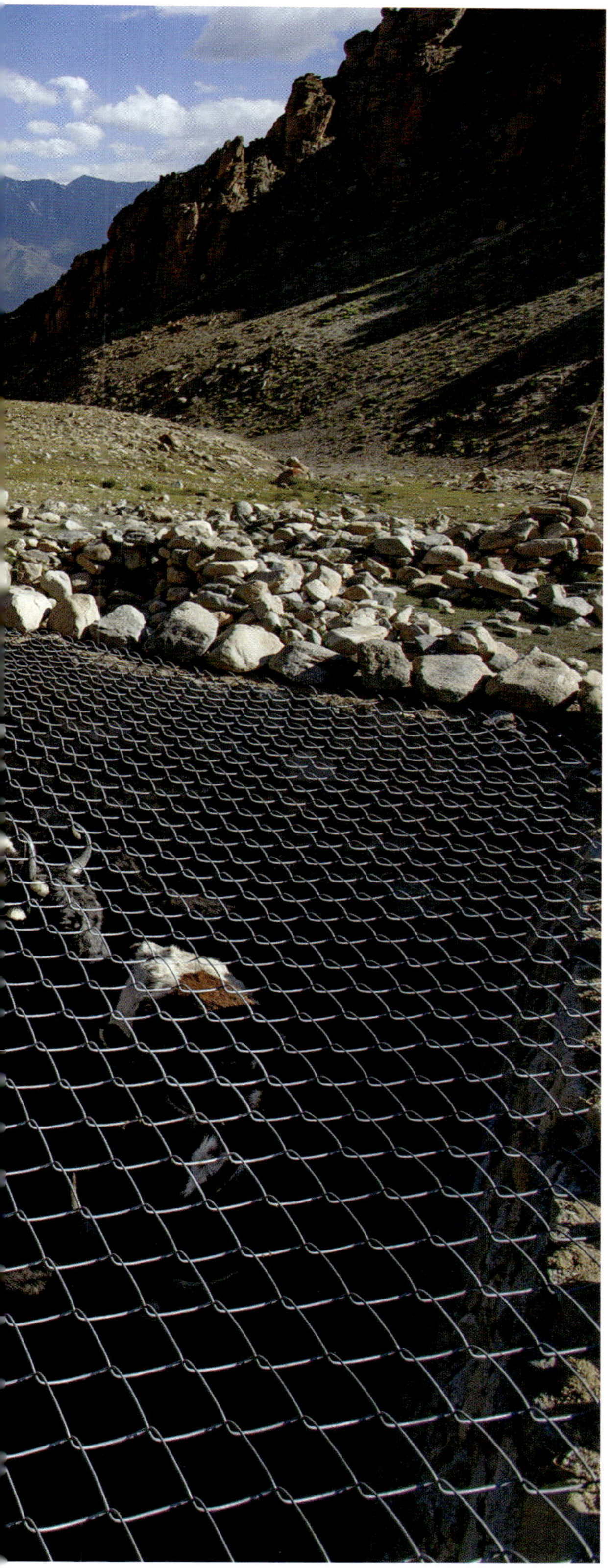

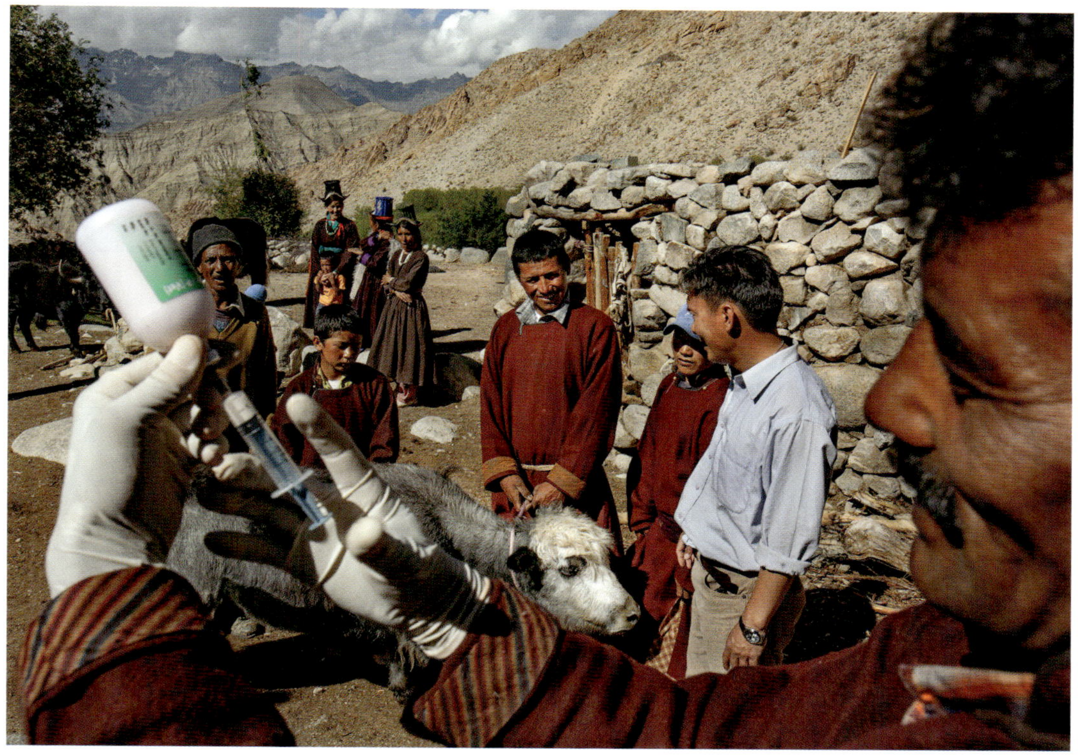

ABOVE: Residents of the Himalayan village of Ulle get free vaccination for their livestock if they agree to protect snow leopards. It's an economic boon: Disease takes down far more of their animals than the cats do.

LEFT: A predator-proof corral keeps livestock safe from hungry snow leopards. Building these structures is just one of several programs that economically benefit communities and foster coexistence with this cat.

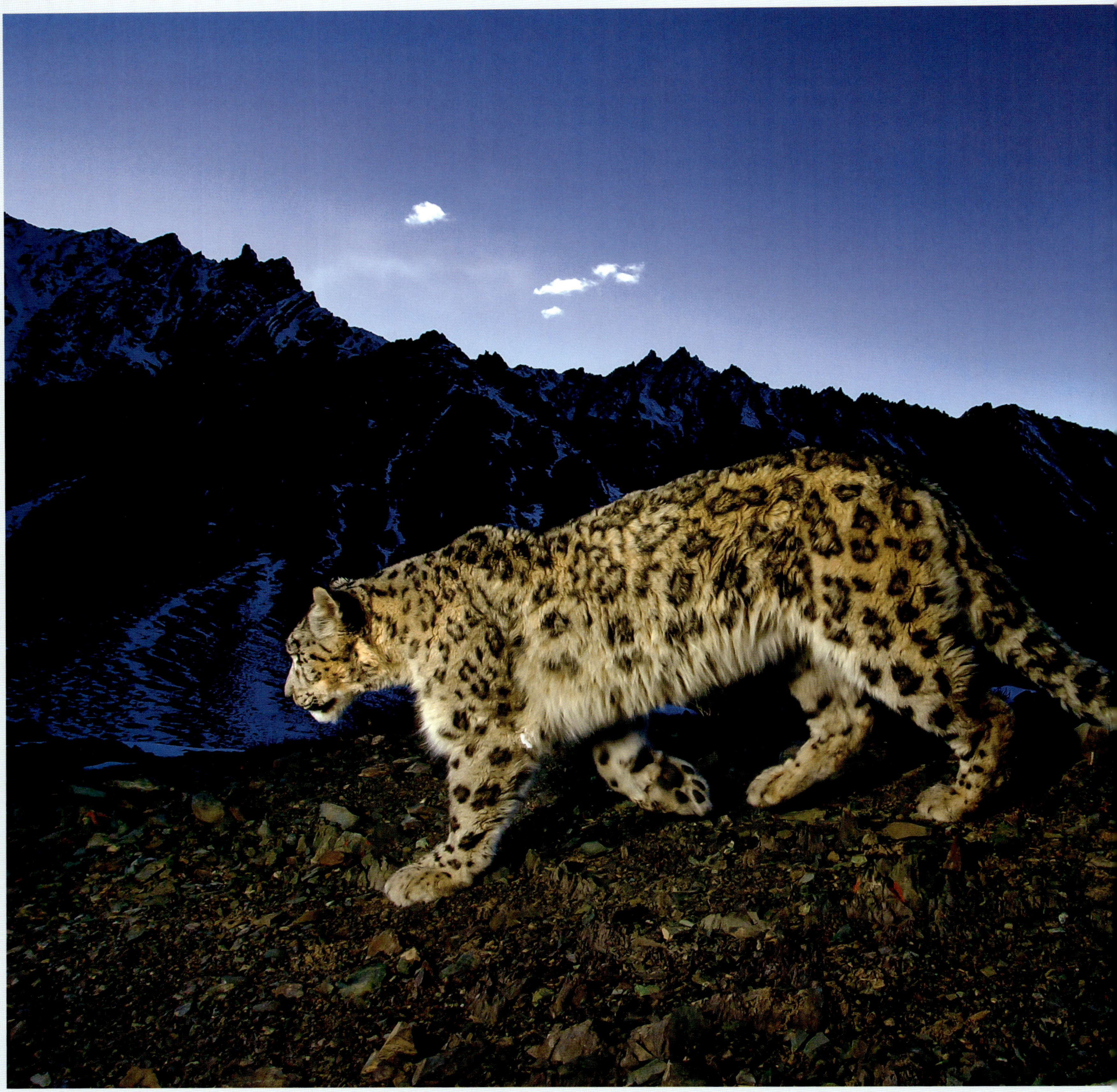

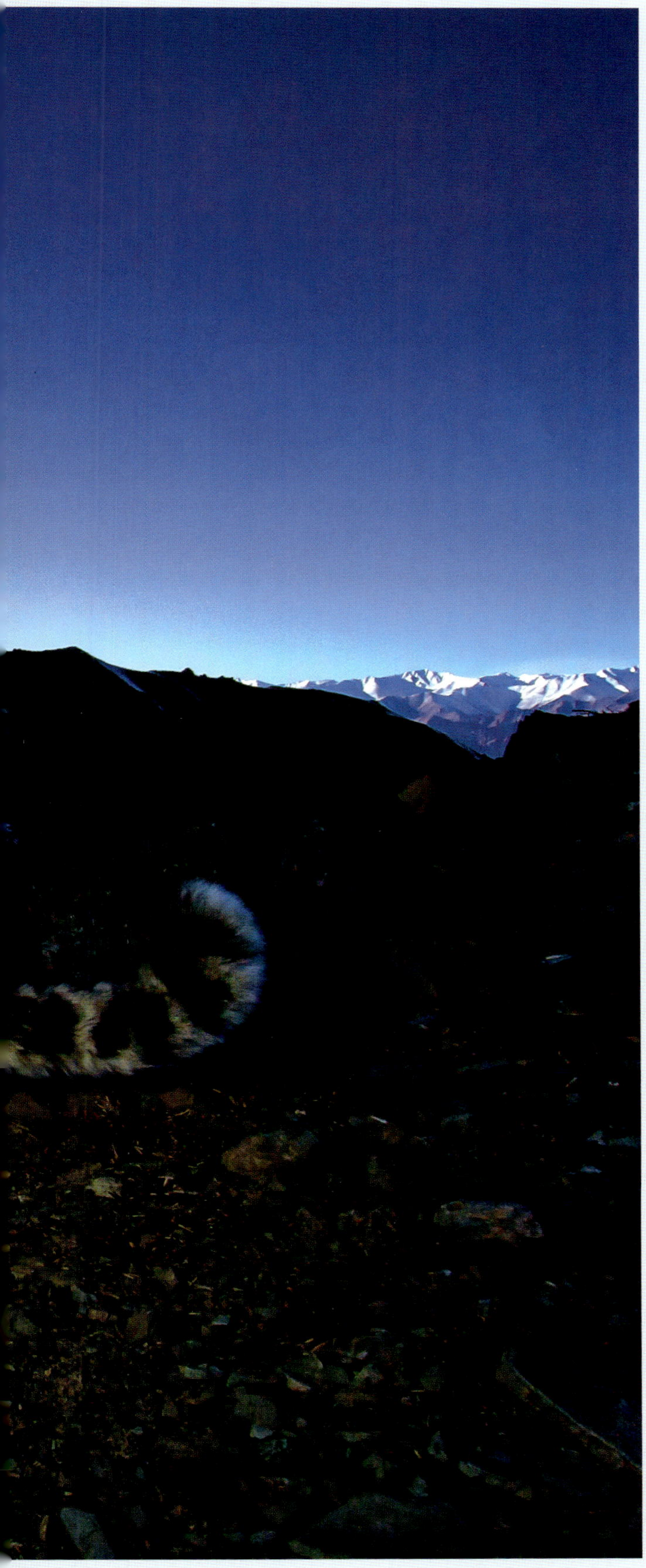

In northern India, experts helped me set up camera traps where snow leopards walked or marked their territory. In some places, it still took months to get a single image.

(Continued from page 120)

shoreline and no place to set up camp. By the time we found a flat spot, my pants were frozen, and I was shaking uncontrollably, bordering on delirium. While some of the team erected the tent, others stripped off my wet clothes and wrapped me in my sleeping bag. The next thing I knew, it was morning.

That was strike two. It would be impossible to get our equipment in and too dangerous to work there. But there was another option. The Snow Leopard Conservancy India, a nonprofit launched by pioneering snow leopard biologist Rodney Jackson, was having success with a camera trapping study in Hemis National Park. This rugged 1,700-square-mile expanse, situated on the northern flank of the Himalayan Range, is believed to harbor the highest density of snow leopards in the world.

Sharon and I explored the Zanskar Valley region on a two-week trek that summer, guided by Tashi Tundup, who worked for Rodney and would be my local assistant. The trip gave me enough confidence to tell Kathy I could pull this off. I hired her former assistant, Emilene Ostlind, to come with me. Emilene hailed from Wyoming, so for her, high mountains were home. Until that point, I'd only worked in jungles. I was a sea-level, tropics kind of guy.

In addition to camera gear, I had to outfit an entire camp for seven people. We'd need to survive three months in Himalayan winter. Emilene and I flew to Delhi and then north to Leh with 33 bags: cameras, lenses, 14 camera traps, flashes, batteries, tools, tents, sleeping bags, headlamps, severe weather clothing, cots, a propane heater, a kerosene stove, my satellite phone, solar chargers, etc., etc.

We shopped for food and other provisions in India. The last things we bought were a pile of DVD movies and a case of Old Monk rum—our only entertainment. After shopping and acclimating to the altitude for a couple of days, we loaded everything into trucks and 4×4s and drove to the end of the road. It was a rutted single-car-wide mud track, strewn with rocks and boulders, etched into a peak with a 1,000-foot drop-off to the valley below. There were times when a car or truck came the other way. Passing was terrifying.

From there, we loaded everything on packhorses and walked in. I wondered what Canon would think if they saw the 1200mm lens they'd lent me strapped on a horse. It was one of just 10 in the world, so expensive they had to be special ordered. We camped at 15,000 feet in a valley surrounded by saw-toothed ridges. It looked like we were on the moon: It was a world of rock and scree.

On our first day there, the cook was scanning the ridges with binoculars and spotted two moving dots. I grabbed my megalens and got pictures of two snow leopards, minuscule in the frame. I was psyched! The next day we glimpsed them again, in the same spot. But that was the last time I saw one. I never met one face-to-face, never saw its golden eyes or its coat, which writer Doug Chadwick later described as "the color of dappled moonlight on frost."

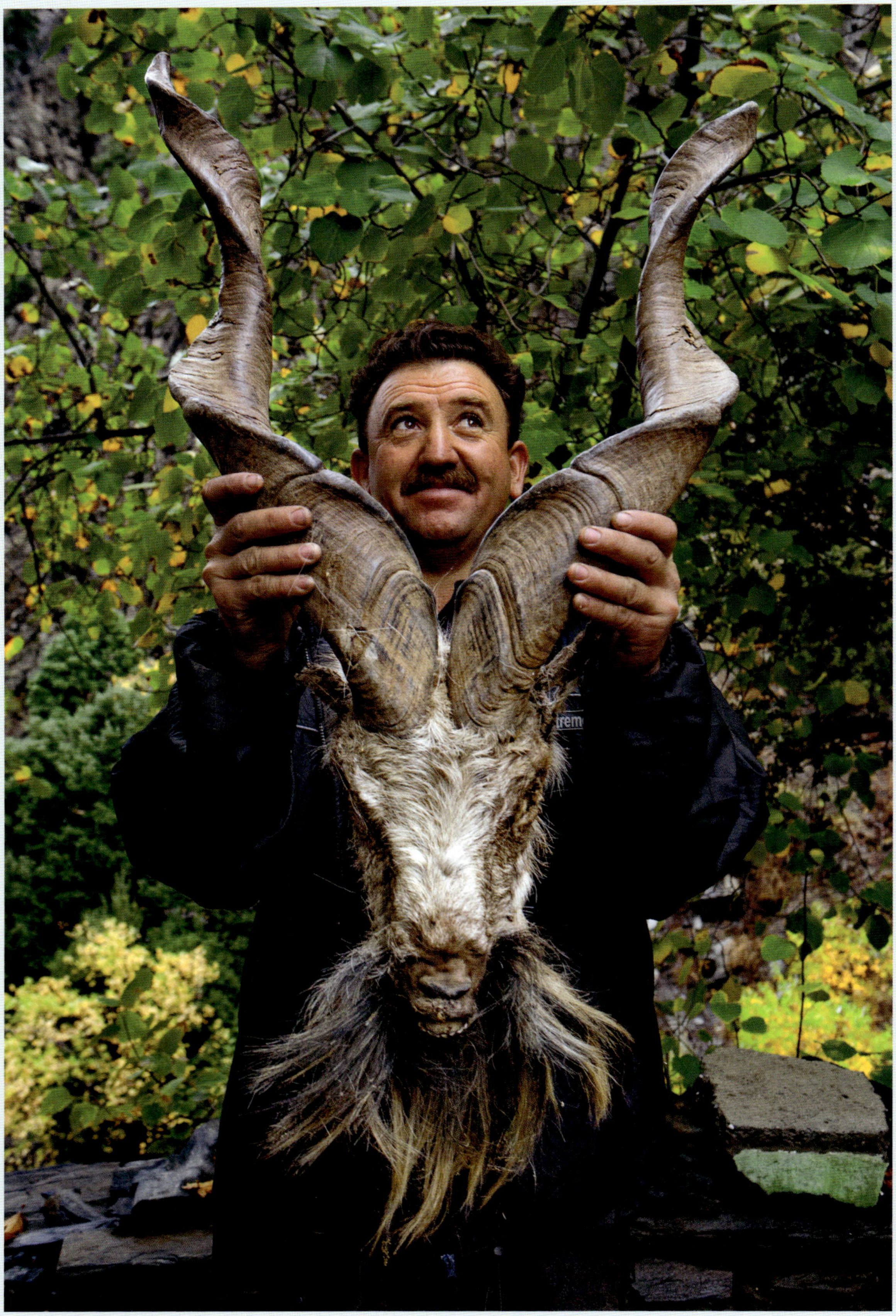

A ranger from Chitral National Park in western Pakistan holds up the head and horns of a markhor, a wild goat, which was killed by a snow leopard.

PAGES 128–129: A snow leopard walks amid a nighttime snowstorm in the high peaks of India's Hemis National Park. I camped out in those remote mountains for three months in winter to produce the story.

Snow leopards can travel 25 miles in a single night, but they have regular habits, making the rounds in their massive territories. In this vertical landscape, they pad silently along trails etched into mountainsides, their steps muffled by broad paws with extra fur between their toes. They're ambush hunters that careen downhill after blue sheep or other hoofed prey, steered by their impossibly thick, long tails, nearly as long as their bodies.

Snow leopards have bounced back far enough to change their conservation status to vulnerable, but they're not yet safe.

We scouted for the locations where they marked their territory, which are perfect places for camera traps—I consider them my movie sets. I composed my frame, set up multiple flashes to light the cat and the scene, and wired it all together. (Now they're wireless.) To make sure all was functioning, Tashi crawled in on all fours to break the infrared beam, triggering the camera and flashes. Then I'd wait for the wild actor to step onstage. When I got pictures, we "mined" that trail with cameras.

Every day we trudged up and down the rocky scree slopes to check, take down, or move the camera traps. They are notoriously temperamental, and the weather posed another challenge: The extreme cold quickly killed the batteries. With this much effort, the failures and malfunctions really hurt.

It sometimes warmed up to freezing during the day, always sunny, but it dropped down to 30 to 50 below zero at night. The propane heaters I'd brought didn't work: There wasn't enough oxygen at 15,000 feet to feed a flame. I slept fully clothed, in my parka, wore gloves and two hats, and hunkered down inside three sleeping bags. The cook put a hot water bottle inside the top bag; it was frozen solid by morning.

Given the rugged terrain and this cat's singular nature, it was impossible to assess populations, which ran from 4,000 to 7,000. Despite protection under the Convention on International Trade in Endangered Species in 1975, their numbers were halved during the 20th century. They were hunted for their luxurious pelts and coveted for the fashion trade, and their parts were used as ingredients in traditional Chinese medicine. All fetched a small fortune on the black market. And since they hunt yaks, goats, sheep, and other livestock, deadly conflict with herders was far too common.

I always photograph the people who live with my wild subjects, so I visited tiny settlements in the region, accessible only on foot. I photographed the Ladakhis in their daily lives. Through an interpreter, I sometimes heard gripes about snow leopards killing the animals they relied on for survival in this cold, dry, unforgiving place. They were not personally in danger, though. Snow leopards don't attack humans.

Conservation only works if it benefits local people economically, and the Snow Leopard Conservancy established ways to do just that. One program was inexpensive and genius. They built predator-proof corrals to protect livestock overnight: There was no shortage of rocks for the walls—and they supplied chain link roofs. The Conservancy also helped launch Himalayan Homestays, a program that helps trekkers find lodging with families.

A friend of Doug Chadwick's at the U.S. Embassy told him that a pile of livestock vaccines was sitting in a warehouse in Delhi. Since herders lost far more animals to disease than to predators, this could offer another incentive. I arranged to have the vaccines brought north and hired a veterinarian to administer them.

And then the Dalai Lama made a proclamation as part of teachings that emphasized respect for all life, urging followers to protect snow leopards and stop adorning festival clothing with their fur. His message resonated across this Buddhist land.

During my three months in camp, my equipment failed, broke, and was damaged—but I miraculously continued to get pictures. A snow leopard gazing at the camera, tail up, marking its territory on a boulder. Another mesmerized by my flashes, gazing upward in what looks like wonder. An ethereal image of a cat on the prowl at night during a snowstorm.

I flew home and left the traps up. Tashi kept them running in my absence. I got one final image: a cat walking up a trail amid pink spring blooms—a different world than the winter rockscape I'd worked in.

The snow leopard story ran in *National Geographic* in June 2008, and while I was away on another shoot, Sharon entered these photos in the Wildlife Photographer of the Year competition. And I won! I'd never entered contests or really promoted my work once it was done: It was always on to the next story. I'm a freelancer. I don't get a regular paycheck and need to be out in the field, paying the bills. But the recognition gave snow leopards global attention.

Soon, travel to the Zanskar Valley increased. Today, at least 200 families host trekkers, and another 550 families indirectly benefit from the visitors. The vaccination program has continued. Now, outfitters guarantee snow leopard sightings to visitors in Hemis National Park.

Snow leopards have bounced back far enough to change their conservation status, according to the IUCN Red List, from endangered to vulnerable—great news, but they're not yet safe. In some places, their numbers are stable or slightly increasing, but in other places, they continue to slip. In today's world, protecting wildlife requires steady attention, especially for big cats and other predators.

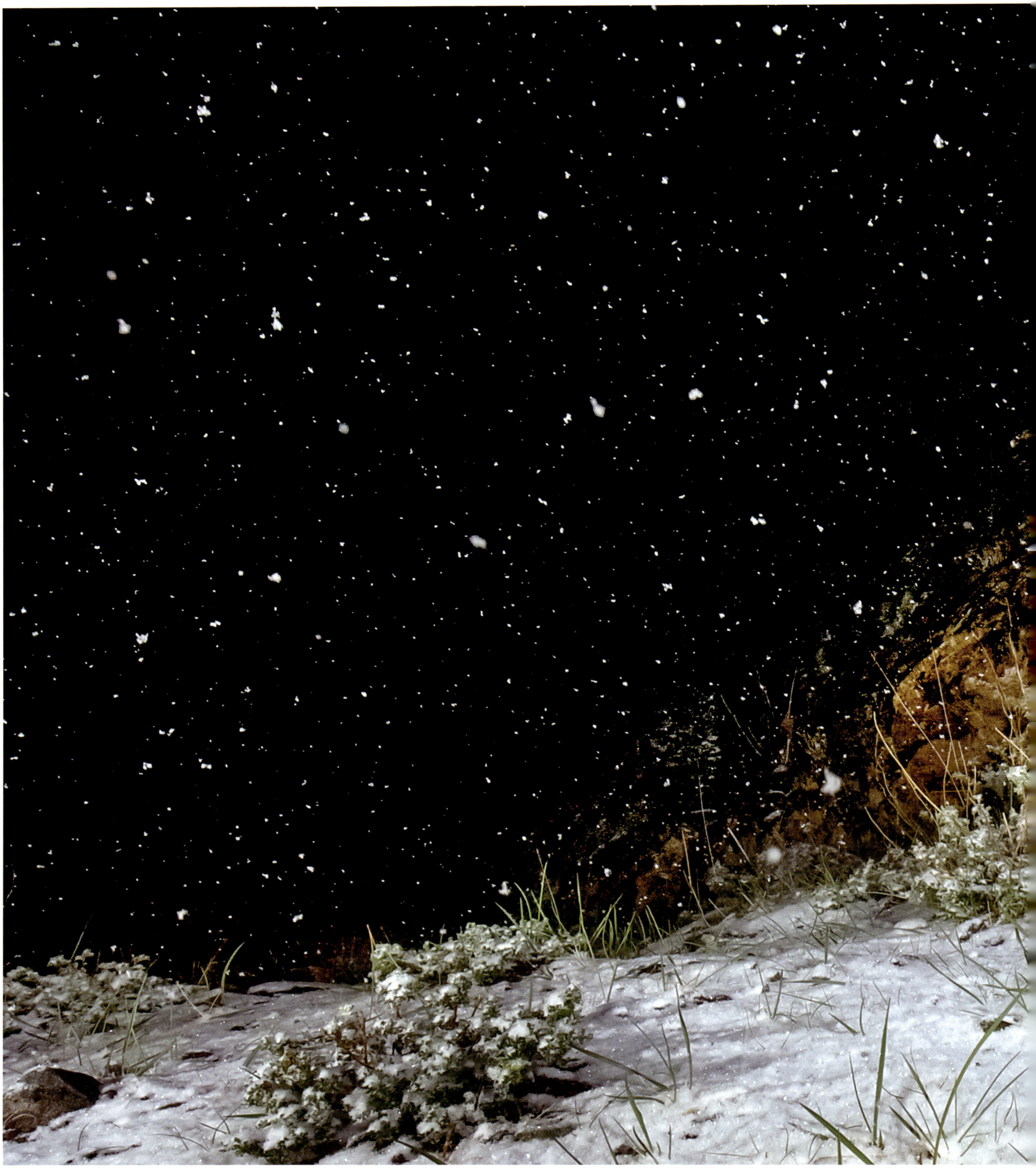

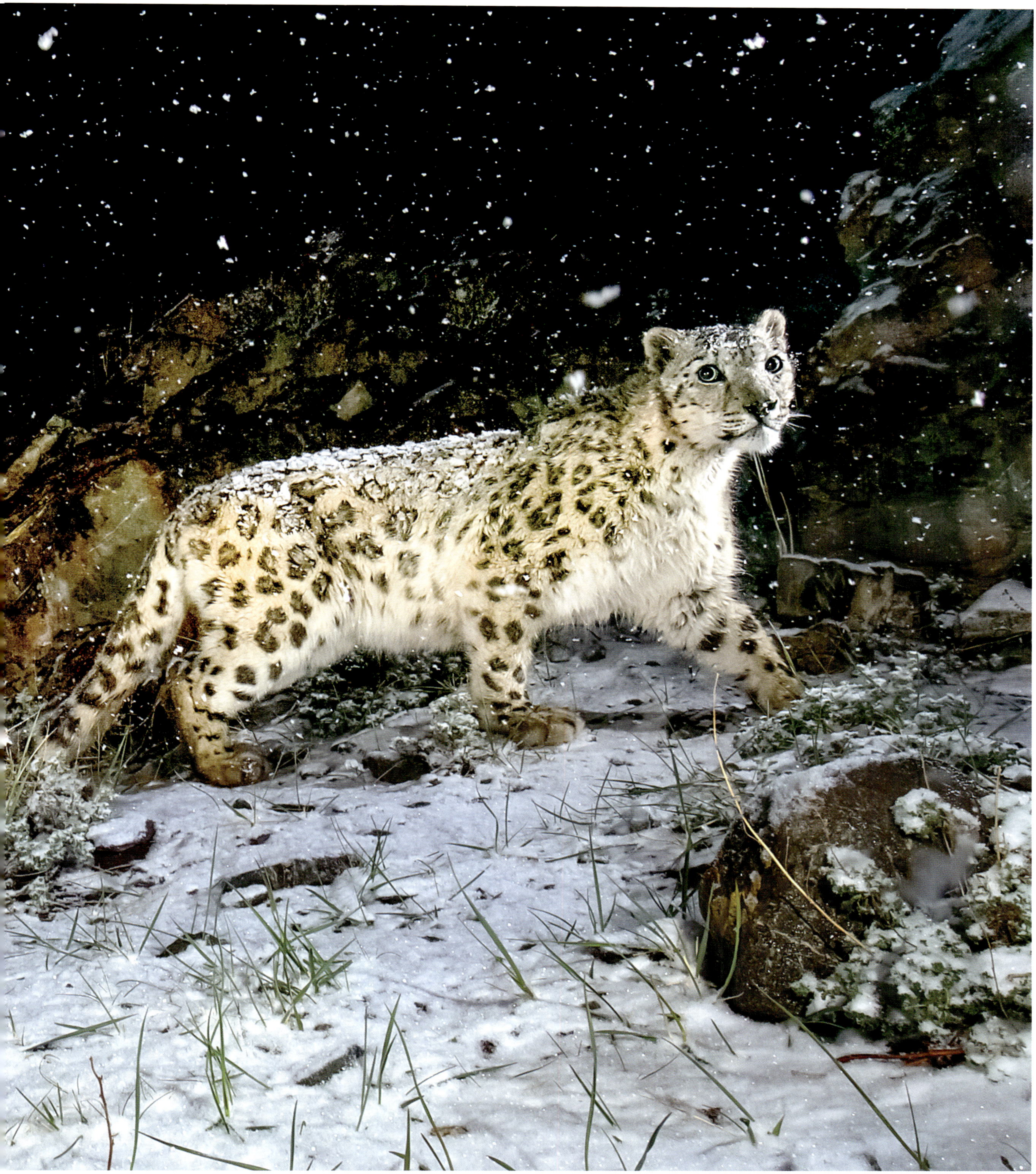

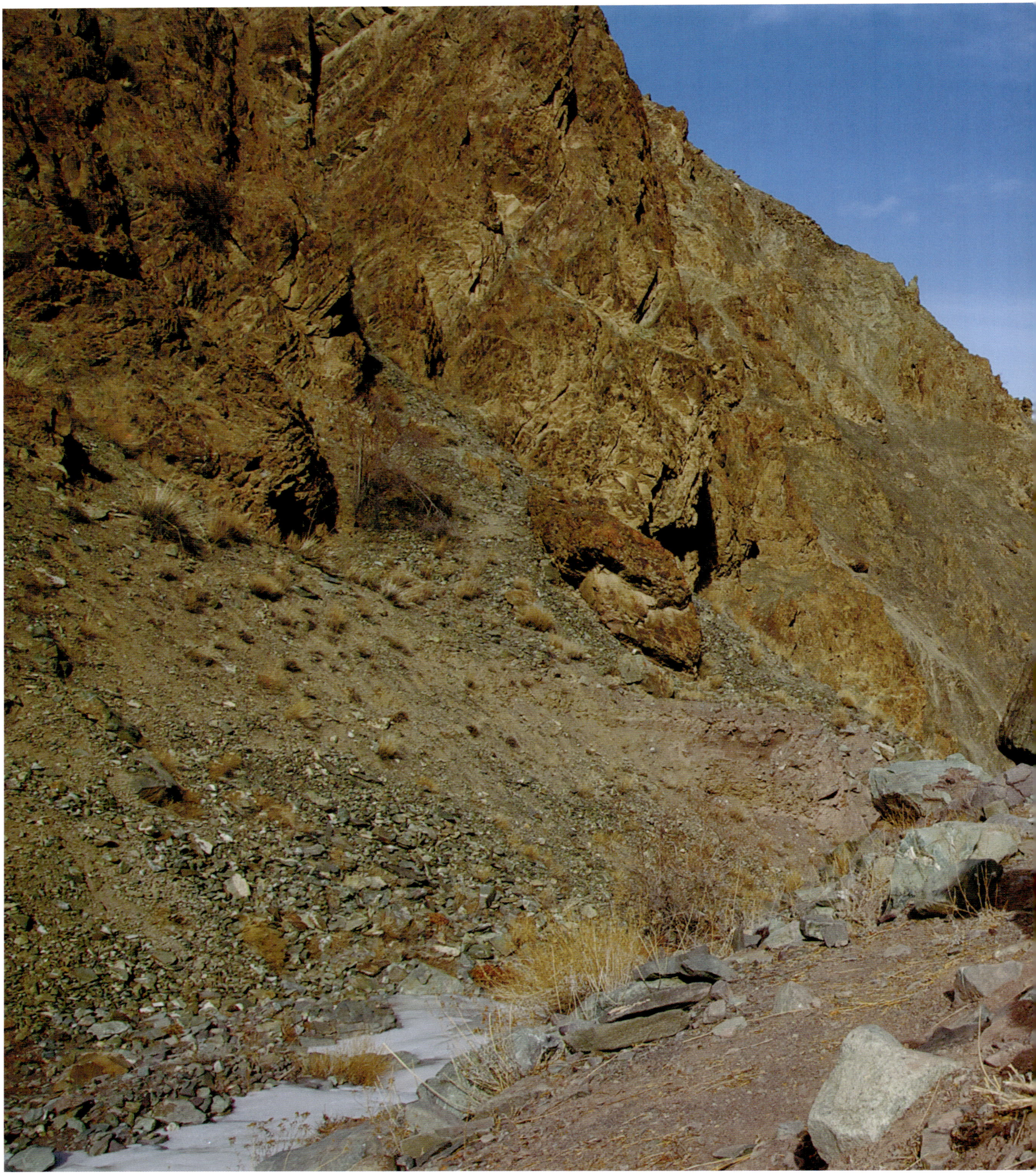

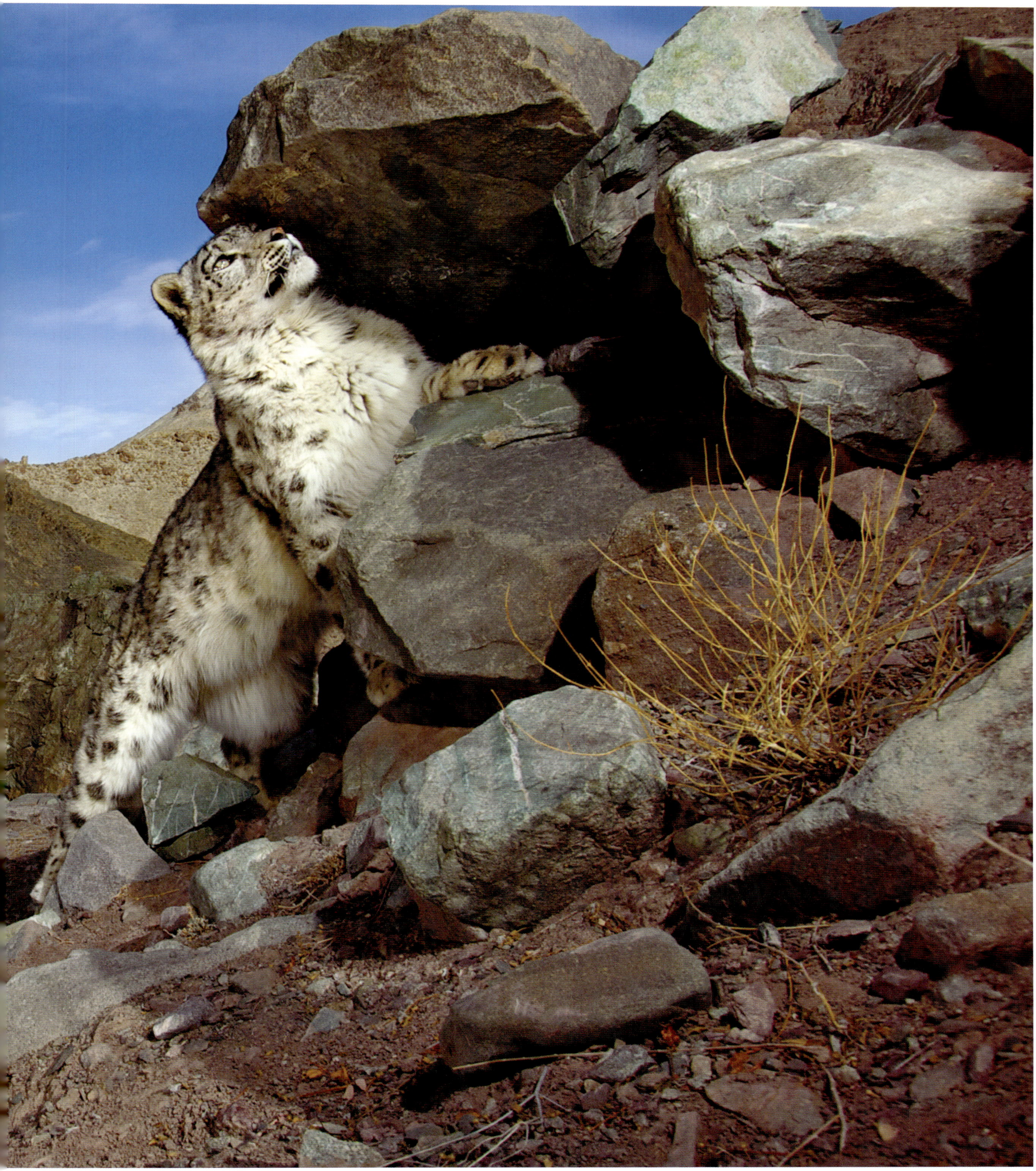

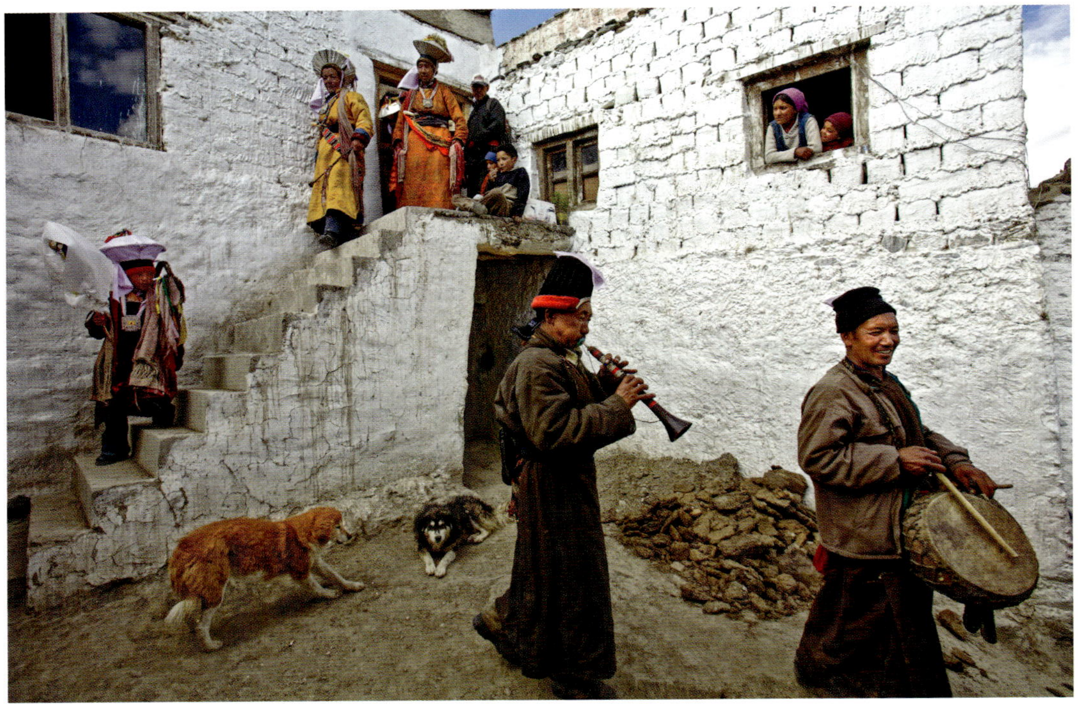

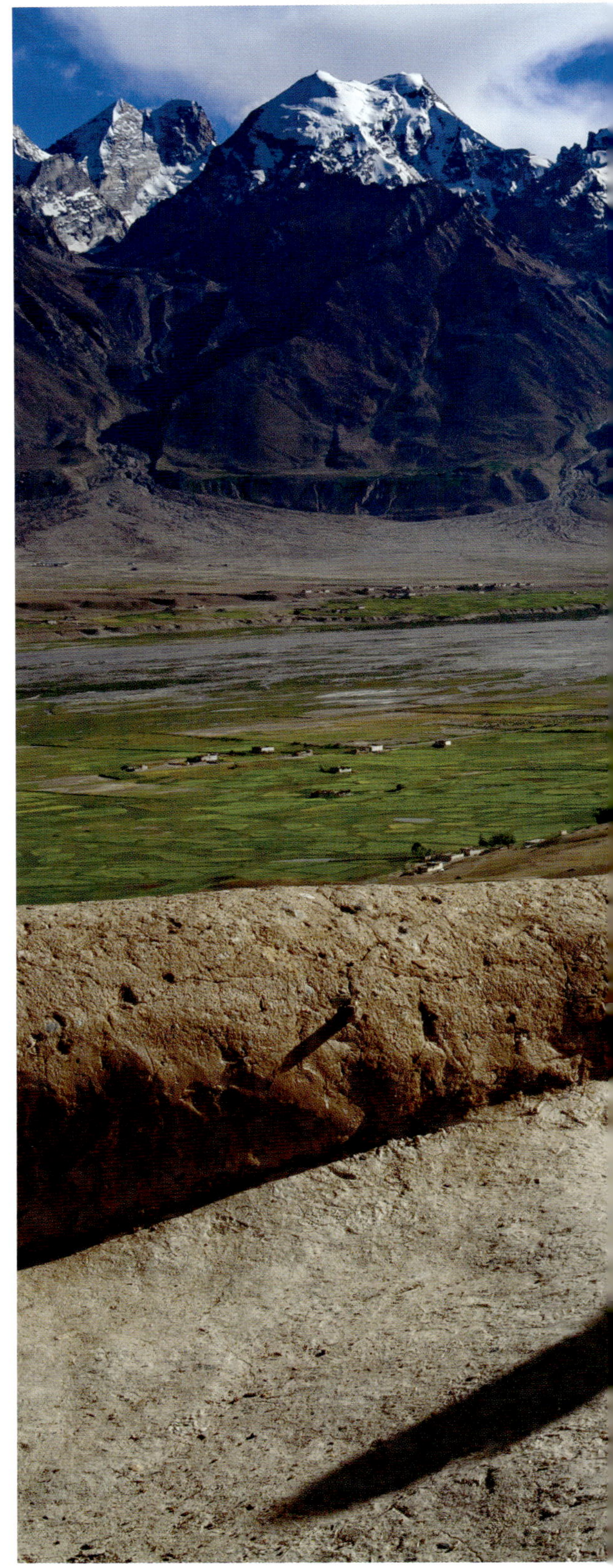

ABOVE: A musical group leaves a wedding ceremony in Karsha, a village with a population of 13,000 in northern India, where people share the landscape with snow leopards.

RIGHT: Karsha Monastery's head abbot looks out over the Zanskar Valley. In 2009, the Dalai Lama issued a proclamation calling for people to cease killing snow leopards.

PAGES 130–131: Snow leopards leave each other messages and delineate territory with scent markings. They rub or spray their scent from glands on their cheeks, paws, and other parts, as this cat is doing.

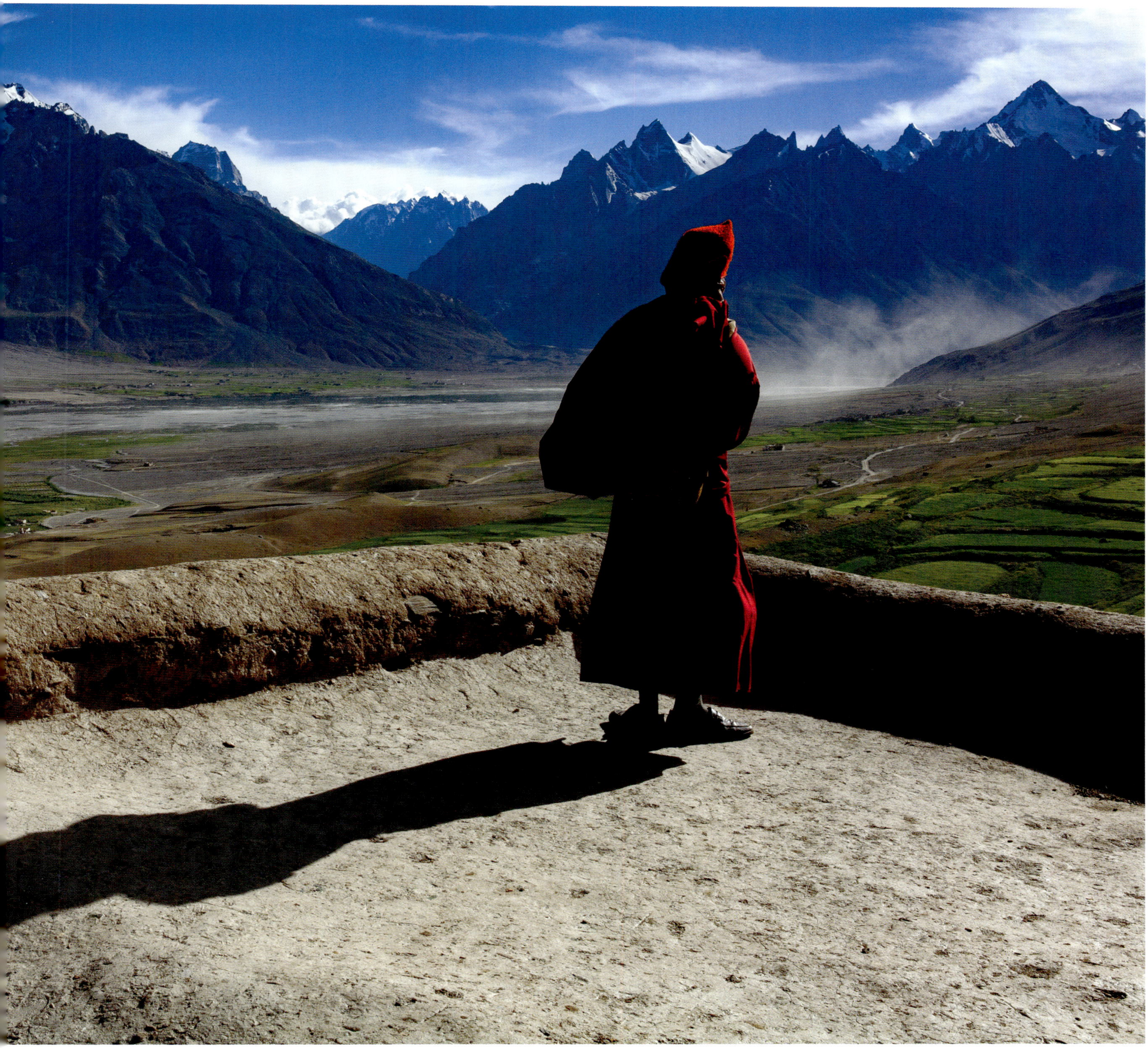

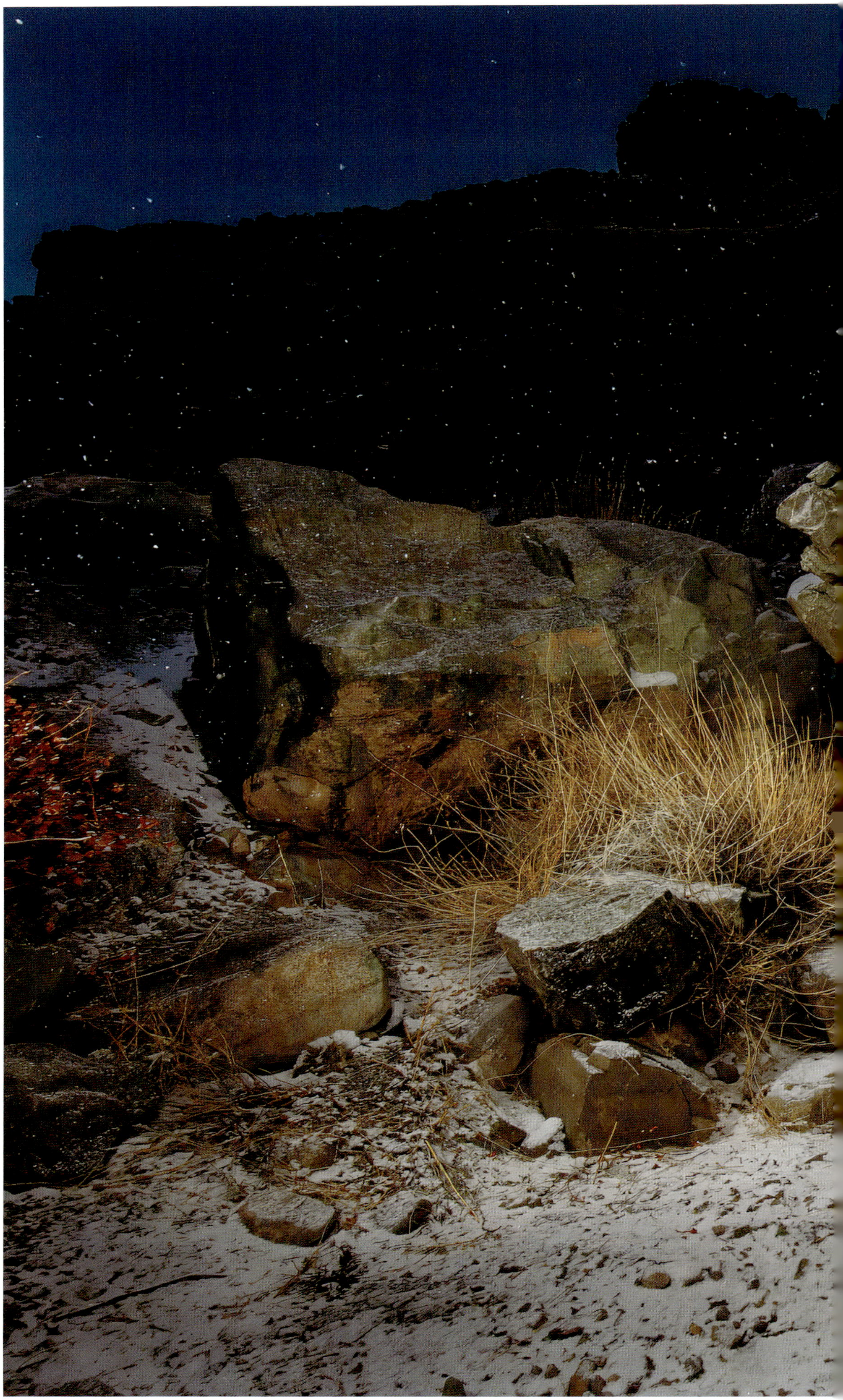

Camera trapping at altitudes of 15,000 to 18,000 feet in the dead of winter was challenging: Batteries died quickly and equipment regularly malfunctioned. It took months to capture this image of a snow leopard marking its territory.

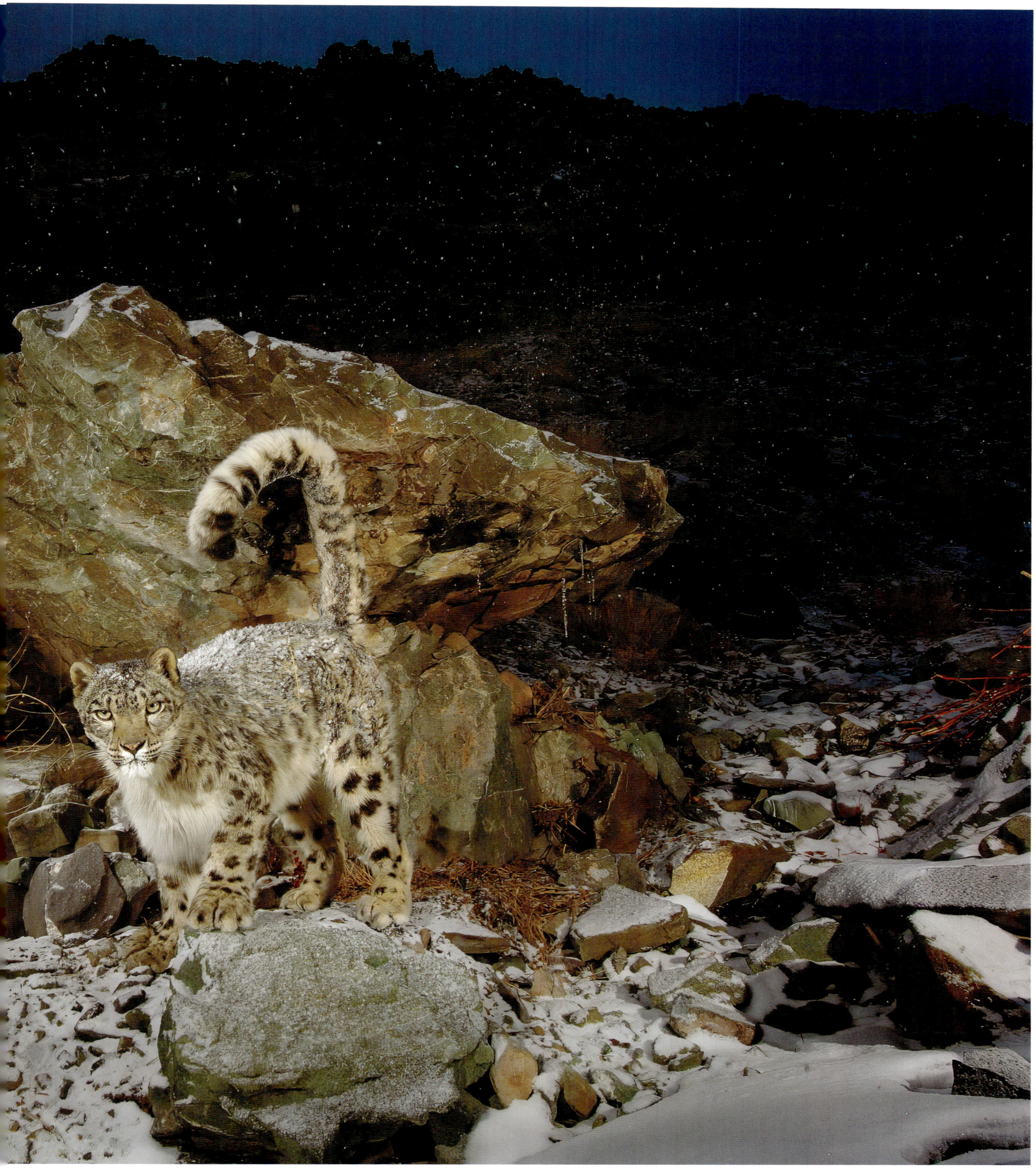

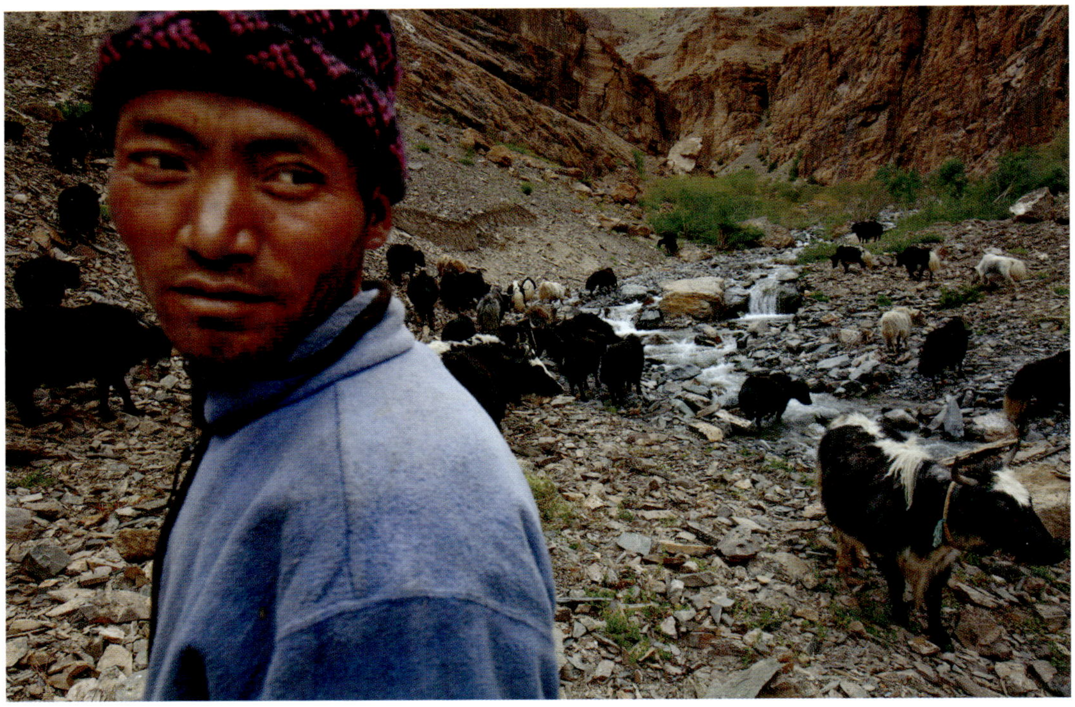

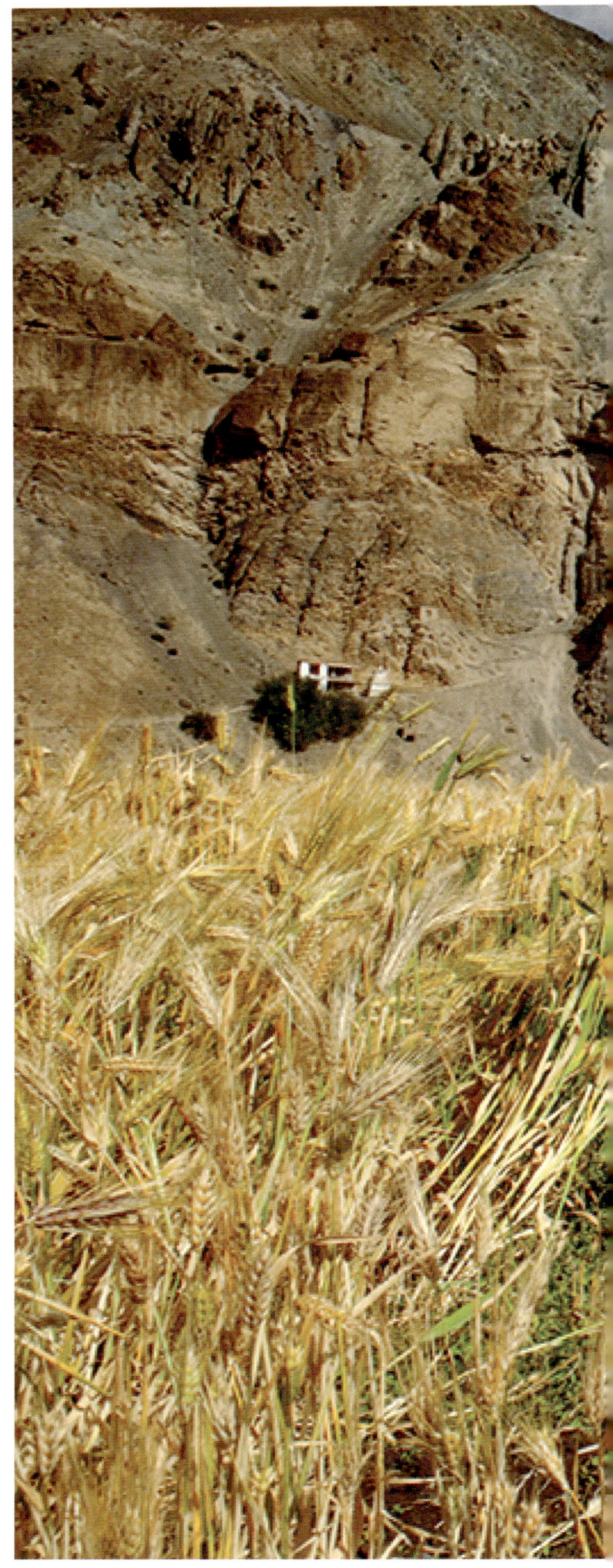

ABOVE: For people living in India's unforgiving Zanskar Valley, livestock are key to survival. This herder, here with his yaks, sold his goats and sheep because they were more vulnerable to attacks by snow leopards and wolves.

RIGHT: Local farmers harvest barley in the Zanskar Valley, with the 700-year-old Phuktal Monastery in the background.

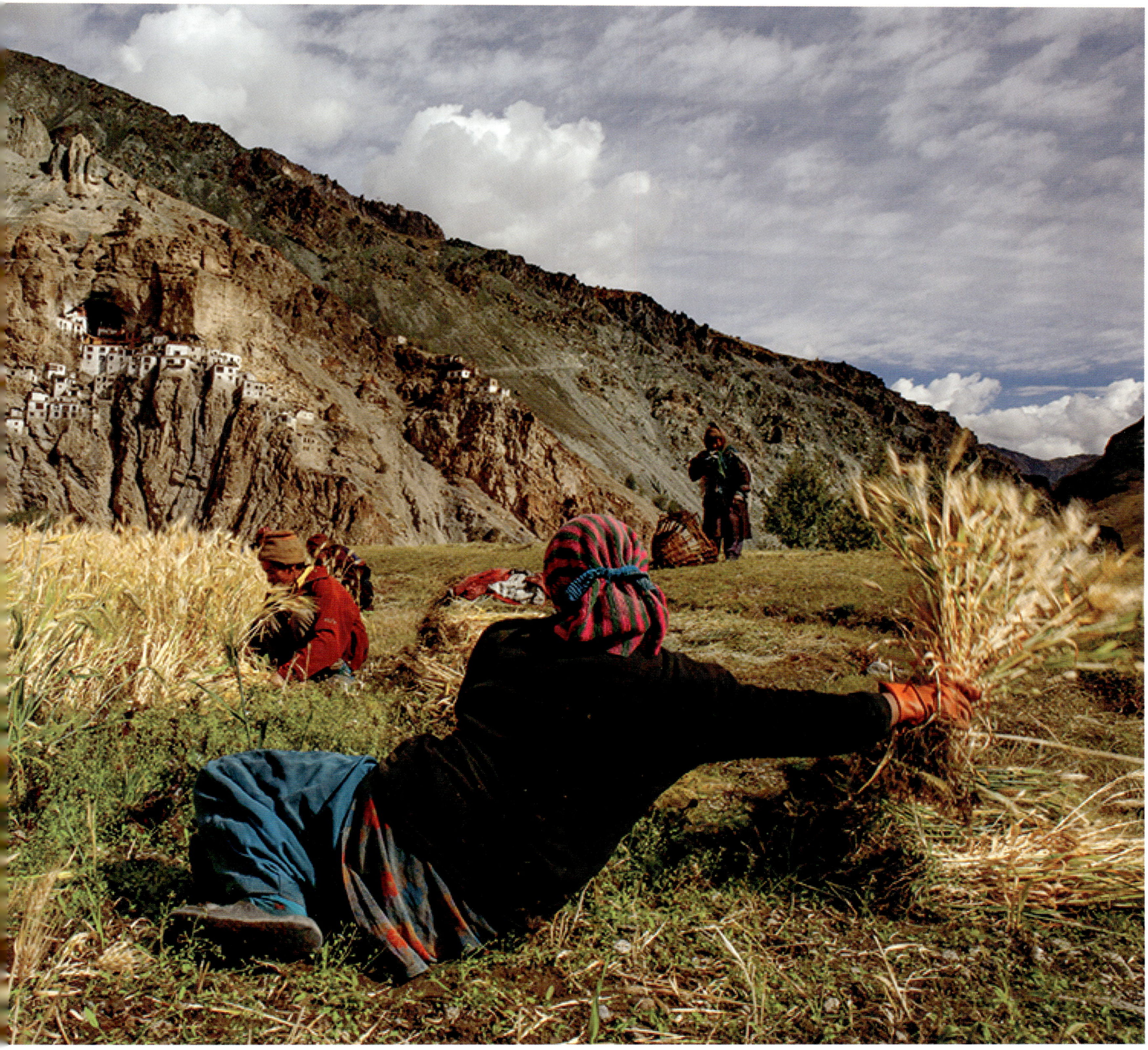

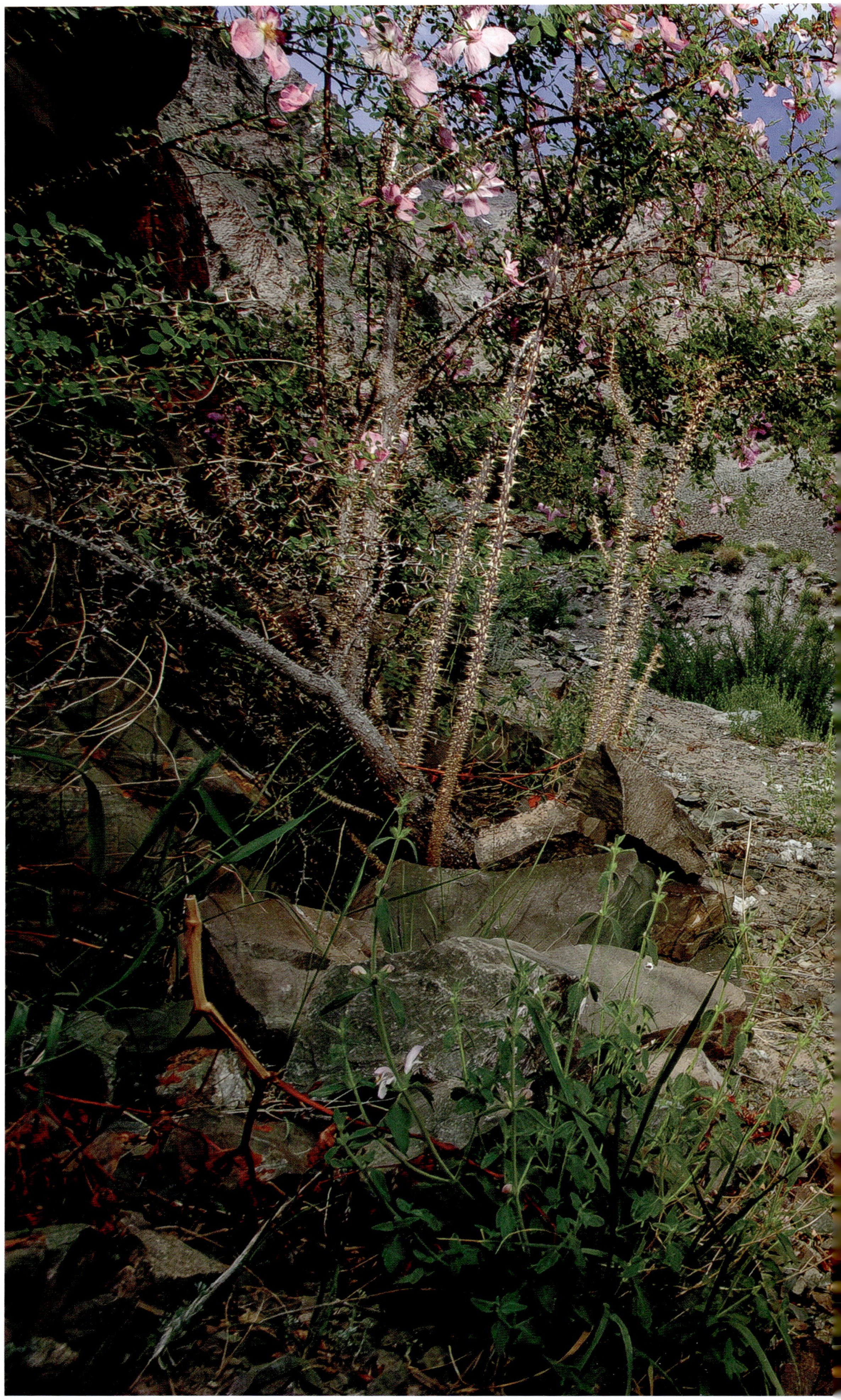

A snow leopard walks through the Zanskar Valley in June. Spring comes late to this high-altitude region, briefly bringing color to a mostly barren landscape.

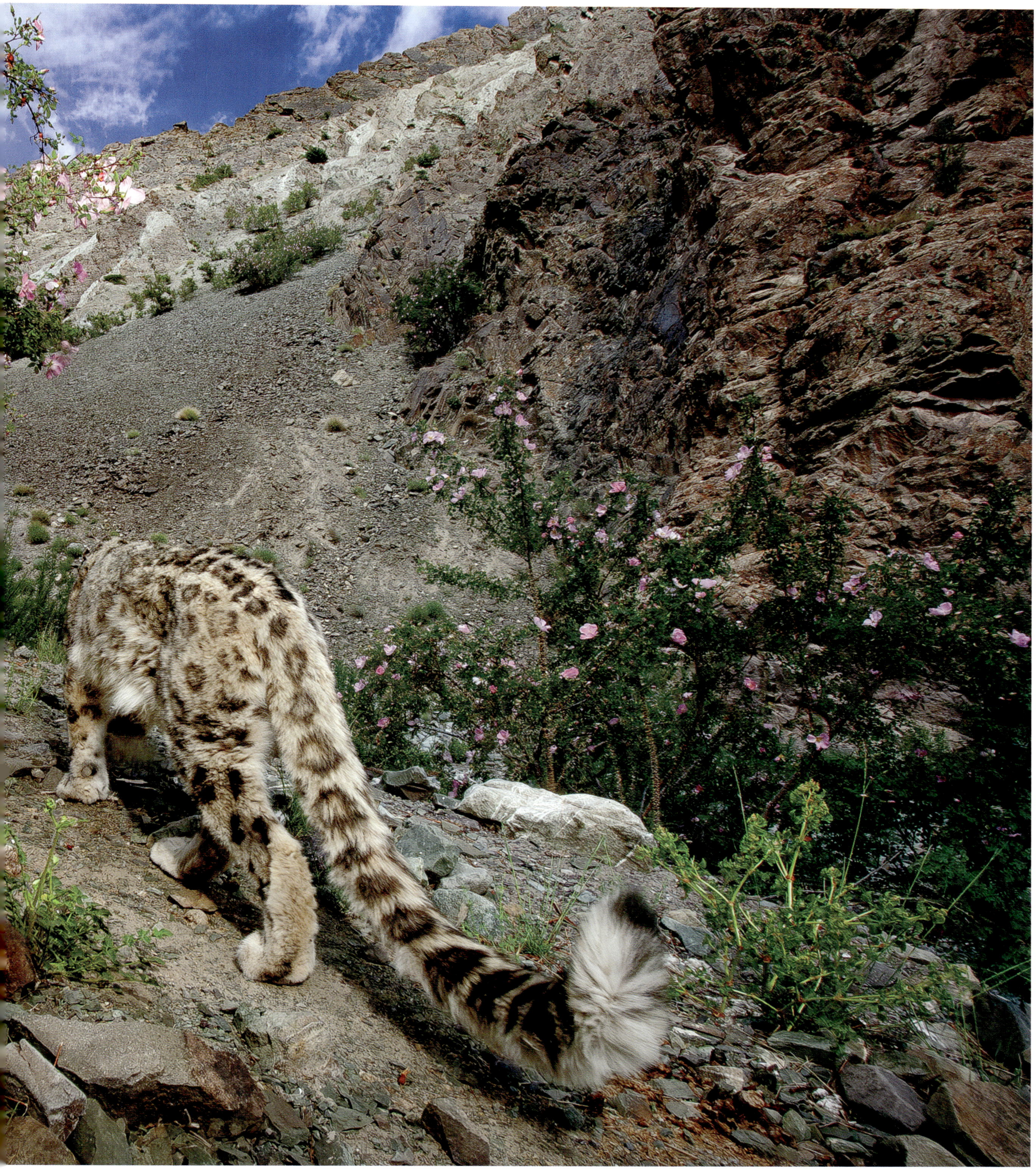

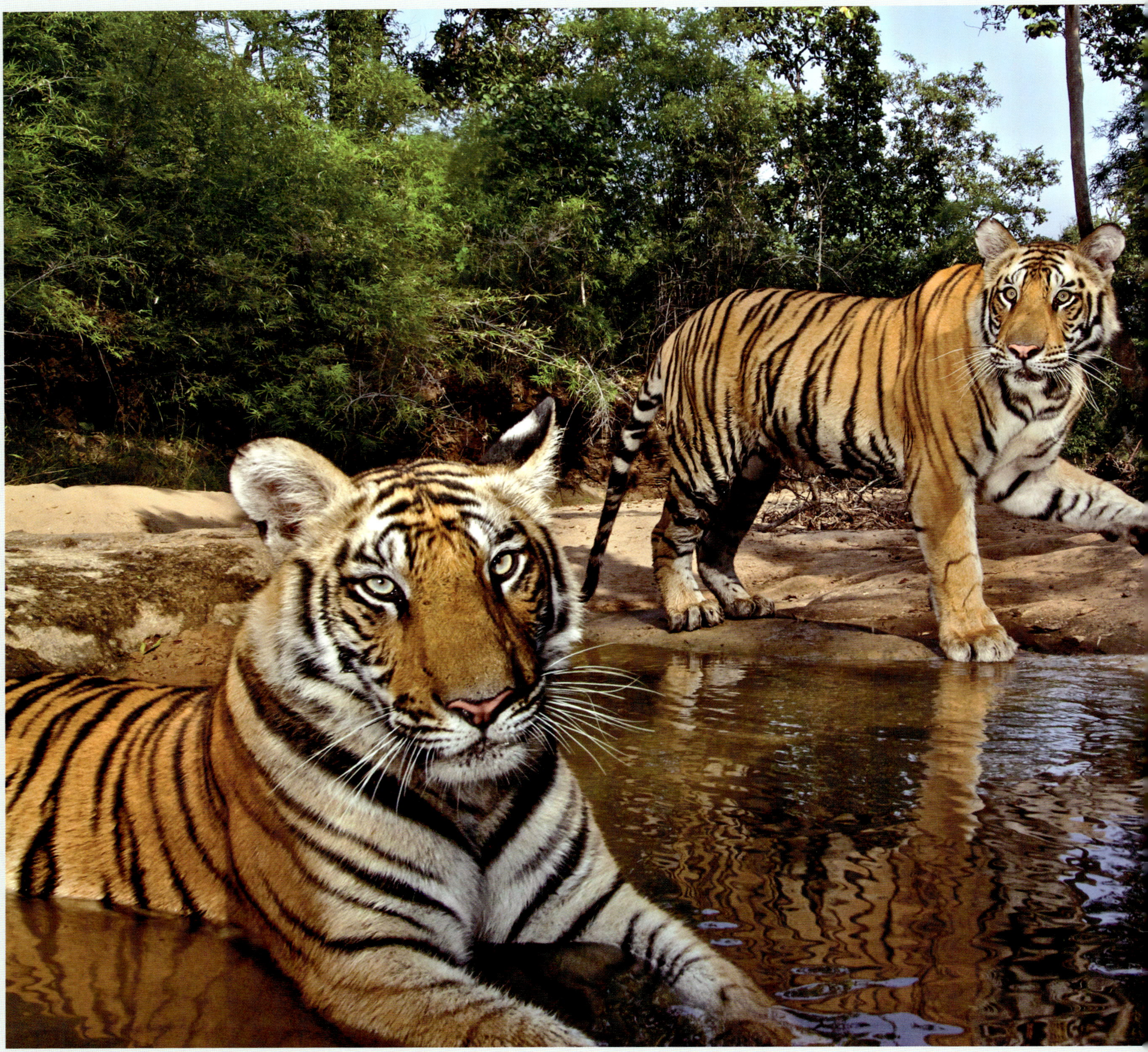

CHAPTER FIVE

Wild Tigers Feared, Revered, Threatened

The resurgence of India's Bengal tigers has been hard-won; the fight to save all the tiger subspecies continues across the nations they call home.

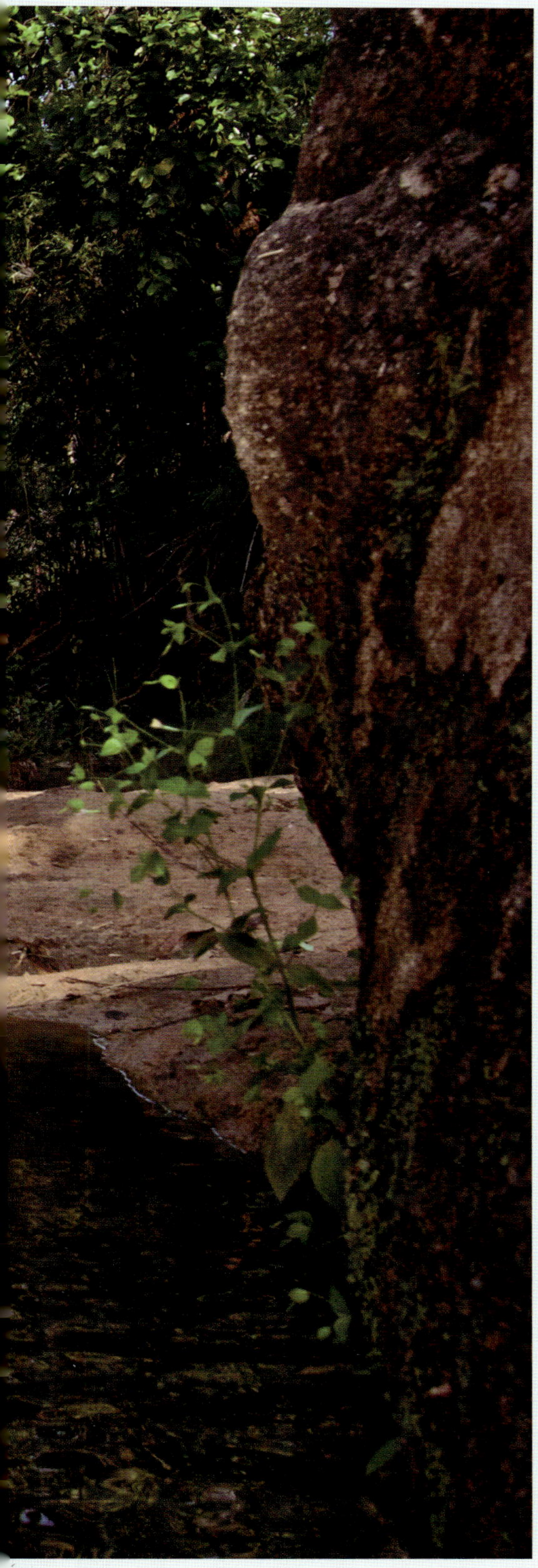

Almost grown tiger cubs cool off in their favorite water hole in India's Bandhavgarh National Park. I captured this image on the first day I set up a remote camera.

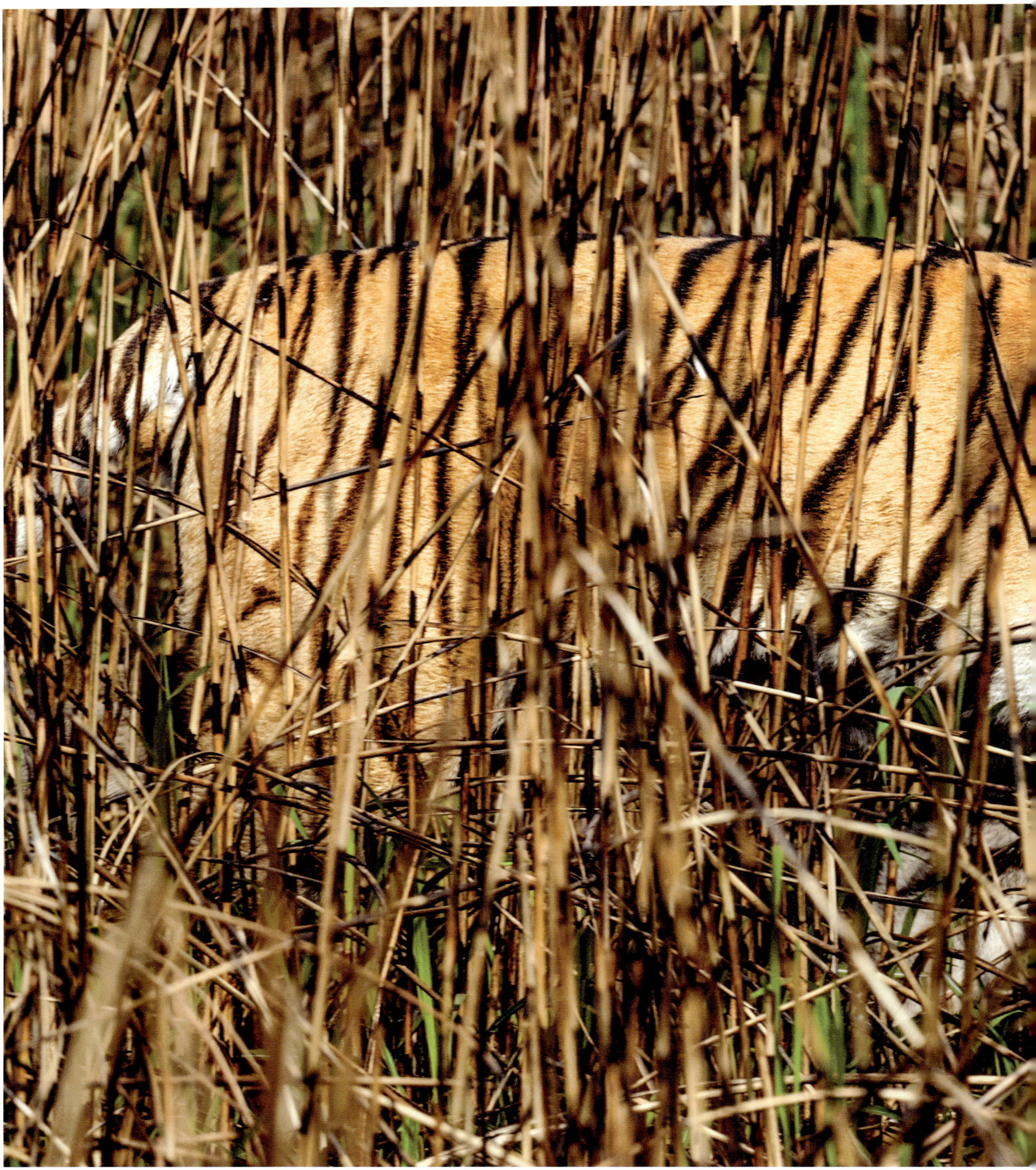

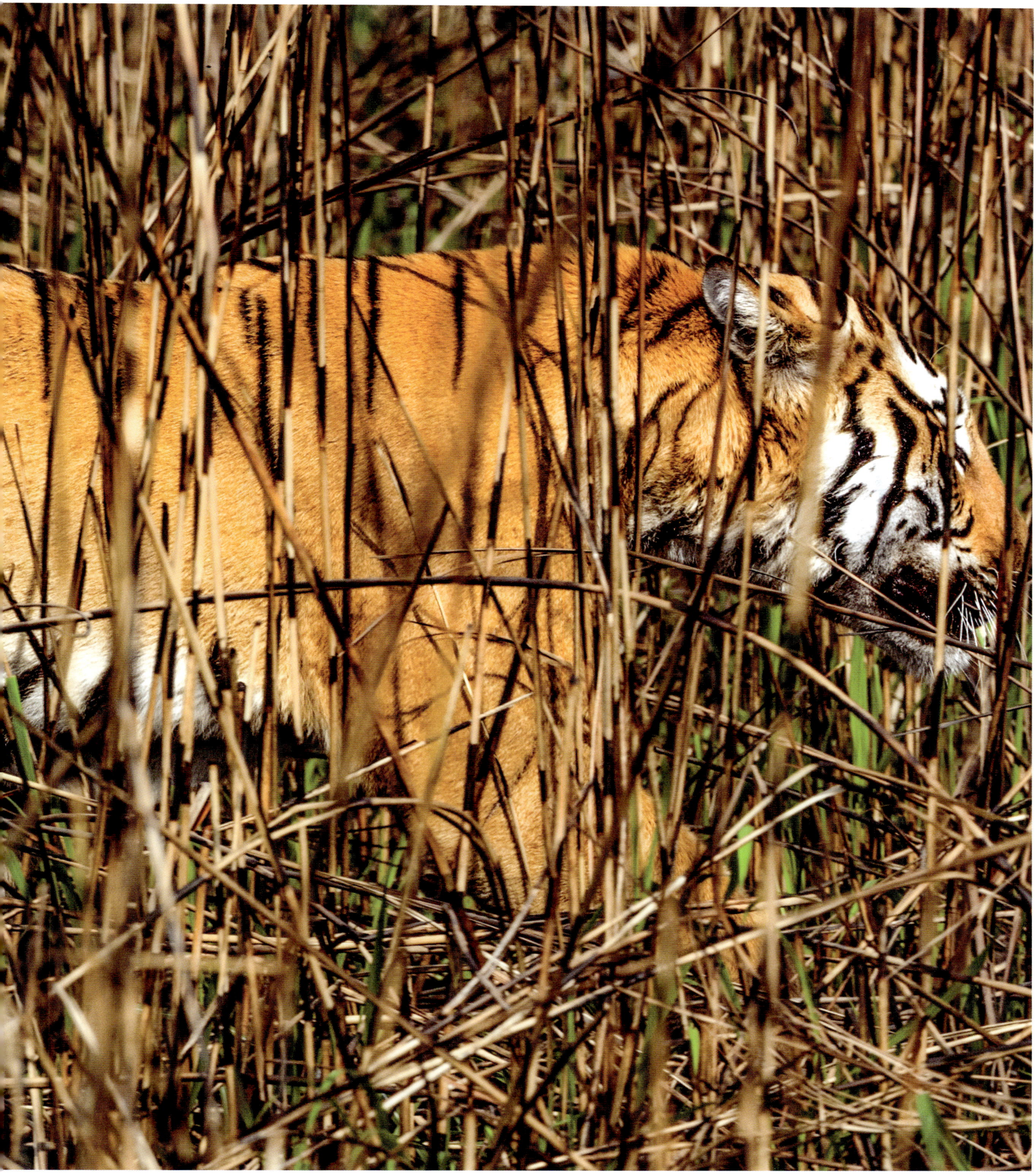

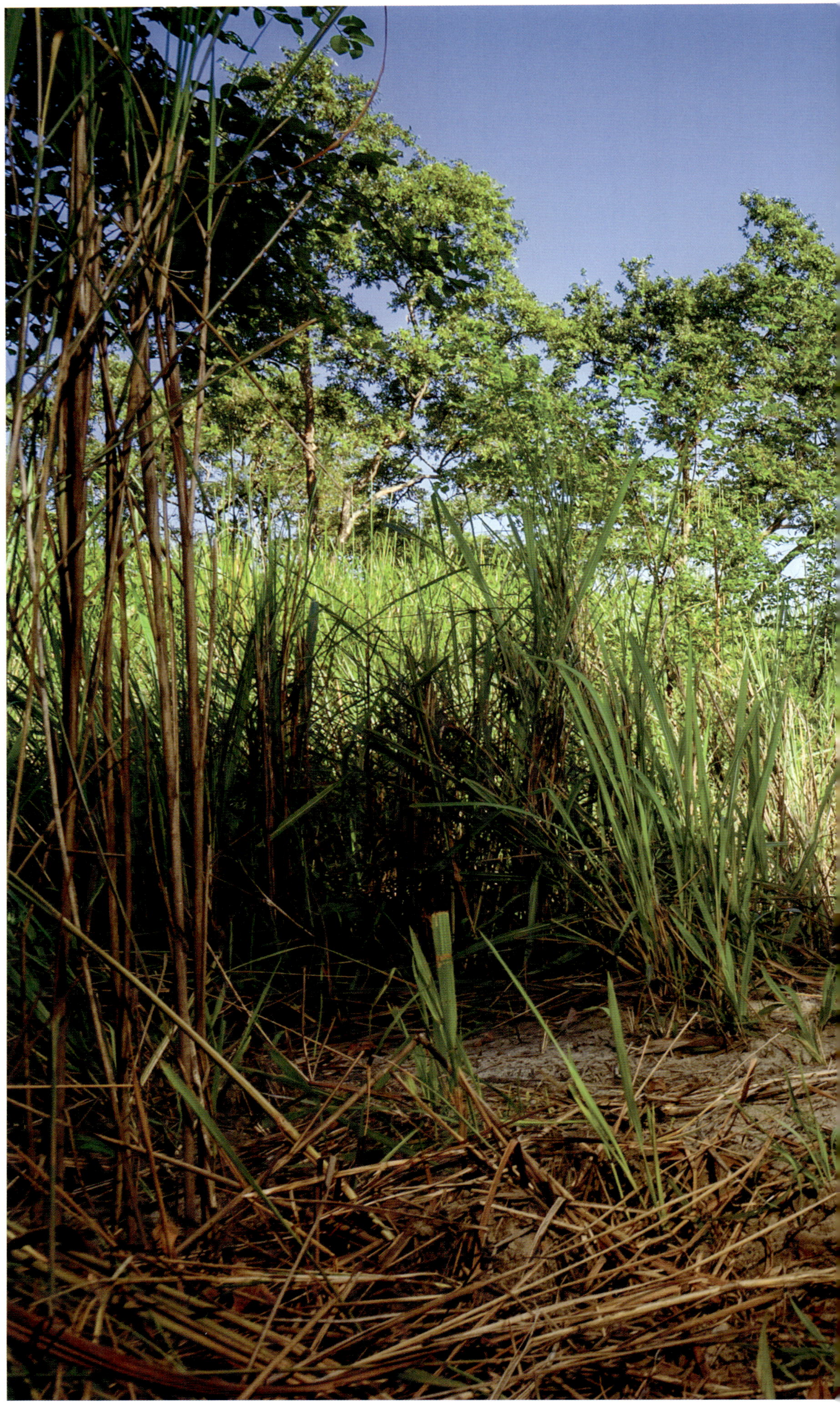

A curious young tiger stumbles upon a camera trap while walking in the elephant grass of India's Kaziranga National Park. He stopped, looked side to side, and then backed away.

PAGES 142–143: A Bengal tiger in Kaziranga National Park moves through tall burnt grass, which provides excellent camouflage for hunting.

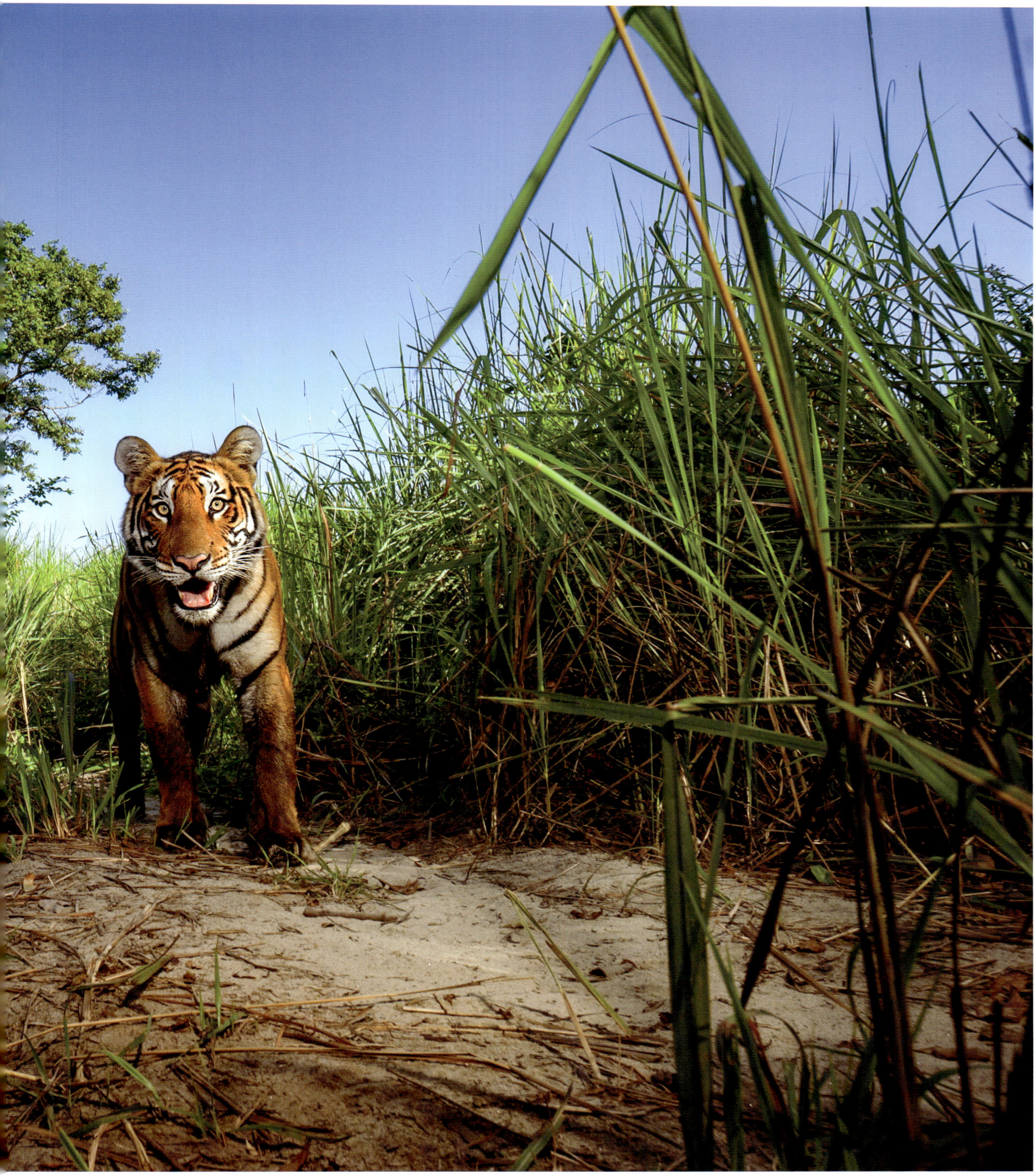

My first attempt to photograph tigers

My first attempt to photograph tigers was in northwest Myanmar in 2002. It was a story on the new Hukawng Valley Wildlife Sanctuary, created to protect some of the last Indochinese tigers, one of the most endangered tiger subspecies. Whenever possible, I work with biologists, and Tony Lynam and his Burmese partners were assessing the state of the region's wildlife. I joined them in the jungle and set up camera traps to track the tigers. Dubbed the "Valley of Death" during World War II, this remote, nearly impenetrable floodplain is choked with bamboo and razor-sharp rattan, canopied by forests that block the sun. The only way to navigate this forbidding land is along riverbanks.

We followed paw prints that inevitably disappeared into thick jungle. One day, I stepped on a glistening patch of sand and sank to my waist, saved by a huge fanny pack that held a telephoto lens. It was quicksand. After laughing hysterically, my guides (who didn't speak English) hauled me out with a bamboo pole. Back in camp that night—and every night—I picked off dozens of leeches. We slept on platforms in the trees.

In five months, I never captured a single tiger image. A massive influx of gold prospectors had pulsed into the valley on new roads, hunting and poaching anything alive, selling it openly or shipping it across the border to China. That became the story.

Tigers have been feared and deified for at least 8,000 years, since Neolithic people drew pictures of them on cave walls in what is now India. They're an iconic symbol of power and courage woven into culture, religion, and folklore, sometimes assigned otherworldly superpowers. And no wonder: Tigers are the largest big cat, stunningly beautiful, fearsome, and stealthy, with formidable claws and teeth and a bone-chilling roar.

They once roamed across 30 Asian nations, from Turkey to Eastern Russia, and from China and Southeast Asia to Indonesia. They're adaptable enough to survive Siberian winters, swim through the Sundarbans mangroves in Bangladesh, and thrive in India's steamy jungles. But they've disappeared from 93 percent of their historic range, and today breeding populations hang on in just 10 countries.

At the turn of the 20th century, when Rudyard Kipling penned *The Jungle Book,* there were roughly 100,000 tigers. In the decades that followed, the Javan, Bali, and Caspian tigers were lost forever. By 2002, when I began the Hukawng Valley story, roughly 3,200 tigers remained, split among six subspecies—all of them in trouble. The last South China tigers are gone from the wild, living only in zoos. Bengal and Siberian tigers are endangered; Sumatran, Malayan, and Indochinese tigers are critically endangered. Four of the five wild subspecies hang on with just a few hundred remaining each.

Tigers disappeared for the same reasons that all big cats are embattled: Humans engulfed their habitat and hunted them, killing them in huge numbers in extravagant *shikar* (hunting) parties; placed bounties on their heads; retaliated when they preyed on livestock; and later, slaughtered them to supply pelts for the 1960s fashion craze for fur in Europe and the United States. They were also killed for their parts: Tiger teeth have been worn as talismans for millennia.

Then, in the 1980s, tigers became walking gold. From whiskers to tail, their parts suddenly commanded a fortune on the black market for use in traditional Chinese medicine. Poaching skyrocketed.

By the 2000s, a high-level report found that demand for medicine had dropped while clamor for tiger parts used in exotic luxury products soared. These items became a way to flaunt wealth and prestige among those who were powerful enough to circumvent federal and international laws. The cats' beautiful pelts and bottles of tiger-bone wine were (and still are) bestowed upon high-ranking military or police officers or public officials, or gifted as part of major business deals.

Tourist venues—from zoos and circuses to cub-petting attractions in China and other countries—were stoking this trade, entertaining tourists out front while selling tigers out the back door. This deadly trade is masterminded by international criminal networks that also smuggle guns, drugs, and humans. Comparatively, trafficking wildlife is a low-risk, high-profit enterprise.

I finally glimpsed a tiger in 2007 while shooting a story about Kaziranga, a small national park cradled by India's Brahmaputra River. Better known for its Indian one-horned rhinos, it's also home to the world's highest concentration of tigers. I photographed a male wading through mud up to its elbows and a curious young cat emerging from head-high elephant grass.

After witnessing this cat's beauty, power, and unchallenged rulership of the jungle, I fell under its spell. Tigers were (and still are) in crisis. I was driven to tell their story, hoping to reinvigorate passion to save them.

I talked to experts, including Alan Rabinowitz and George Schaller, and decided to focus on India's Bengals, which have the best

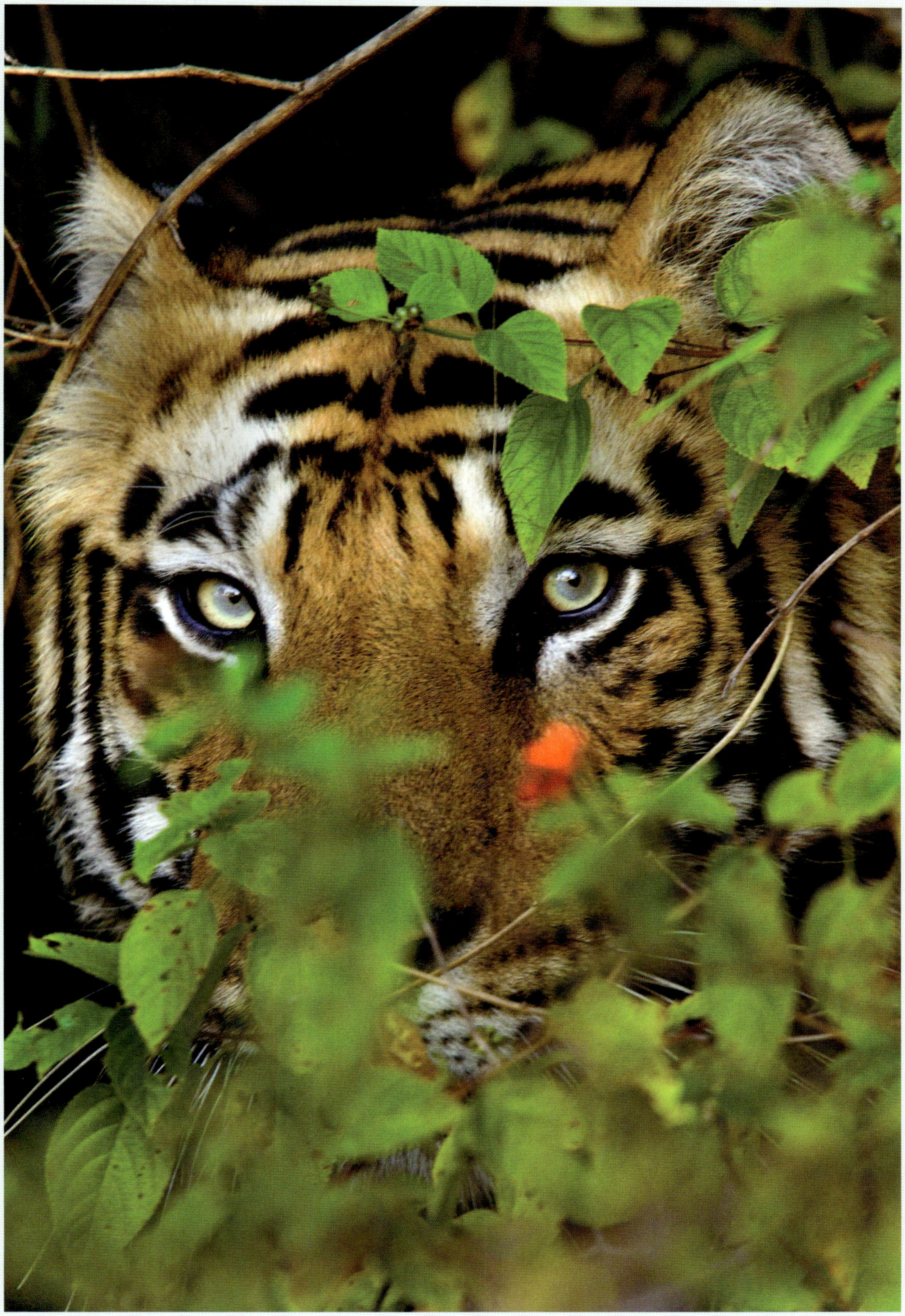

A male Bengal tiger peers through foliage, eyeing a herd of water buffalo. It wasn't a productive hunt. A hunting tiger's success rate ranges from 5 to 20 percent.

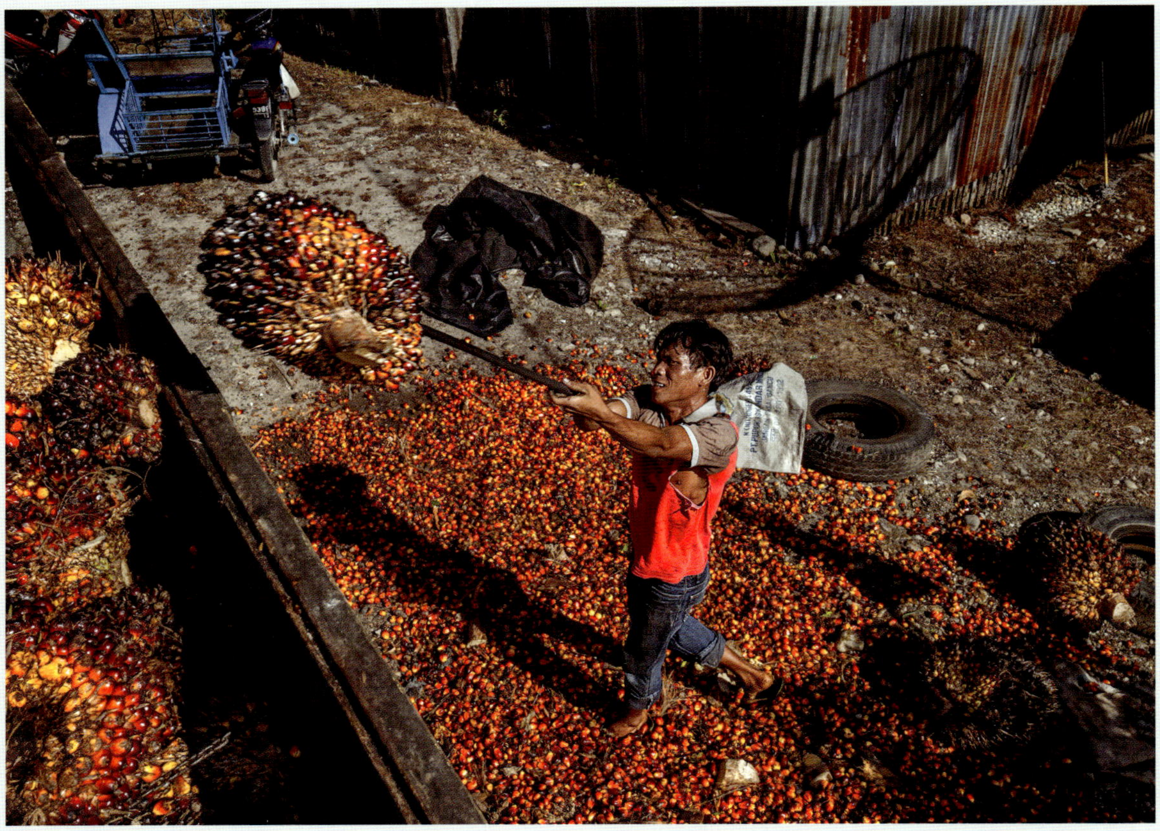

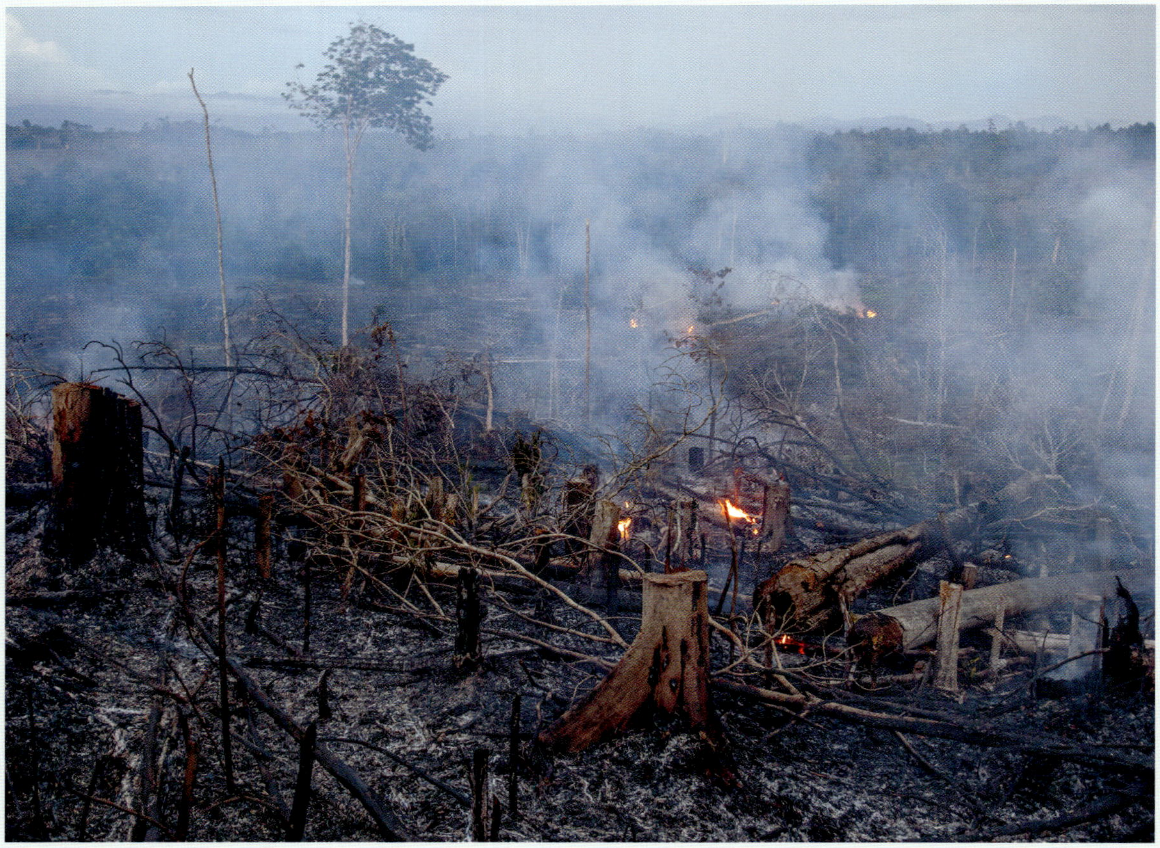

ABOVE: A man loads palm oil fruit harvested from his small-scale farm. As of 2025, 42,000 square miles of Sumatra's rainforests have been leveled for palm oil plantations.

LEFT: With so much rainforest cut and burned for palm production, critically endangered Sumatran tigers are left with little habitat. Now, they are mostly relegated to Sumatra's central mountain spine and a handful of national parks.

outlook for long-term survival, as well as two critically endangered cats, Indochinese tigers in Thailand and Sumatran tigers in Indonesia.

Dutch colonists in Sumatra once described the tigers there as a "plague," plentiful and bold. Today, though, their numbers have dwindled to perhaps 400, threatened by relentless poaching and palm oil plantations that continue to strip the island of rainforest. Flying in, I saw that the island's mountain spine was still cloaked in native forest, but little else remained wild.

When I landed in Medan in summer 2009, Munawar Kholis met me at the airport. He was a veterinarian with a wildlife response unit, and he'd just received a text alerting him that a poacher had been arrested. We loaded my 18 bags and cases into a van and drove to the offices of Indonesia's elite forest police unit, where an officer showed us a confiscated tiger skin. It revealed this species' unique qualities. Sumatran tigers are the smallest, with a darker, rusty coat, slashed with many stripes. Its face is framed by a slim white mane.

That experience was a prelude of what was to come, all of it horrific. I photographed illegal logging in protected parks and a forest leveled and burned for palm oil plantations, leaving blue-smoke ruin. I photographed a cub that was caught in a snare and had to have its front paw amputated, and a crime scene where a beloved tigress was poached out of her cage in a zoo.

In two months traveling across the island, I didn't get a usable tiger image. I left my camera traps up, and hired people to change batteries and memory cards for me. Sharon asked me how I didn't have a nervous breakdown. The pressure to produce unique images for every National Geographic story is intense, and so far, I'd failed in Sumatra. But somehow, I can shrug it off: For me, every morning starts a new day.

Five weeks later, I got an email saying I'd gotten two pictures from my camera traps. One was a stunning Sumatran tiger, caught midstride at night in front of a buttressed tree, gazing directly into my lens. Ironically, a former poacher had pointed out that location. He'd come to know the tiger's habits well.

After Sumatra, I joined the Tiger Team in one of Thailand's last gems of nature, Huai Kha Khaeng Wildlife Sanctuary. A couple, Achara and Saksit Simcharoen, headed what is the longest-running tiger research program, started by Alan Rabinowitz in 1986. Achara was studying how much space a tigress needs for herself and her cubs. Her team baited and snared them, letting males go and outfitting females with radio collars to track their movements.

My assistant, Joe Riis, and I camped with Achara's eight-person team, sweltering in 110°F heat. In the dead of night, the team captured a female, accompanied by a five-month-old cub. The team took samples of their blood and fur, then weighed and collared both of them. There was a moment when I paused: It was amazing to see these cats up close.

After setting up camera traps, I requested permission to go out with the SMART patrol, named for the Spatial Monitoring and Reporting Tool technology they use. This elite force, 190 strong, used digitally mapped field and intelligence data and carried serious firearms to protect against poachers. I spent two days with Thai military and police officers who conducted intense boot camp training exercises with new recruits.

Then we went on patrol. That morning, the unit investigated a tip-off. They discovered and dismantled six traps, each fashioned from a small pipe, a bullet, and gunpowder attached to a trip wire. Later, they followed footprints and found a camp littered with garbage and shotgun shells. They recorded each bit of evidence, including latitude-longitude coordinates, into a handheld GPS device while others scouted the area and a ranger photographed the scene.

I'd thought camera trapping in Huai Kha Khaeng would be relatively easy, but that wasn't the case. We captured only a few tiger images—and then they stopped altogether. And each time I downloaded my memory cards, I was shocked by the number of people who walked through. My standard line became, "Are they pilgrims or poachers?" With Buddhist shrines scattered across the sanctuary, it was impossible to know.

There were active poachers. Guards reported that a male tiger was killed, then a tigress and her two cubs. All were killed by poachers who poisoned animal carcasses with carbofuran, a cheap, easily available pesticide.

Since I wasn't getting pictures, I headed home, but I stopped en route to photograph at a tiger tourist trap, the Tiger Temple: Tigers used in entertainment were an important aspect of this story. I had no clue that I would return to investigate the place.

A few months later, I heard that SMART patrol officers arrested the alleged poacher after a shootout. Photos on his cell phone showed a notorious poaching ring posing with a dead tiger. It was IDed as a cat from the sanctuary. Two gang leaders were convicted of killing four tigers, and they received the stiffest jail sentences for wildlife crime in Thailand's history. The patrols in the park increased, and there's been perhaps one tiger-poaching incident in the last 15 years.

And then in India, it was all tigers, all the time.

Every day, we drove into Bandhavgarh National Park in the blush of first light. A tourist hot spot, it's heavily patrolled by rangers and antipoaching patrol guards who track tigers on elephant-back. Many of them didn't speak English, but they taught me bad words in Hindi, we laughed a lot, and we became friends. They helped me find locations for camera traps—and install them without being eaten.

A trap we set up over a water hole was a prime-time spot for a tigress and her three nearly grown cubs: a female and two males. The female was by far the bravest, sniffing the camera and shooting selfies. I nicknamed her brother Smasher because he regularly swatted the camera, ripping it down at least four times.

I spent weeks with them. It was amazing (and incredibly rare) to see so many big tigers together. Tigers are loners except when mating or raising young. I photographed the massive cubs swimming, napping together, and chasing each other with the playfulness of kittens. I also photographed cats mating, using trees as giant scratching posts, stalking, hunting, and eating. One day we surprised a sleeping tiger. He startled awake, shot at the jeep like a cannonball, then halted and sauntered off. I was terrified, but I got the picture.

Tigers are among the most photographed animals on the planet, so I wanted to find a different approach. I'd glimpsed a remote-control car in the basement at National Geographic headquarters and asked the guys in photo engineering to outfit it with a camera. It was like an oversize toy, with tank-style treads. The female approached, flinching a little with each click of the shutter, but gaining confidence. I got a tiger's-eye view that I could never have gotten unless I was in a tiger's mouth! Then she trashed the camera.

My camera traps also captured images I couldn't have gotten any other way. I hit the jackpot at that water hole: There were hundreds of tiger pictures that I later made into an animation, just for fun. I also got an important image illustrating one reason why saving wide-ranging, dangerous wildlife is complicated: a photo of "the family" slipping through a chain-link fence that bordered the park, headed for a nearby village.

The resurgence of India's Bengal tigers is a hard-won success story. They were nearly hunted into oblivion. Historian Mahesh Rangarajan wrote that "over 80,000 tigers ... were slaughtered in 50 years from 1875 to 1925." Some became the ultimate trophy in elaborate hunts by Indian nobility and the British Raj; others were exterminated as livestock killers.

When Indira Gandhi was elected prime minister in 1966, she spearheaded a fight against the growing crisis with Project Tiger, the world's most ambitious initiative to preserve the species. She established nine tiger reserves, and protections paid off: Tigers were on the upswing, numbering 4,000 in 1984, when she was assassinated. Iconic tiger conservationist Valmik Thapar called her "India's greatest wildlife savior." When her son Rajiv Gandhi became prime minister, he passed a seminal law, the Environment Protection Act of 1986.

Then political will evaporated, creating India's second tiger crisis. Forests were felled. Wildlife habitat drowned beneath dam floodwaters. The land was carved up by miners and ruined by industry and haphazard development. Human populations soared, poachers swarmed, and tigers vanished. By 2008, after a 34-year, $400 million effort, just 1,411 remained.

This news shocked the nation and jump-started new protections. By 2022, the year of the most recent census, the tiger population had doubled, reaching 3,682. Today, 82 percent of all wild tigers live in India. Now, 58 tiger reserves span Himalayan foothills, grasslands, jungles, deserts, and the Bay of Bengal's mangrove swamps. The next challenge is preserving the corridors linking them, preventing them from becoming inbred "island" populations.

My story still needed images of young cubs, and two Bandhavgarh tigresses had recently given birth. But tragedy struck. One mother was hit by a park vehicle, the other poisoned by a farmer after she killed his buffalo. Losing a female of cub-bearing age is particularly tragic for an endangered species: Her future offspring are collateral damage. The cubs were lost to the wild, sent to a zoo. Again I went home without a photograph I needed.

Another litter was born, and I returned to Bandhavgarh a few months later, this time under tight deadline and budget pressure. The head of the park's poaching patrol, E.A. Kuttappan, agreed to take me to the tigress's den.

It was deep in the jungle, too dangerous to walk. We could only get there by elephant. The afternoon heat was brutal, hitting 120°F, too hot for an elephant to be saddled and carrying us around, so we had to set out at dawn and head back before midday.

The tigress had returned to her natal den, high up in a narrow gorge with a creek running through. The cat had known Kuttappan and his elephant all her life, so she tolerated our presence. For the next three weeks, I shot photos of her from below. I couldn't get any closer, so I was using a 600mm lens in a dark place, sitting on an elephant that never stood still. And the cubs didn't appear.

On my 24th morning there, as the tigress lay amid a tangle of ferns and tree roots, I thought I glimpsed an ear. I snapped a few frames and then wondered if I'd imagined it.

Back at camp, I downloaded the memory card—and cried. The cub was cheek-to-cheek with its mother. So much of my two years on this story had been dark: poaching, devastated habitat, conflict with humans. But this was an image of hope. Tigers are a resilient species. There's enough habitat left for them, and a female can raise six to eight cubs over her lifetime. They can bounce back.

I got my wish to reinvigorate love for tigers. The images went around the world in *National Geographic,* but also in exhibits when I was named Wildlife Photojournalist of the Year and won the Pictures of the Year International Global Vision Award. My tiger photos were exhibited at the Newseum, where a 40-foot-long image of a tiger in Kaziranga's elephant grass hung on the facade, gazing out on Washington, D.C.

When you see and learn about these magnificent animals, how can you not care about their welfare? Today, poaching and loss of habitat remain the major threats to their survival, and their fate varies in the countries where they still roam. But tiger numbers in India and Huai Kha Khaeng continue to rise, and where there's life, there's hope.

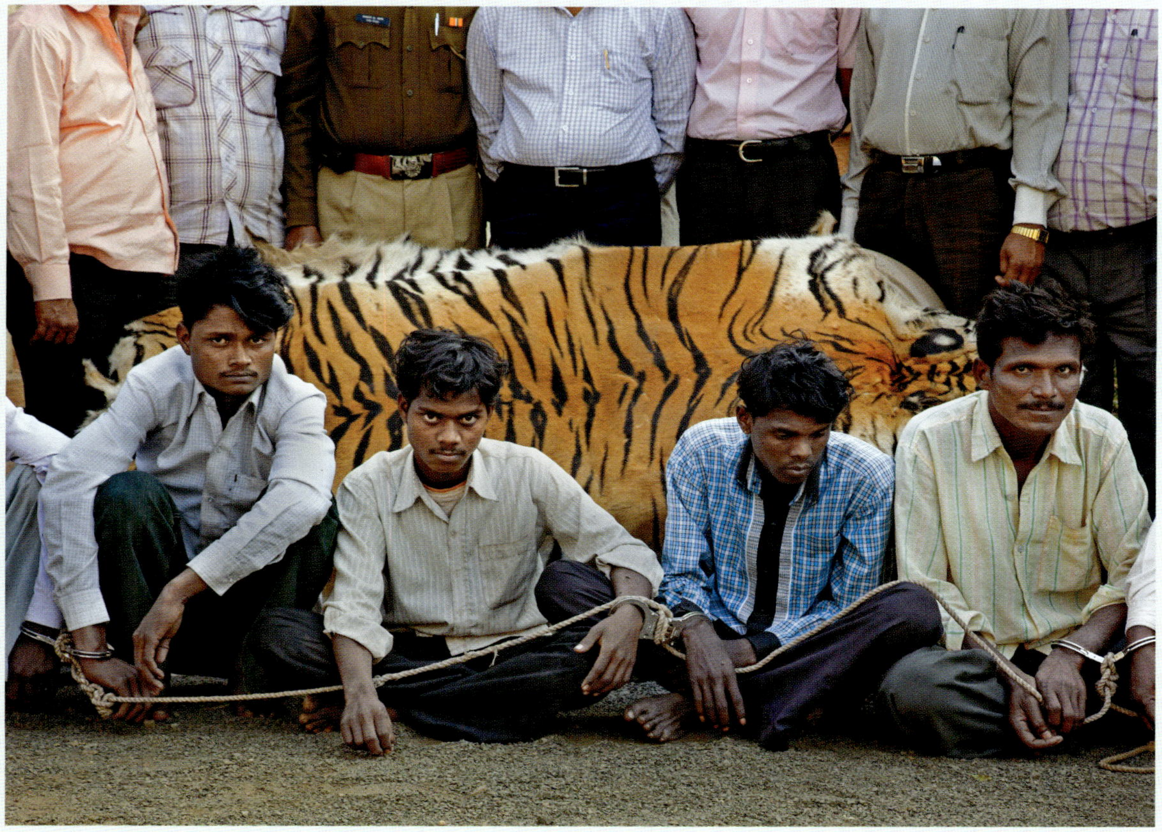

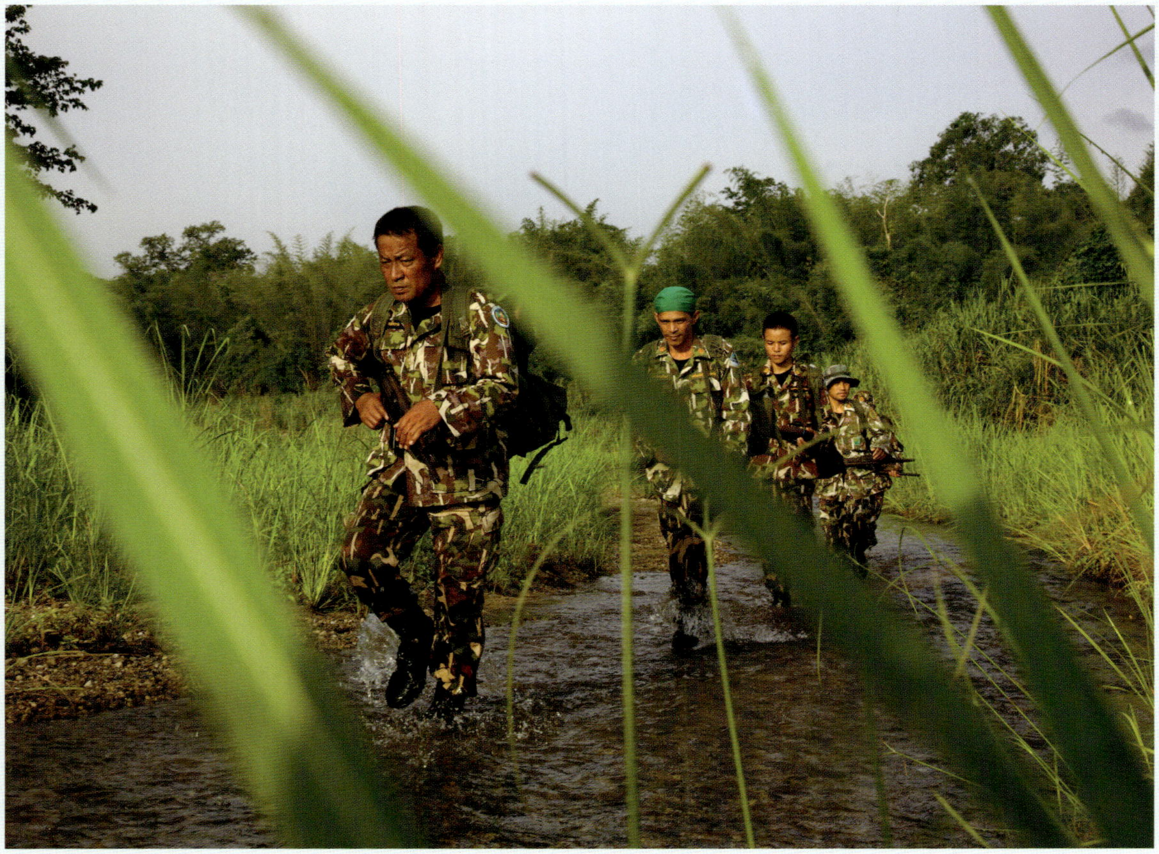

ABOVE: These poachers were apprehended while attempting to board a bus with a Bengal tiger pelt stashed in a burlap bag. Killing tigers or trading in their parts is illegal: All six subspecies are endangered and protected under an international treaty, and tigers are protected under Indian law.

RIGHT: Thailand's SMART patrol rangers march through Huai Kha Khaeng Wildlife Sanctuary, the country's last stronghold for critically endangered Indochinese tigers. This force has essentially ended poaching here, making this area the last best hope for saving this cat.

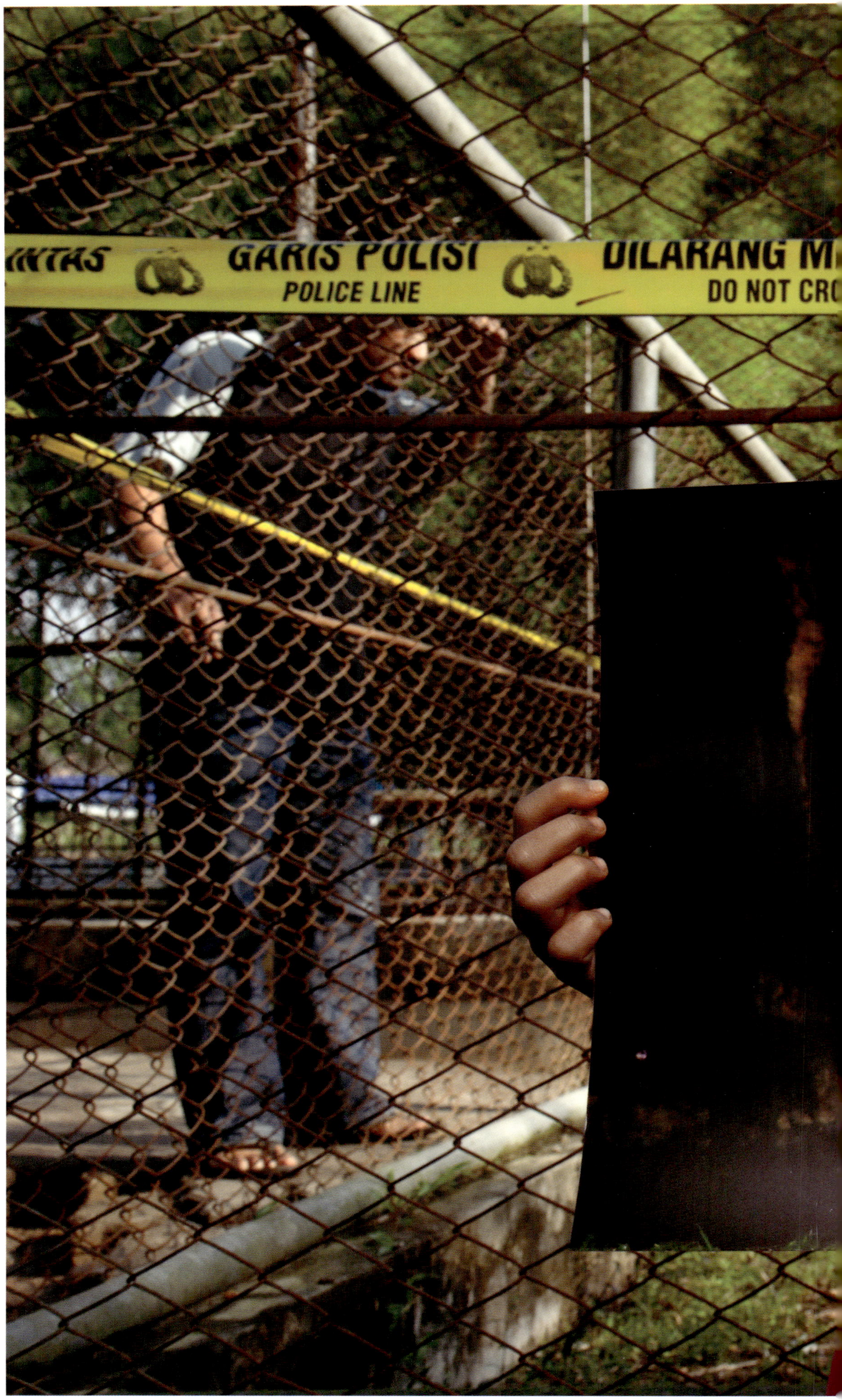

This zoo near Jambi, Sumatra, is a crime scene: A child holds a photograph of a tiger that was killed in its cage here for its pelt, bones, teeth, and other body parts.

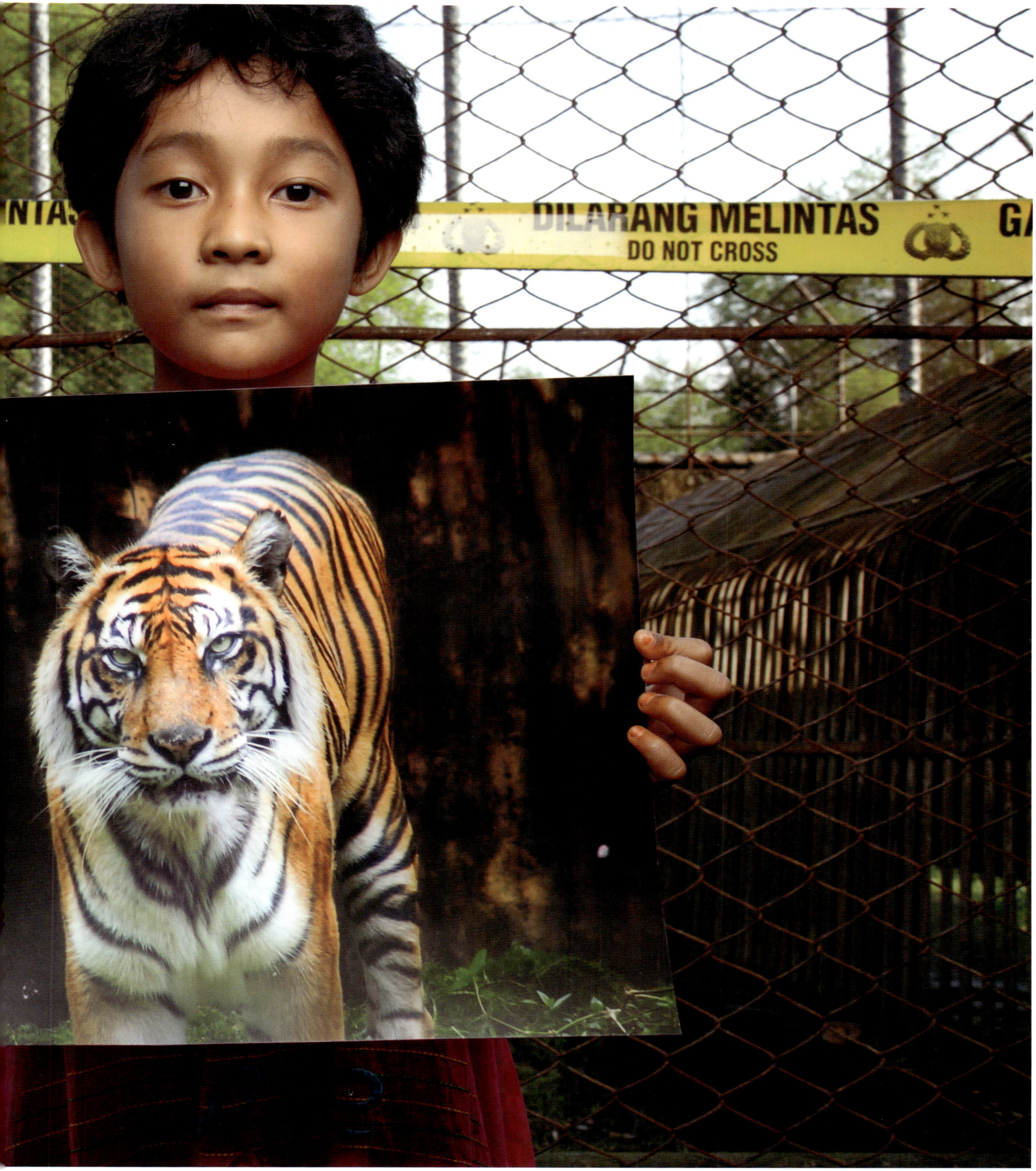
DILARANG MELINTAS
DO NOT CROSS

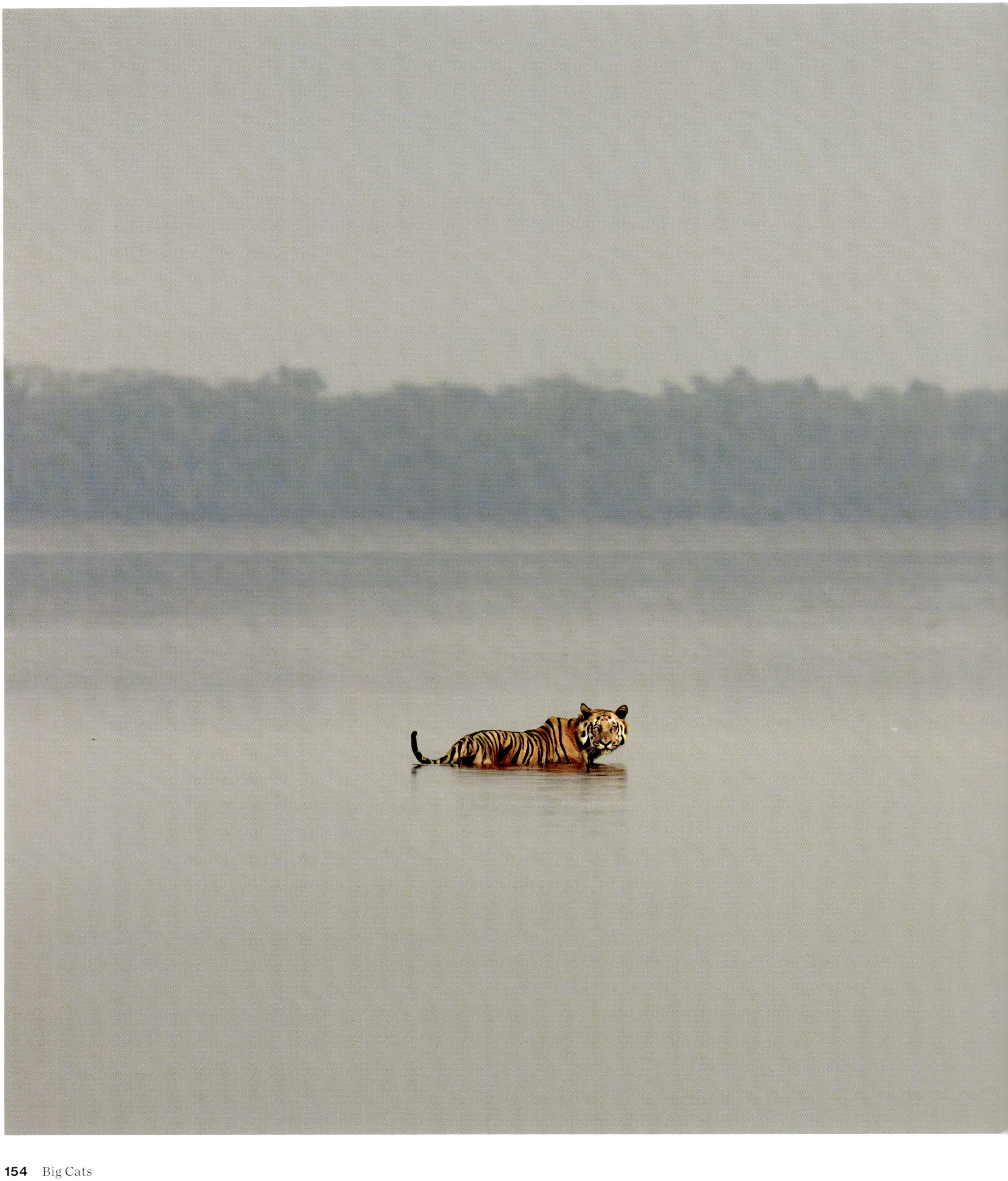

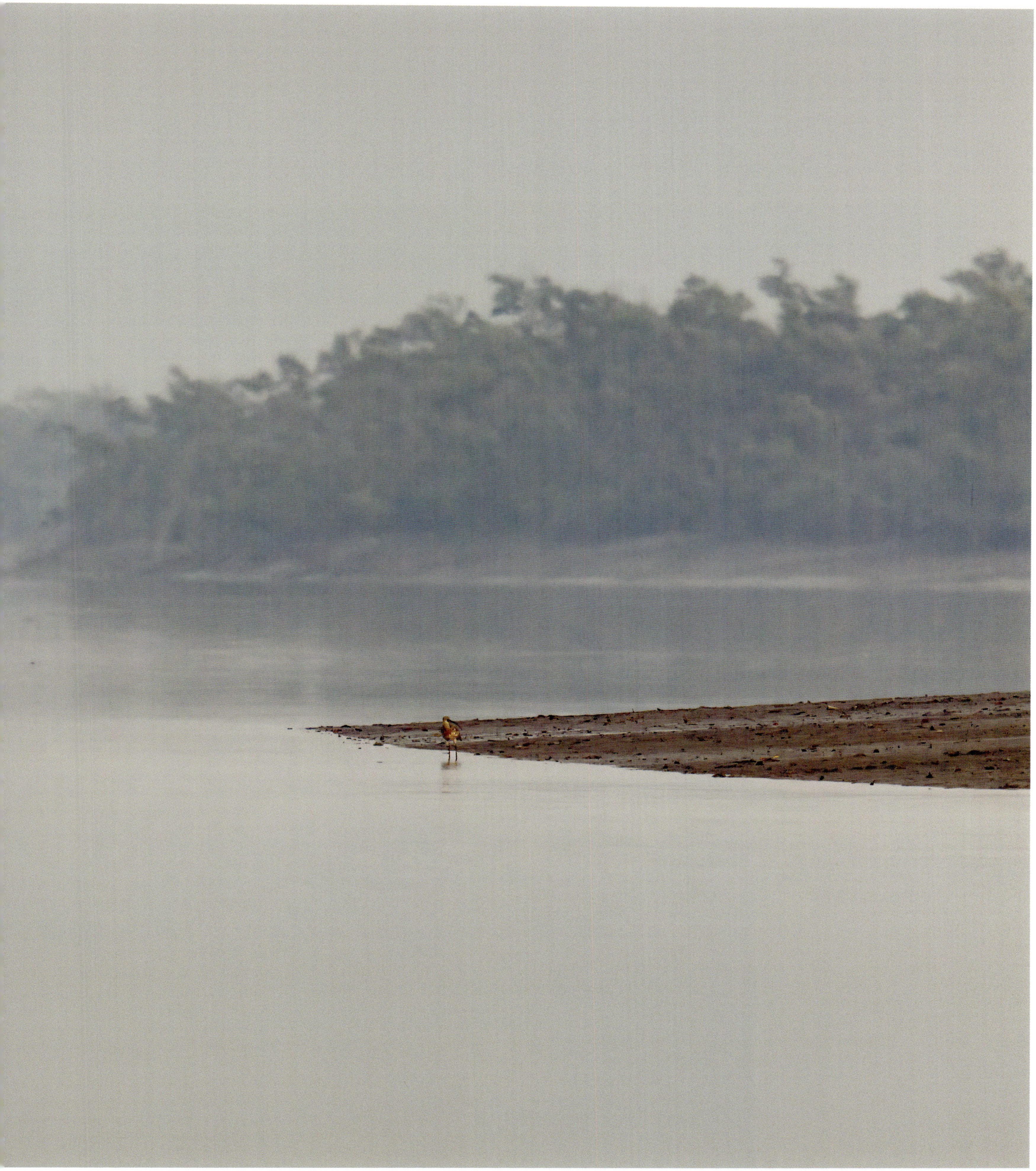

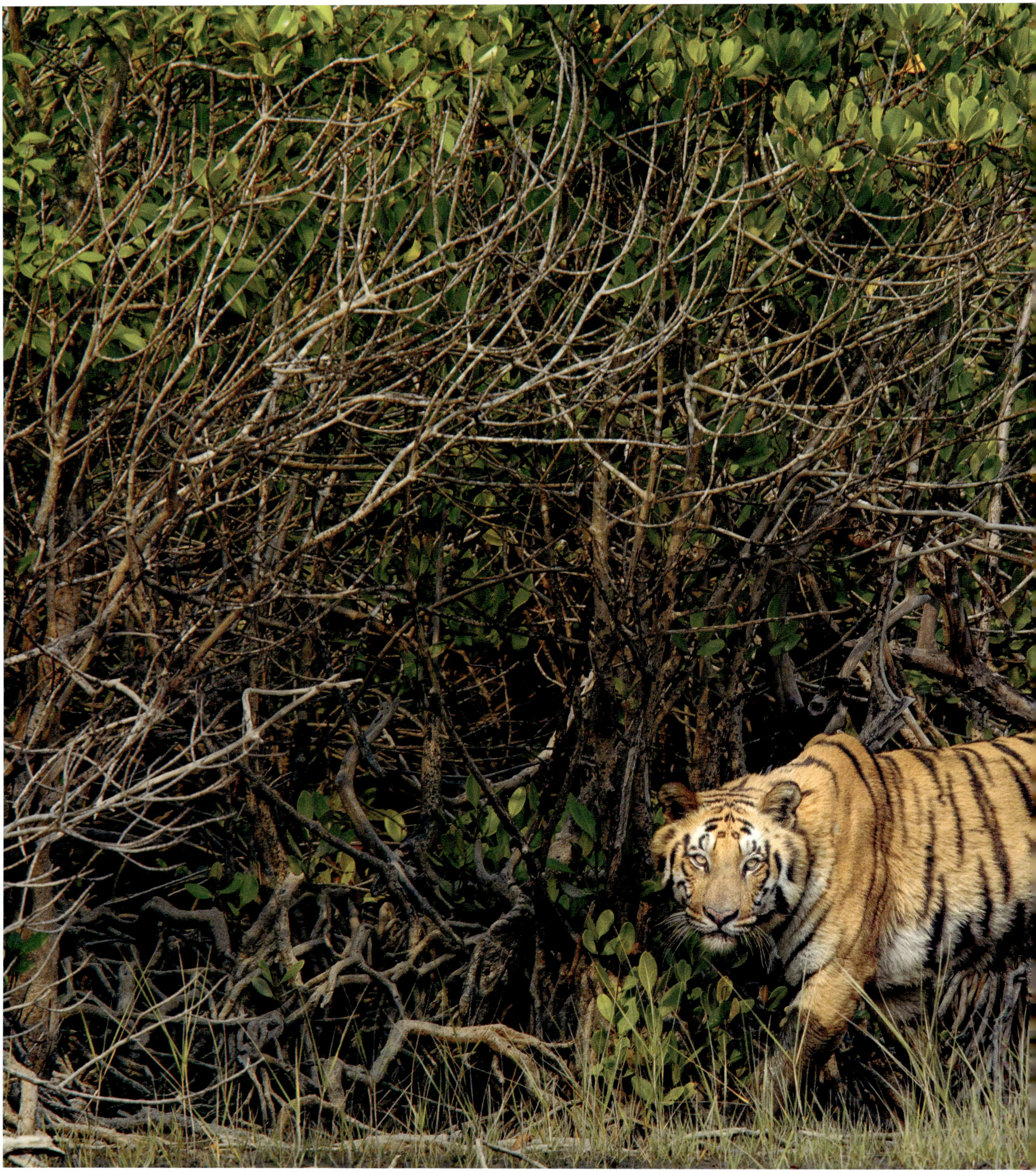

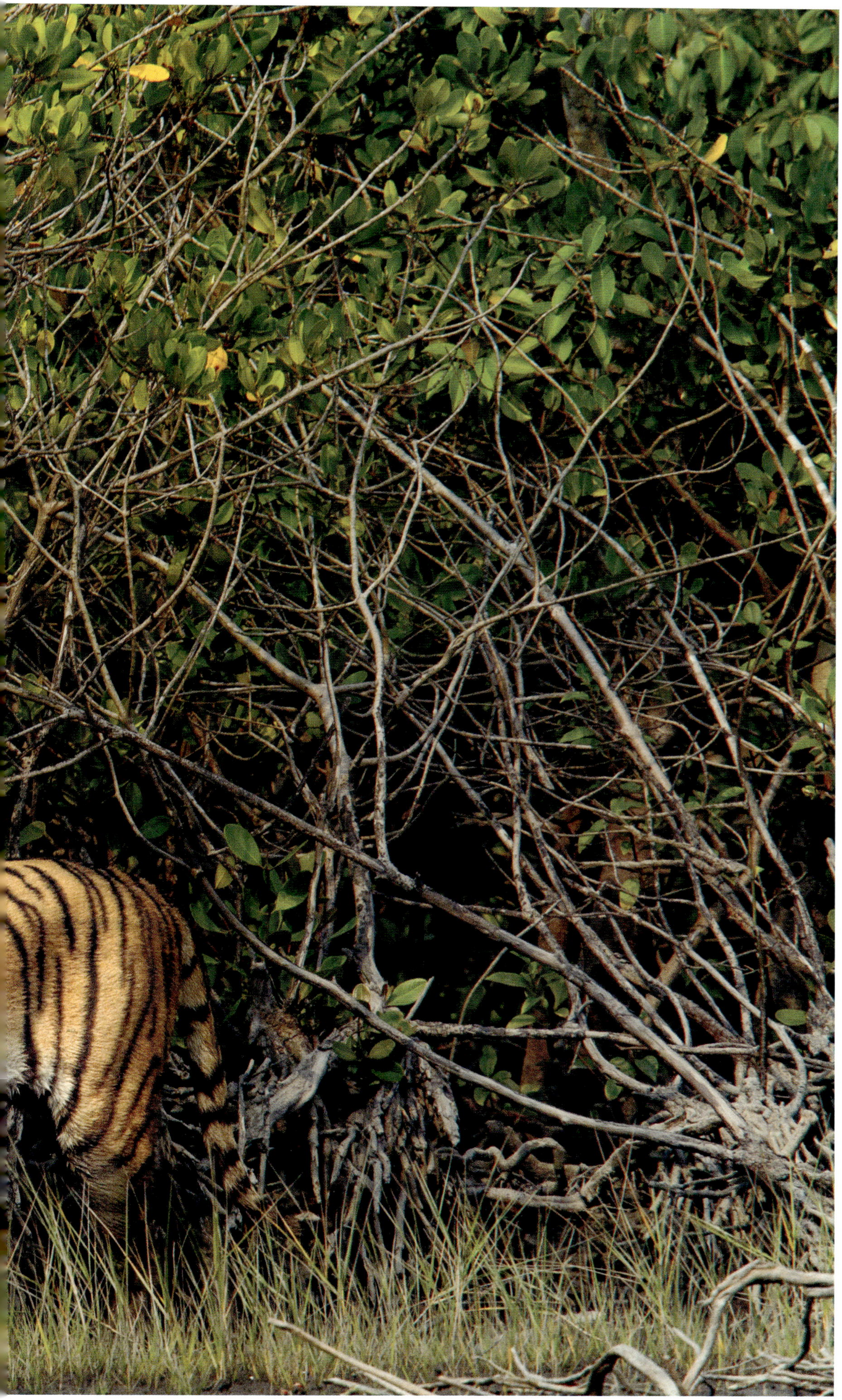

A Bengal tiger moves through the Sundarbans, mangrove swamps that straddle the India-Bangladesh border. These cats are strong swimmers and are known to be particularly aggressive.

PAGES 154–155: A male tiger pauses mid-swim in India's Sundarbans wetlands as he listens to another tiger's roar. After hearing that warning, he swam back the way he had come.

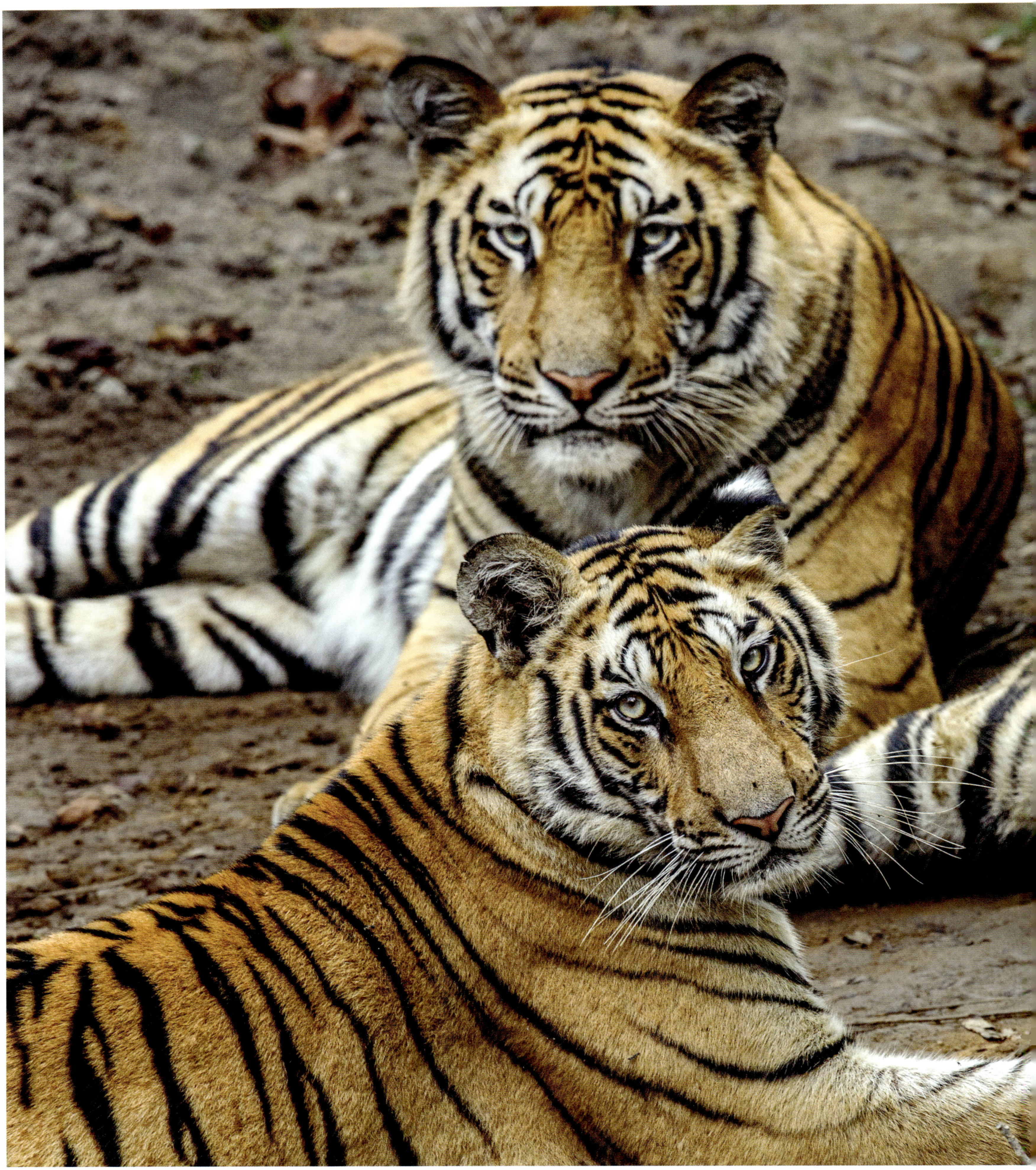

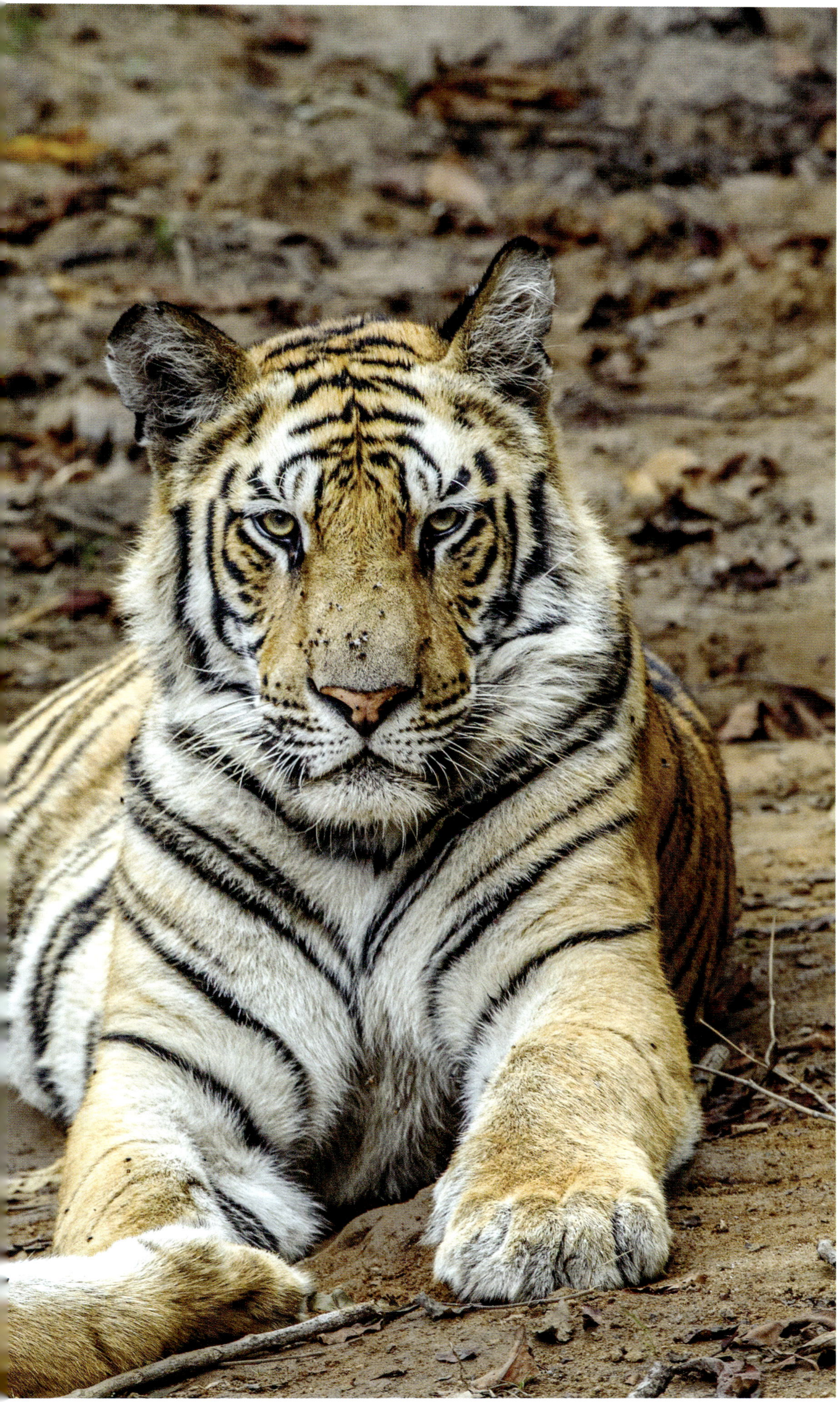

At 16 months old, these three siblings are nearly ready to leave their family and establish their own home territories. Tigers live solitary lives, coming together only to mate and, for females, to raise their cubs.

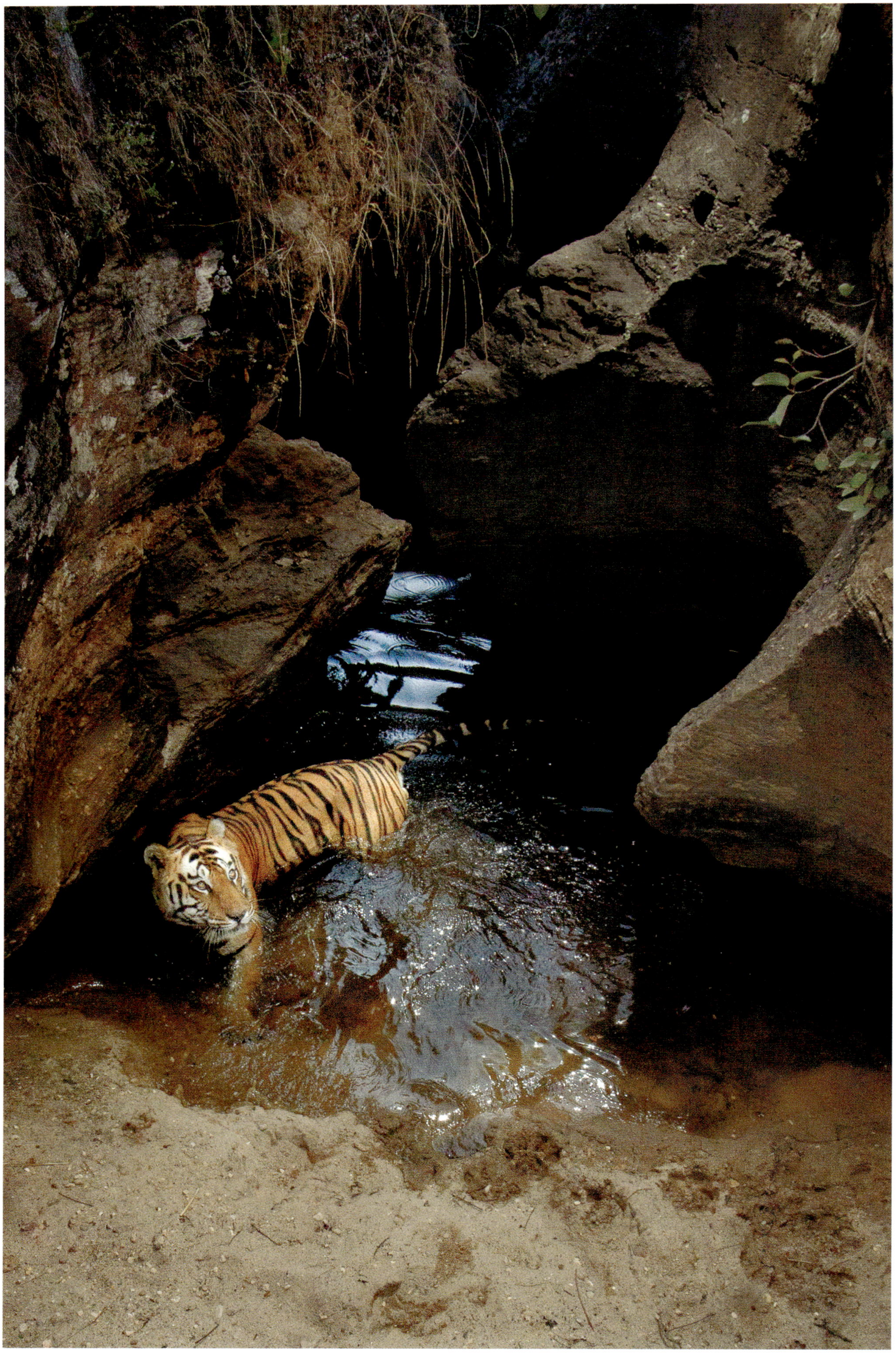

When they come to this water hole to drink, tigers back into the water, always vigilant for rival cats. More than 75 tigers live in Bandhavgarh National Park (an area of about 150 square miles), among the highest densities of tigers in the world.

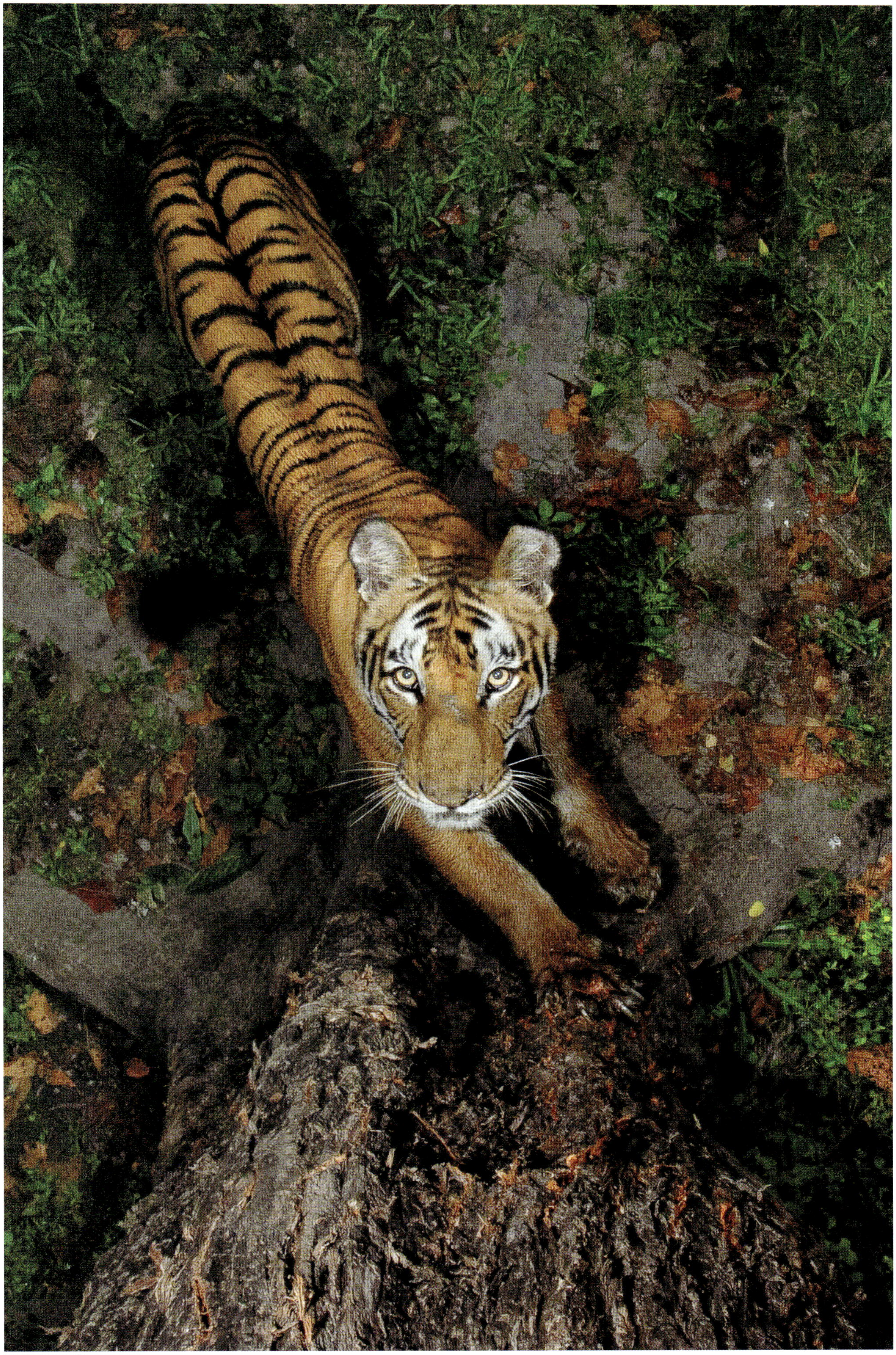

A remote camera captures a Bengal tiger raking scratches into the base of a tree, marking his territory with scent glands between his claws, in Kaziranga National Park, India.

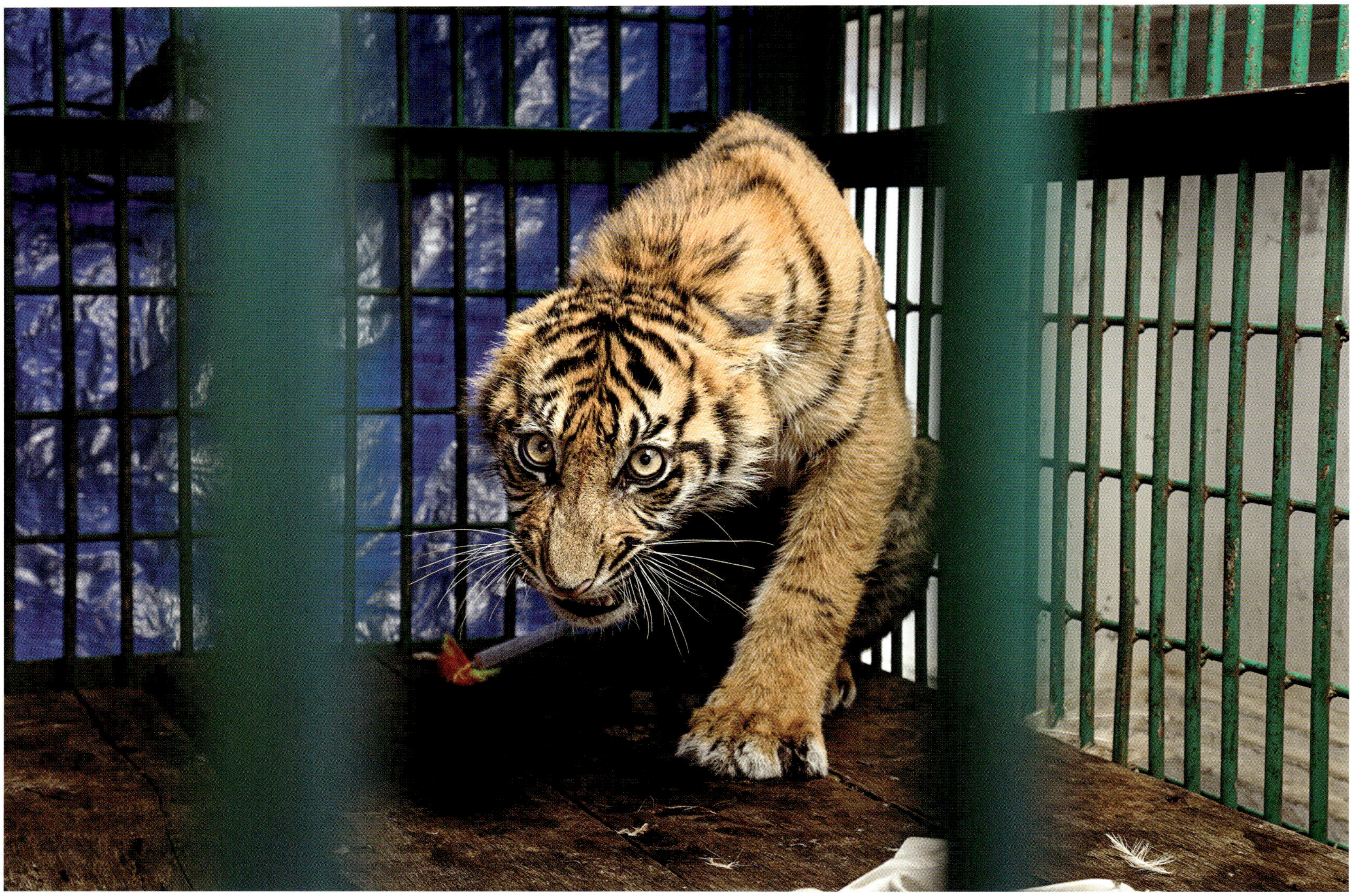

Since palm oil plantations take five years to bear fruit, new farmers set snares to capture wildlife for food. But snares are indiscriminate: This Sumatran tiger cub, trapped for three days, lost its right front leg and will live out its life in a zoo – a devastating loss for a critically endangered species.

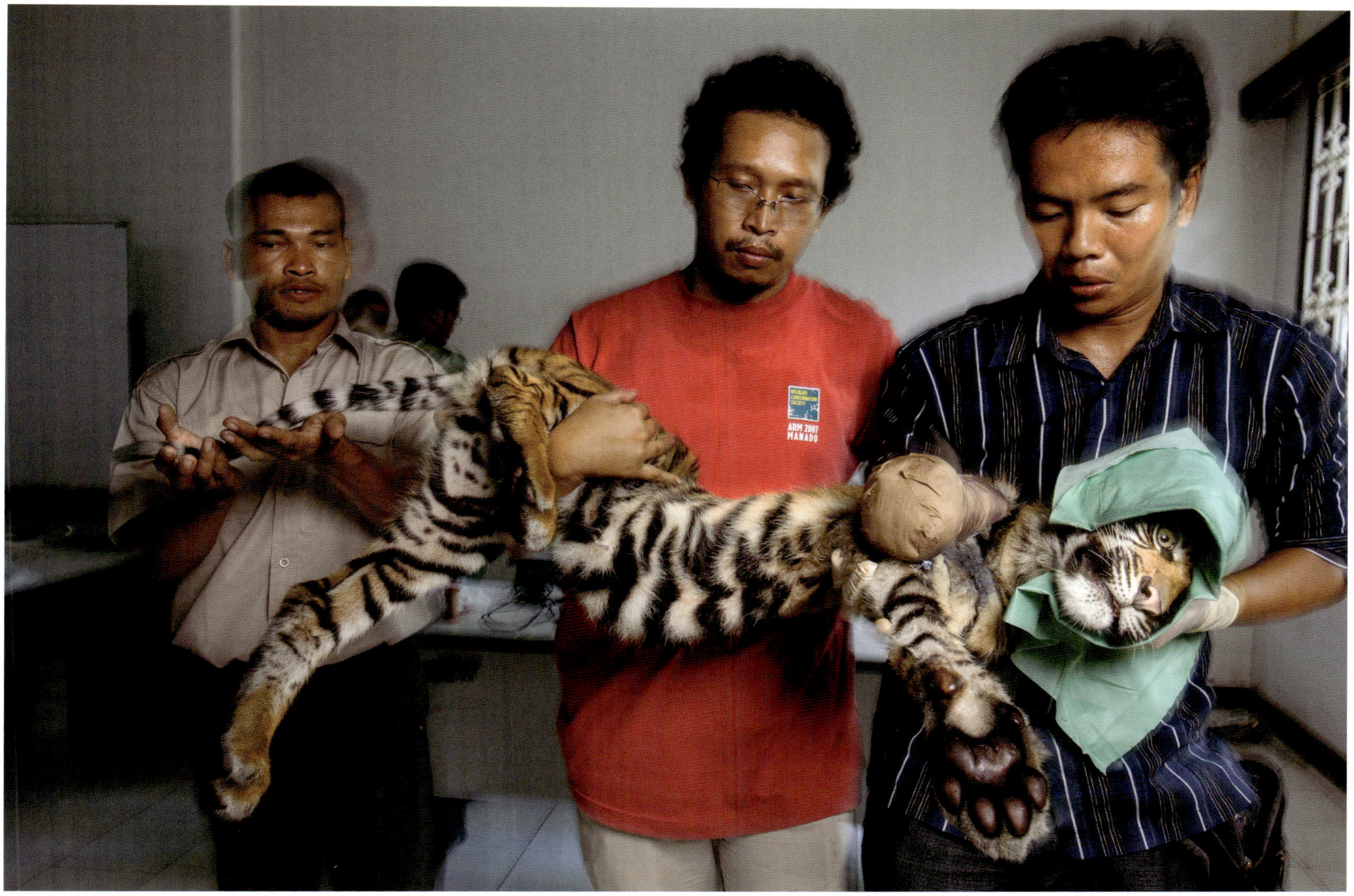

Veterinarians and a ranger cradle a tiger cub after it underwent surgery to amputate its leg in Banda Aceh, Sumatra. Sumatran tigers hang on the brink: Perhaps 400 remain in the wild.

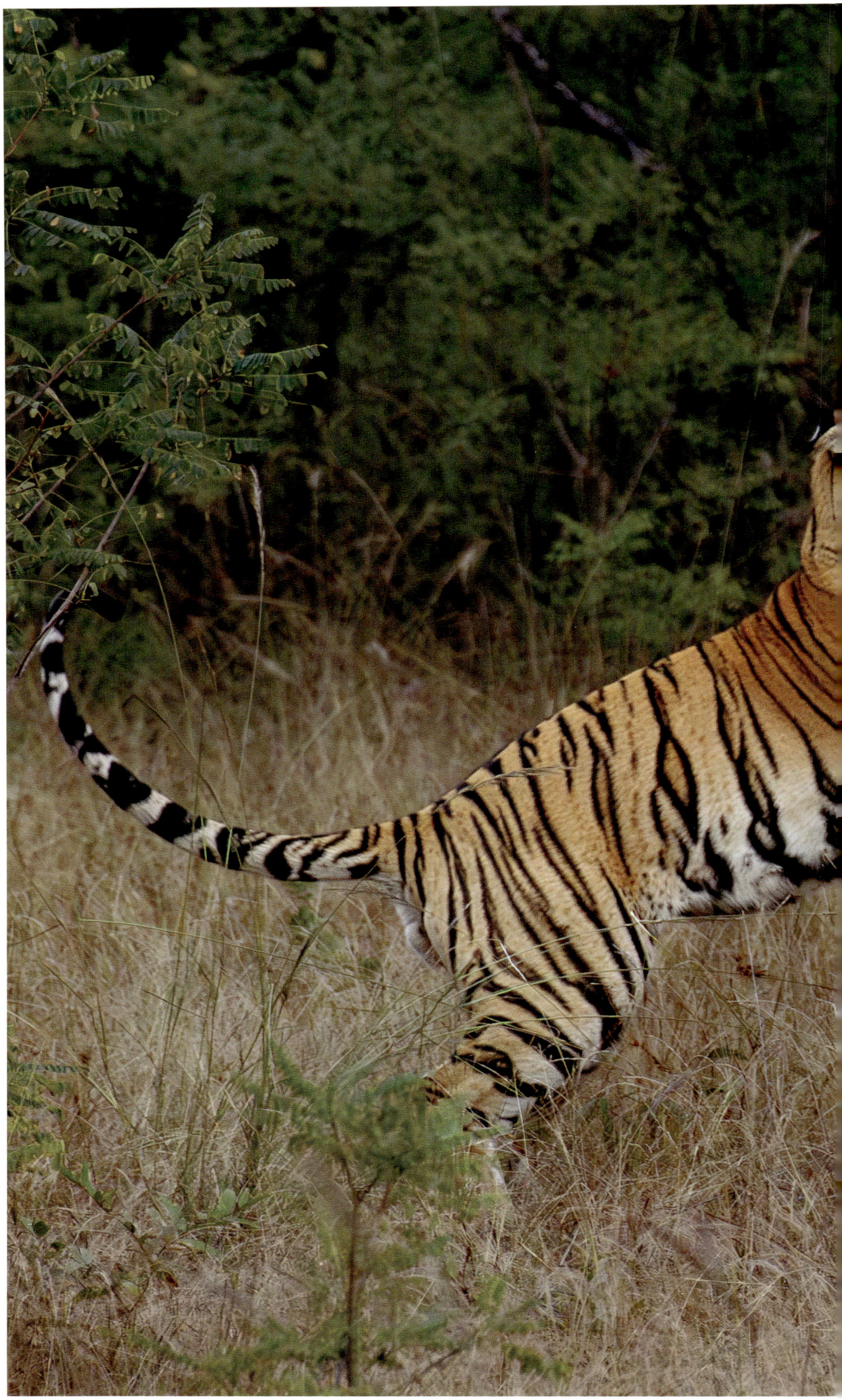

In Bandhavgarh National Park, these nearly grown tiger siblings spar – play that hones their fighting skills. Cubs stay with their mothers for up to two years before finding an unoccupied territory to claim as their own.

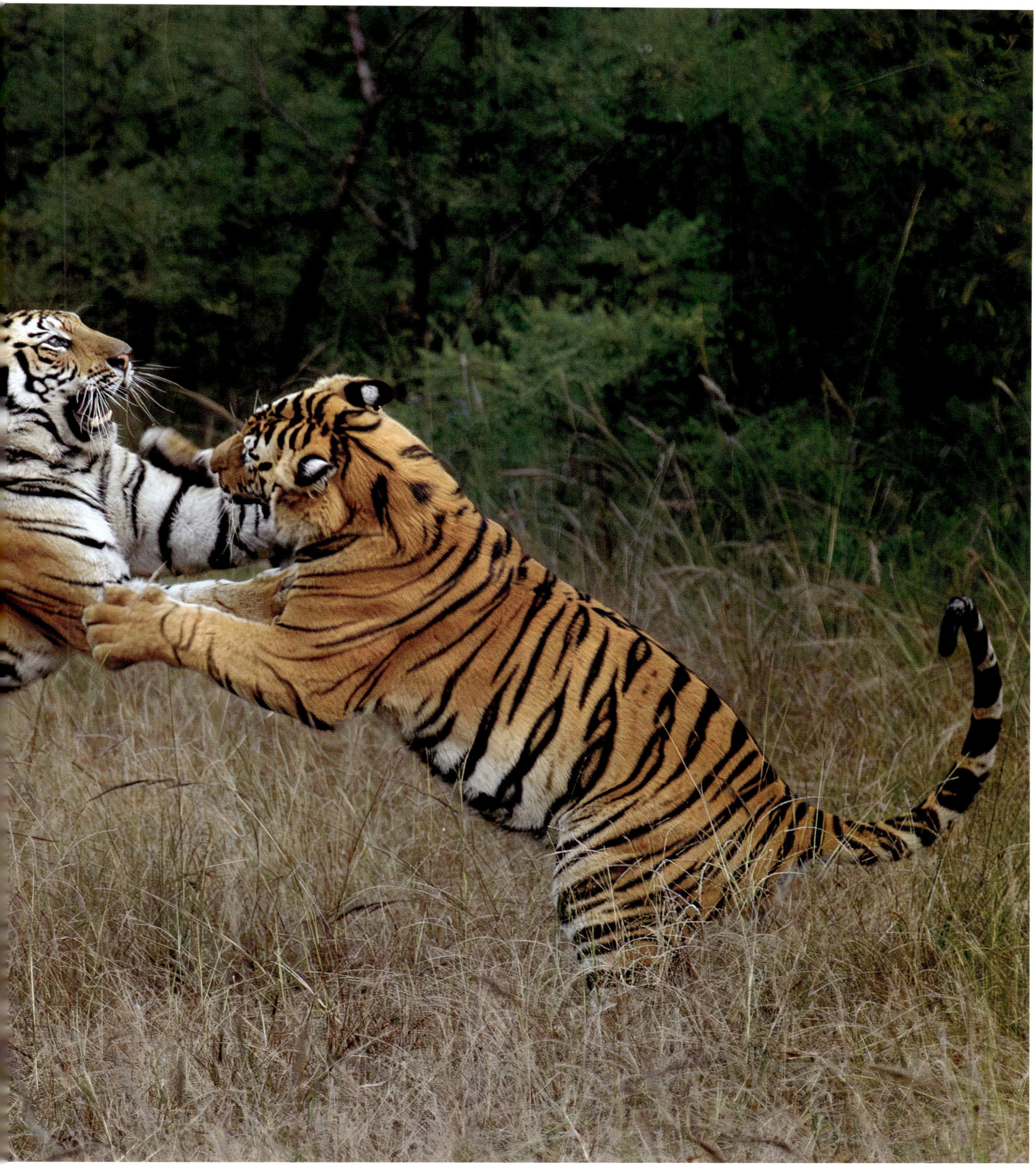

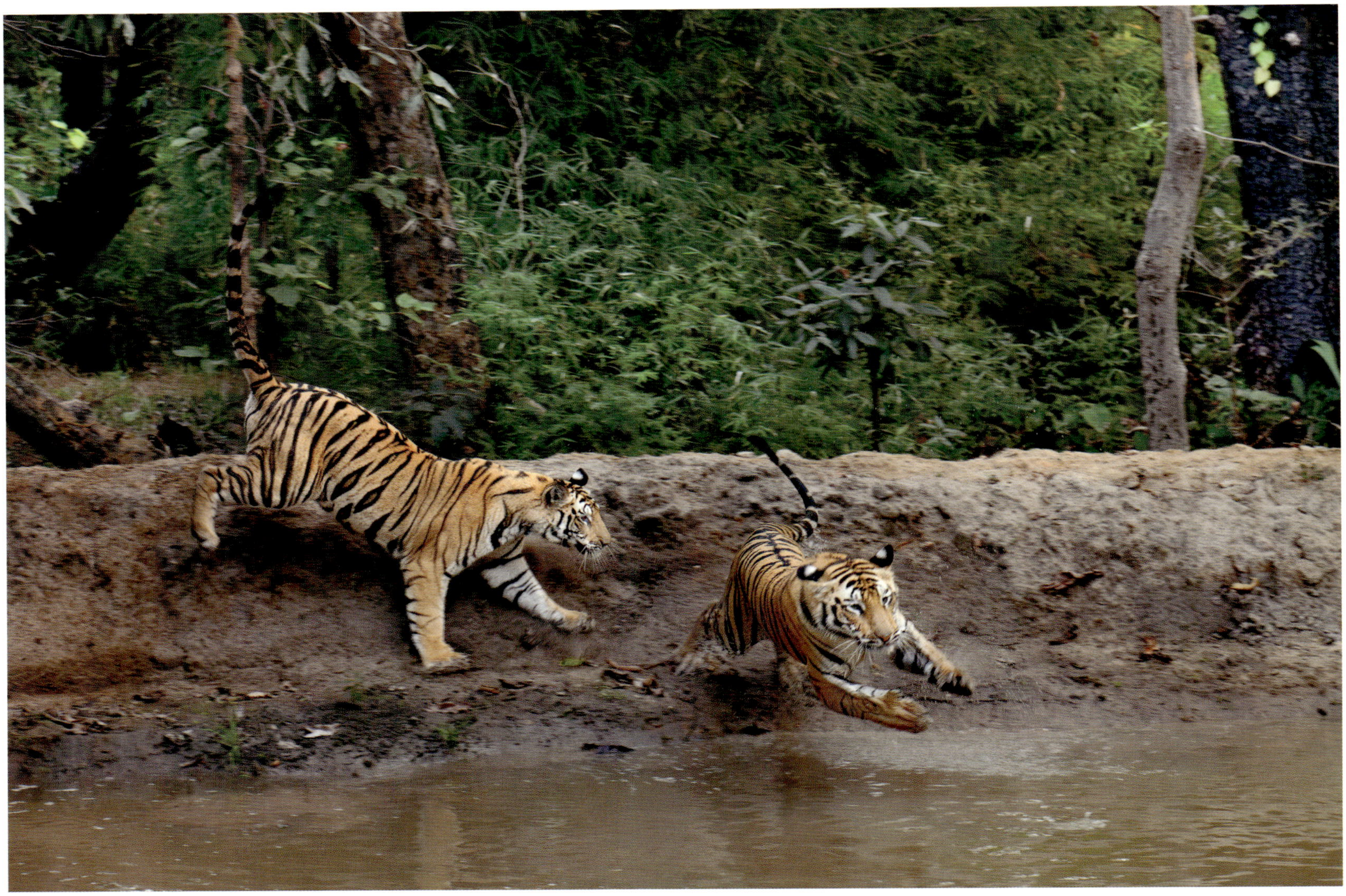

Bengal tiger cubs chase each other at an artificial water hole. India's 58 tiger reserves require intensive management: At last count, 3,682 tigers live in the world's most populous nation.

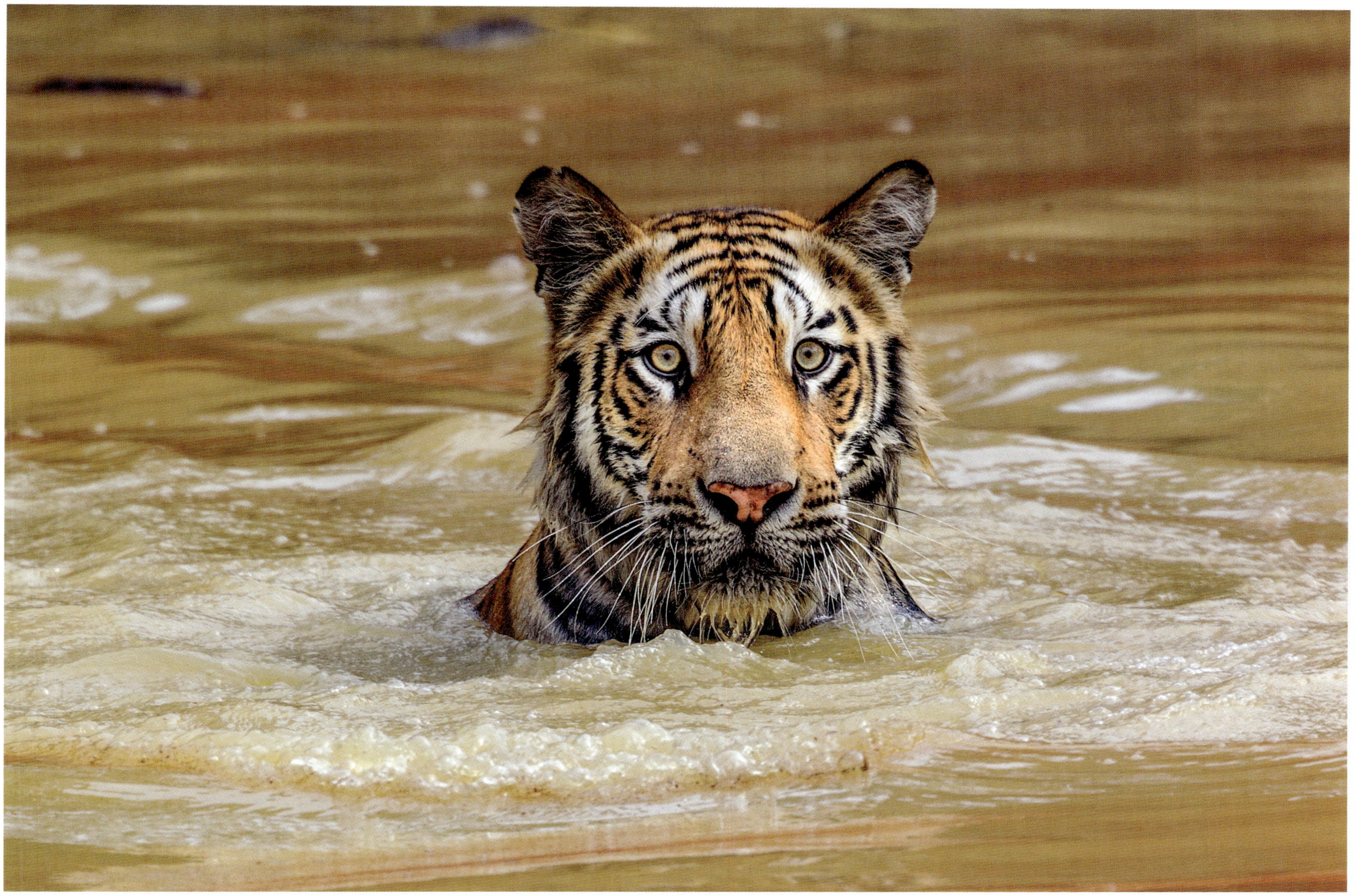

A young tiger fixates on a herd of chital – spotted deer – as it cools off in a water hole in Bandhavgarh National Park. Sambar deer and chital are Bengal tigers' favored prey.

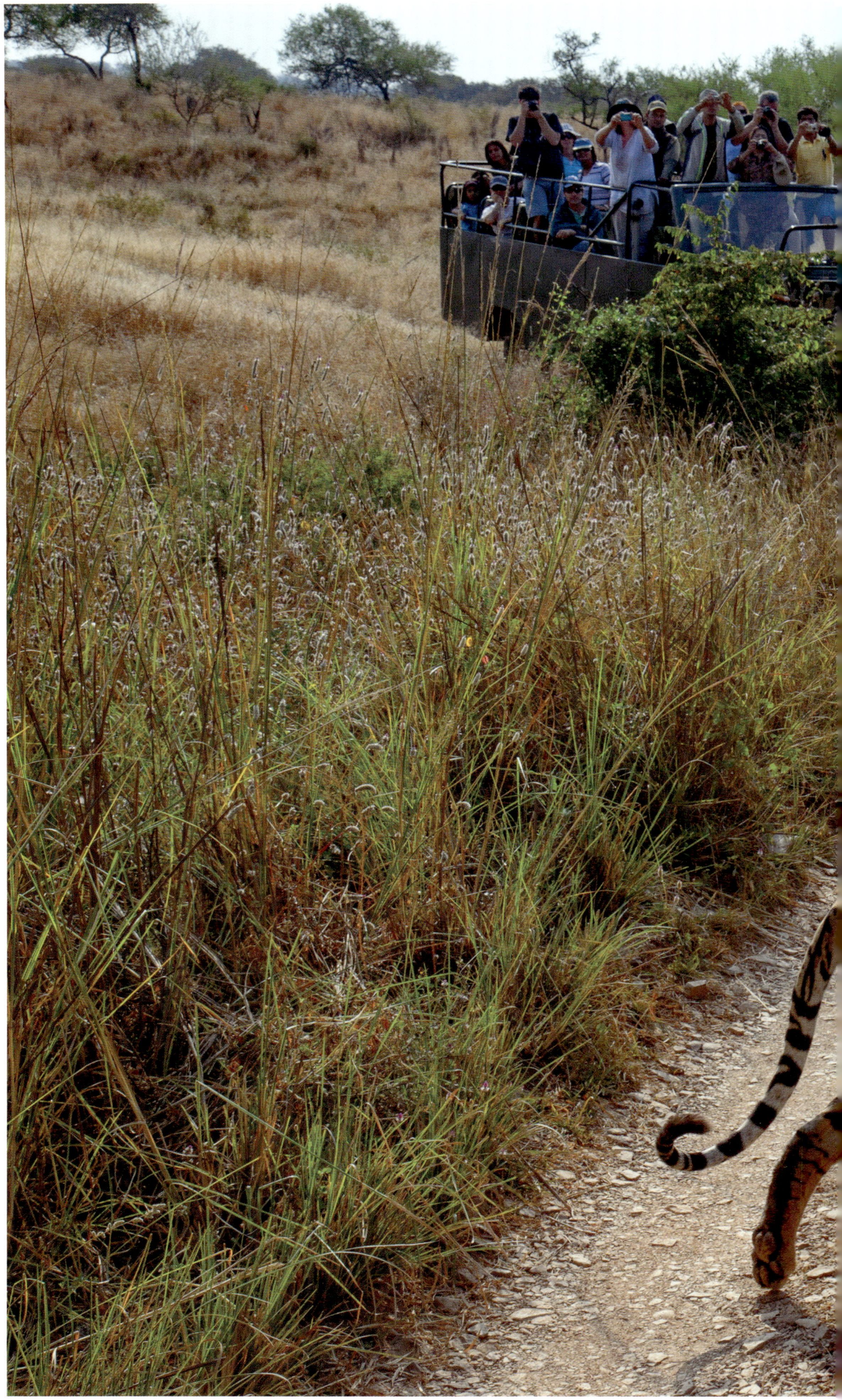

Tourists from around the world flock to India's parks to see tigers like this collared cat in Ranthambore Tiger Reserve. Tigers rule here, and they generally ignore visitors, sometimes hunting or mating in full view of safari jeeps.

PAGES 170–171: A Bengal tiger rests in the shade of a bamboo forest in Bandhavgarh Tiger Reserve, with the parched, premonsoon landscape providing perfect camouflage.

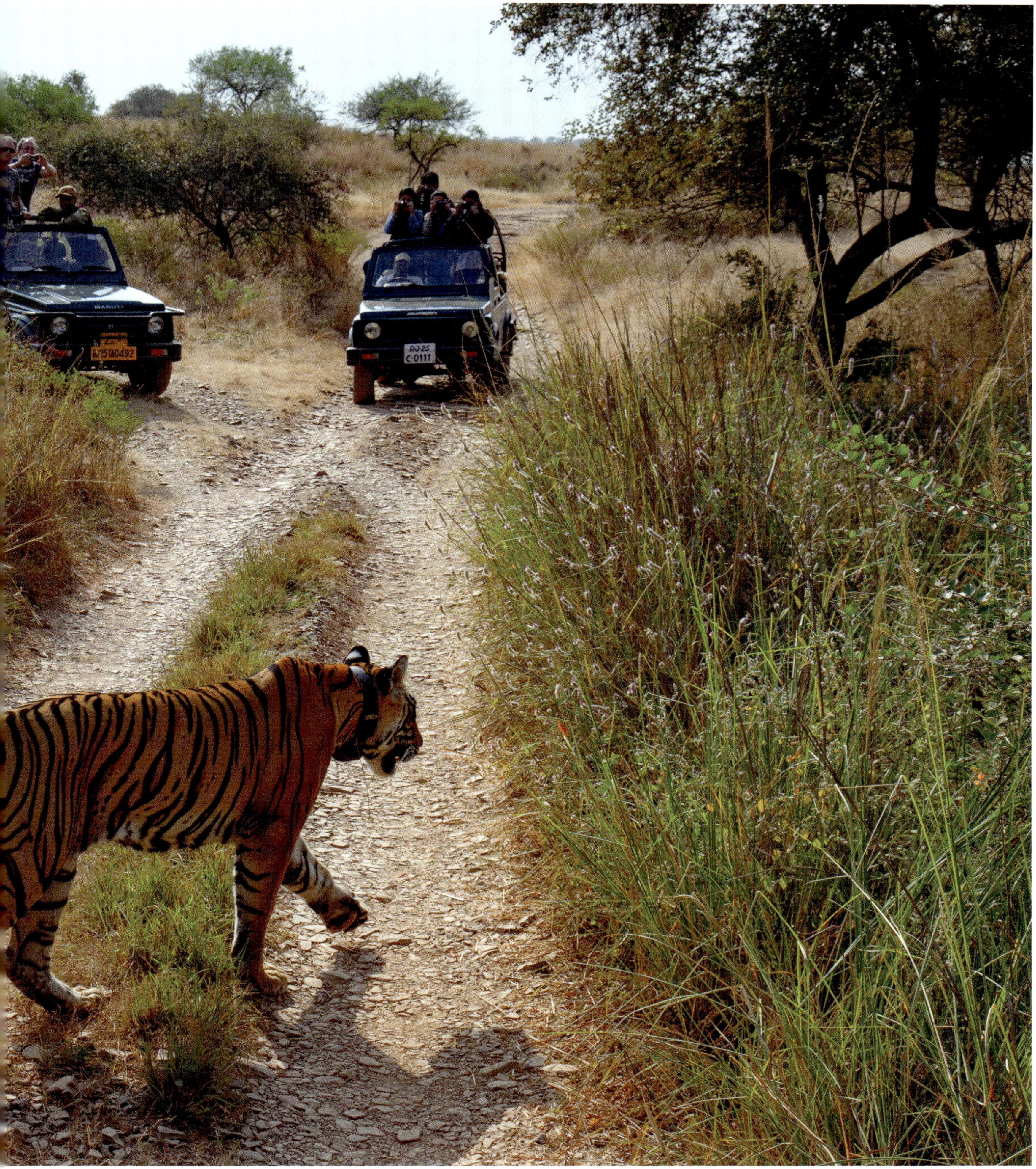

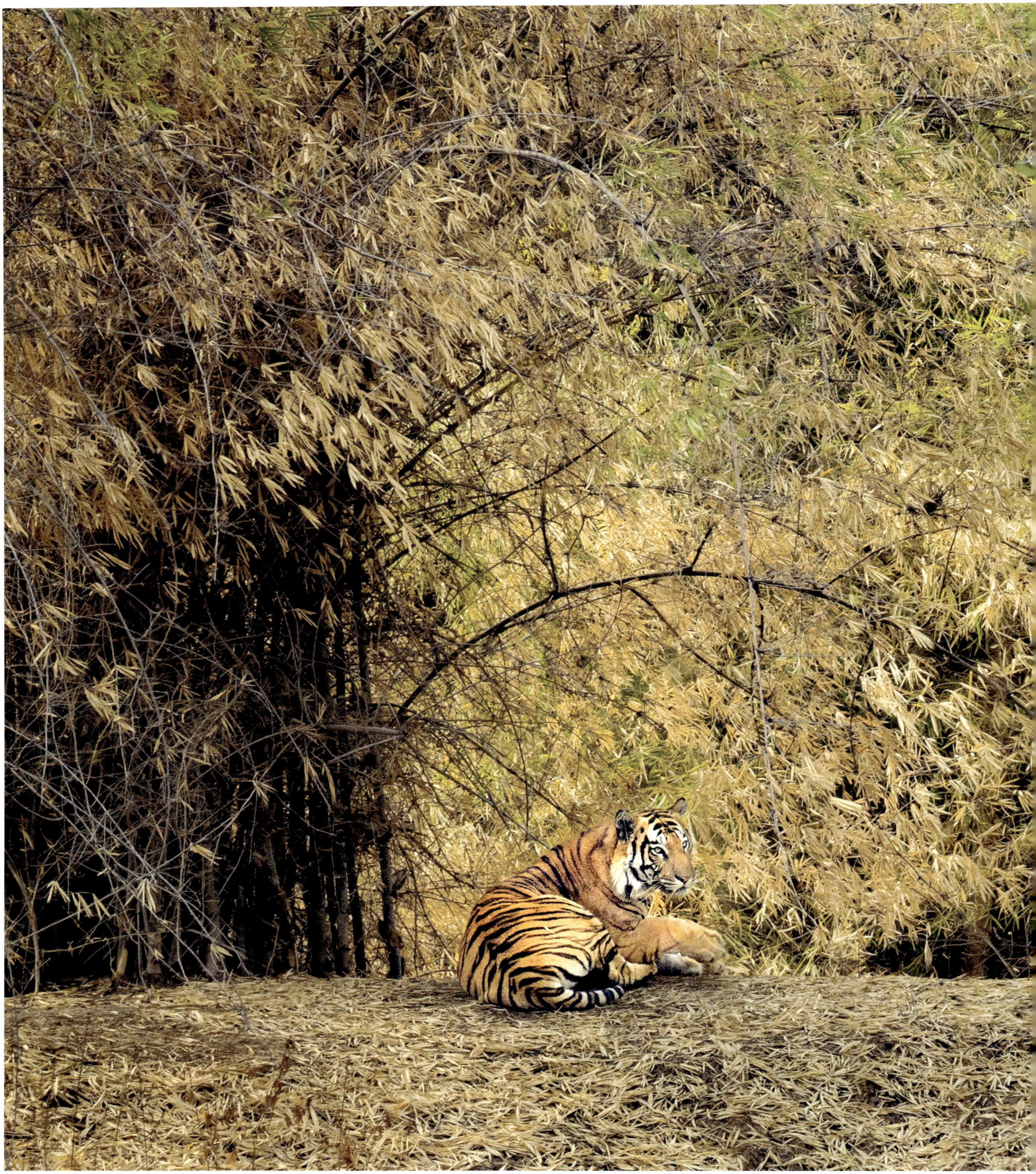

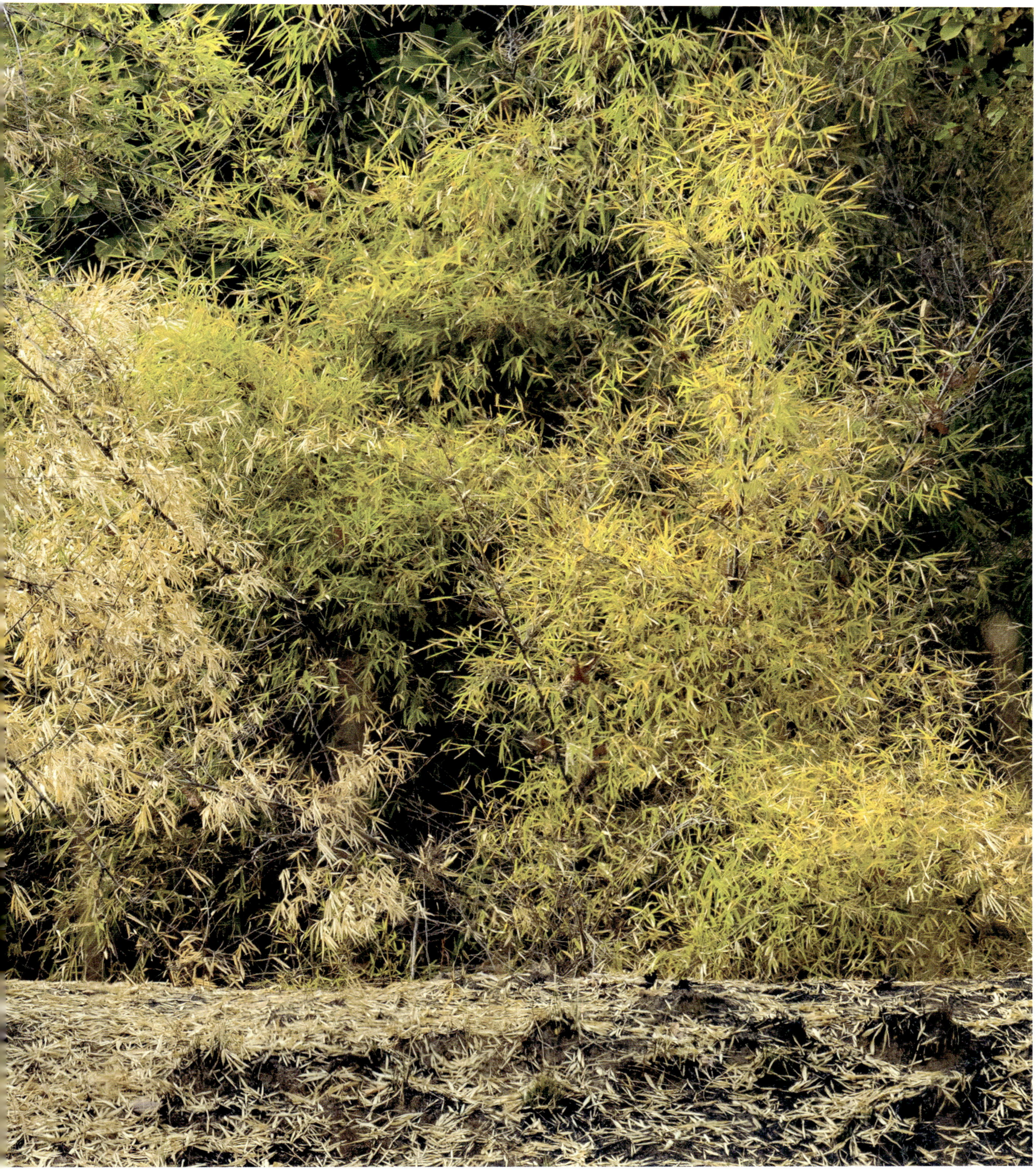

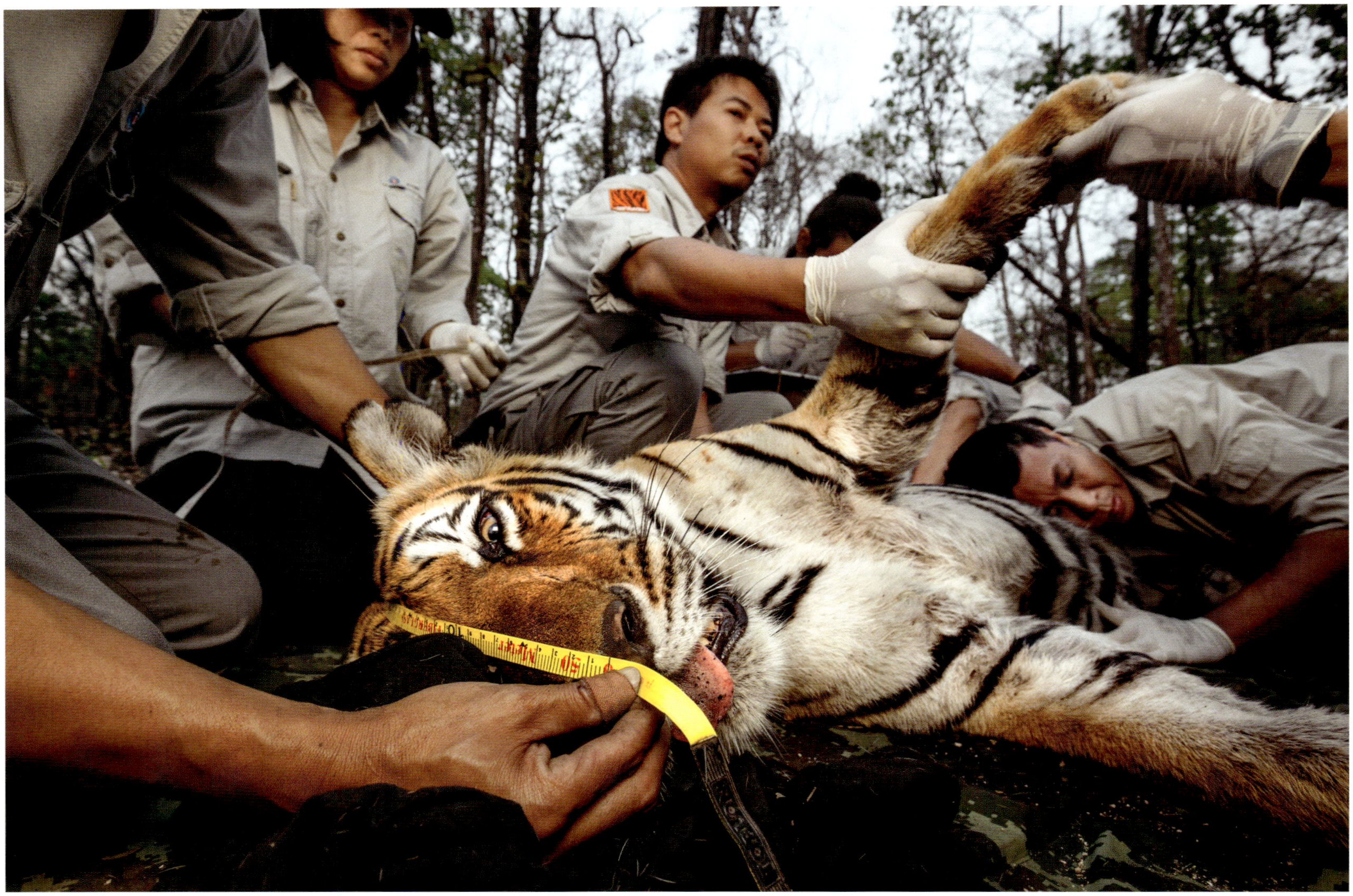

Researchers with the Thai Tiger Team, led by Achara and Saksit Simcharoen, measure, weigh, and fit an Indochinese tiger with a radio collar in Huai Kha Khaeng Wildlife Sanctuary.

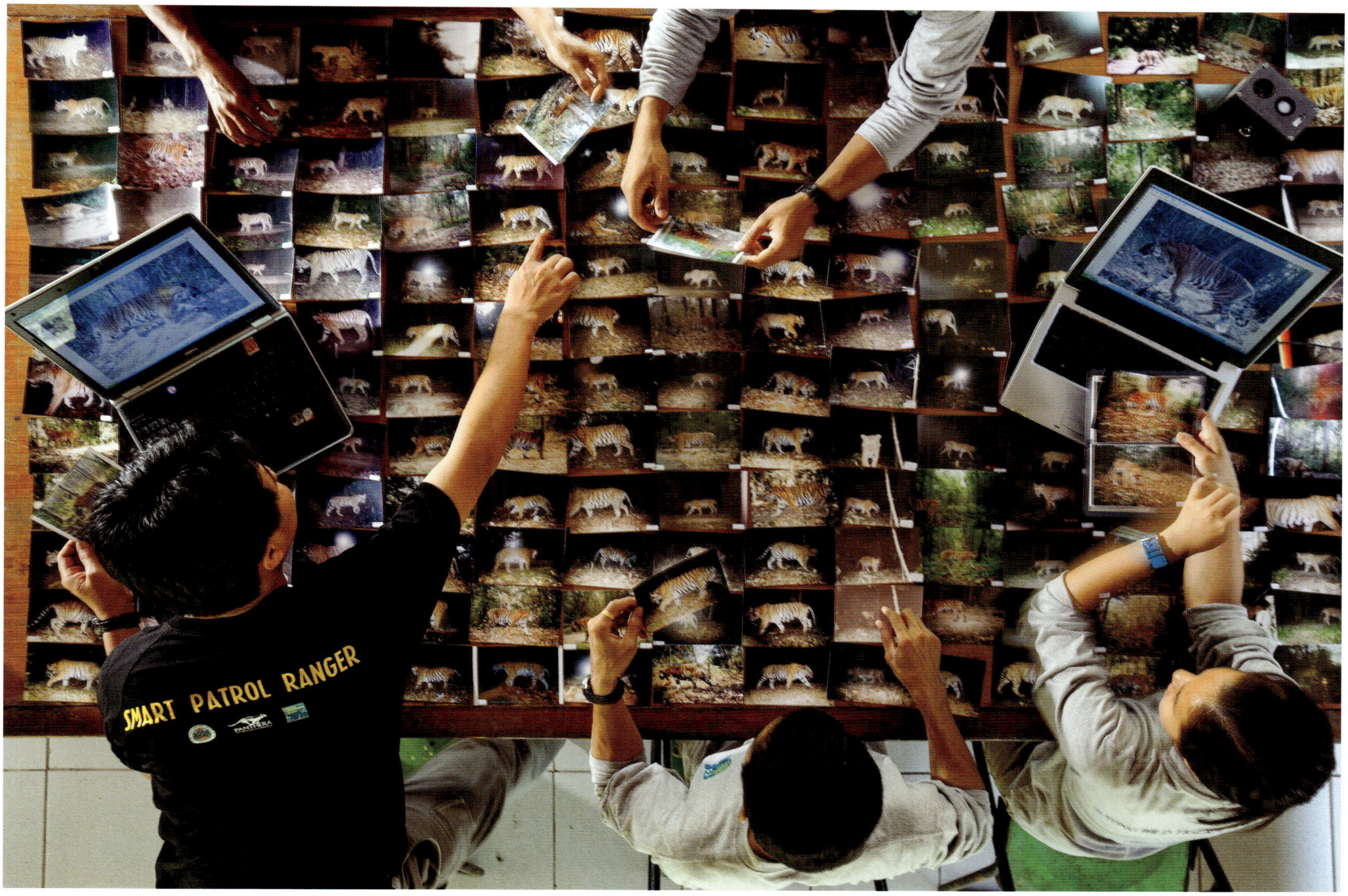

Biologists examine camera trap images of tigers made in Huai Kha Khaeng Wildlife Sanctuary, where the world's longest-running tiger research program, launched by biologist Alan Rabinowitz, has been ongoing since 1986.

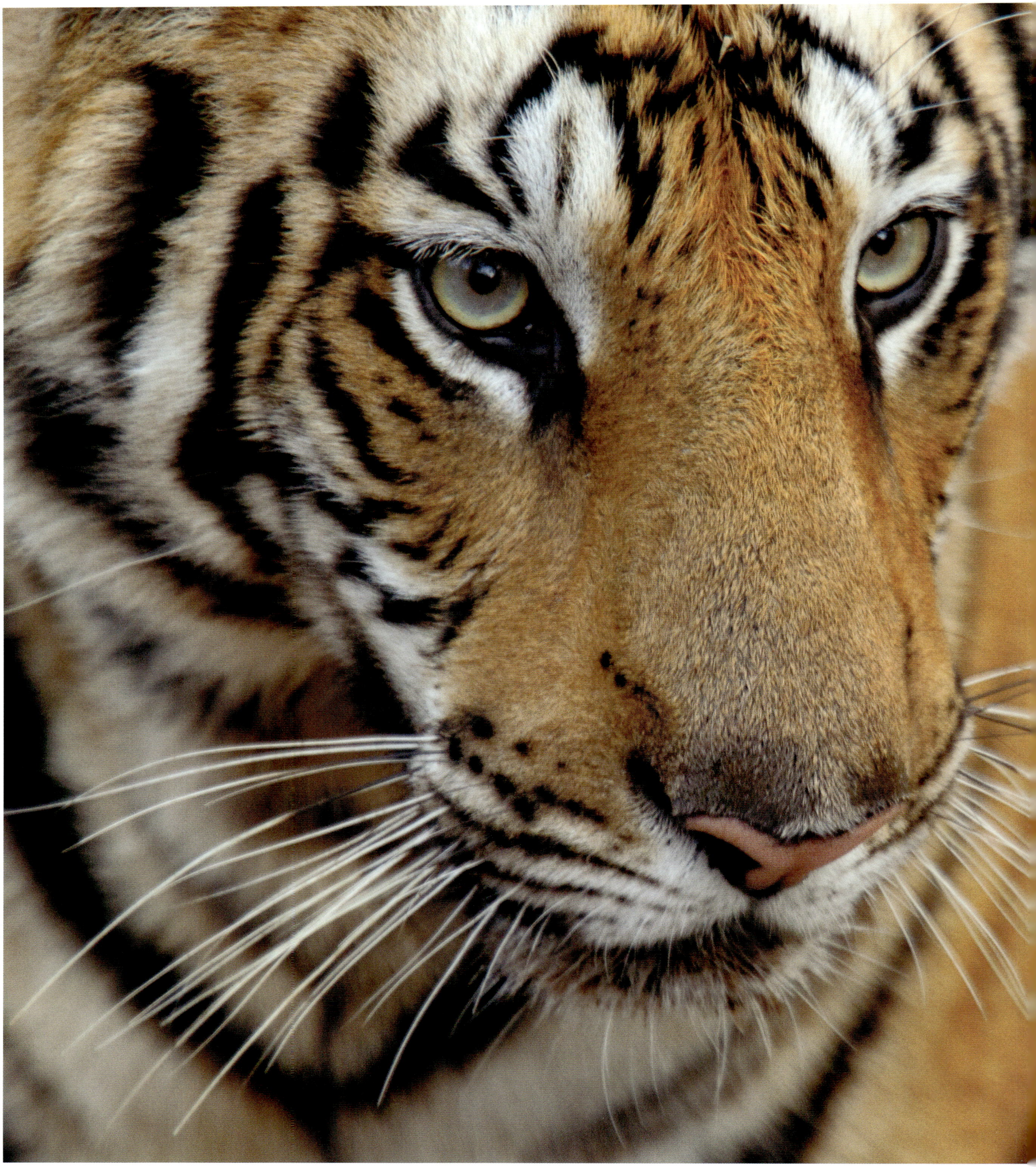

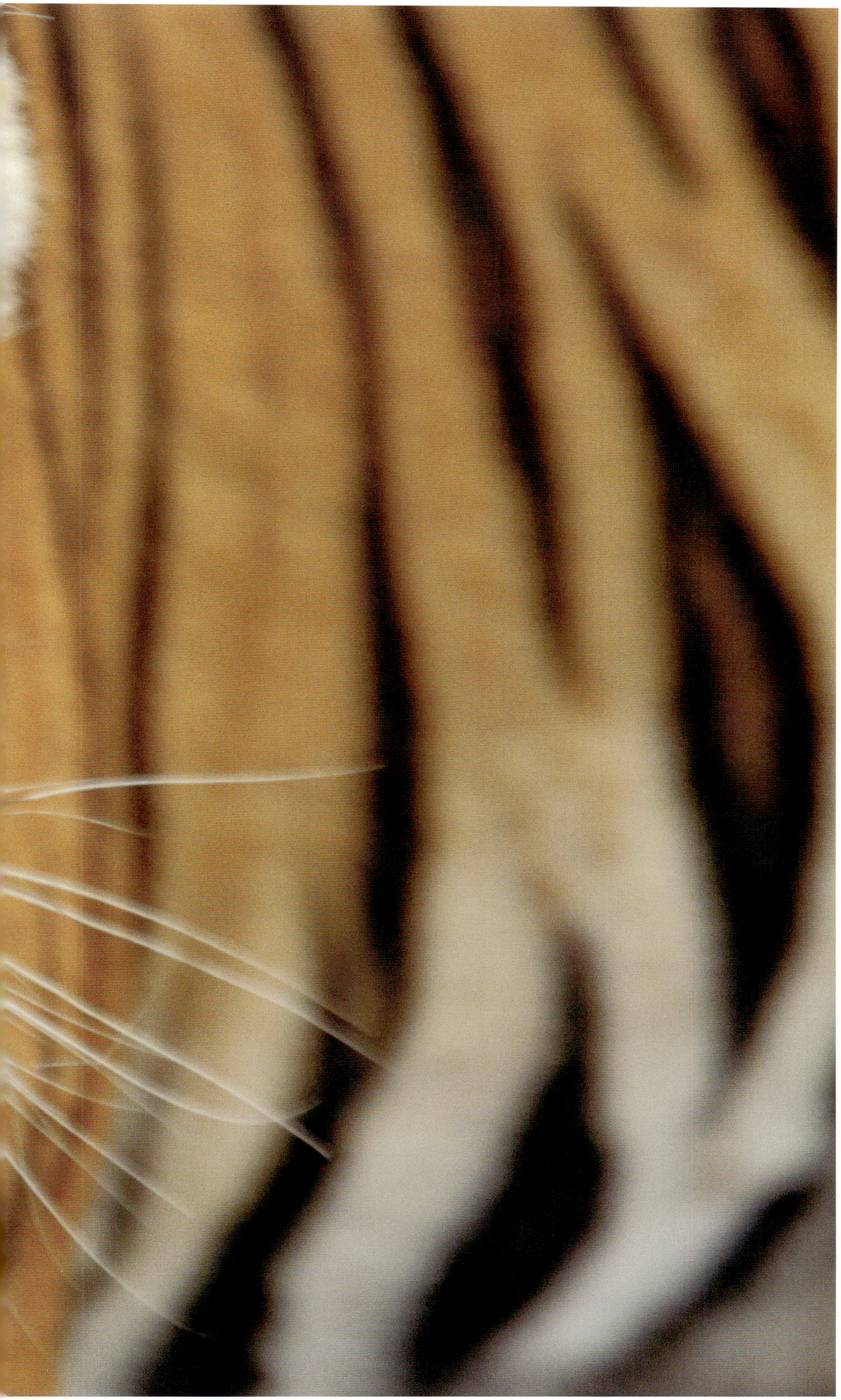

This male tiger in Bandhavgarh National Park represents a small recovery for the local population. In 2008, just 1,411 tigers were left in India. After dedicated conservation, the most recent census counted two and a half times that many.

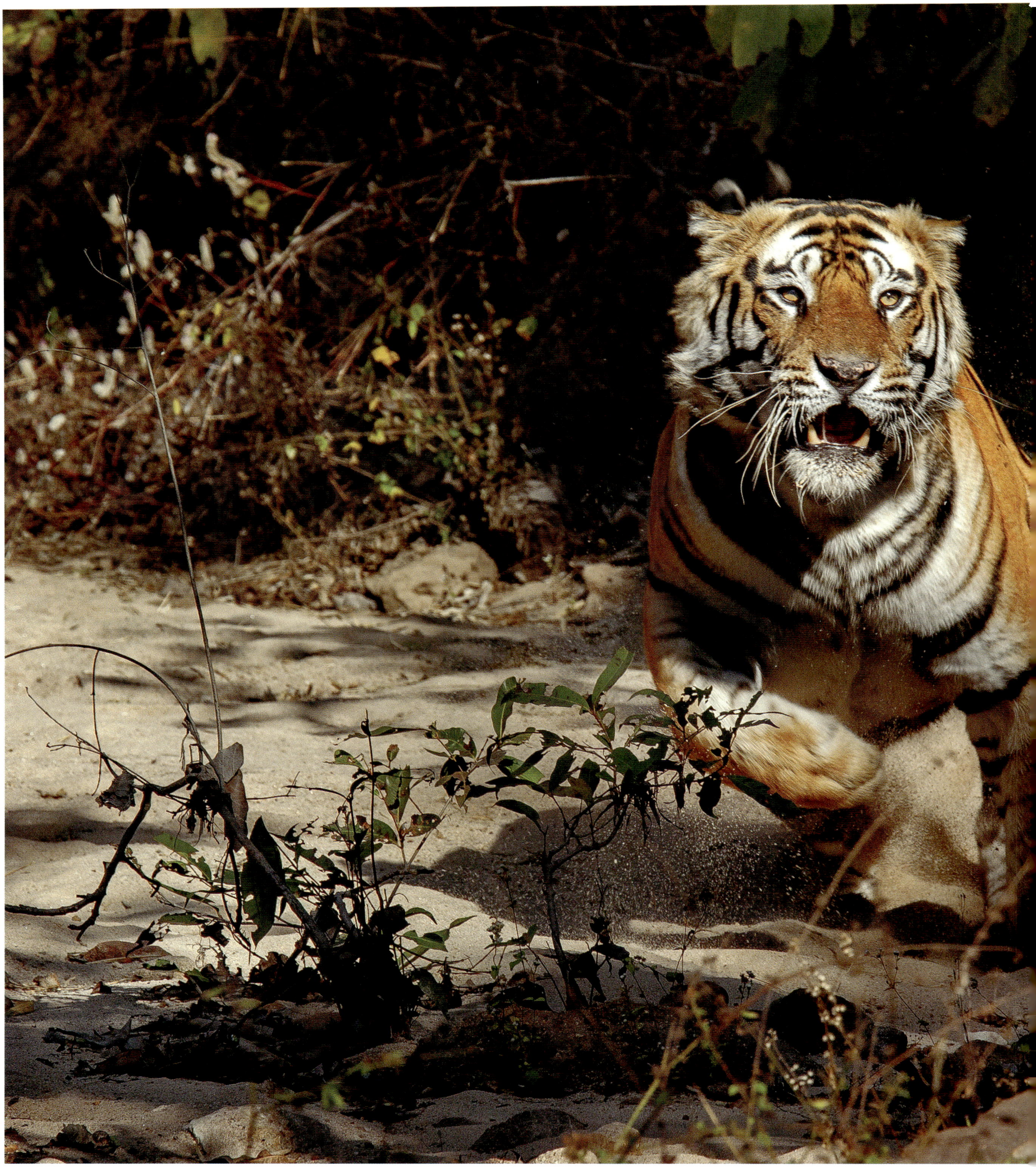

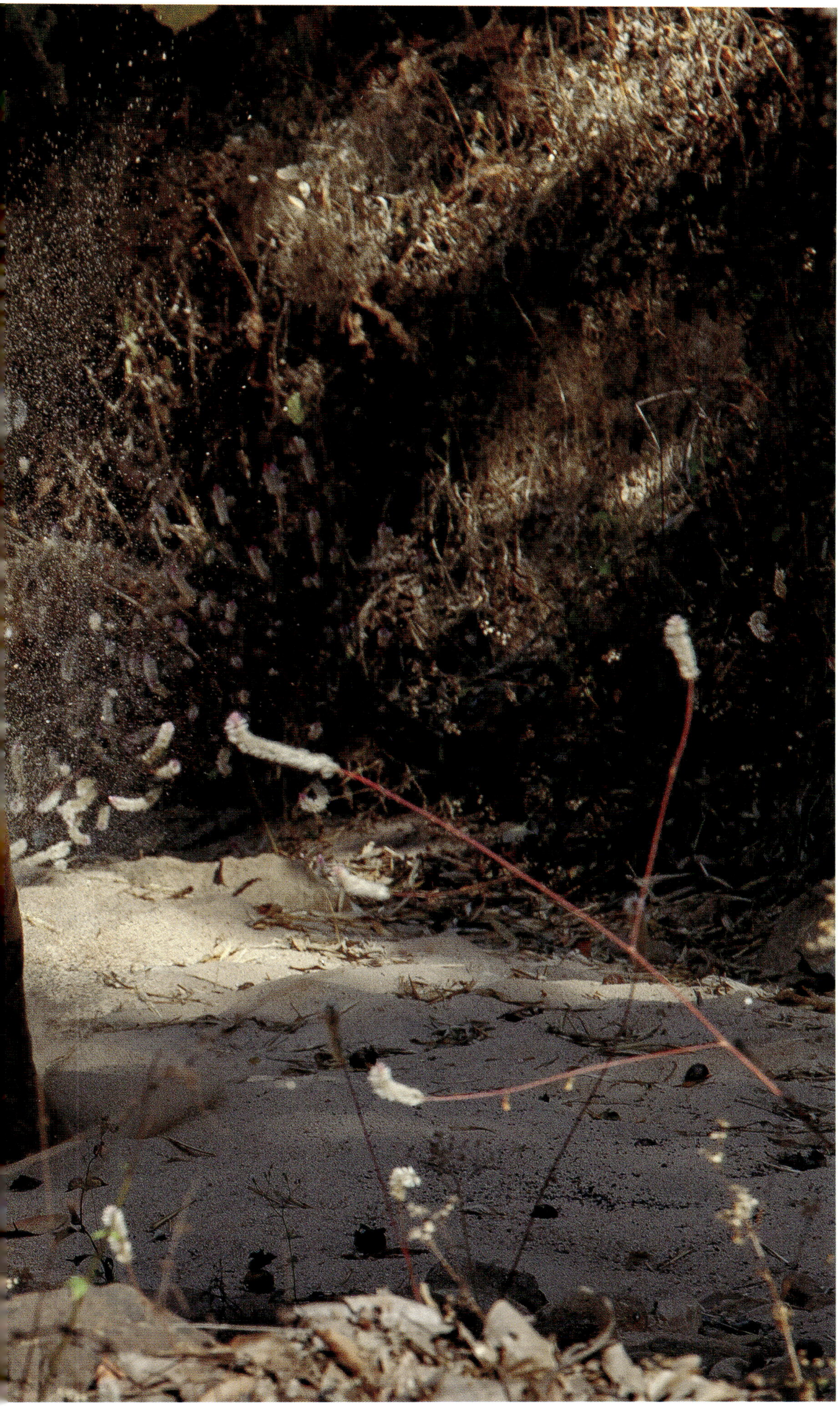

We startled this alpha male tiger from a nap as we drove through Bandhavgarh Tiger Reserve. He charged our vehicle, then stopped just before he reached us.

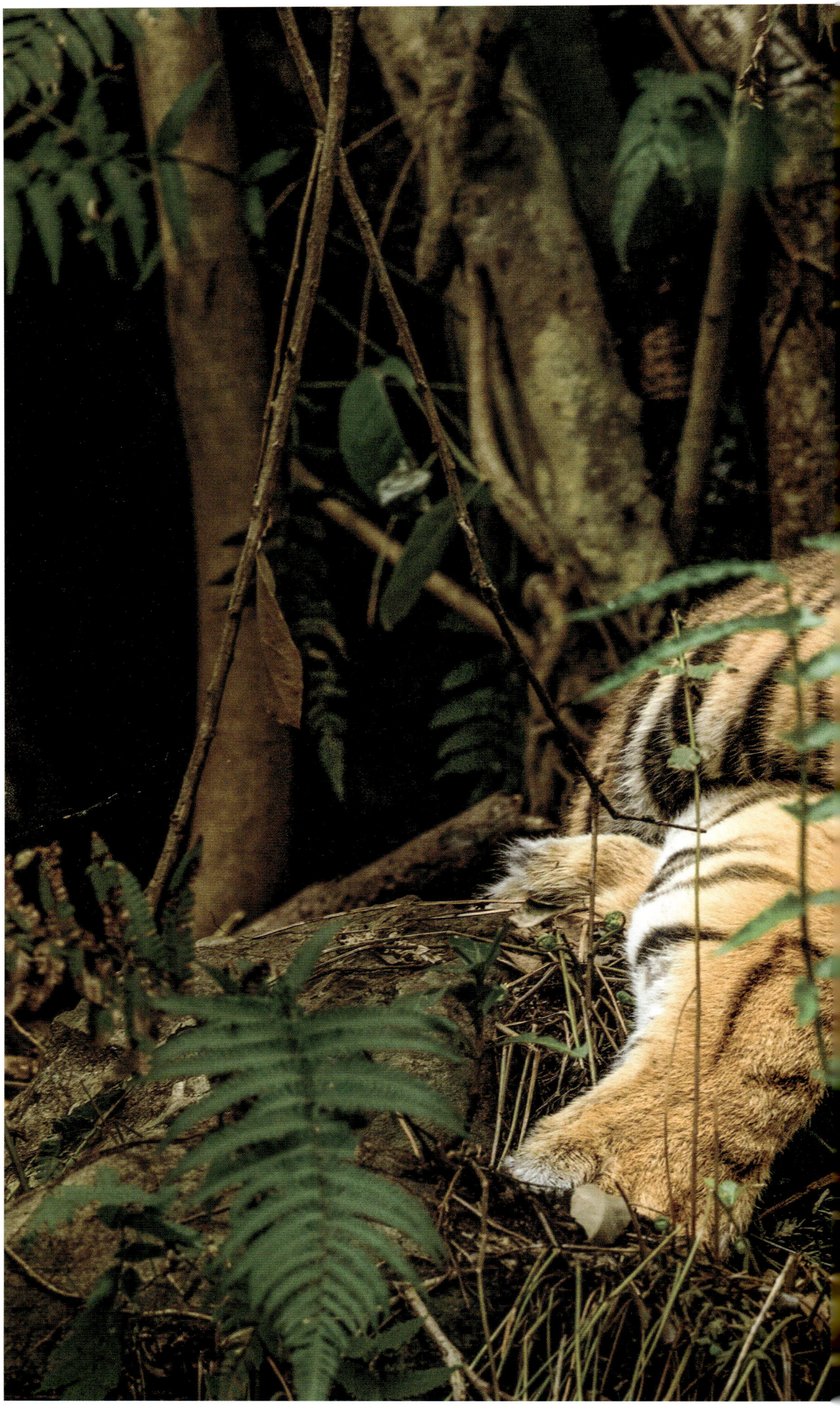

I needed a photo of a mom and cubs for my tiger story. I waited 24 days until, finally, this cub emerged from its den. I snapped off just five frames before it disappeared.

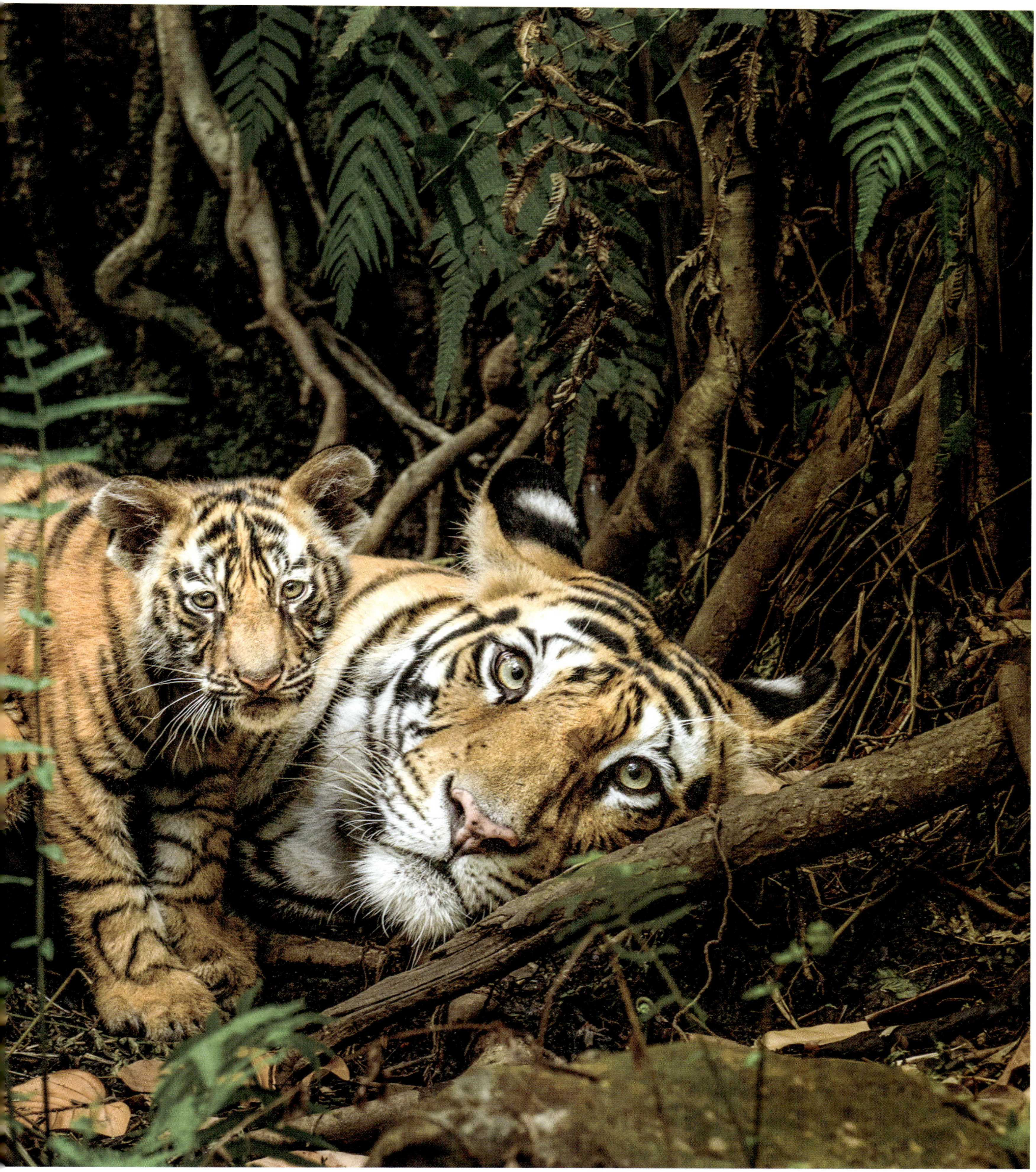

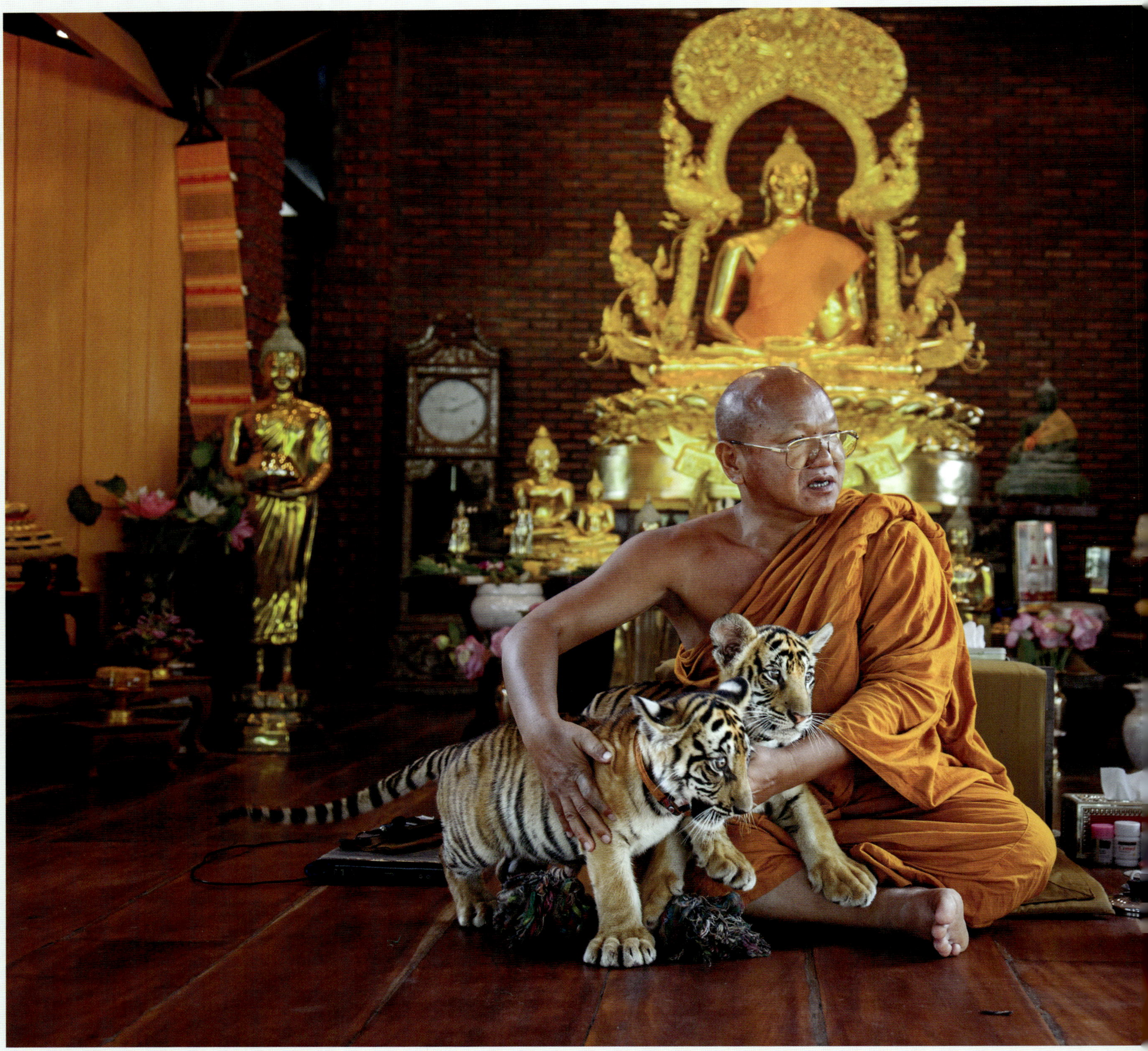

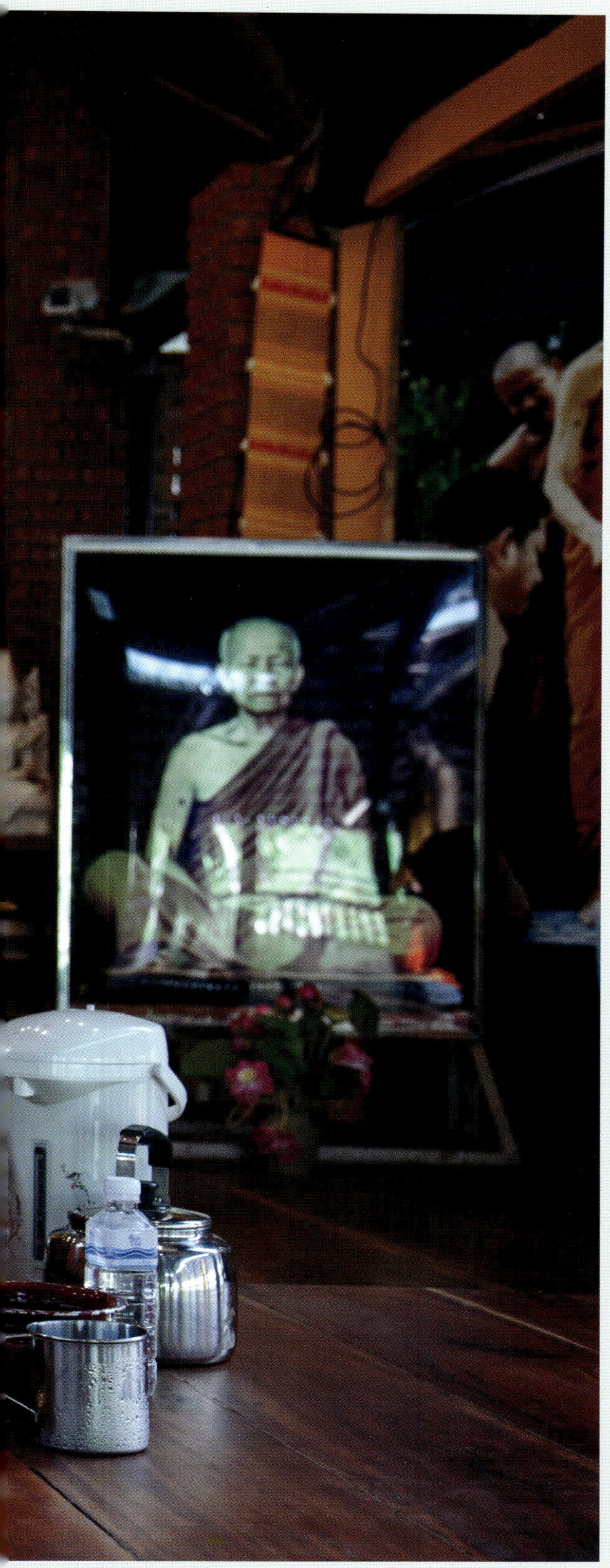

CHAPTER SIX

Thai Tiger Temple

Petting Cubs, Trafficking Tigers

A Buddhist monastery became a popular tourist attraction where visitors handled tiger cubs—and was eventually accused by activists and the Thai government of mishandling and selling them into the illegal wildlife trade.

The abbot of the famed Thai Tiger Temple plays with two tiger cubs at this Buddhist monastery that once doubled as a tiger-petting tourist attraction.

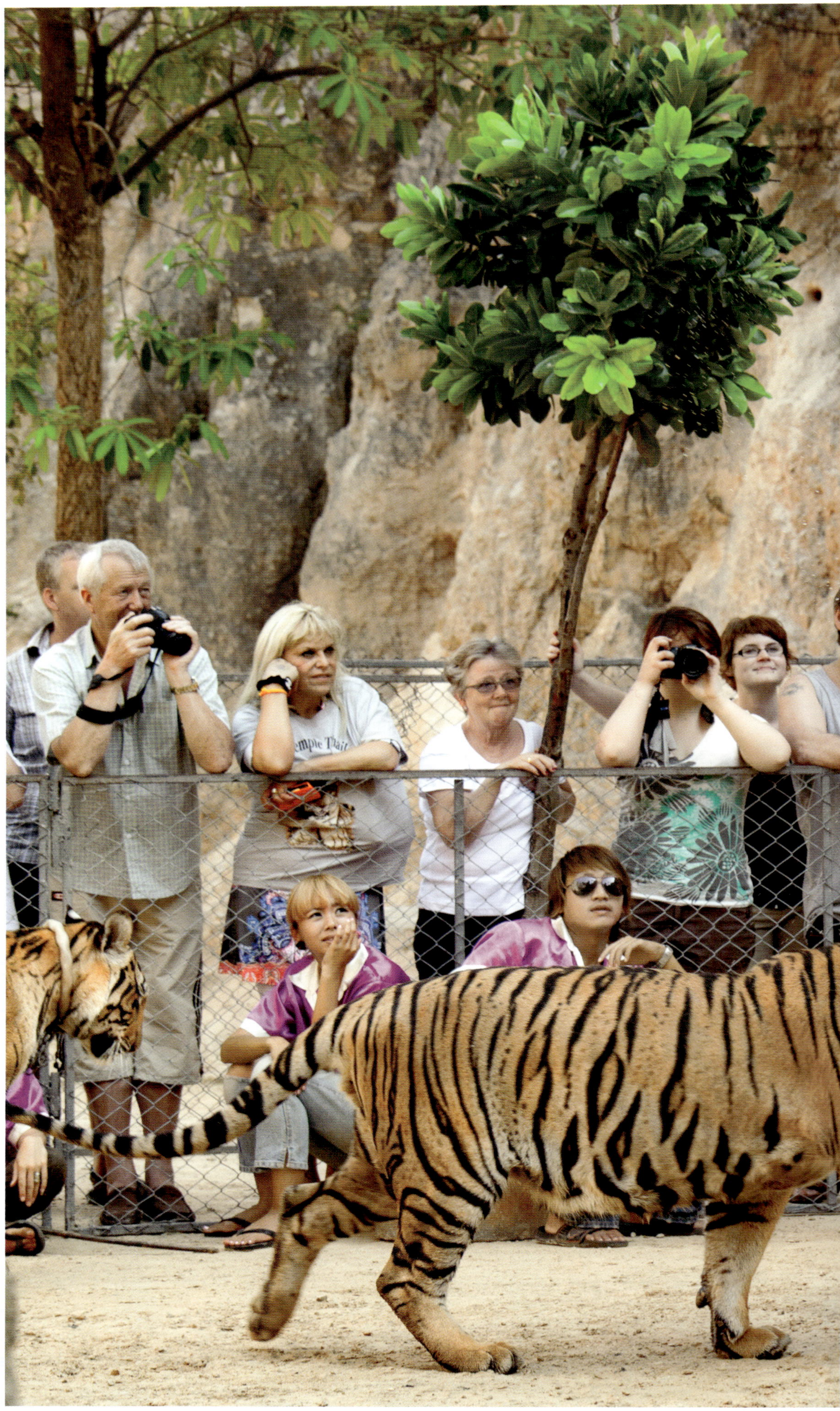

Adult tigers perform for visitors at the Tiger Temple, where tourists walked, fed, and snapped selfies with these cats. The venue generated an estimated three million dollars a year until our investigation and others revealed evidence of tiger trafficking. After government raids, the temple closed to tourists.

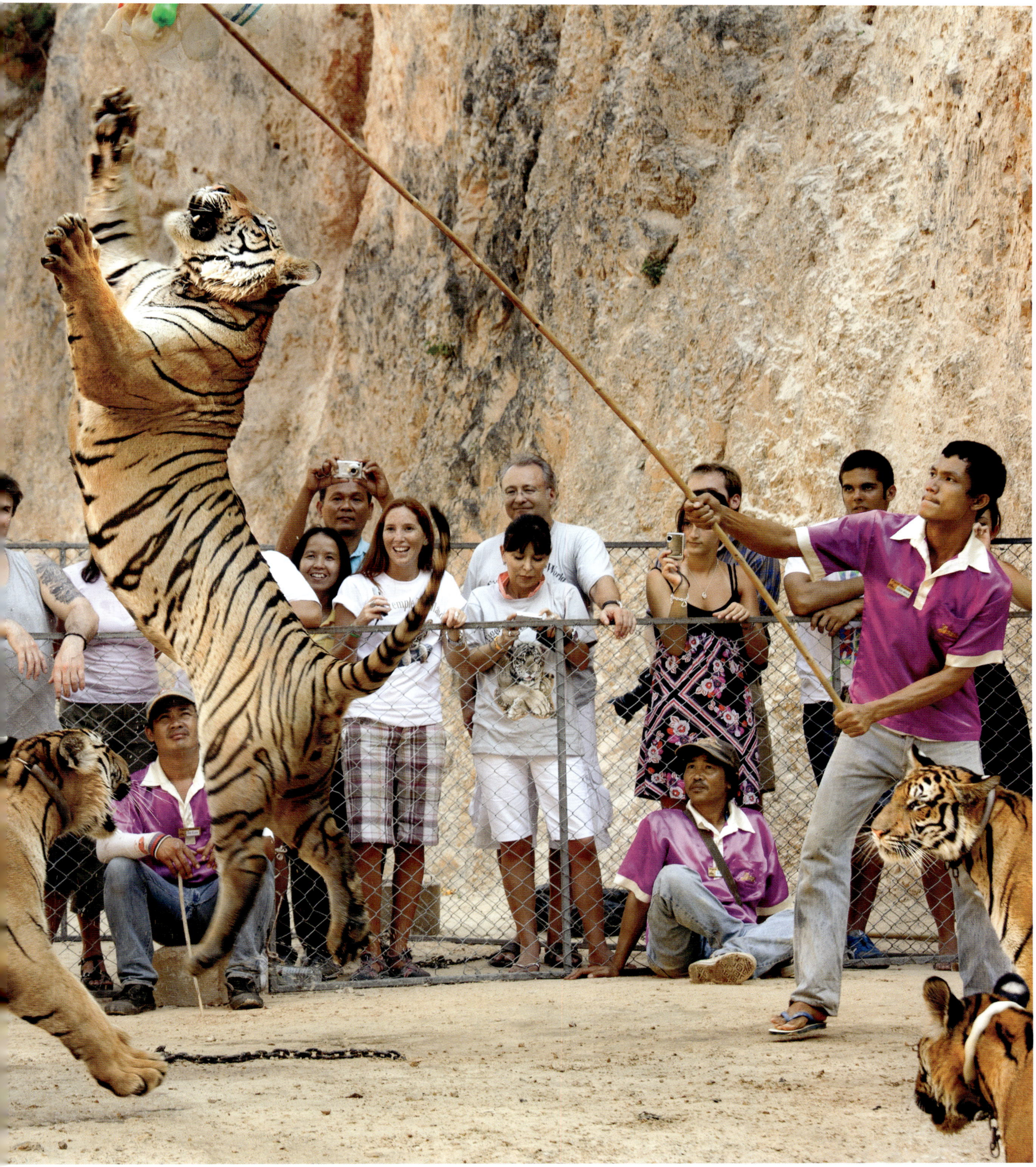

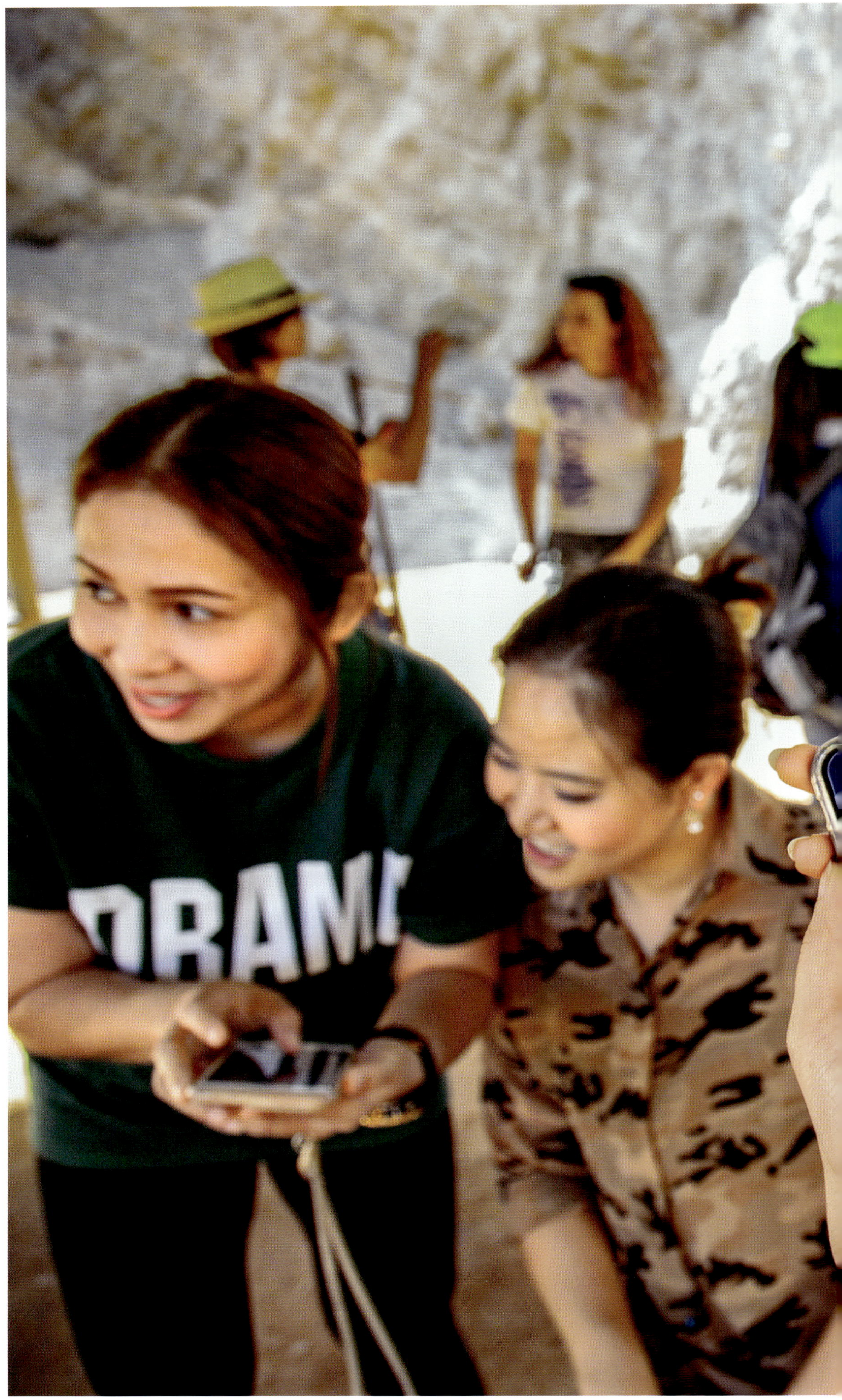

Tourists flocked to the Tiger Temple to take selfies with cubs and grown tigers that they often posted on social media. Our investigation found evidence that tigers at the temple were repeatedly bred and that some were sold, both alive and dead, into the illegal Asian wildlife trade.

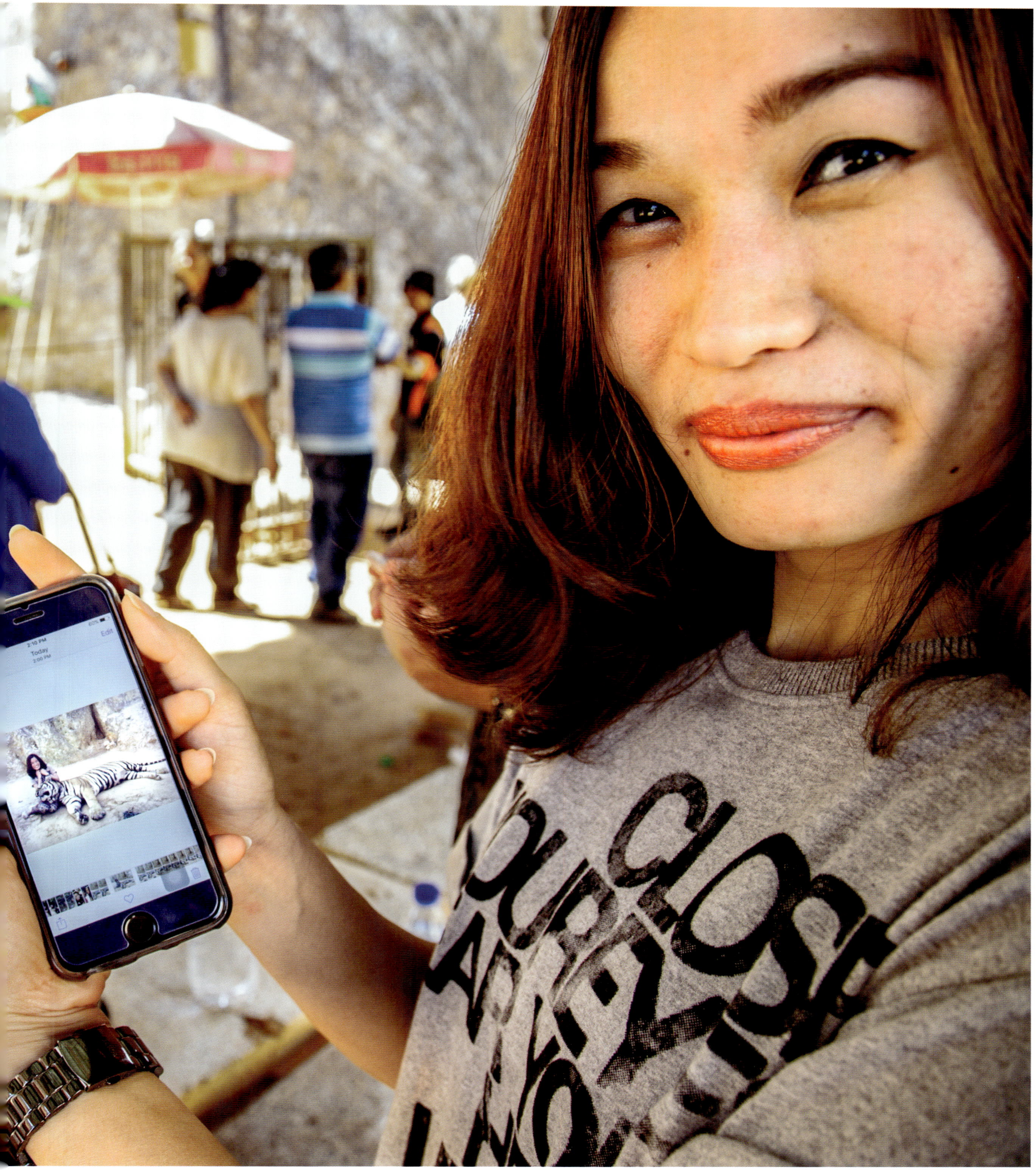

Rumors swirled for years about dark dealings at the Tiger Temple, a Buddhist temple that doubled as a tiger tourism attraction. So when I was in Thailand in 2011 working on my wild tiger story, I asked around. Busloads of tourists came to pet cubs and take selfies with some of the temple's 137 tigers—photos that were plastered all over Facebook and Instagram. Meanwhile, there were whispers of exploitation, abuse, and wildlife trafficking.

I arranged press access and spent four days shooting at the temple. A sign at the front gate claimed it was a sanctuary that saved tigers. I photographed tourists bottle-feeding and cradling cubs, some tiny babies, others big enough to do damage if they bit or clawed. People walked grown tigers on leashes and posed with an adult tiger's head in their lap. I photographed Phra Acham Phoosit (Chan) Kanthitharo, the abbot, with cubs, seated before a huge, opulent altar: a gold-leaf-plated Buddha framed by a pair of elephant tusks. I wasn't allowed behind the scenes.

On the last day, they escorted me to an outdoor area where 15 tigers leaped after plastic bags attached to bamboo poles, which handlers waved above them—oversize cat toys that doubled as long clubs to whack the cats if they got too close. They put me in a flimsy cage that could easily be knocked over by these massive animals. It was stupidly dangerous for the visitors, too. They were "protected" by a waist-high, movable fence. An adult Indochinese tiger can weigh up to 400 pounds and can easily jump atop an Asian elephant.

It was intensely depressing. Infuriating. The whole experience made me feel dirty. It raised serious questions, but this was just a small part of my tiger story. I'd hit my deadline and headed home.

Fast forward. I used some of those photos in a book published in 2014, *Tigers Forever: Saving the World's Most Endangered Big Cat,* and Sharon's text mentioned the temple. After it came out, an Aussie, Sybelle Foxcroft, reached out. She said she could prove that the temple was trafficking tigers into the illegal wildlife trade.

Sybelle had lived at the temple in 2007, doing research for a master's thesis on Indochinese tigers, which gave her access to areas that were off-limits to foreigners. She quickly realized something was wrong. Females were constantly pregnant, cubs and grown tigers appeared and disappeared, but overall numbers remained the same. When she asked questions, she was threatened—and she fled.

We decided to do the story because the effects of trafficking from captive-breeding facilities like the Tiger Temple extend far beyond the welfare of individual cats. They impact wild cats, says Debbie Banks, a tiger expert with the London-based Environmental Investigation Agency. That's because underworld sales stoke demand for tiger pelts (used in luxury home decor), bones (used in expensive tiger bone wine), and body parts (used in traditional Chinese medicine). This deadly demand drives the poaching of the world's last wild tigers.

For the next year, Sharon and Sybelle teamed on a complicated investigation, digging into criminal activity in a foreign country (from the U.S. and Australia, respectively) and looking into wrongdoing at a Buddhist monastery in a deeply religious country.

They obtained vet records proving that four of the temple's original tigers were wild caught—poached, not rescued, as the temple claimed. They found that a female, Nanfa, was smuggled from Laos in 2004 without permits needed to legally transport an endangered species across borders. They unearthed evidence of missing tigers that were listed in a government database: microchipped cats that had a paper trail with Thai authorities.

A former adviser for the temple turned informant. We named him "Charlie" to conceal his identity. If his involvement came out, he'd be in serious danger. Influential people, including an admiral and a former police colonel, were staunch Tiger Temple supporters, and the venue raked in three million dollars a year in tourist income. Who knows how much it made trading in tigers.

Charlie gave us recordings of the abbot discussing the missing tigers. In those tapes, the abbot said that cubs were smuggled alive and adult tigers were killed before they were sold on the Asian black market. Charlie obtained closed-circuit TV footage showing vehicles coming and going after the temple gates were locked for the night and the receipt for a cub sold to a "tiger farm" (aka breeding factory) in Laos.

Former volunteer Sylwia Domaradzka revealed that when monks got rid of tigers, the temple covered its tracks by renaming cats with the missing cats' names so it looked like none of the tigers were ever really "gone." She corroborated what Sybelle had seen: Tigresses were speed-bred and their newborn cubs snatched so they'd quickly produce another litter. Sharon and Sybelle also uncovered evidence of egregious abuse: The cats were beaten and fed poorly. Many were sick and in need of veterinary care.

We waited until the end of the investigation to go to Thailand.

(Continued on page 191)

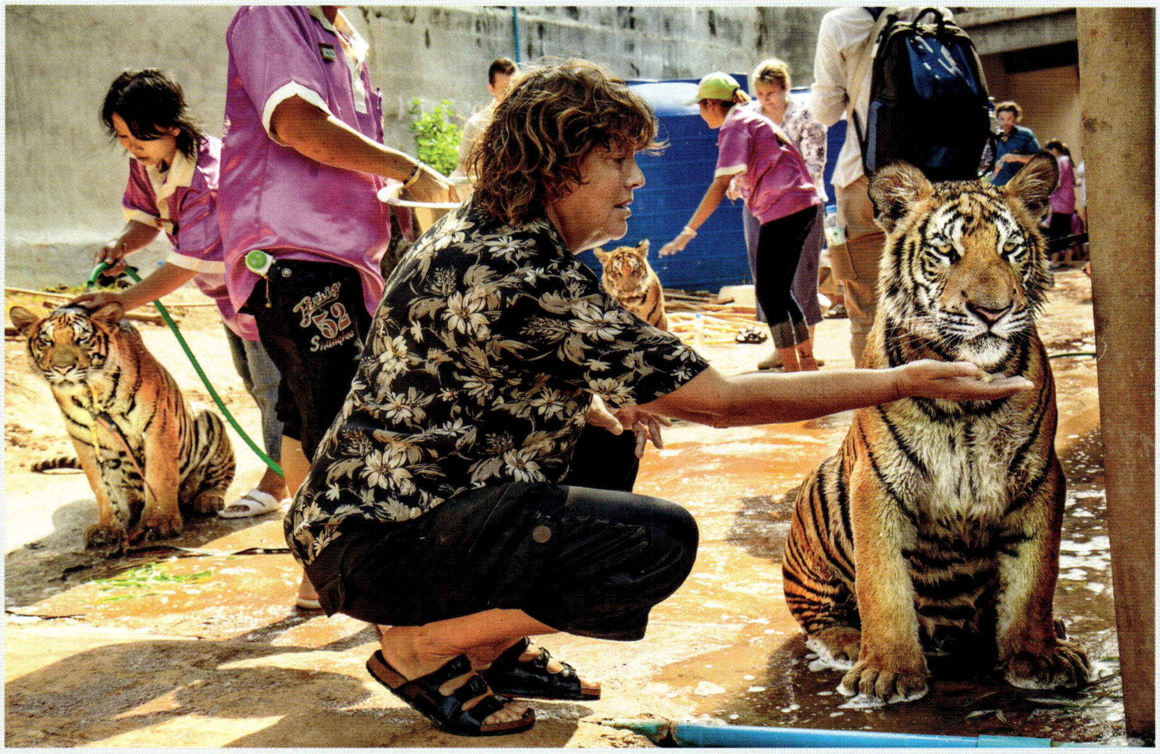

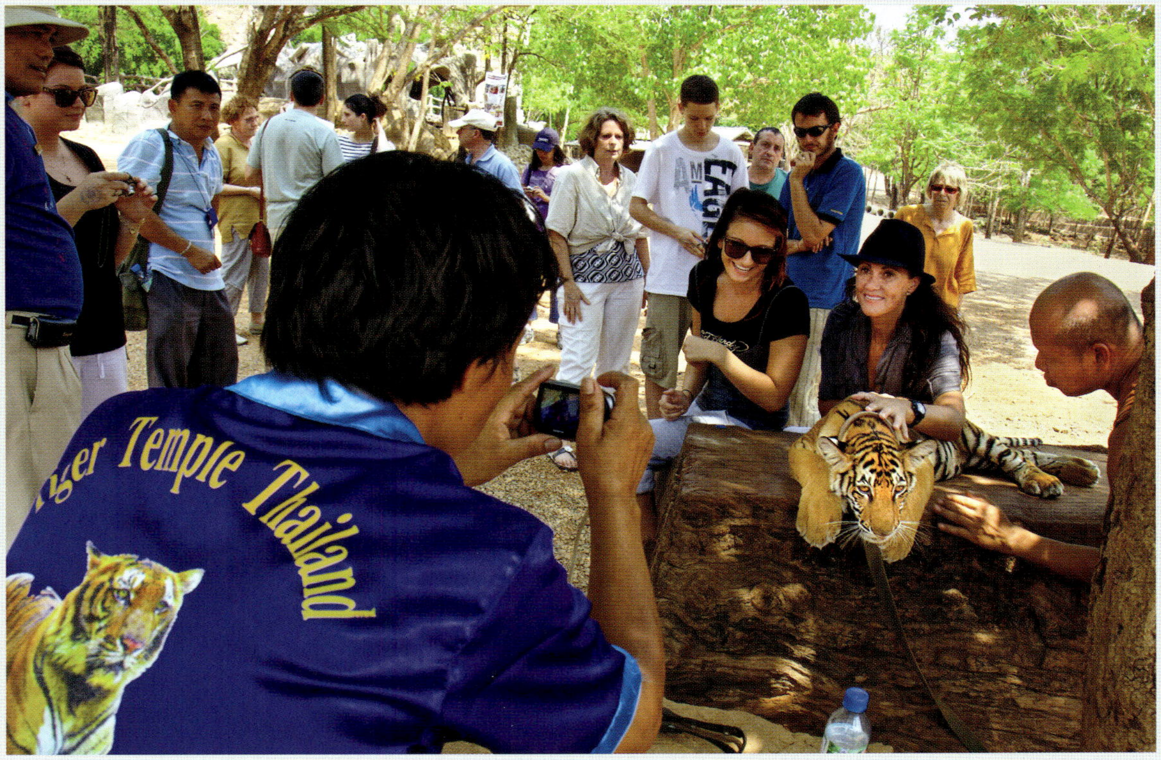

ABOVE AND RIGHT: Tourists paid $200 or more to bottle-feed cubs, bathe a yearling, walk a tiger on a leash, or take selfies with the cats. The Tiger Temple once housed up to 178 cats at a time in abysmal conditions, speed-breeding and trafficking them and their parts to Laos and China.

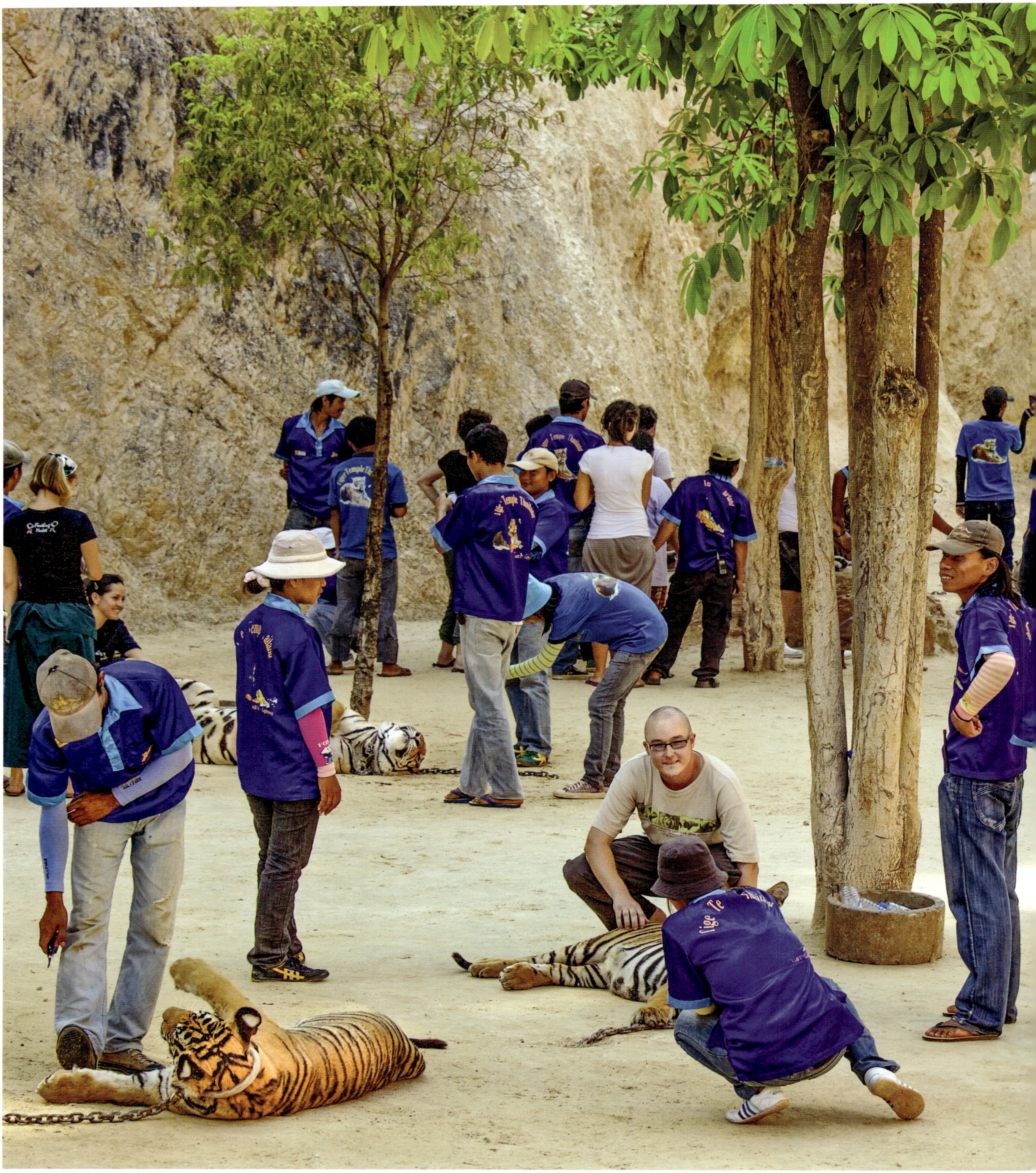

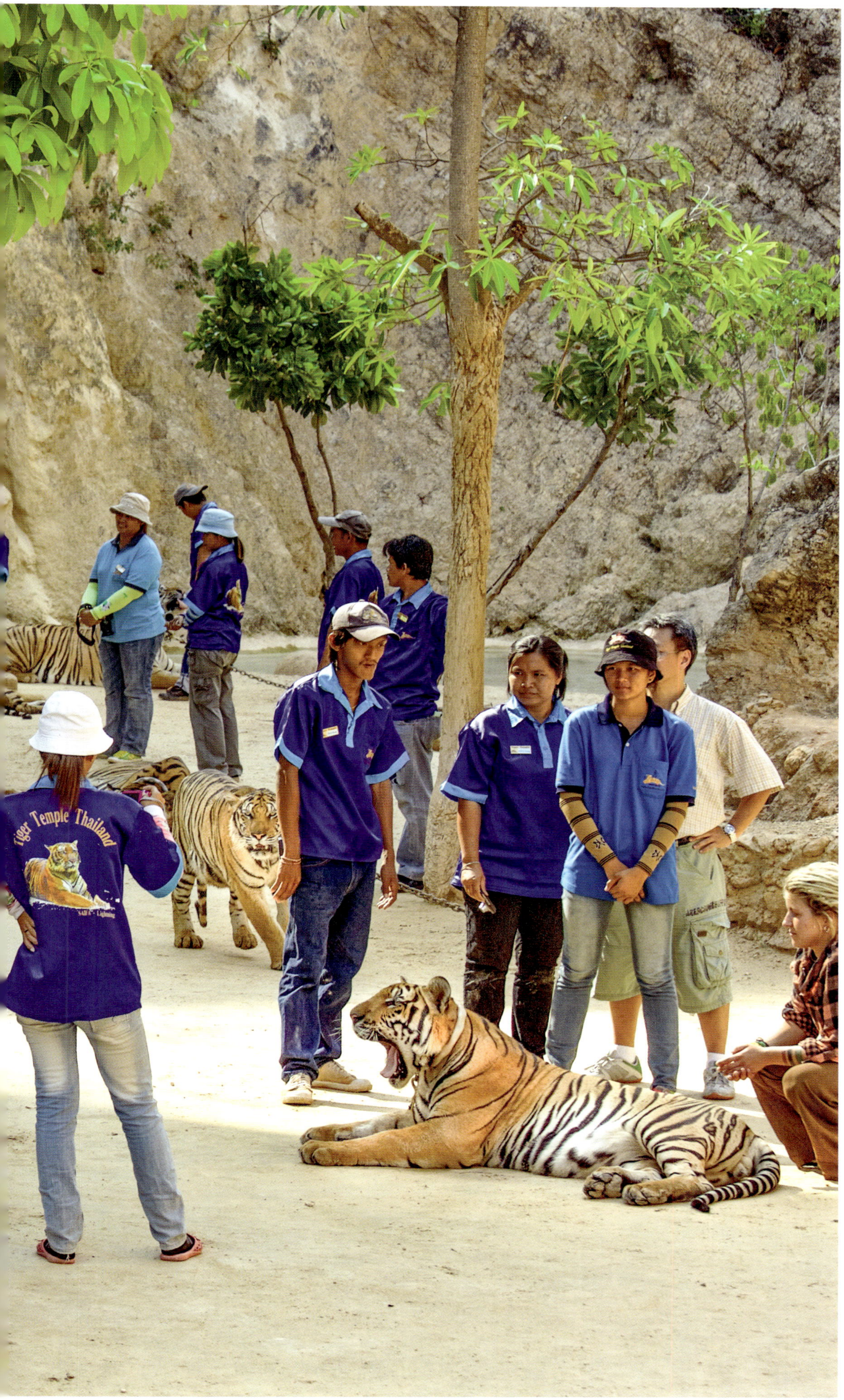

In one area of the temple, tourists could pose with huge adult tigers that were chained so close to the ground that they could barely lift their heads and couldn't stand. The behavior of these tigers led us to believe that the temple drugged them in order to make these encounters possible.

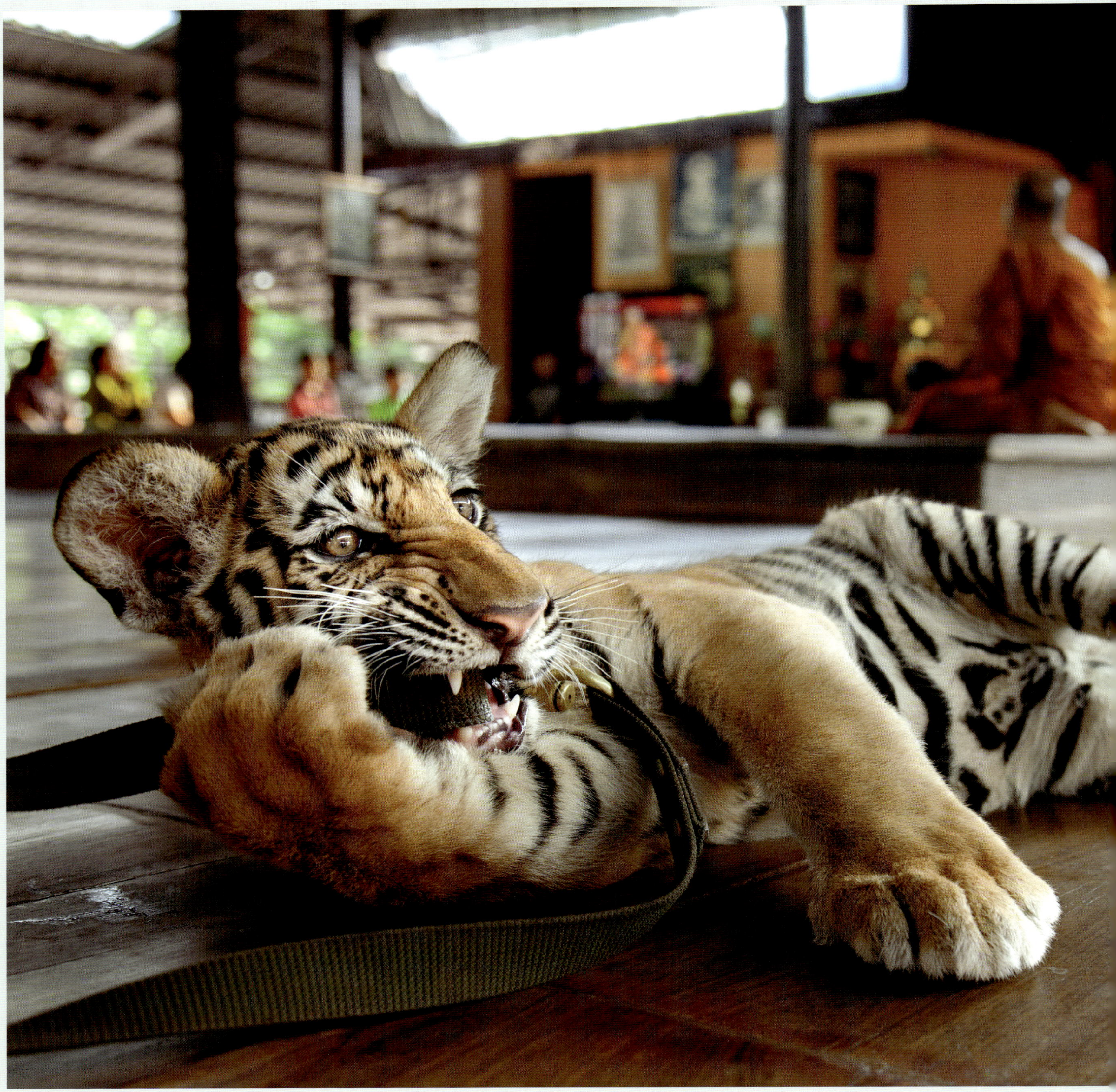

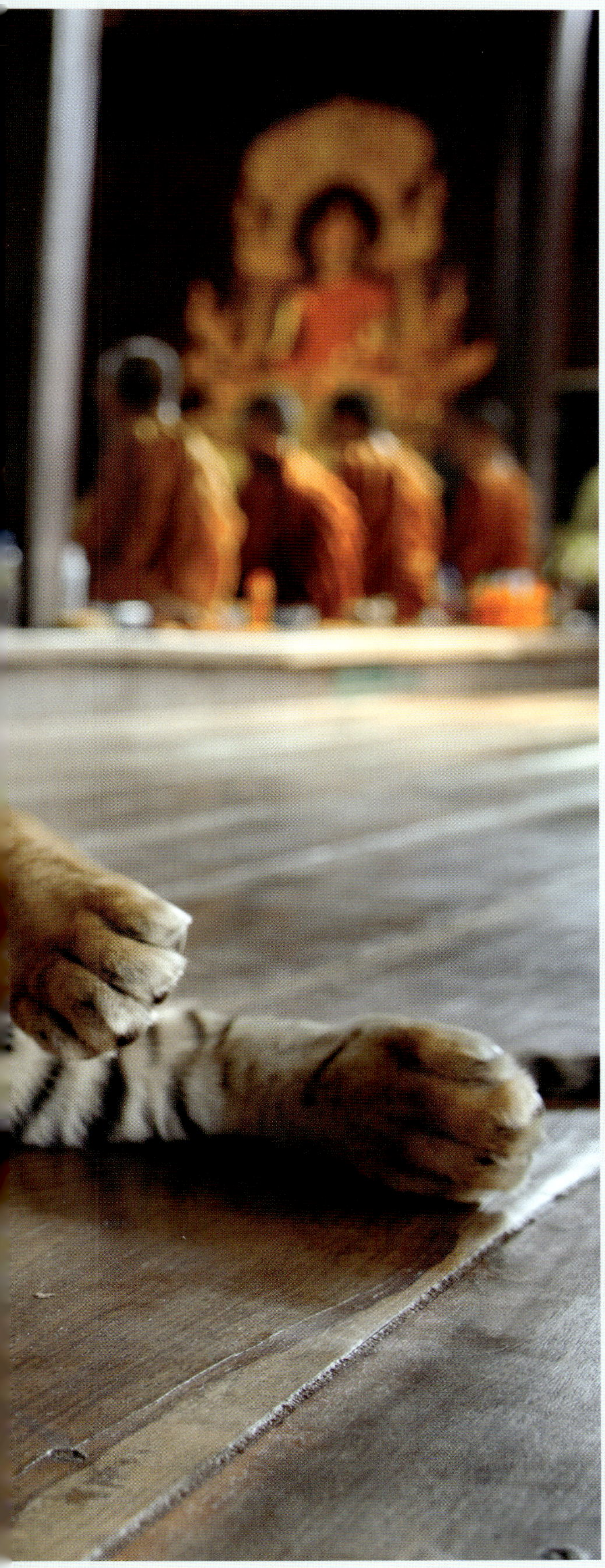

A tiger cub chews its leash in the worship area of the temple, formally known as Wat Pa Luangta Bua Yannasampanno. In an audiotape shared with us by a fellow researcher, the abbot admitted he was aware of the large-scale breeding of tigers at the temple and of activities that investigators later identified as illegal trafficking.

(Continued from page 186)
Sybelle stayed at a hotel; some monks knew her. Sharon and I went to the Tiger Temple as tourists, paying a cash-only entrance "donation" of $139. With other activities, our visit cost $200-plus. These tigers were cash cows.

We carried minimal equipment and tried to fit in as we filmed and photographed. Toward the end of the day, we slipped into an unauthorized area to document rows of jail-like enclosures. Just as we emerged, a monk recognized me from my previous visit. We went with him to the office, which was convenient, because our next step would be to return with press access. Weirdly, they welcomed us. They'd liked the two photos that ran in my tiger story. "You came to save us," someone said. They'd had a lot of bad press. We smiled and said nothing.

The next day, I photographed and shot video as staff showed us around. Everywhere. In one exhibit, a dozen adult tigers walked free; a handler "guarded" us with a stick. Tourists bathed tigers, the cats' skin raw from constant shampoo. The abbot walked a tiger on a leash with a parade of monks and tourists behind. Visitors posed for pictures with huge male tigers chained so close to the ground they could barely lift their heads. They were lethargic, likely drugged, possibly with opium.

The staff even took us into the back, where we discreetly documented small herds of tigers in filthy, cement-floored cages. In the wild, tigers are solitary animals that roam massive territories. This was bleak.

We interviewed the abbot's right-hand man, Jakkrit, who insisted their tigers were safe and happy. Then Sharon pulled out an iPad and showed him a bill of sale he'd signed for a cub trafficked to Laos. Flustered, he ended the interview. We left and met Charlie at a safe house. We blurred his face and altered his voice in the video we produced.

***National Geographic* published** the story and posted our seven-minute video online in January 2016. It kicked up a national and international uproar. Bangkok's largest TV station ran the video on the nightly news. The *Bangkok Post* and dozens of outlets across the globe covered it. Sanctuary Asia, an Indian nonprofit, launched a #TigerTempleTakedown campaign, which flooded the environment minister's office with letters and emails.

Four months later, in an operation involving 500 people, Thai authorities raided the temple. They tranquilized and removed six tigers that had been set loose in the kitchen. They found jewelry made from skins and teeth and seized numerous tiger pelts, some of which were hanging on the wall inside the abbot's residence.

Acting on a tip from a temple volunteer, they made a truly macabre discovery: 40 cubs in an industrial-size freezer and 20 large jars with cubs floating in formaldehyde, all one to seven days old. Ten tigresses were lactating, but no live cubs were found. Officers confiscated 137 tigers and shut the venue down.

We published three follow-up stories. This was a revelatory moment for wildlife tourism, showing the horror behind those selfies.

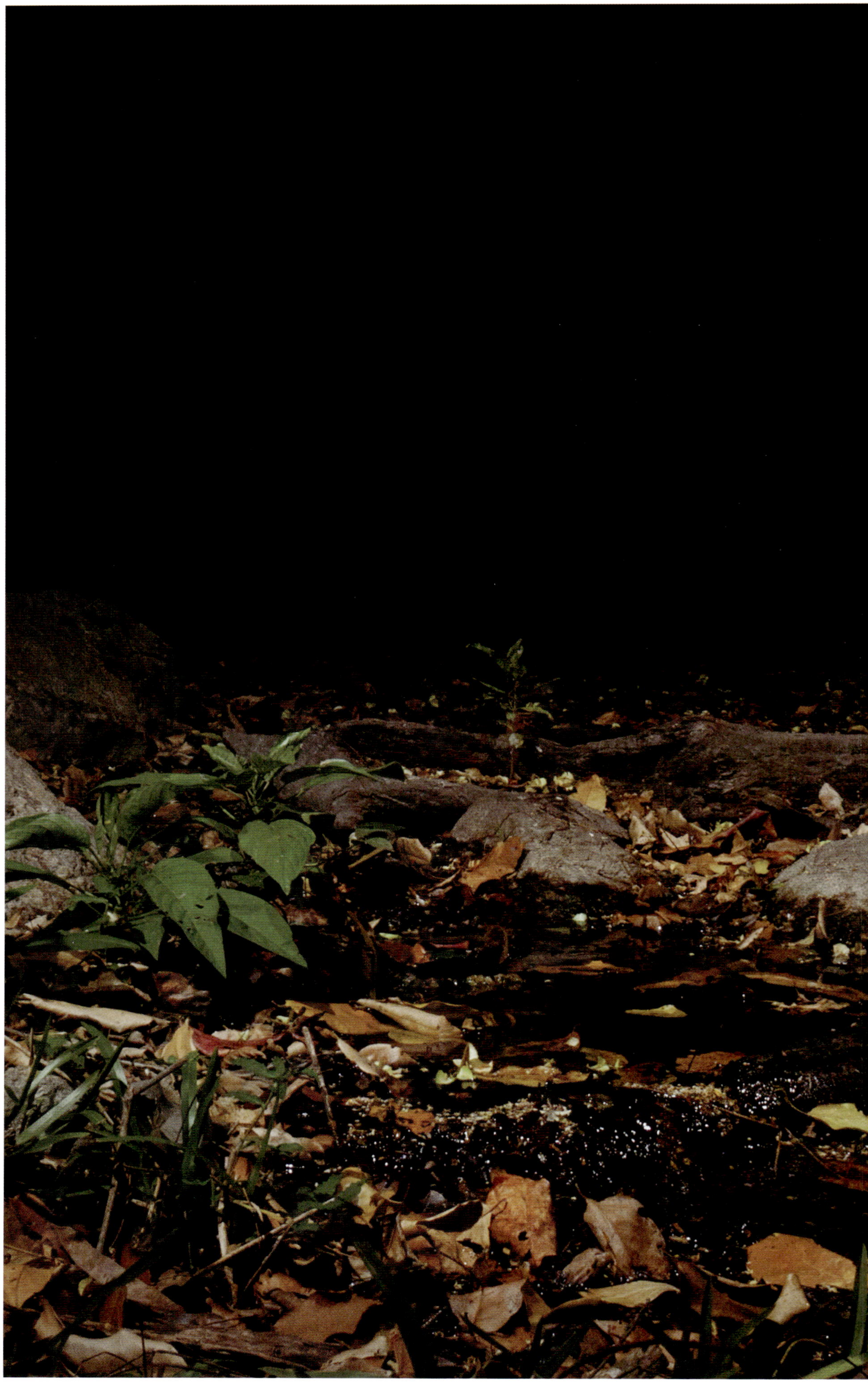

My remote camera captured an image of an Indochinese tiger in Thailand. This species is one of the most endangered of the five tiger subspecies that remain in the wild.

PAGES 194–195: Indochinese tigers once roamed six Southeast Asian countries. Some may remain in Myanmar, but their last best hope for survival is in Thailand's Western Forest Complex, where this photo was made.

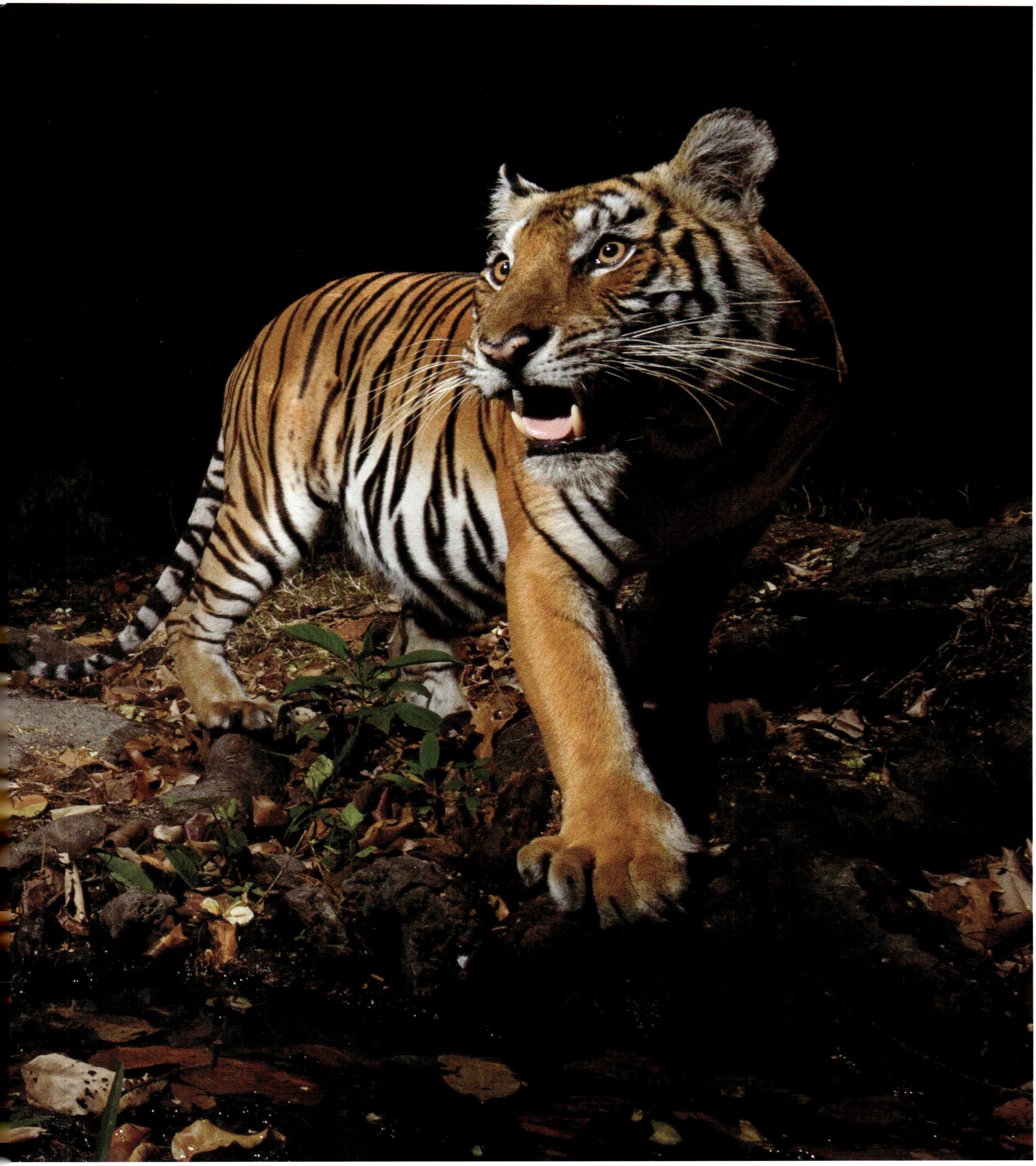

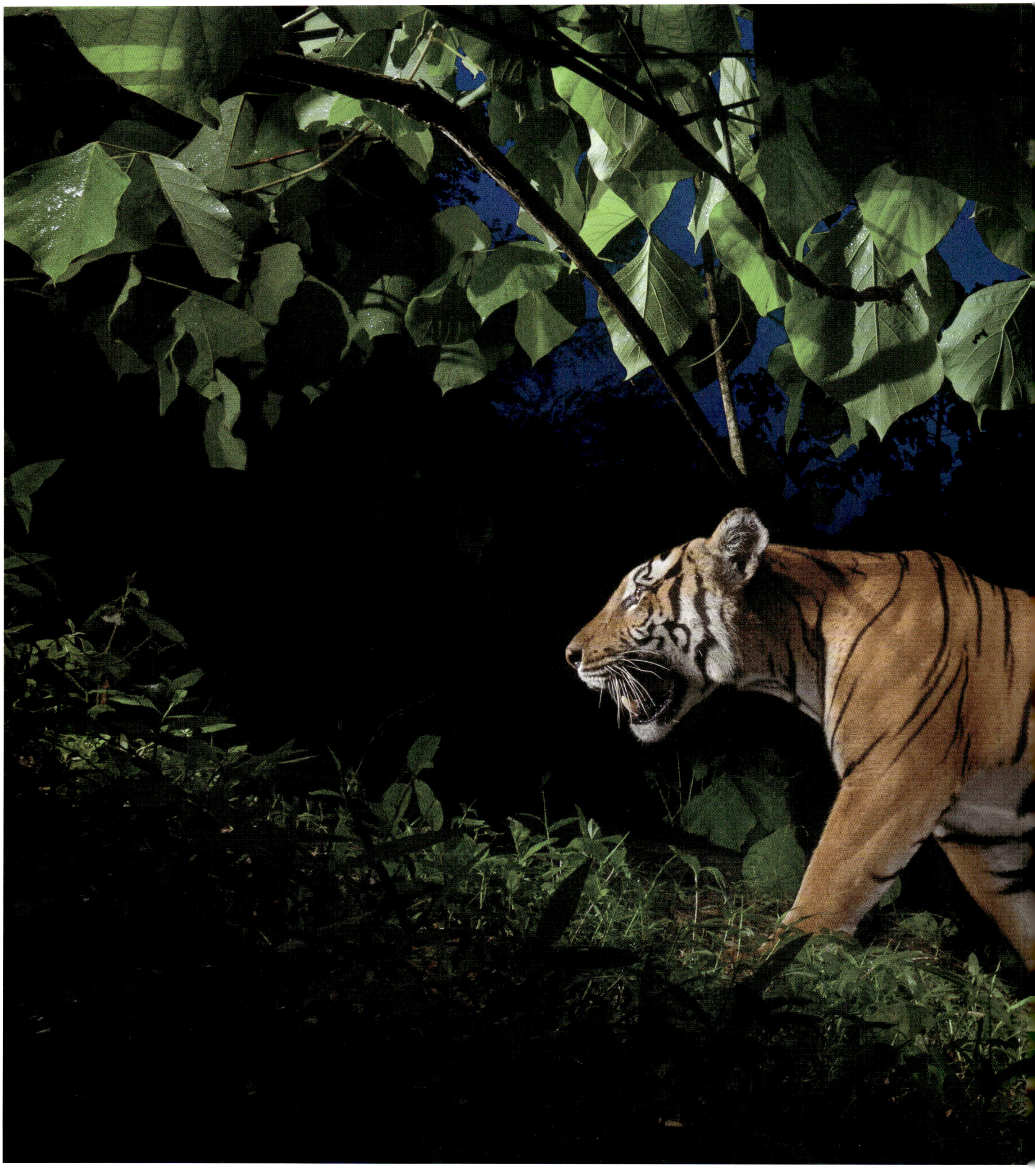

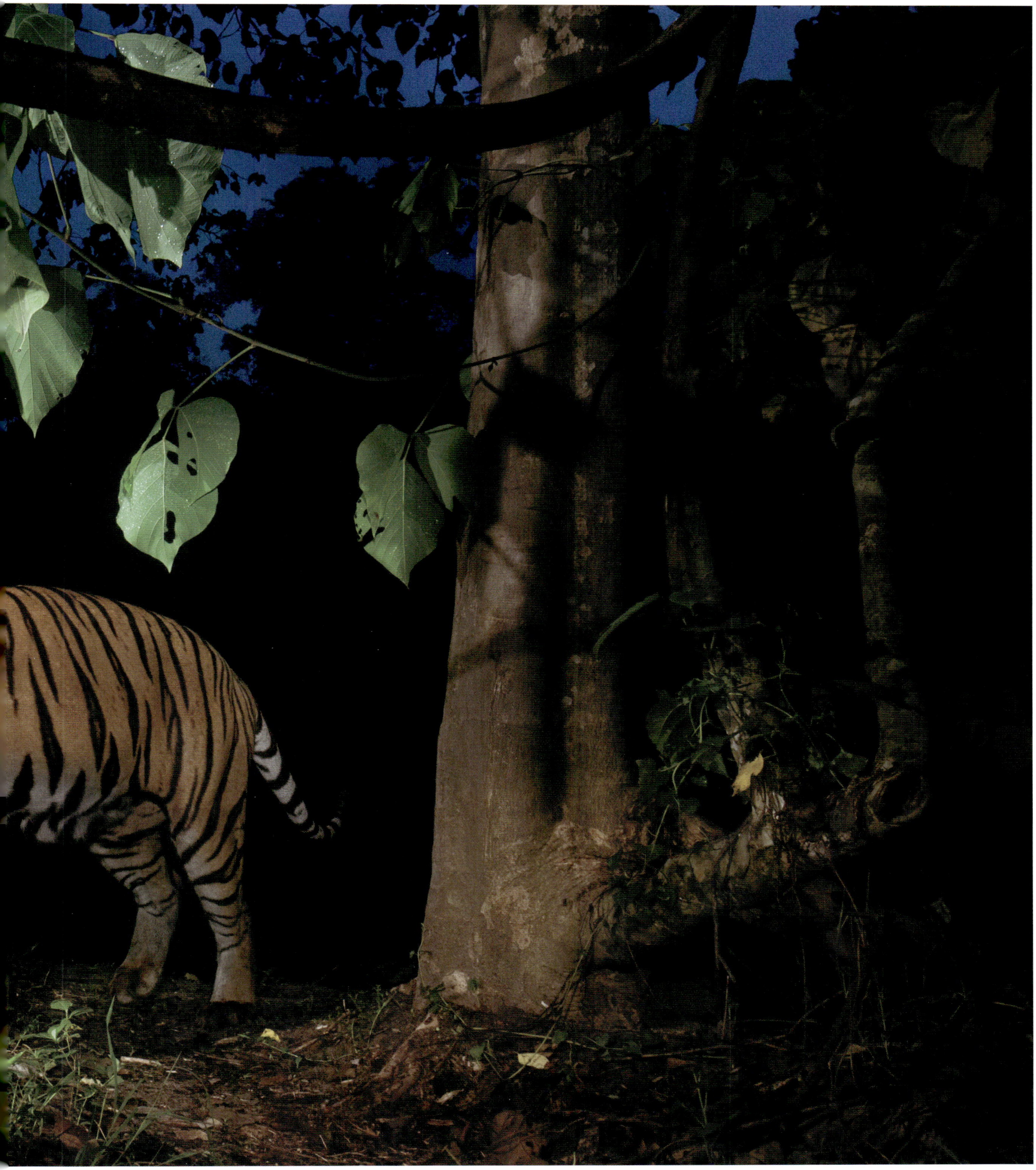

CHAPTER SEVEN

Tigers in the U.S.A.

Investigating Captive Tigers in America

For years, more tigers lived in cages in the U.S. than roamed their Asian homelands. Our investigation helped shut down a lucrative cub-petting industry.

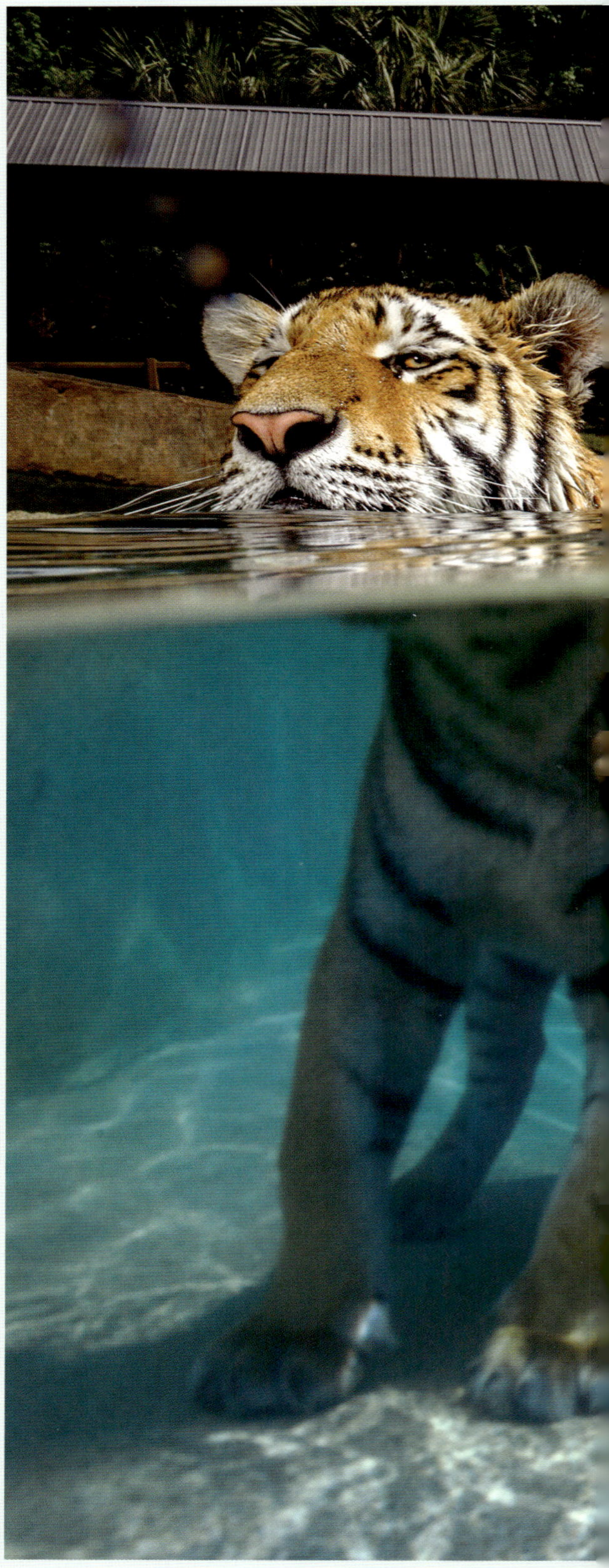

Kevin "Doc" Antle (far right), also known as Bhagavan, poses with his staff in a pool used for tiger shows at South Carolina's Myrtle Beach Safari. In June 2025, Antle was convicted on federal charges including wildlife trafficking and money laundering and sentenced to 12 months in prison. His zoo remains open.

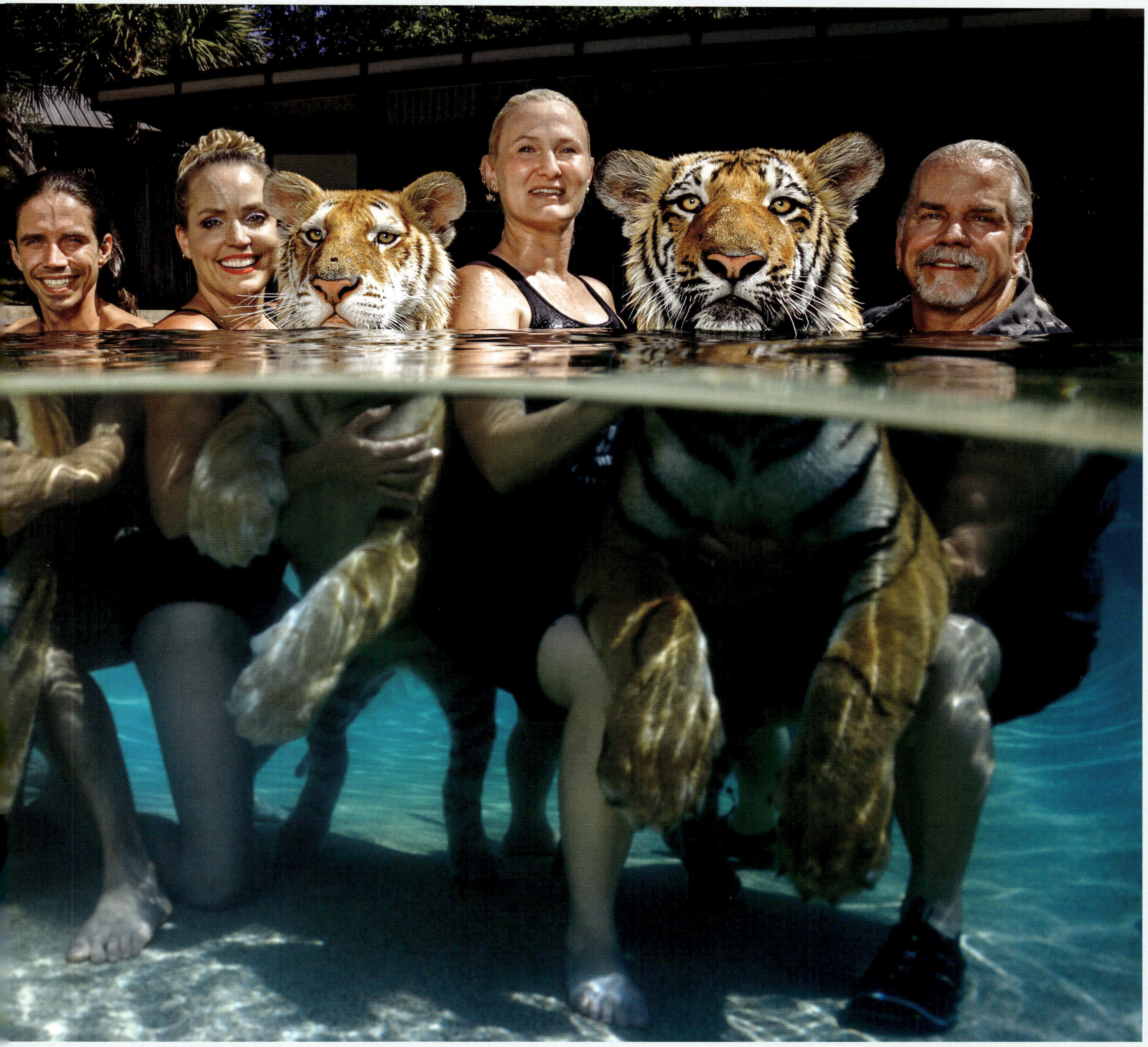

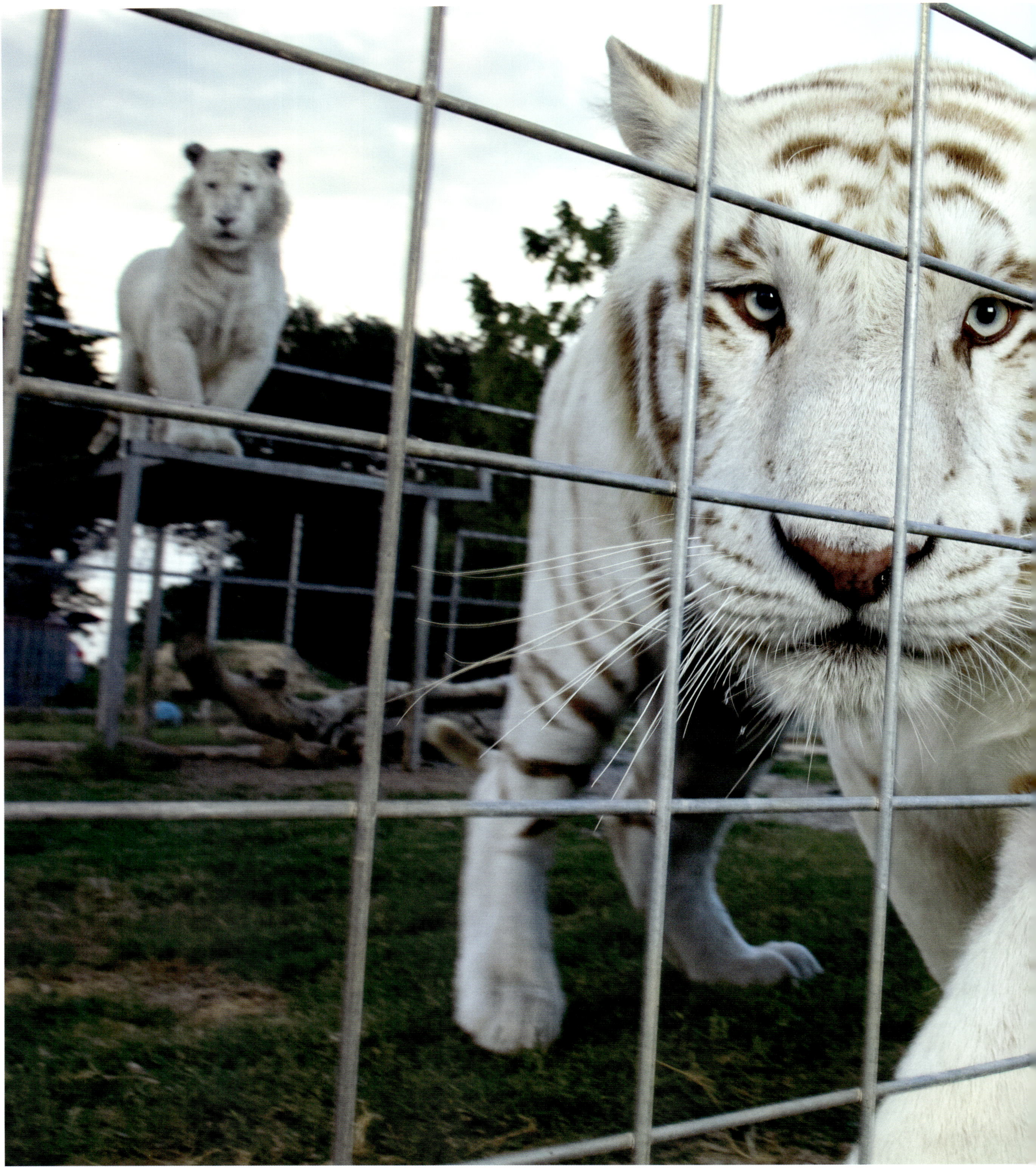

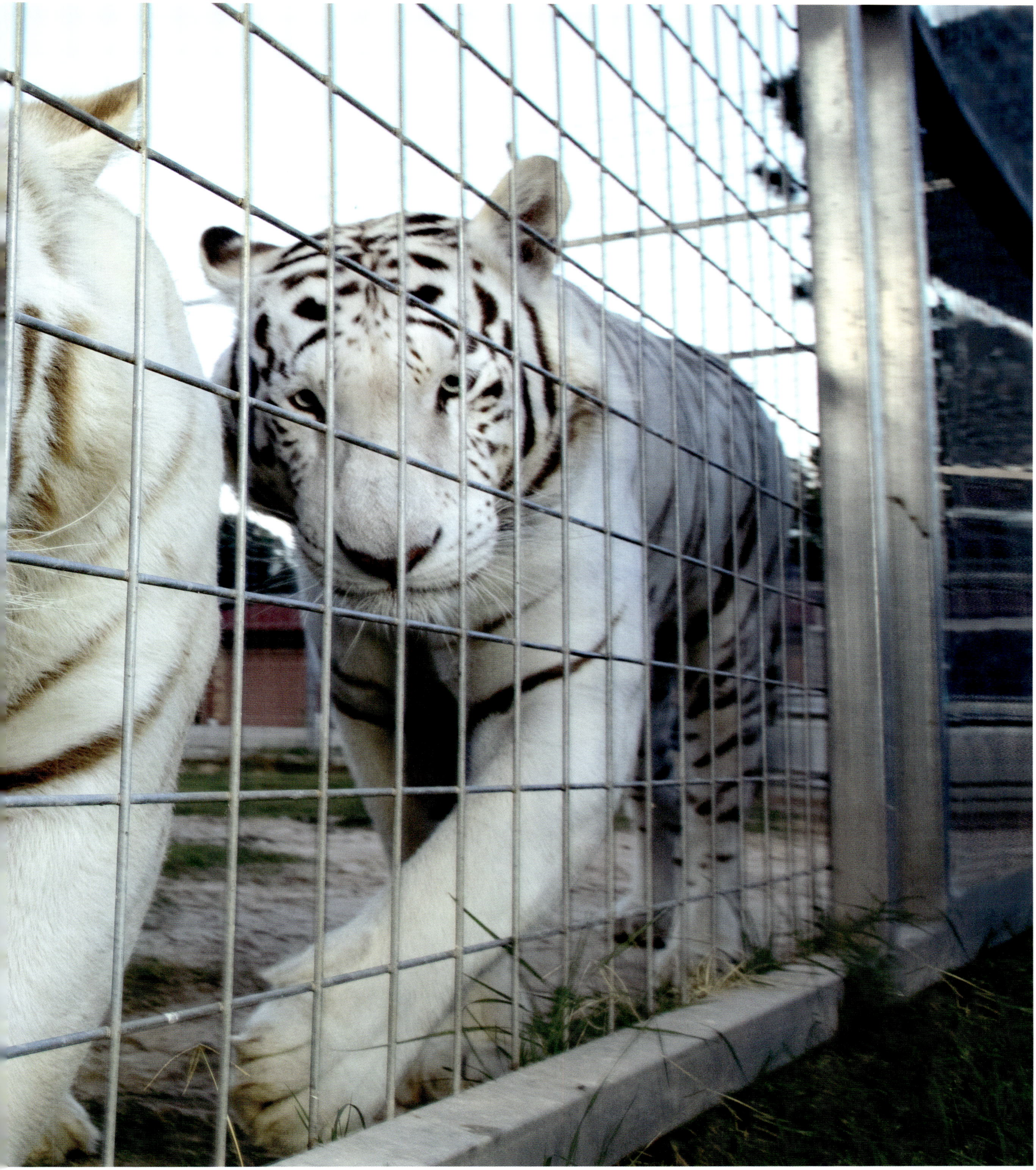

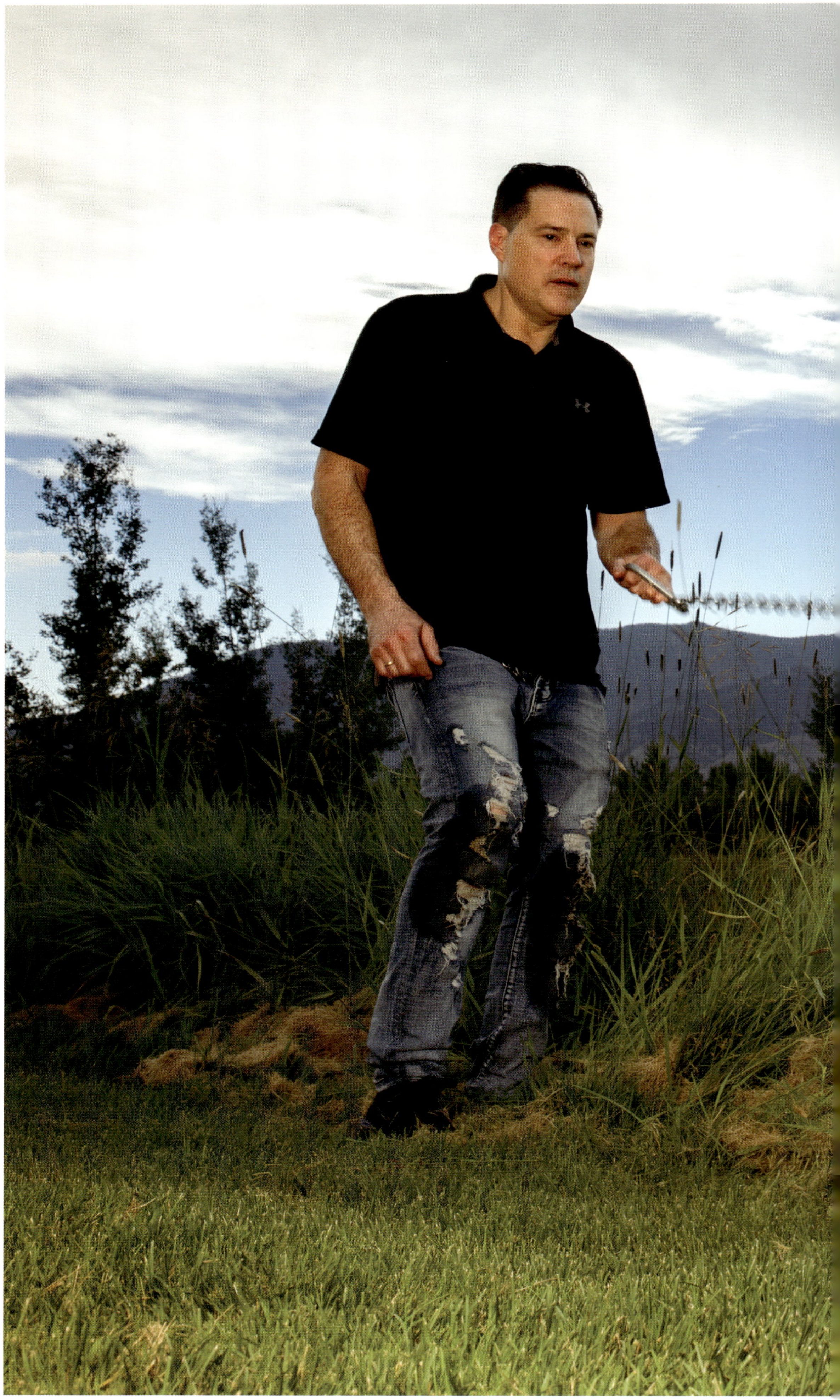

Illusionist Jay Owenhouse walks Selah, one of three tigers he used in his touring magic act, near his home in Bozeman, Montana.

PAGES 198–199: White tigers, made famous by Siegfried and Roy's Las Vegas act, are not a distinct species. Their snowy coat comes from a pair of recessive genes, one from each parent, so they must be inbred with direct relatives to produce white cubs.

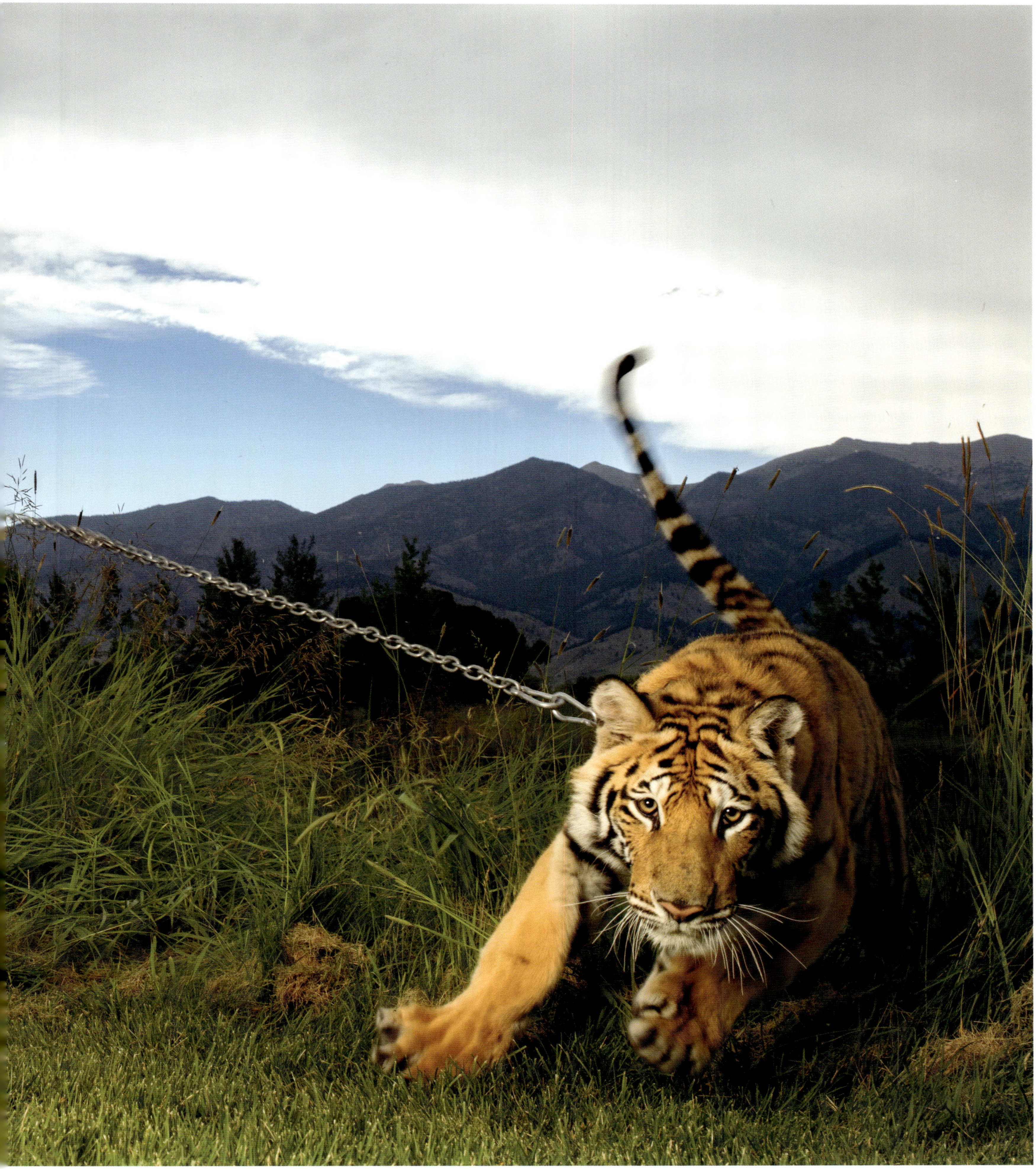

One of our earliest forays

into the U.S. tiger cub–petting industry was in 2018, at a rented ranch-style home in Oklahoma. James Garretson carried Hulk into his living room, where the McCabe family waited on the couch. He placed the squirming cub on nine-year-old Ariel's lap and shoved a baby bottle into the tiger's mouth. "Hold it just like that," Garretson instructed. Ariel nodded.

Everyone beamed, stroking the three-month-old, cocker spaniel–size cub. Hulk clutched the bottle in his massive paws, slurping away, then wandered onto the coffee table. Garretson coaxed him back with another bottle, and five-year-old James got his turn.

Then Hulk leaped off the sofa and went for my son, Nick Ruggia, who was shooting video. The cub swiped and bit—leaving cuts and massive bruises—and grabbed Sharon from behind. He ripped five-inch-long scratches into her thighs and sank his claws in, holding on with surprising strength. Garretson quickly peeled the cat off her legs and joked about his playful little kitty. The family laughed nervously. They'd paid $200 for a half-hour session with Hulk.

Garretson had two more cubs in a back room, and he'd built chain-link enclosures for six full-grown tigers outside. When the landlord found out four months later, he evicted Garretson, and a team from Turpentine Creek Wildlife Refuge drove from Arkansas to rescue the six adult tigers. One, named Diesel, quickly died of a treatable tick-borne infection. Hulk and the two other cubs disappeared, possibly sold.

We later learned that Garretson had a history. He'd been a strip club owner, he had a felony firearms conviction, and seven tigers under his care at another facility had killed a volunteer in 2003. Court documents noted the cats were "extraordinarily hungry" and had reached through flimsy cattle fencing to rip a volunteer's arm off "in a feeding-like frenzy." He was fined $32,560 and ordered to never again exhibit, breed, buy, or sell animals that required federal licensing—including tigers. In 2017, he opened this new operation, licensed in his girlfriend's name.

We'd recently completed our Thai Tiger Temple story when Sharon heard a mind-blowing statistic at a conference. Carson Barylak, a policy expert for the International Fund for Animal Welfare (IFAW), told the audience that between 5,000 and 10,000 tigers were living in cages in the United States, though just 3,900 remained in the wild in Asia. No one, including government officials, knew exact numbers, and there were no federal regulations. IFAW and others had been trying to pass a federal law, the Big Cat Public Safety Act, for years.

That conference launched a two-year investigation. For the first 18 months we remained undercover, posing as tourists. We filmed and photographed cub-petting and tiger exhibits at run-down zoos, a high-end safari show, an Illinois county fair, and a Vegas-style magic act. I drove through 32 states.

Roadside zoo owners were jumpy and aggressively vigilant. The Humane Society, PETA, and other groups had begun to expose animal abuse at cub-petting operations. Some places didn't allow cameras. I shot furtive pictures, sometimes on my phone. Sharon audiotaped animal shows and conversations, and Nick shot video where he could.

Conditions ran the gamut. Many of these places were poorly regulated and understaffed, sometimes employing addicts, criminals, and other desperate people, with full-grown tigers pacing the perimeters of tiny, squalid enclosures. At a few high-end venues, visitors interacted with cats in lush habitats—with the breeder cats housed beyond public view in chain-link cages. Many of the tigers we saw were ill, skeletally skinny or morbidly obese, listless, scarred, or had festering wounds. Some were covered in biting black flies, and some limped from declawing—a process that amputates the first joint of their toes. None of these cats radiated the power or confidence of tigers we'd seen in the wild.

Signs of inbreeding were rampant, especially among the white tigers. Owners bred mother to son, father to daughter, and sister to brother to pass on a rare genetic mutation that sometimes spawns a snowy-coated, blue-eyed cat. Many of them were cross-eyed, with spinal deformities and distorted faces.

Roadside zoos were often in remote rural areas, with few services. Too often, we stayed in run-down motels and lived on gas station food. At night, we showered off dust that reeked of cat urine and slumped on our beds, our heads in our hands.

The American tiger craze was ushered in by the Siegfried and Roy magic show, which dazzled Las Vegas audiences for decades. Commercial sale of "mutt" tigers was legalized by the U.S. Fish and Wildlife Service (USFWS) in 1998. It ruled that only sales of purebred tigers required endangered species permits. Virtually every tiger outside of accredited Association of Zoos & Aquariums facilities was a crossbred mutt, often Siberian-Bengal mixes.

Suddenly tigers were everywhere. They became a sought-after pet.

Tiger and liger cubs were cash cows at Myrtle Beach Safari: Playtime and photographs cost between $339 and $689 per person. At 12 weeks old, cubs become too dangerous to pet, at which point the business kept some as future breeders and sold others. Many simply disappeared.

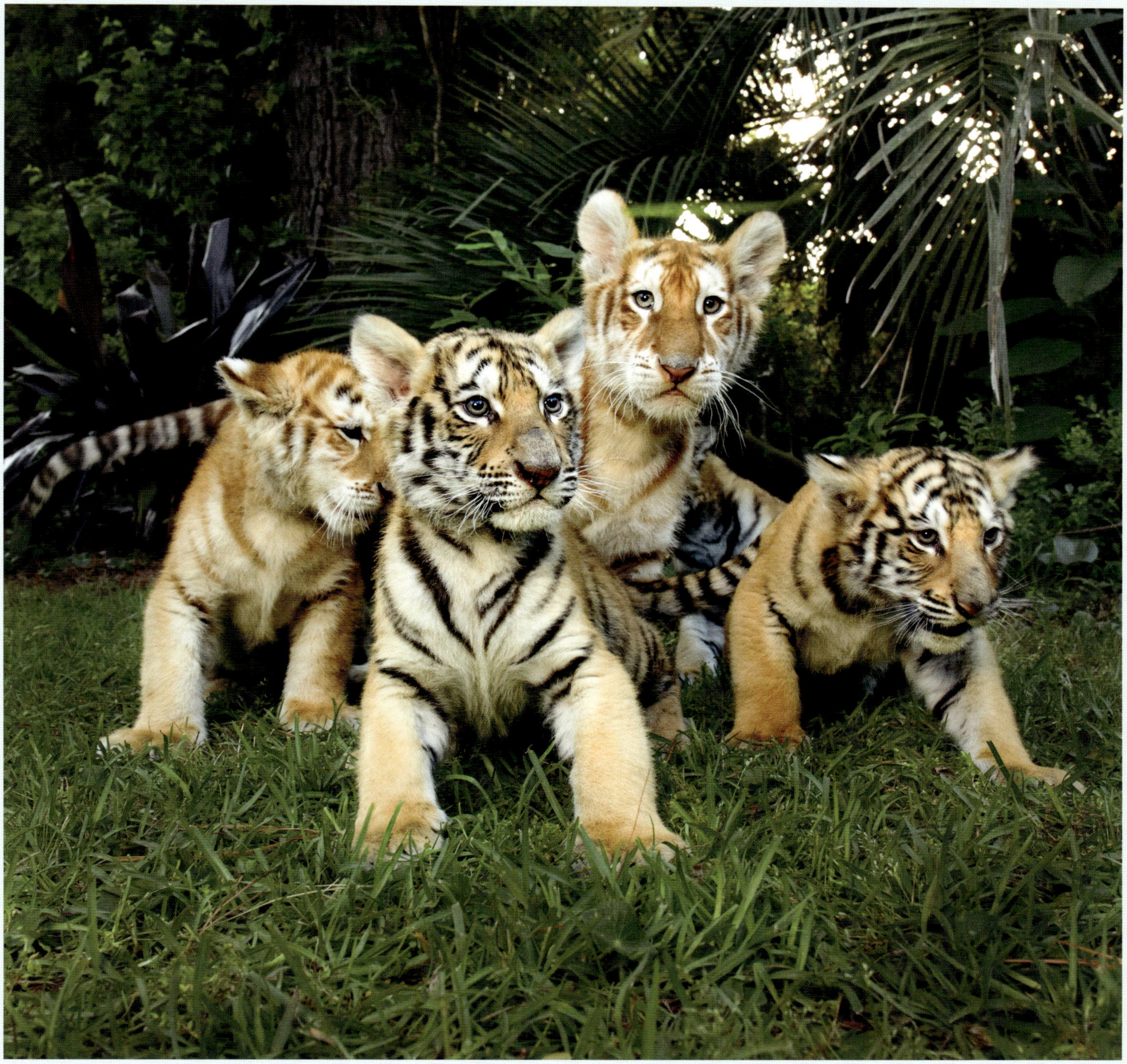

You could buy cubs at auctions or from newspaper ads. They were on late-night TV with celebrities like Jack Hanna. You could play with one at a flea market, in a parking lot, in a mall—or at a roadside zoo.

> **Joe Exotic tried to get his archnemesis murdered. She'd made it her mission to reveal his mistreatment of big cats.**

We knew this was popular in Southeast Asia, but in the U.S.? We'd never done stories on captive cats before our Tiger Temple story. Captive breeding feeds black market trade, and that demand is poaching the last wild tigers toward extinction. The out-of-control situation in the U.S. was providing cover for tiger farming and trafficking in China, Laos, Thailand, and Vietnam.

However, the incentives were different in North America. Many roadside zoos were registered as nonprofits, but these "pseudo-sanctuaries" were lucrative operations. Some raked in millions. Tiger cubs were a gold mine, especially the white ones. I photographed tourists fondling, feeding, and posing with young tigers, some tiny babies and others as big as golden retrievers (and they could do serious damage). The price tag for a quick photo op with a baby tiger ran from $10 to $100; special encounters could cost $700 per person. Myrtle Beach Safari offered a $5,000 swim-with-a-tiger experience. These venues passed themselves off as rescue operations. Staff at these places enthusiastically told guests that they were contributing to tiger conservation.

To maintain a year-round supply of babies, some roadside zoos were commercial megabreeders. At these tiger mills, females were forced to churn out two or three litters a year, compared with one every two years in the wild. Staff wrenched newborns from their mothers at birth. They became photo props once their eyes opened, passed around to hundreds of people for up to 10 hours a day. But cubs have a short shelf life. At three months old, they're too big and dangerous to handle. Some females stayed on as future breeders; others were sold off. Many simply disappeared.

That USFWS ruling was later overturned, making interstate transport of any tiger illegal without permits, but the cub-petting business couldn't function without smuggling cubs nationwide.

We documented the underbelly of this industry, but we also needed to show the best life these magnificent cats could have in captivity. At Pat Craig's Wild Animal Sanctuary in Colorado, I photographed Clay, Daniel, and Enzo—three of 39 tigers from the infamous Joe Exotic's former zoo. They lazed on a steamy July day; two cooled off in a stock tank. The third settled beside them on the ground, its head resting against the tank, and I had the story opener: three contented tigers, and they were framed by a fortuitous rainbow.

We also photographed at Scott and Tanya Smith's Turpentine Creek Wildlife Refuge in Arkansas and joined the Smiths on the largest tiger rescue in U.S. history, one that moved 78 tigers from Serenity Springs Wildlife Center to sanctuaries across the country. They'd bought the place from Nick Sculac, one of America's biggest tiger breeders, who long ago had lost his exhibitor's license for egregious violations. When they arrived, Scott said, the place was "a true horror show": shin-deep excrement, dead birds floating in water tanks, and tigers with tumors and missing limbs.

Joe Exotic, the "Tiger King," craved the spotlight. He sported a signature horseshoe mustache, a dyed-blond mullet, and tattoos depicting tigers, past lovers, and bullet holes dripping blood. He ran low-budget, Vegas-style tiger shows at malls and at his former zoo, the Greater Wynnewood Exotic Animal Park in Oklahoma, which was also a cub-petting and large-scale breeding operation. At one point, Joe claimed to own 227 tigers, some of them monstrous, distorted tiger-lion hybrids that don't exist in nature.

Joe Exotic tried to get his archnemesis, Carole Baskin, murdered. She'd made it her mission to reveal his mistreatment of big cats, which triggered escalating, ever more threatening public attacks. In a YouTube video, he screamed menacingly at the camera, "Carole Baskin better never, ever, ever see me face-to-face, ever, ever, ever again," and then he blew the head off an effigy of her with a revolver.

In 2019, Joe was convicted on two counts of murder-for-hire and 17 charges for trafficking tigers and executing those that outlived their usefulness. Sharon corresponded with him in jail and attended his trial. He's now serving a 21-year prison sentence.

We gained special permission from USFWS agents to photograph key trial evidence at their Oklahoma office. Officers pulled five tiger skulls from a freezer, each bagged, tagged, and marred by a bullet hole: Joe executed them to make space for better breeders. I photographed them displayed on the floor, careful to avoid the agents' faces and badges: They were undercover.

After the trial, we decided that I should finally approach people openly. My son, Nick, and I started at Safari's Sanctuary in Oklahoma. It wasn't open to the public anymore. With a history of animal attacks and violations, officials revoked its exhibitor's license. I wanted to photograph one of the cats there. Langley was a nearly 300-pound ti-liger (created by breeding a tiger with a lion-tiger hybrid) that owner Lori Ensign-Scroggins kept in her house on the sanctuary property. Somehow, she overlooked the danger of living with a massive, powerful apex predator.

Joe Exotic bred Frankensteinian hybrids like Langley to draw tourists. He dumped most of his cats when they grew too big to pet.

Langley's face was contorted. He was born nearly blind and walked with a rolling limp. I photographed Lori hand-feeding him chicken in her kitchen, watching TV with him on her bed, and walking him on a leash with her two dogs running free beside them.

On one of those walks I squatted down to shoot pictures at eye level. Langley charged, wrenching the leash from Lori's hand, and jumped me repeatedly, my arm in his mouth. A sanctuary staffer ran after him—though she stayed on a call, phone to her ear, the entire time. Somehow, I ducked away, my left kidney bruised, but otherwise okay. Nick got it all on video, which shows how he was raised as the son of journalists. He later told me he filmed it because he assumed that lawyers and insurance agents would need evidence.

The fact is, tigers are apex predators that cannot be domesticated. In 2012, seven nonprofits petitioned the U.S. Department of Agriculture (which has oversight of zoos) to ban hands-on contact with big cats at licensed animal attractions, an issue the agency never addressed. Other groups lobbied for a law to protect the public, using a long list of tragic stories as evidence: People had lost limbs, been bitten, mauled, and killed—even eaten—at these venues.

The victims included workers, tourists who handled tigers, people who kept them as pets, and first responders who encountered them in the line of duty without training in how to deal with dangerous wild animals. Eight years prior, exotic pet owner Terry Thompson set 56 of his animals loose in Ohio—including 38 big cats—then shot himself. Sheriff Matthew Lutz and his deputies had to kill most of the animals in what became known as the Zanesville Massacre.

Former FBI bomb expert Edmund Kelso, Jr., expressed his concern in a letter to Congress. "If I had a choice of working on an IED [an improvised explosive device, or homemade bomb] or responding to an emergency involving a dangerous big cat, I would definitely select the former. This should underscore how dangerous it truly is for these animals to be left in untrained hands in backyards and roadside zoos."

A lot can go wrong with tigers in captivity: a hurricane, tornado, a fire, or a momentary lapse in attention. But it was a long shot that I'd be in the right place at the right time to capture this part of the story, so I resorted to photo illustration. Tim Harrison, a retired police officer with expertise in exotic animals, trained first responders at Texas A&M's Disaster City, which has a collapsed building, a train wreck, and other simulated disaster areas. It was the perfect place to make that picture.

I bought a life-size tiger stuffy and put it on the plane in a duffel bag. (I wondered if they picked it up on x-ray at the airport.) We placed the toy inside a smashed car, and I made my favorite photo of the entire story. Unfortunately, we misplaced the location release, the university had second thoughts about publication, and the photo didn't run in the magazine, but we can show it here.

Next on the list was Kevin Antle's Myrtle Beach Safari in South Carolina. He went by Bhagavan (a name he took after classes with a yoga guru), called himself Doc (though he wasn't a doctor), and was purportedly among the nation's largest breeders of big cats and hybrids. His tigresses—constantly pregnant—were live-monitored so staff could grab cubs as soon as they were born, still wet with afterbirth.

Doc ran an operation that many called a cult. Much of the work was done by 18-year-old female interns. They had to commit to two years of work, adopt Sanskrit names, and sign nondisclosure agreements. They received no pay and no time off for holidays, weddings, or funerals, and they worked only for room and board and one meal per day. Many of them had blond hair and ended up with buxom, plastic surgery–enhanced bodies.

A few former interns told us they felt they were essentially prisoners and that their intimate relationships with Antle influenced their treatment. He had three so-called wives, though he was married to none of them. These women left broke and broken after years there. They had chronic nightmares, they said, which seemed like Stockholm syndrome—and Doc refused to give them a reference or even say they'd ever worked there.

Since I'm best known for my wildlife photos, Doc welcomed me. I think my arrival was a throwback to the glory days when he trained animals for TV and documentaries, and cinematographers regularly shot wildlife footage on his manicured, junglelike property. That was before ethics changed and wild animals had to be filmed in their habitat, not in captivity. Doc had also brought cubs on *The Tonight Show With Jay Leno*. He took cubs to congressional offices for staffers to play with while lobbying against captive animal laws.

Nick and I shot there for five days. Doc is a tall, broad, commanding man who heavily controlled our photo ops. He choreographed pictures of his son Kody posing with tigers and large ligers. I photographed the main staff members in a swimming pool with tigers.

I also documented the three-hour Saturday morning "wild encounters" tour. Nearly 80 people gathered in the zoo's spacious, safari lodge–style living room. Screens played loops of Doc's TV appearances and the films he worked on. The morning was highly scripted. Attractive young women emceed, peppering circus-style oratory with puns and questionable facts.

Then staffer Robert Johnson told the crowd how they were saving tigers, raising cheers with "You guys are part of it now. Congratulations!" We knew better. A captive-bred tiger that is acclimated to humans has never been successfully released into the wild. Even if it were possible, it wouldn't be these inbred crossbreeds or lion-tiger hybrids.

Outside, guests took turns posing for photos with Hercules, a tawny, lion-ish, 900-pound liger that had faint tiger stripes and an overly massive head. Then the crowd got what they came for: They played with tiger cubs. No cameras were allowed. The safari's photographer snapped cub-cuddle pictures that people later bought in

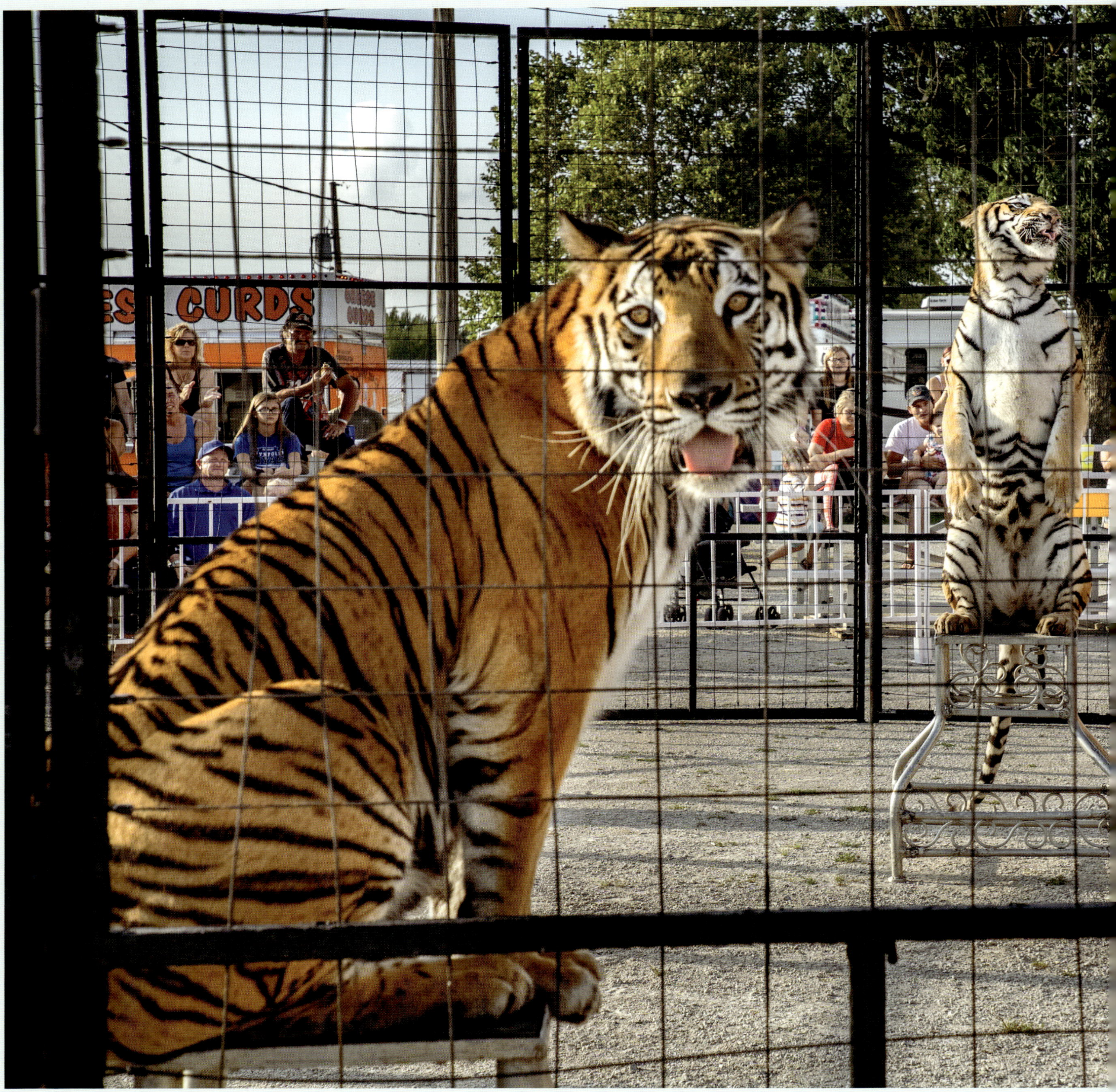
CURDS

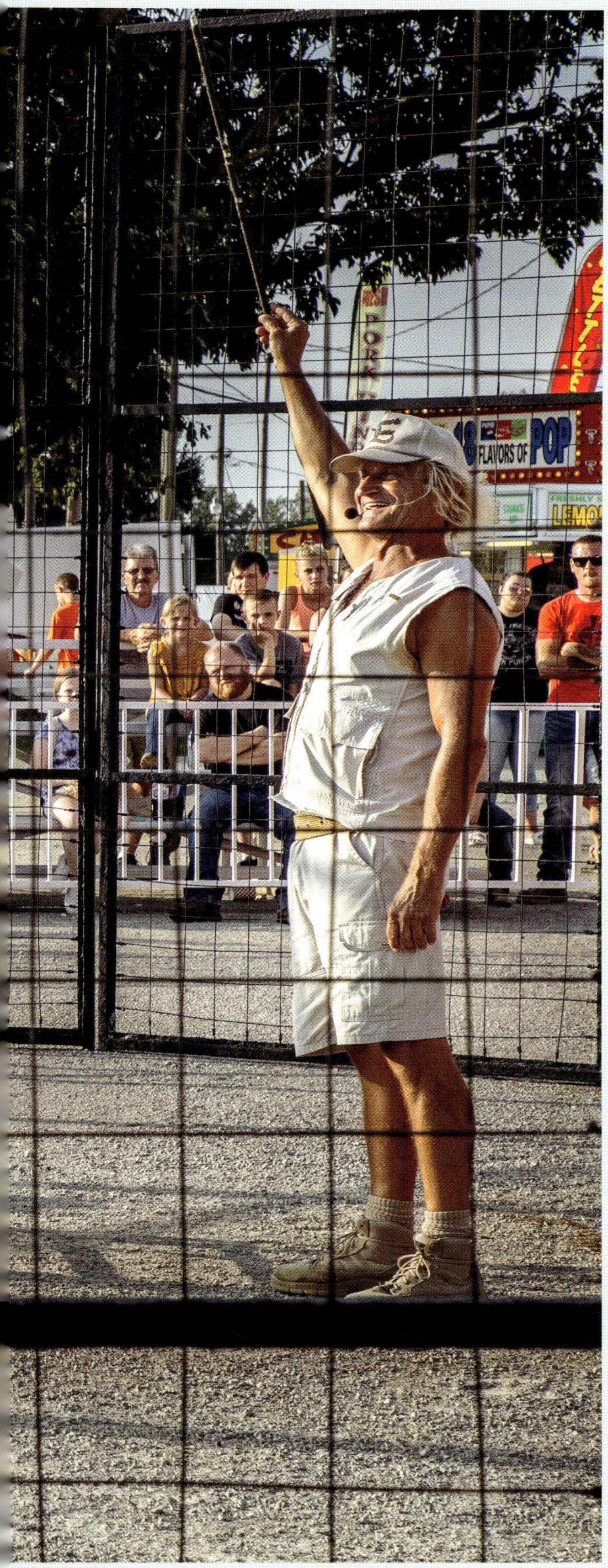

Tiger trainer Brunon Blaszak puts tigers through circus tricks at one of three daily shows at Ohio's Crawford County Fair. Most of the time Blaszak kept the cats in small travel cages. He still performs, but there's growing backlash against shows that use or exploit animals.

pricey photo packages. That night, 30 people joined a "night safari." I conservatively estimated that Saturday's receipts at about $50,000.

Three weeks later, Nick and I returned with Sharon. Doc was hostile from the moment we arrived, which we expected: An internet search of her work brings up the Tiger Temple investigation, which is why we made the first trip without her. Doc tried to intimidate her during their first interview, though a tape recorder was running. He glared at her, bellowing that this better be a positive article or he'd put a bomb in our mailbox.

But Sharon uses what she calls the Hippocratic oath of journalism: Do no harm—unless you're exposing wrongdoing. She dug into Doc's operations, including his nonprofit foundation, the Rare Species Fund, which donated undisclosed amounts to obscure conservation nonprofits; we were told it was a pittance. Supposedly the fund supported "education": Doc hosted tiger/liger shows at Renaissance fairs, dressed in black leather. (He bragged that he'd sold a million photo ops at those fairs.)

The nonprofit also worked "to improve wildlife legislation," or so its mission statement said. But in fact it spent more than a million dollars lobbying against stronger captive animal regulations. Sharon also found that Antle delivered tigers to Thailand's notorious Samut Prakan Crocodile Farm and Zoo, a squalid attraction known for animal abuse.

Our 30-page *National Geographic* story ran in December 2019, and the nonprofit National Geographic Society gave a copy to every member of Congress. But shortly after the COVID-19 lockdowns in March 2020, Netflix aired its *Tiger King* docuseries, garnering 34 million viewers in the first 10 days. It focused on Joe Exotic, turning him into a cult hero, while Carole Baskin was framed as the bad guy. Some people gushed on social media that they wanted to hold a *cute* tiger cub after watching the series. But most were appalled.

Over the next months, we were inundated with requests to speak and publish on captive tigers. The issue gained an even wider audience when I received an award in the Wildlife Photographer of the Year competition for one of my images from the story.

On December 20, 2022, Carson Barylak texted us. "The bill was signed into law!!!" The Big Cat Public Safety Act prohibits hands-on contact and bans owning a big cat as a pet. Current owners were grandfathered in if they registered their animals with wildlife officials.

The tiger cub–petting industry is now a thing of the past, but animal attractions remain big moneymakers. Myrtle Beach Safari and other operations like it have pivoted, now offering photo ops with wolves, chimps, servals, lemurs, bear cubs, or capuchin monkeys.

Doc Antle was sent to prison for wildlife trafficking and money laundering, and we still haven't received any bombs in the mail—although we did check our packages a little more thoroughly for a few months after our story published.

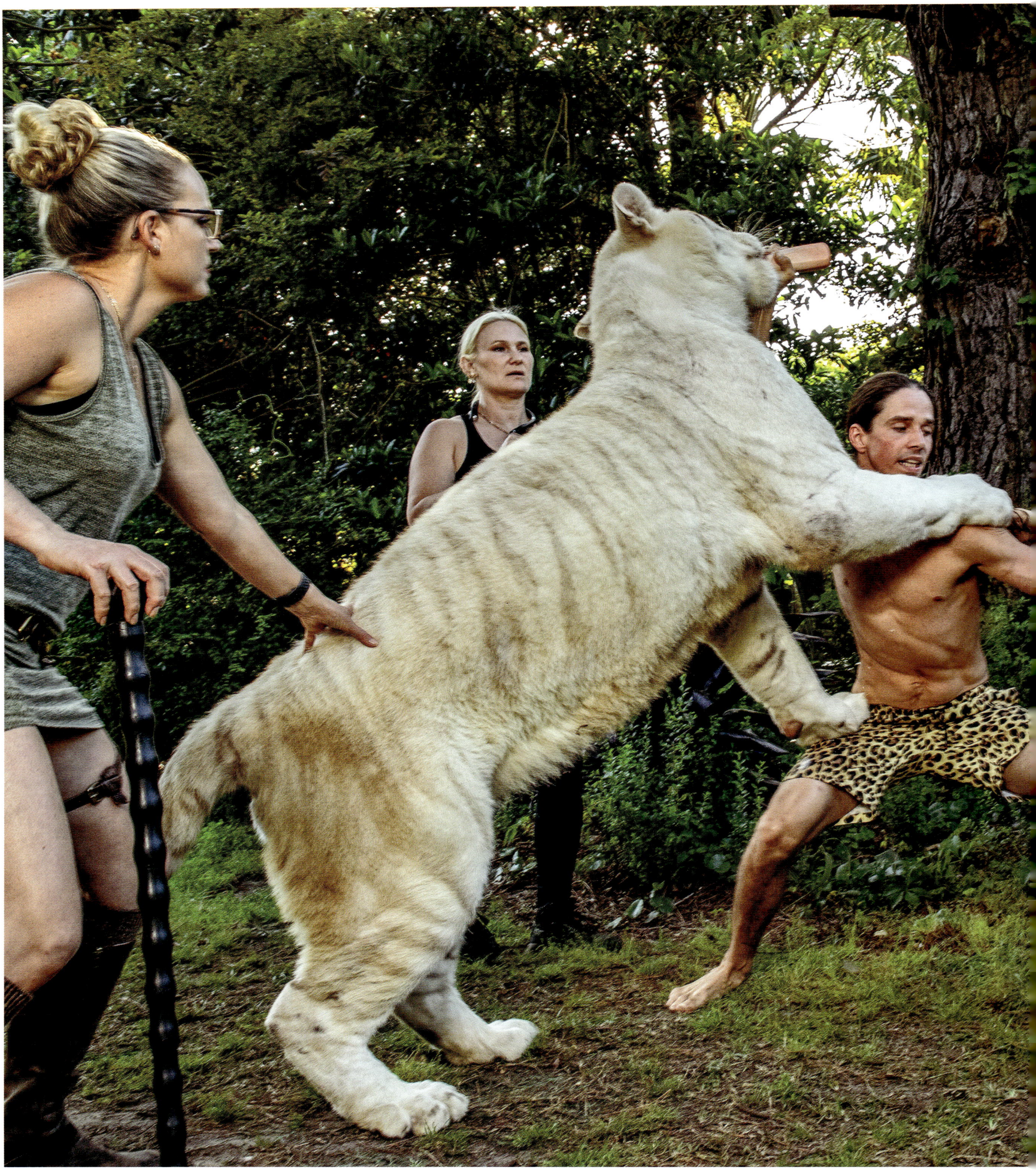

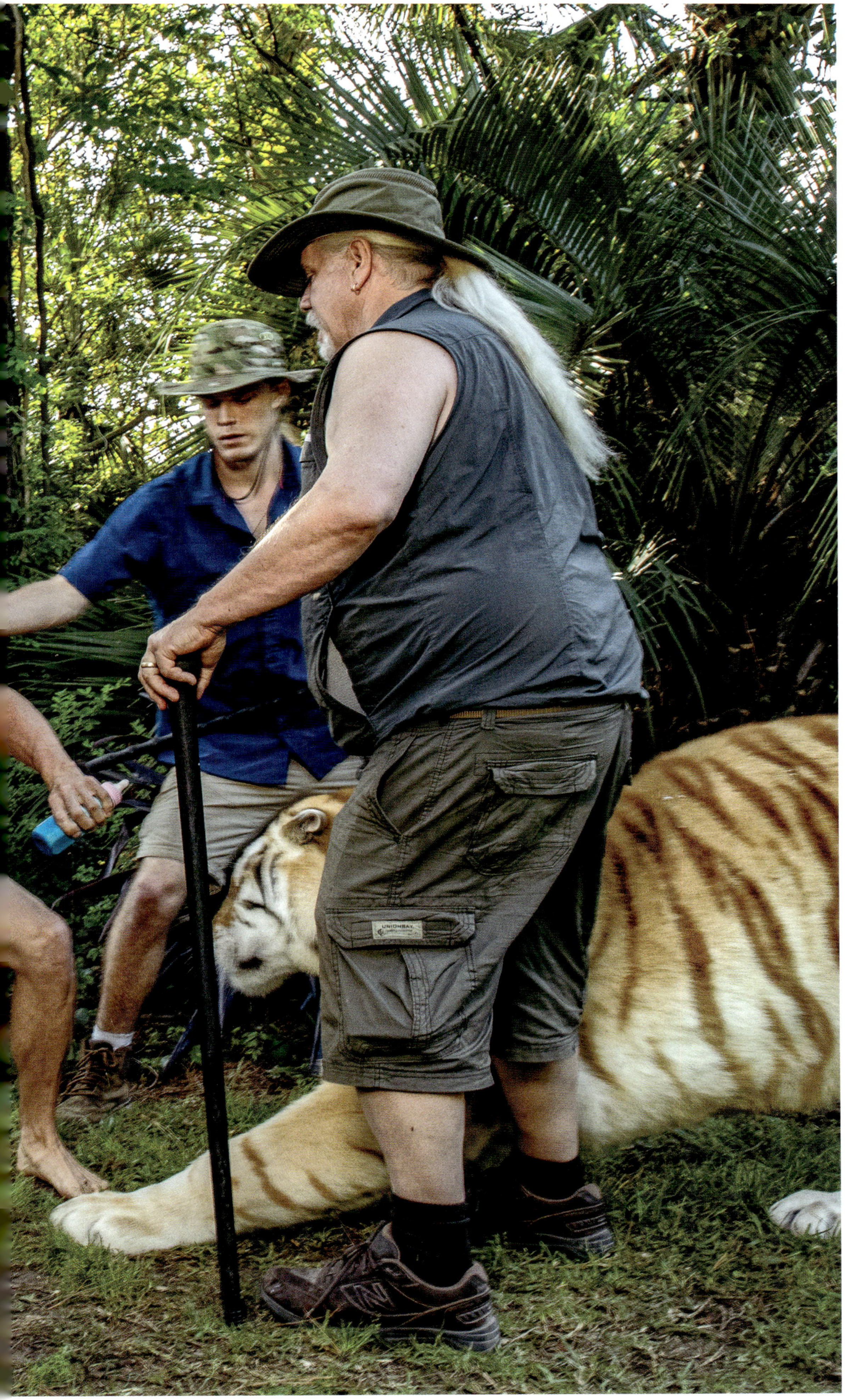

Staff members at Myrtle Beach Safari try to position two massive ligers (tiger-lion hybrids) for a photo session with shirtless Kody Antle, owner Doc Antle's son.

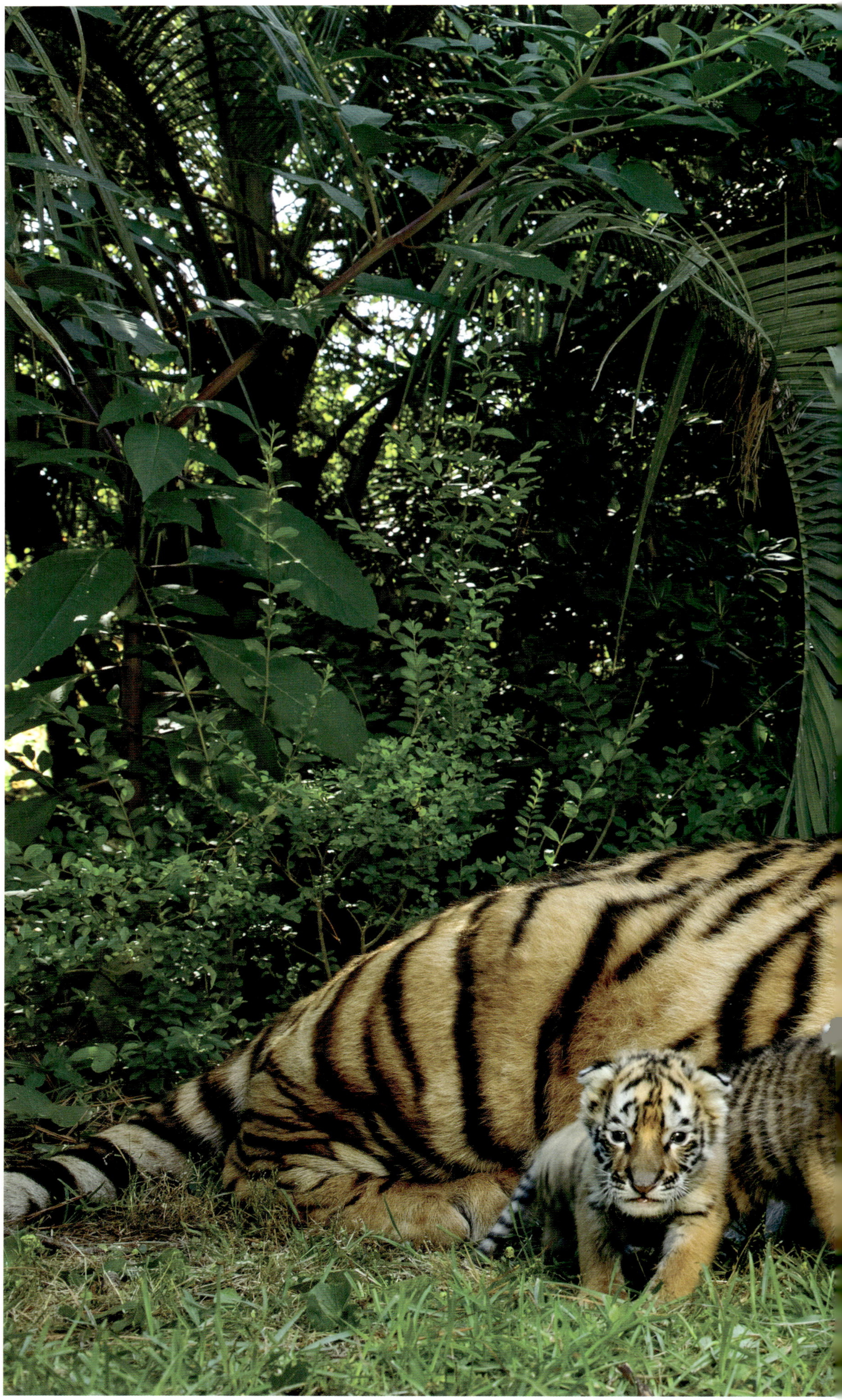

Doc Antle poses with a tiger and tiny cubs at Myrtle Beach Safari. Antle was considered one of the biggest tiger breeders in the United States before passage of the Big Cat Public Safety Act in 2022, which outlawed private ownership and public handling of big cats.

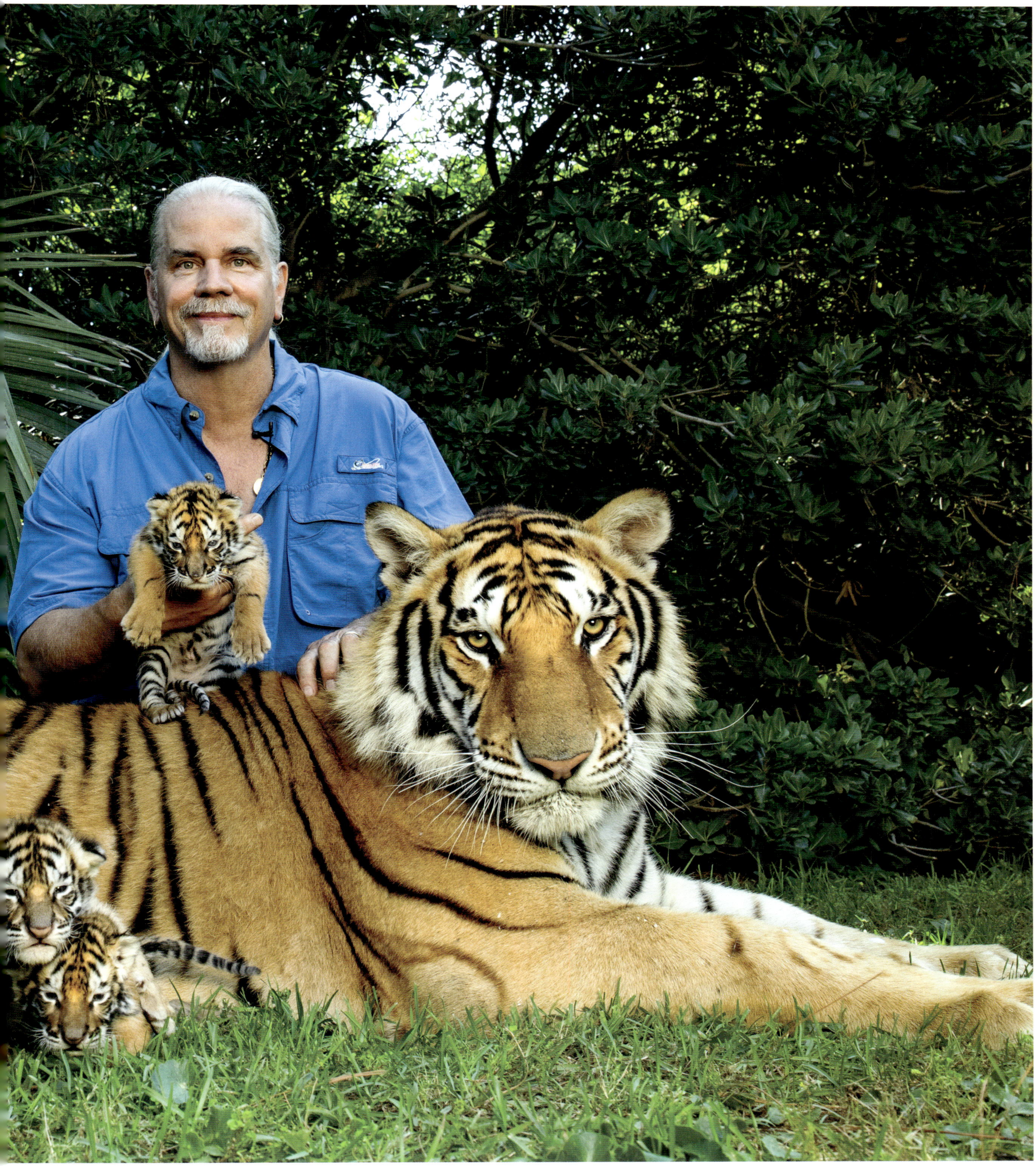

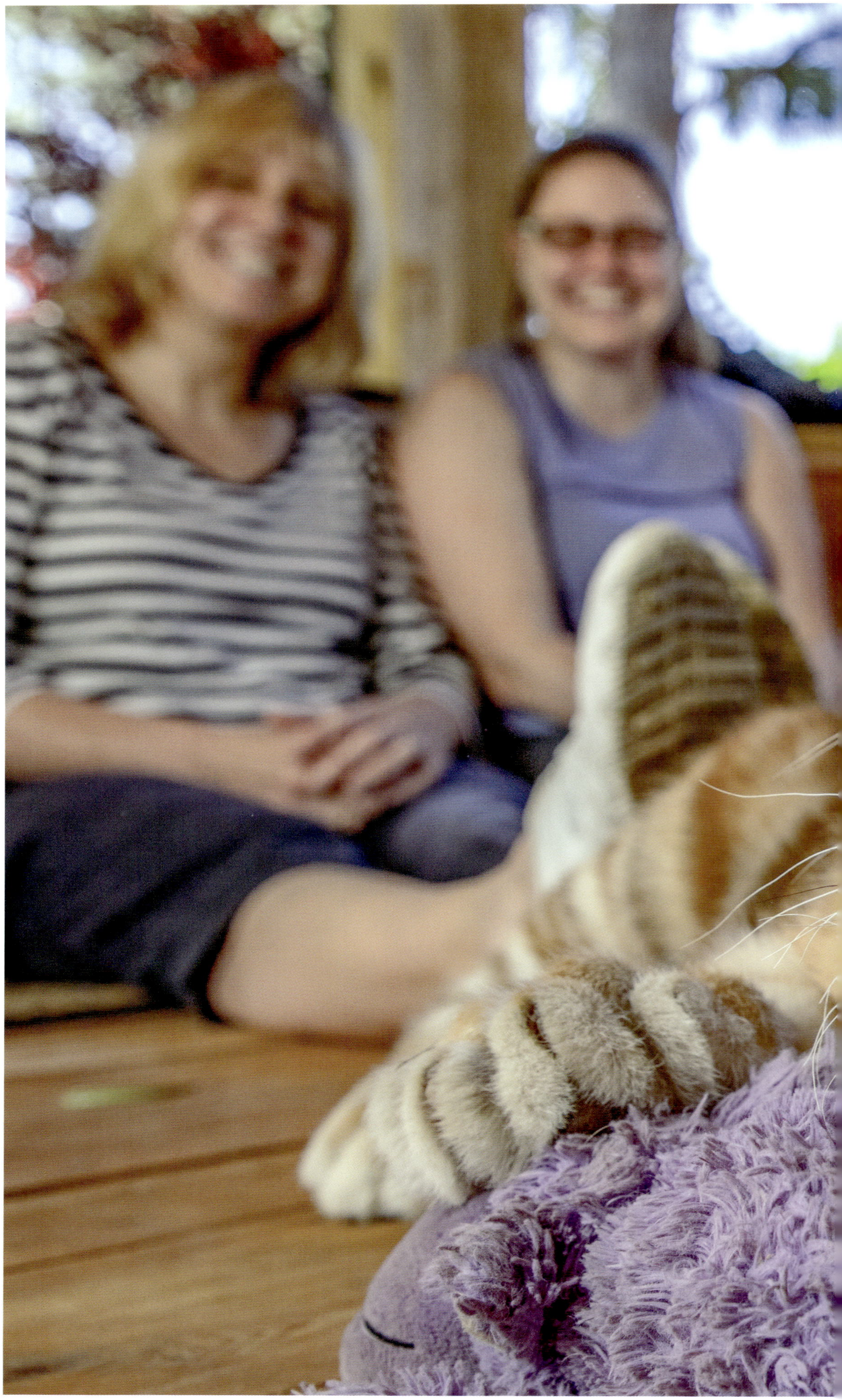

Tourists watch a tiger cub play during a petting and photo session at Myrtle Beach Safari. To maintain a constant supply of cubs, "tiger mills" forced tigresses to produce two or three litters a year – compared with one every two years in the wild – pulling newborns from their mothers at birth.

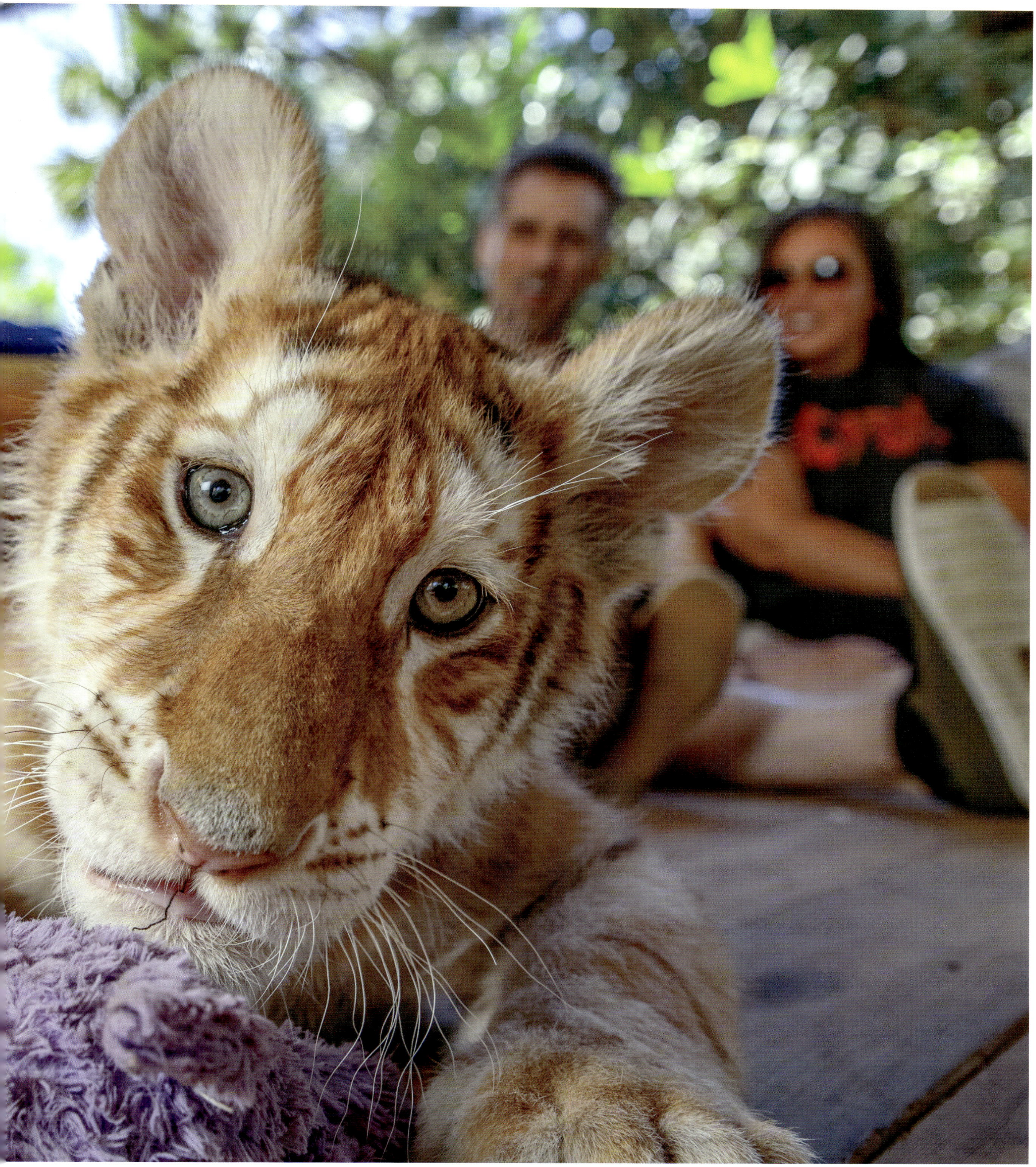

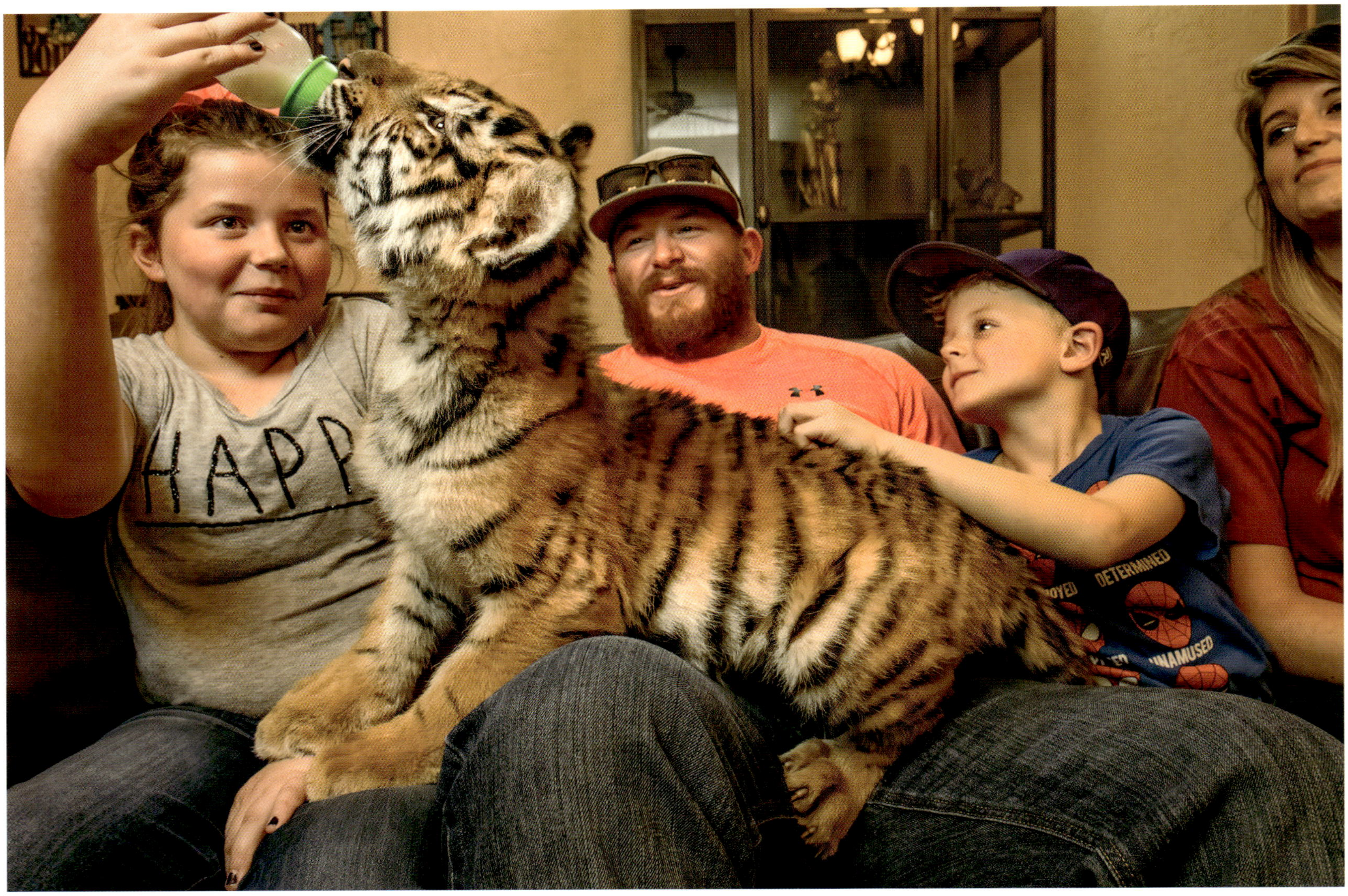

Nine-year-old Ariel McCabe feeds Hulk a bottle as her family watches at the Ringling Animal Care Center in Oklahoma, which offered cub petting in the living room of a rented home until James Garretson, who ran the operation, was evicted. The 12-week-old cub attacked our crew right after this photo was taken, leaving deep scratch wounds in two of us.

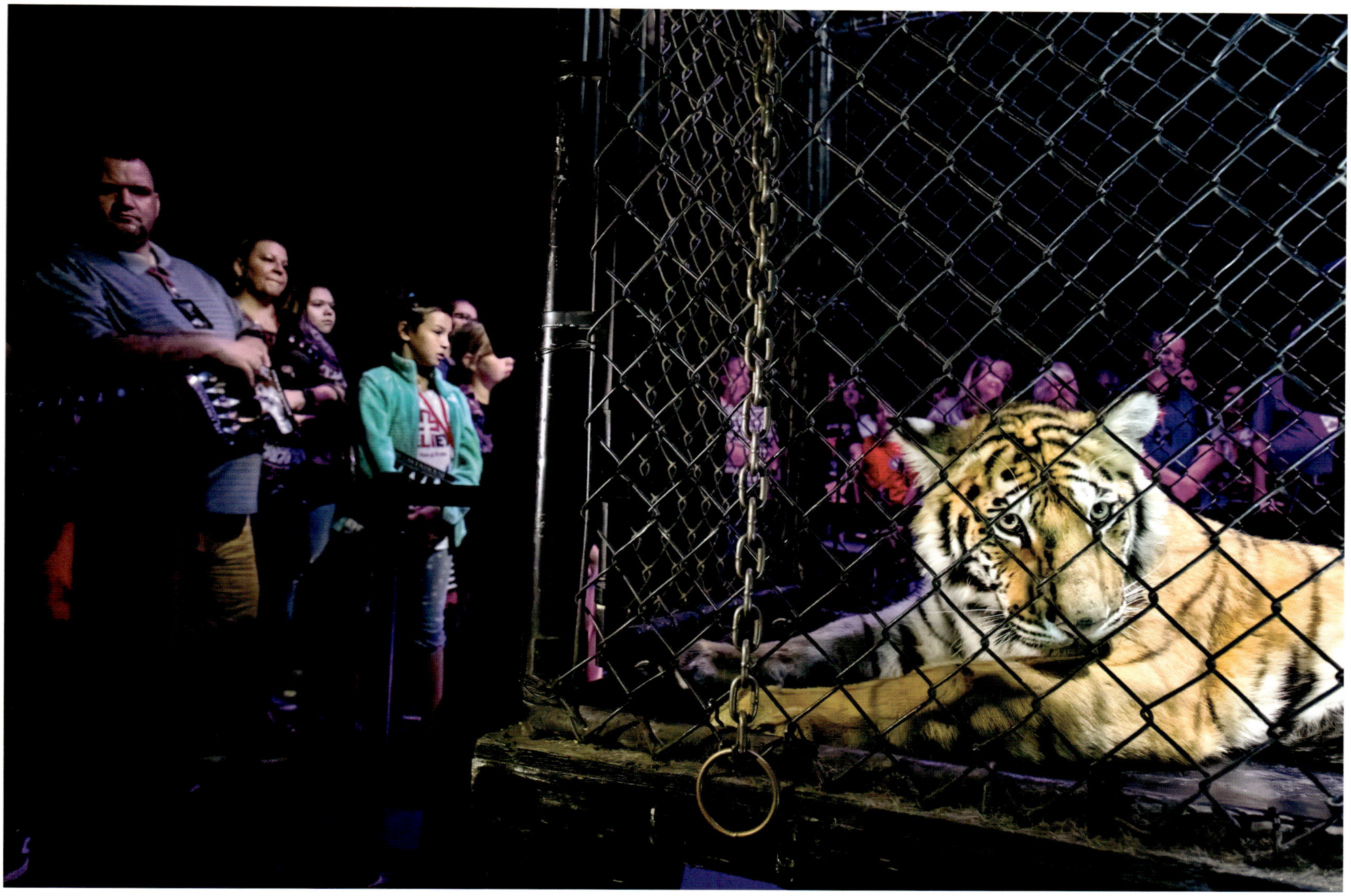

A tiger lies in her transport cage before a performance with Jay Owenhouse, a traveling illusionist who mimicked Siegfried and Roy's famed Las Vegas show.

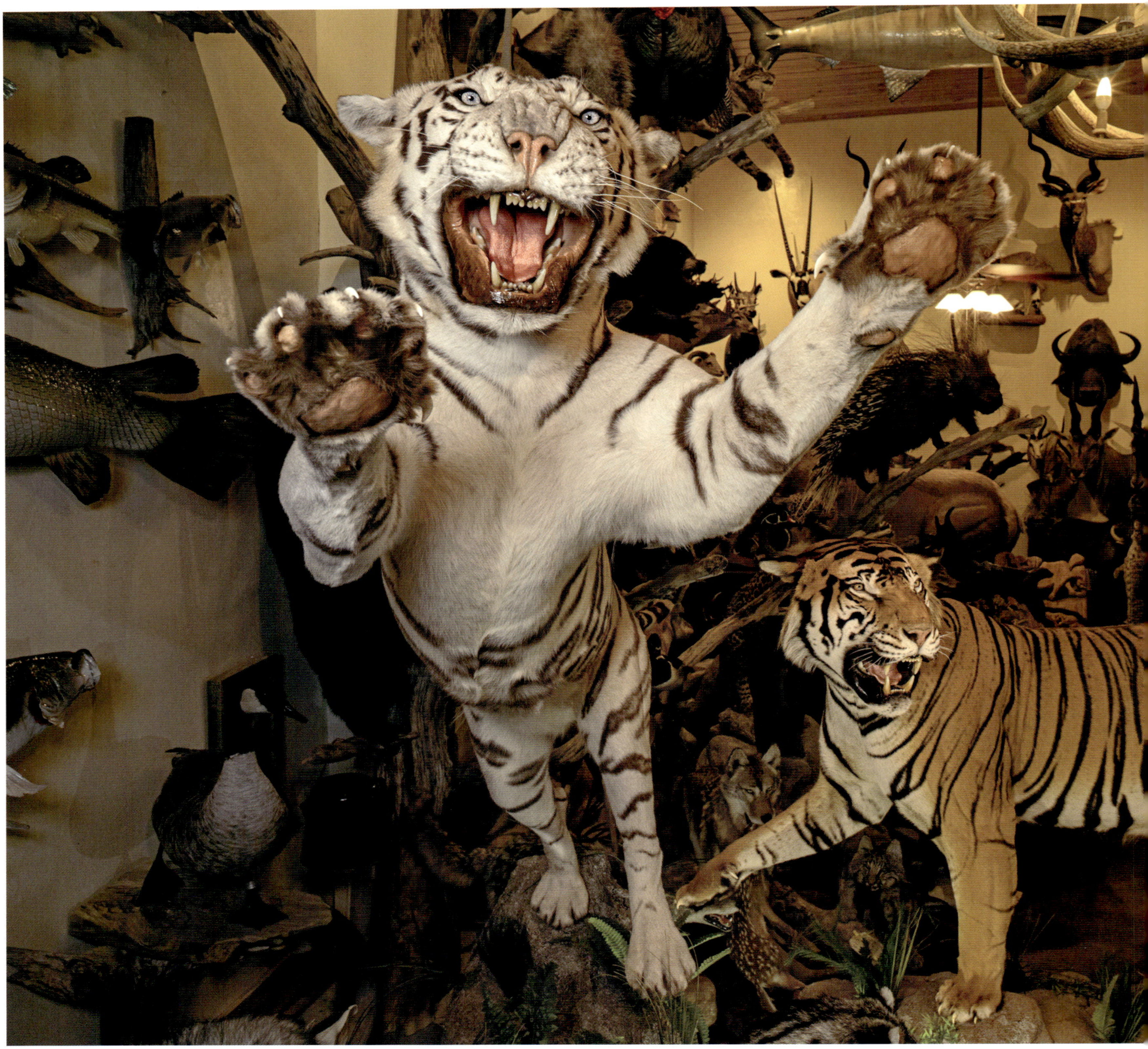

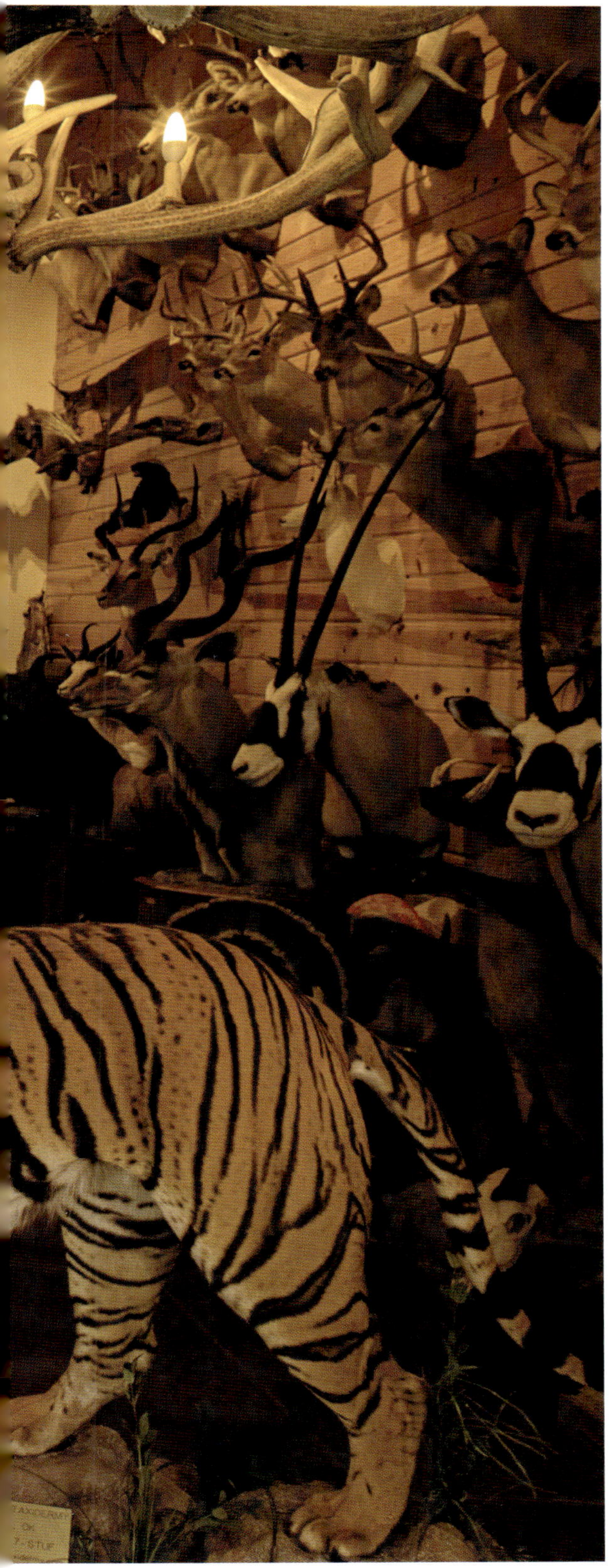

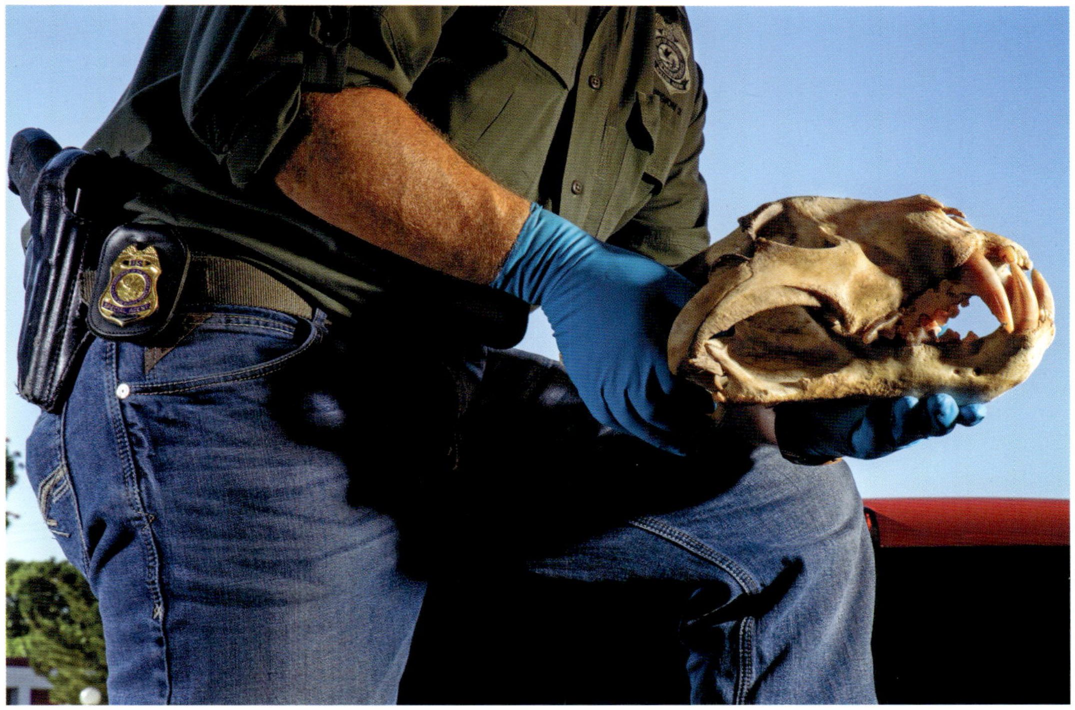

ABOVE: A U.S. Fish and Wildlife agent displays one of five tiger skulls excavated from Joe Exotic's former animal park in Oklahoma. They served as evidence that helped convict Joe Exotic of killing, selling, and trafficking tigers.

LEFT: The proprietor of Terry's Taxidermy in Oklahoma City claims that he acquired these tigers from an Oklahoma animal park – which would make it legal to stuff them, because they were not transported across state lines.

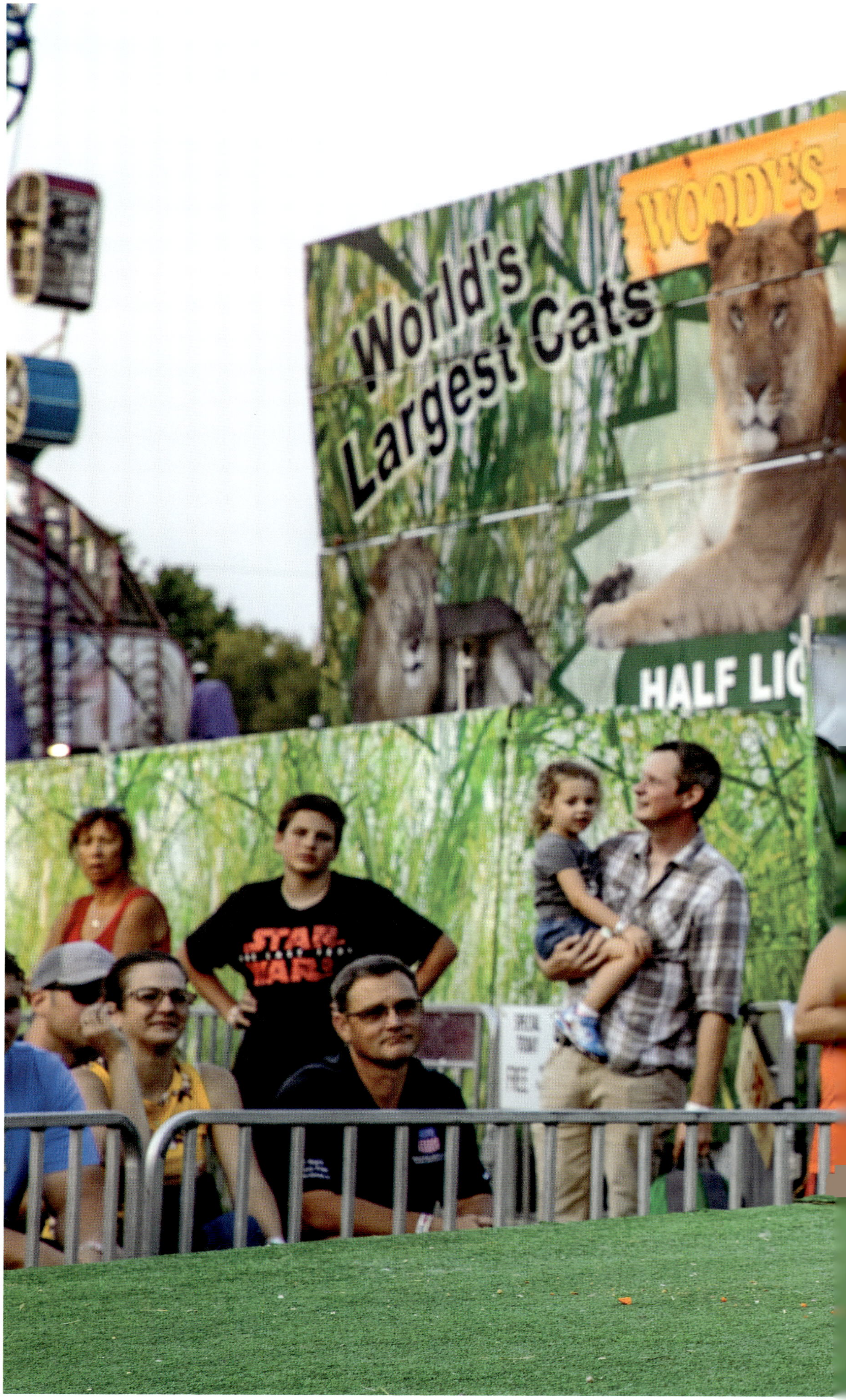

Gregg Woody and his wife parade a tiger cub before a crowd at the Ogle County Fair in Illinois. We watched Woody sell quick photo ops with the cub under cover of darkness.

PAGES 220–221: This plush tiger, photographed at Texas A&M's Disaster City (a training ground for first responders), illustrates the public safety risk posed by poorly run roadside zoos if a big cat is set loose by human error, a hurricane, a flood, or other disaster.

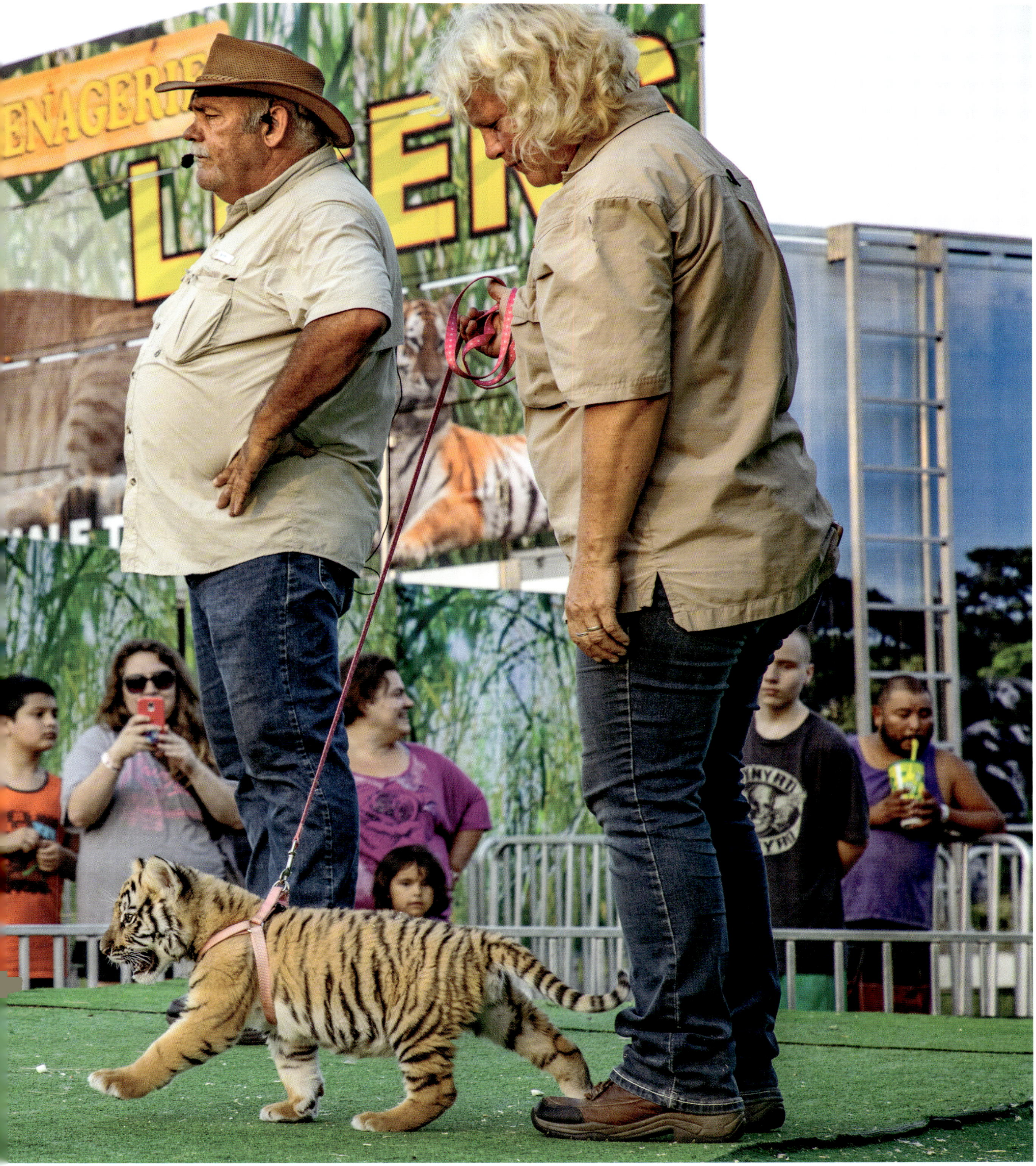

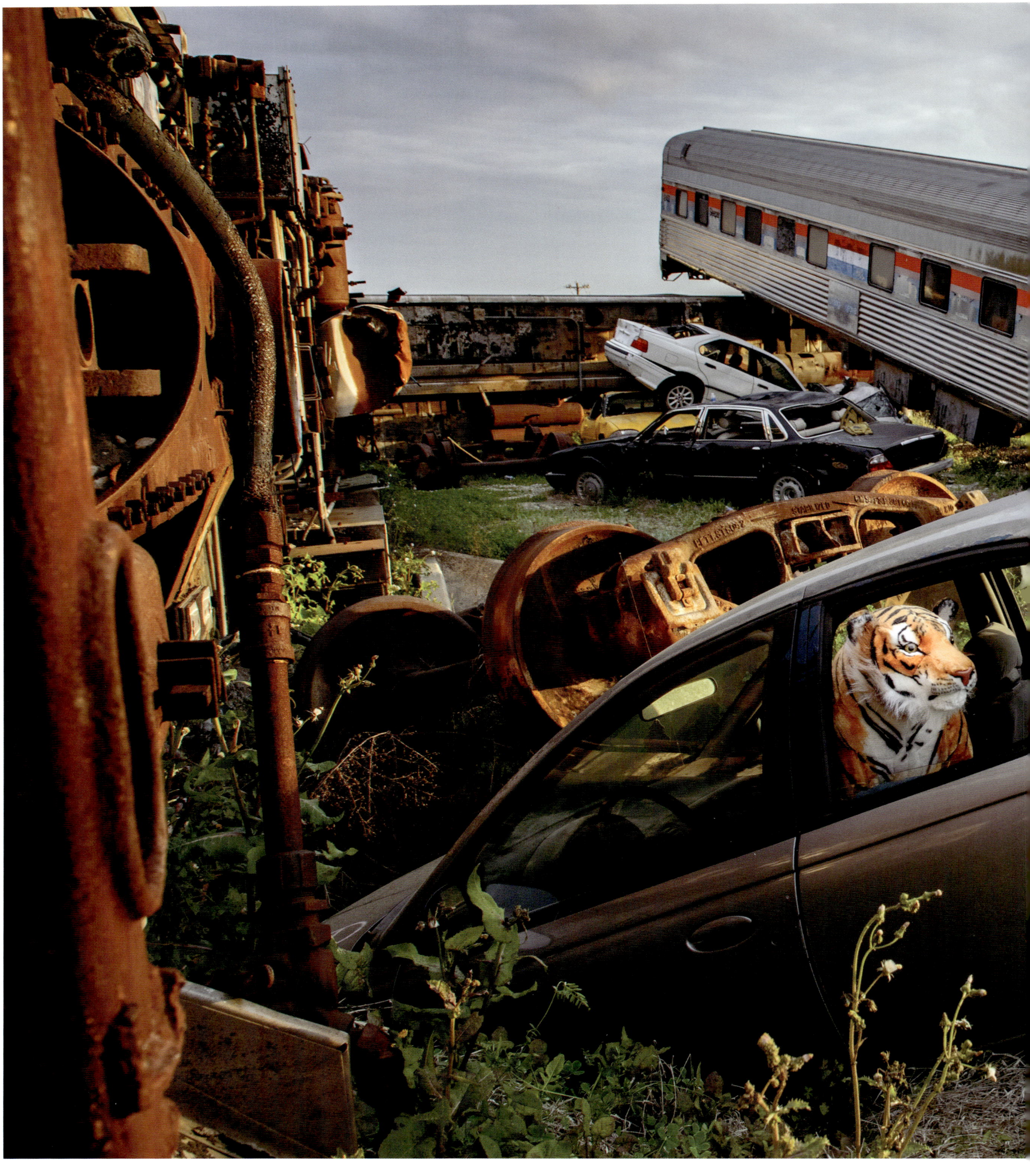

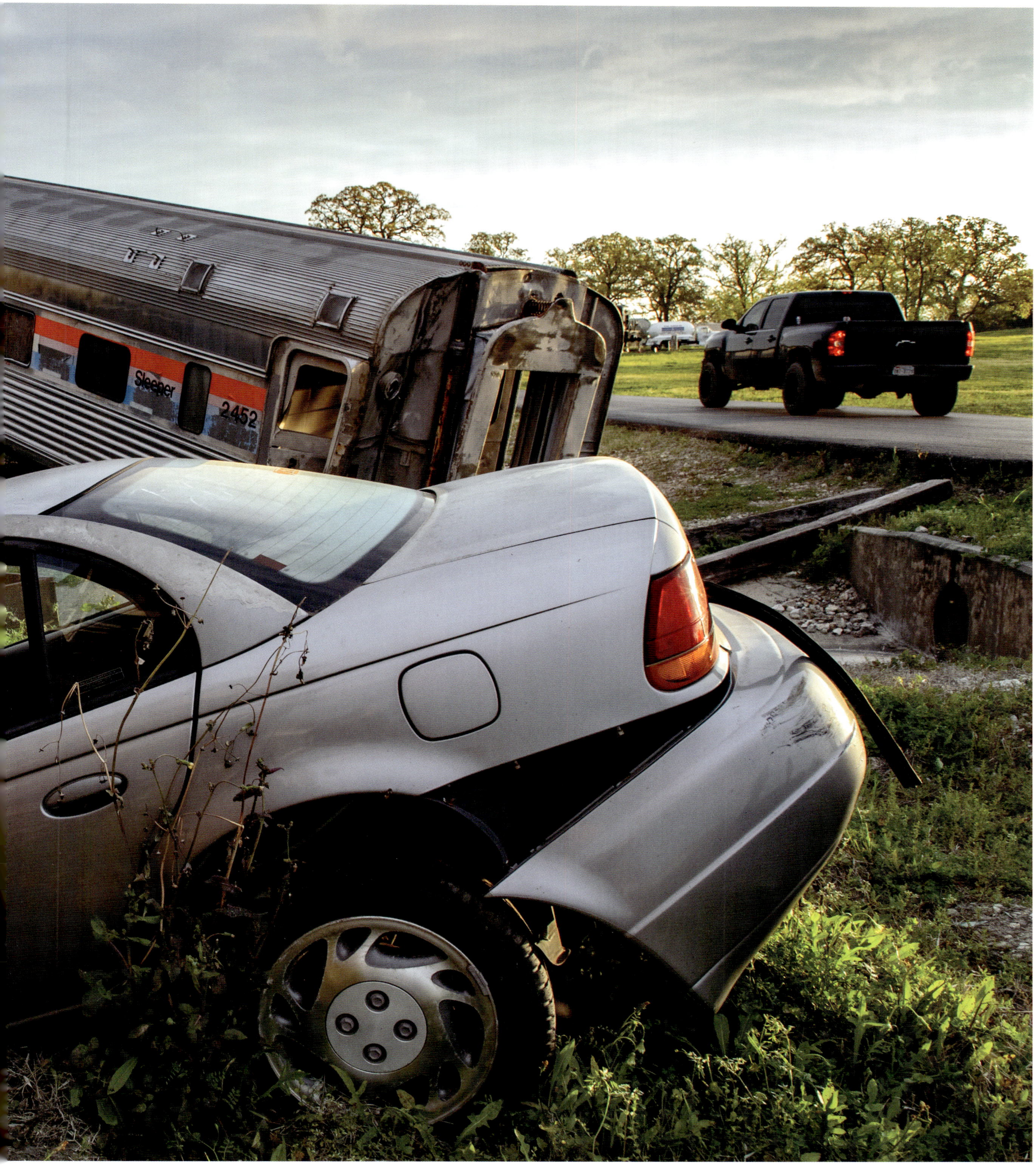
Sleeper
2452

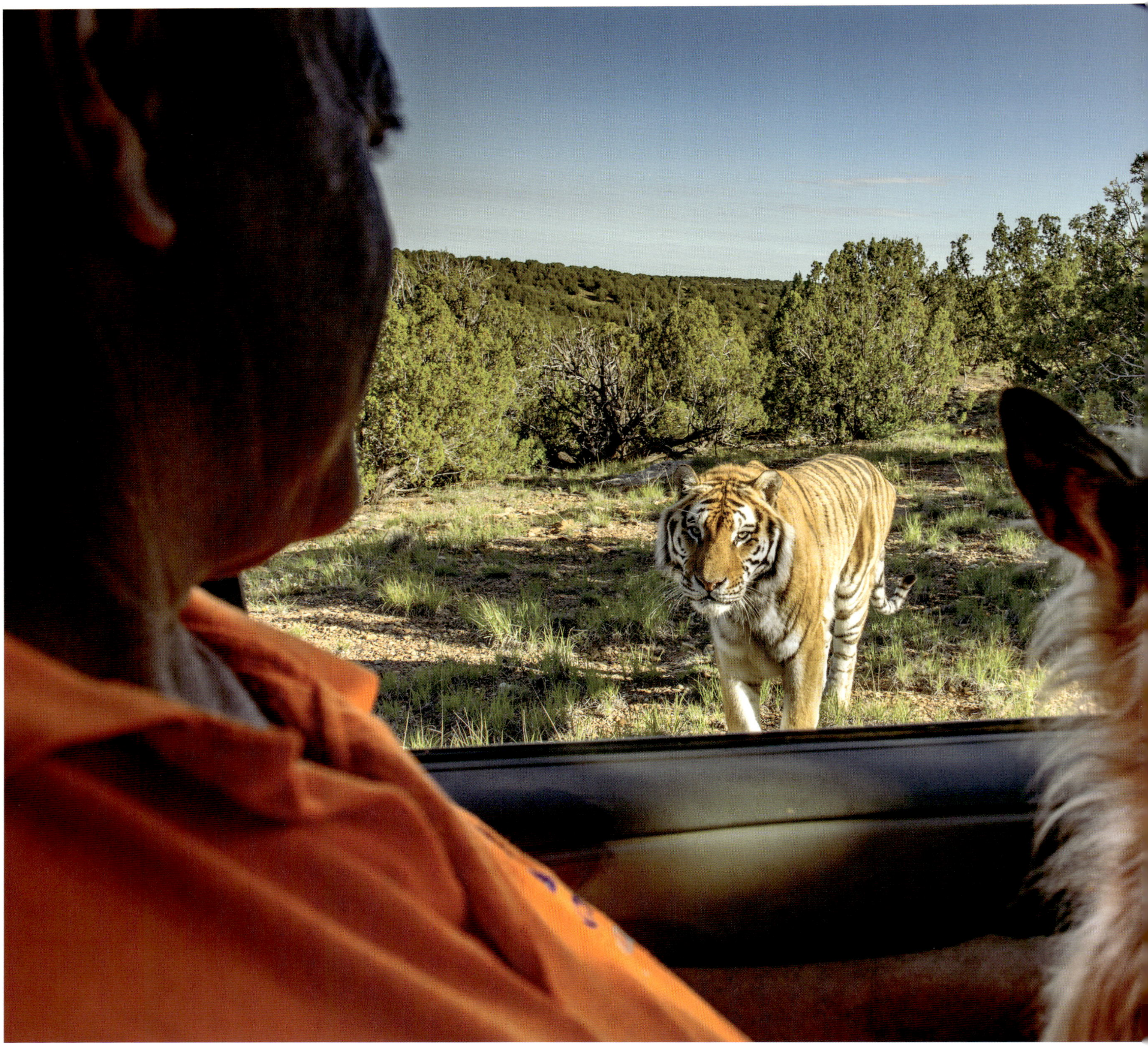

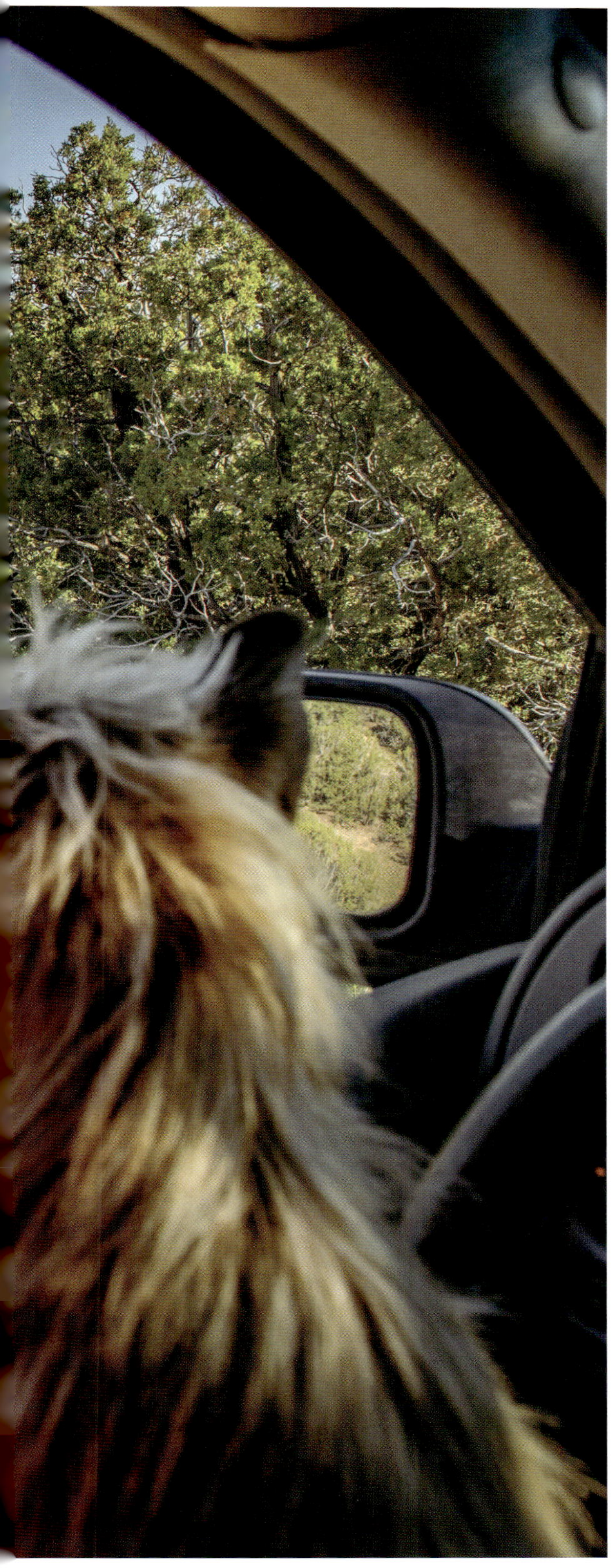

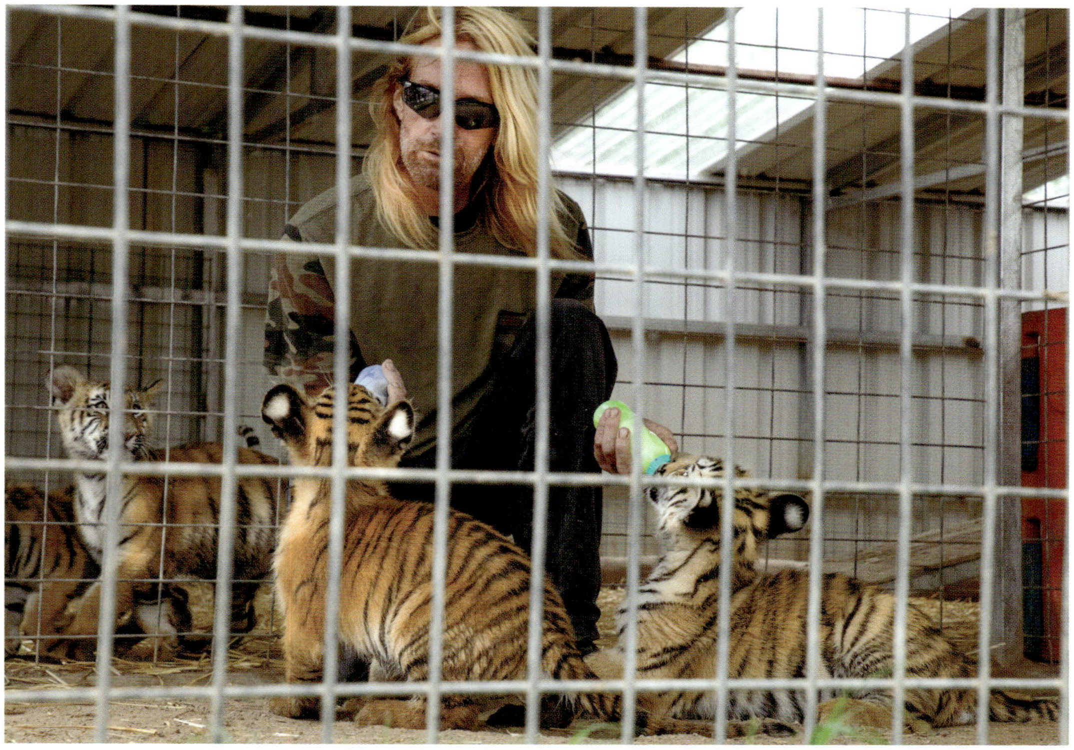

ABOVE: Eric Cowie feeds cubs at Joe Exotic's former zoo, the Greater Wynnewood Exotic Animal Park in Oklahoma. Joe Exotic mass-bred tigers and was convicted of trafficking them across state lines.

LEFT: Pat Craig and Little Bit, his dog, make the rounds at the Wild Animal Refuge in Colorado, a 9,743-acre carnivore sanctuary, the world's largest. One of the sanctuary's first arrivals, Budah, is settling into his 35-acre enclosure.

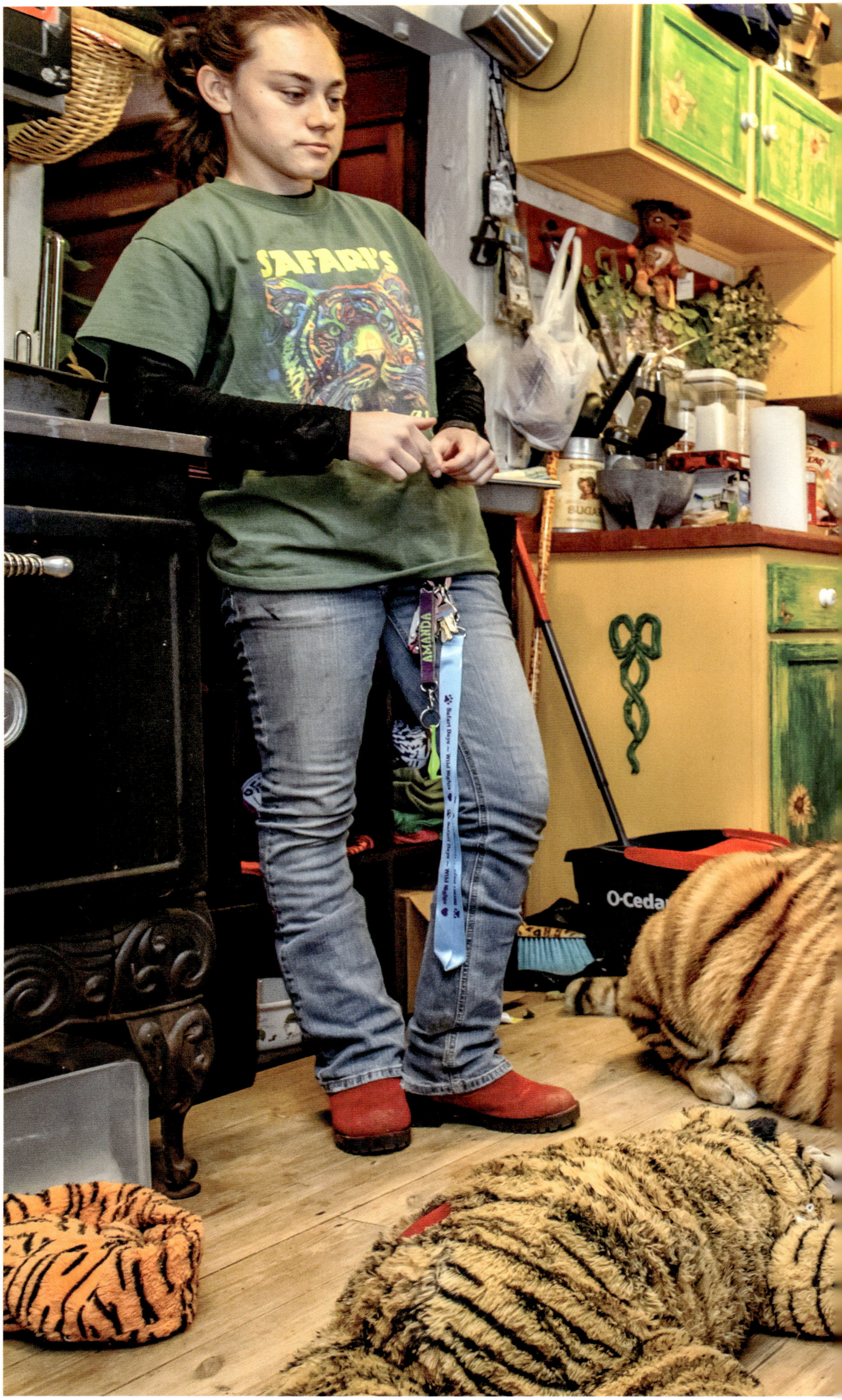

Lori Ensign-Scroggins feeds chicken to Langley, a 300-pound ti-liger (a hybrid that doesn't exist in nature) in her kitchen at Safari's Sanctuary in Broken Arrow, Oklahoma. Langley was born nearly blind, with a limp. Ligers, tigons, and other hybrids were bred by Joe Exotic and Doc Antle to attract tourists to something like a circus sideshow.

PAGES 226–227: A tiger rests on a platform at Exotic Feline Rescue Center in Center Point, Indiana, a facility that houses more than 200 cats.

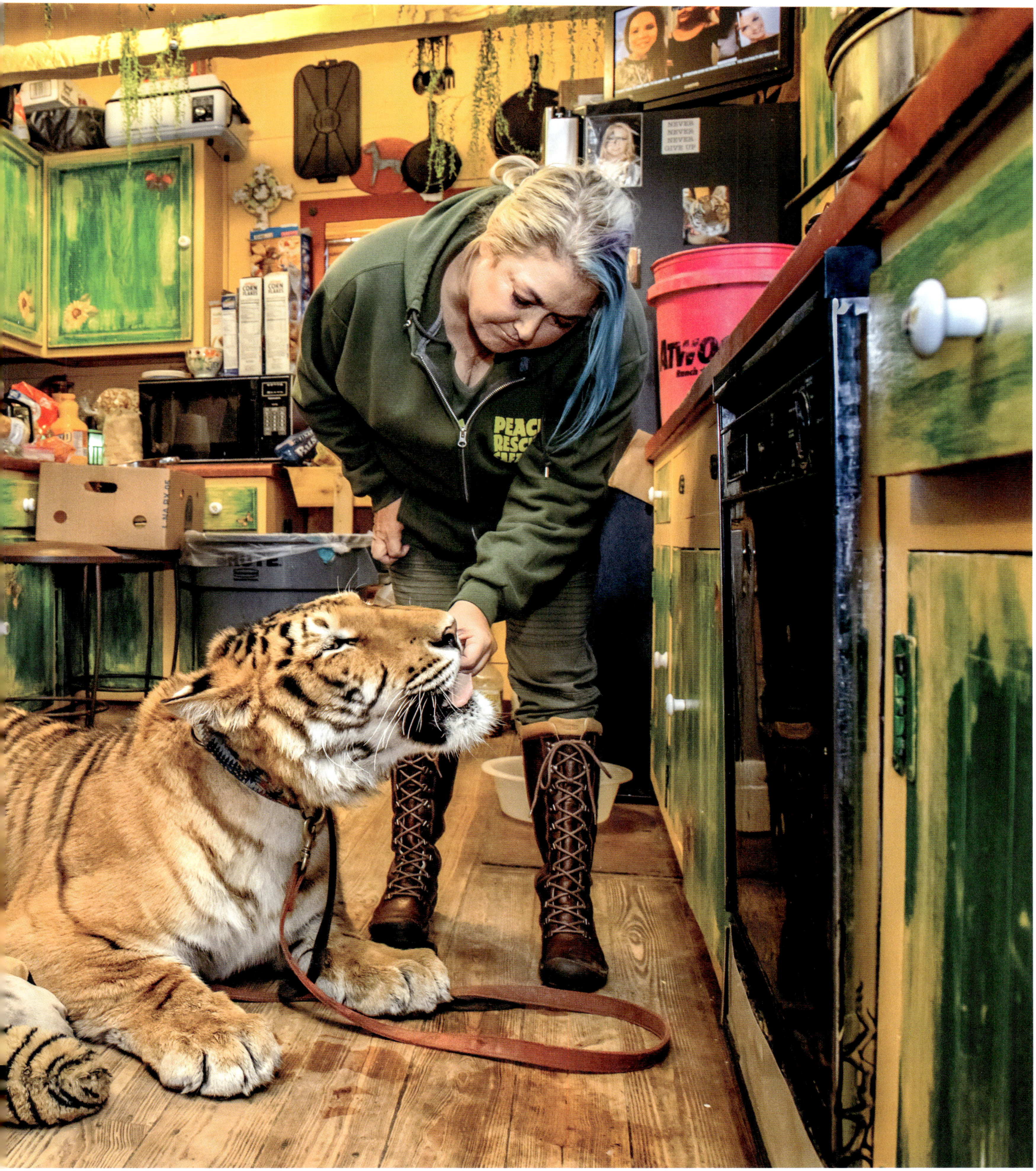
NEVER
NEVER
NEVER
GIVE UP

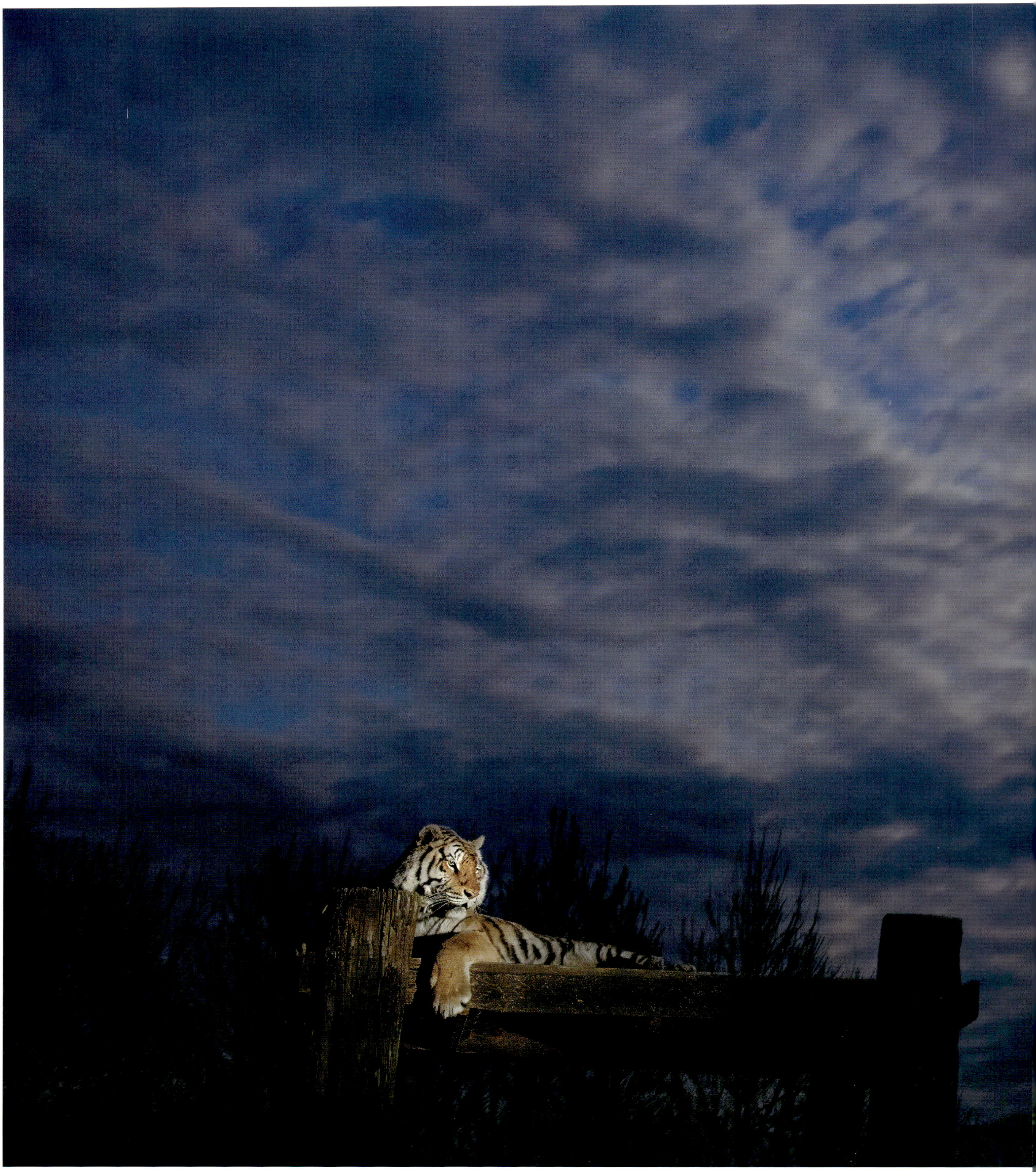

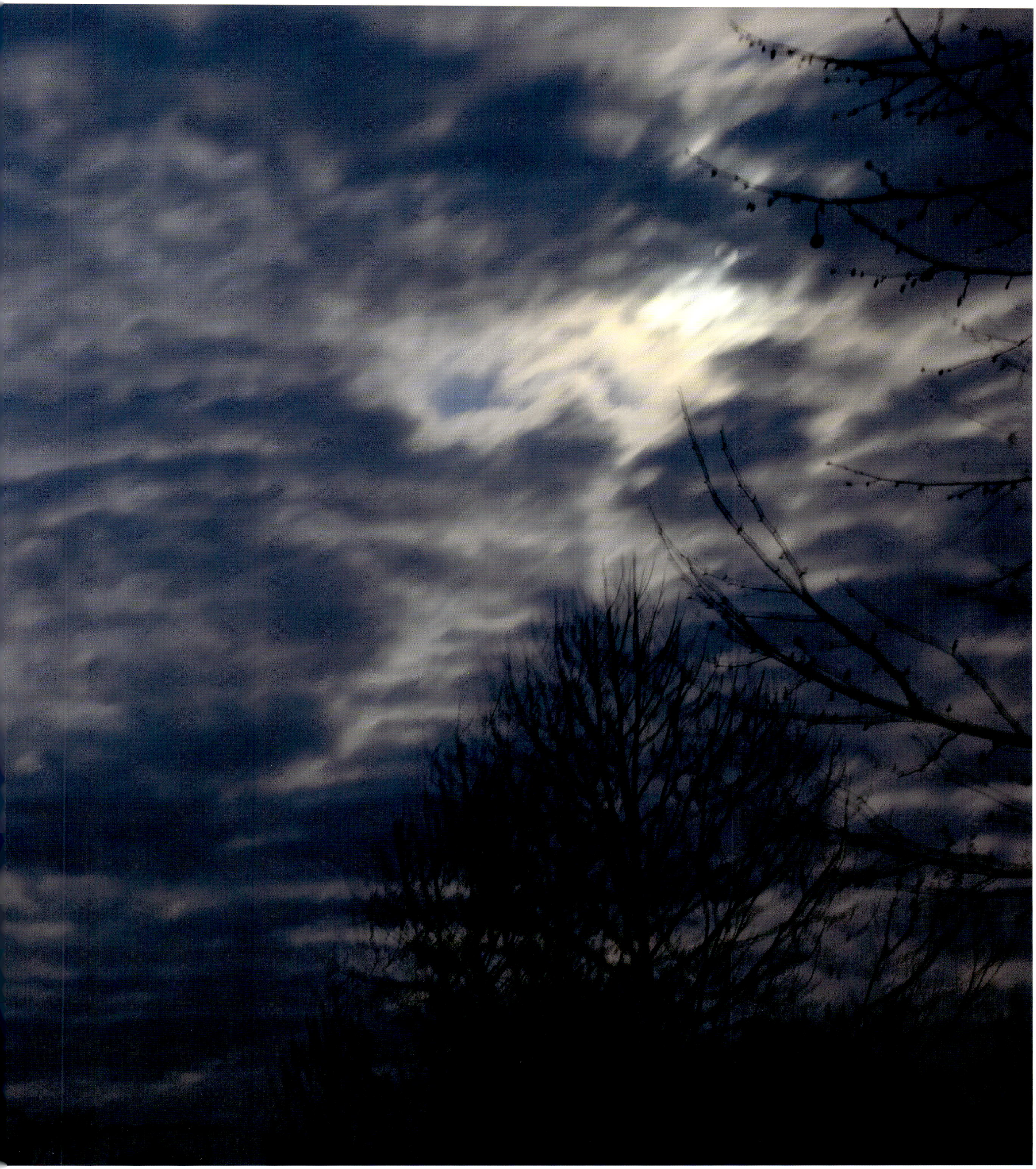

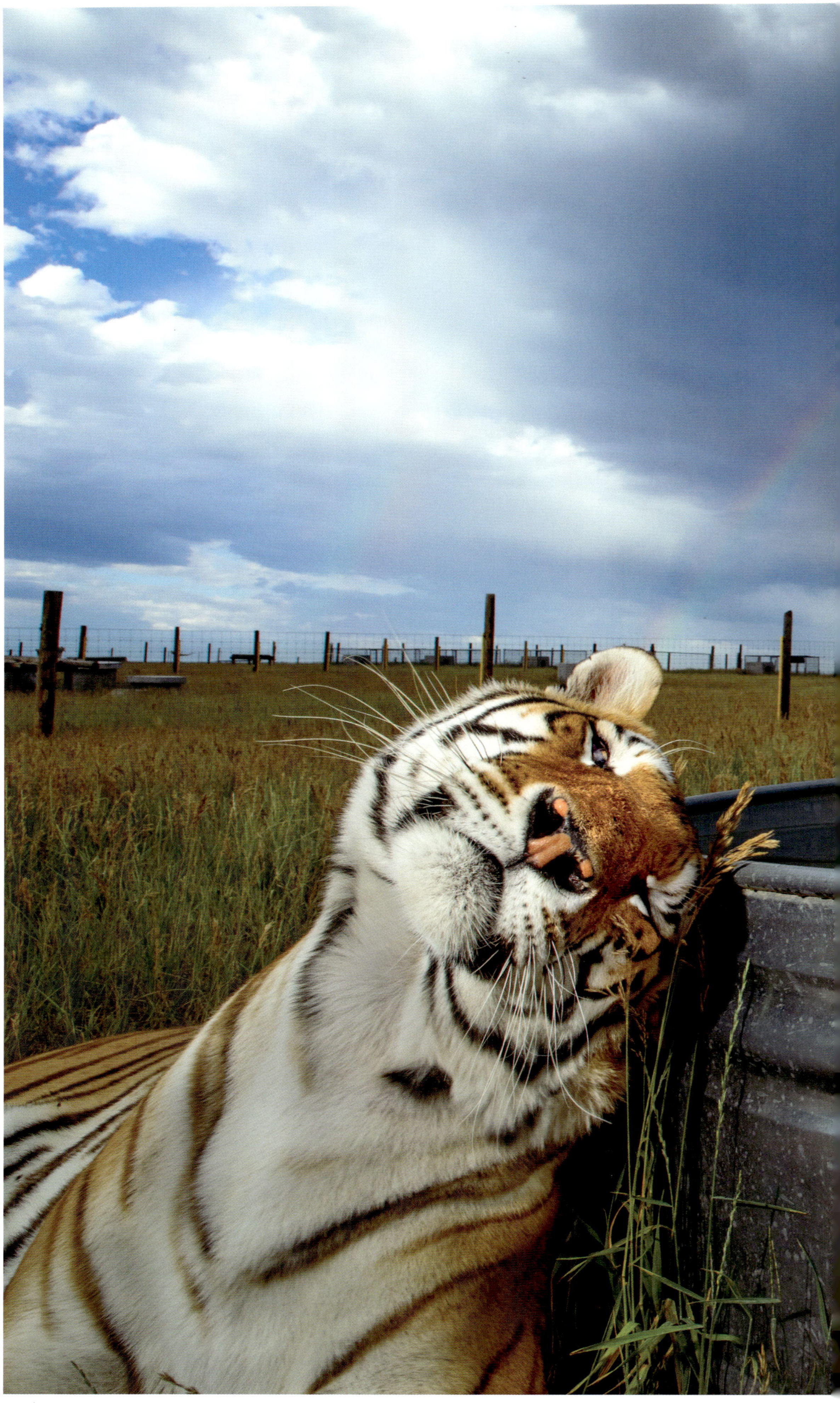

Clay, Daniel, and Enzo, three of 39 tigers rescued from Joe Exotic's now defunct Oklahoma zoo, cool off at the Wild Animal Sanctuary in Keenesburg, Colorado. These cats will live out their lives here, receiving proper nutrition and vet care.

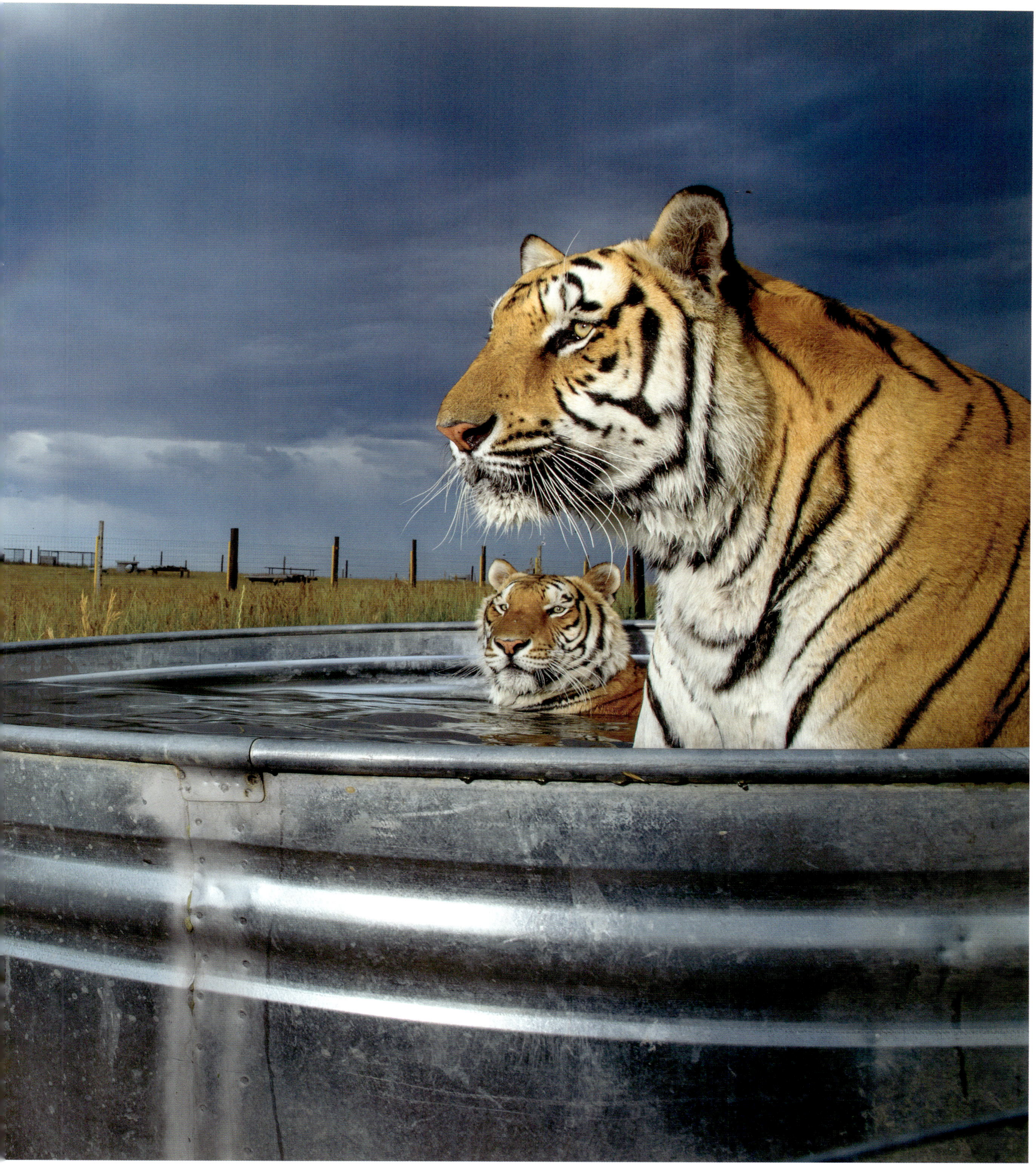

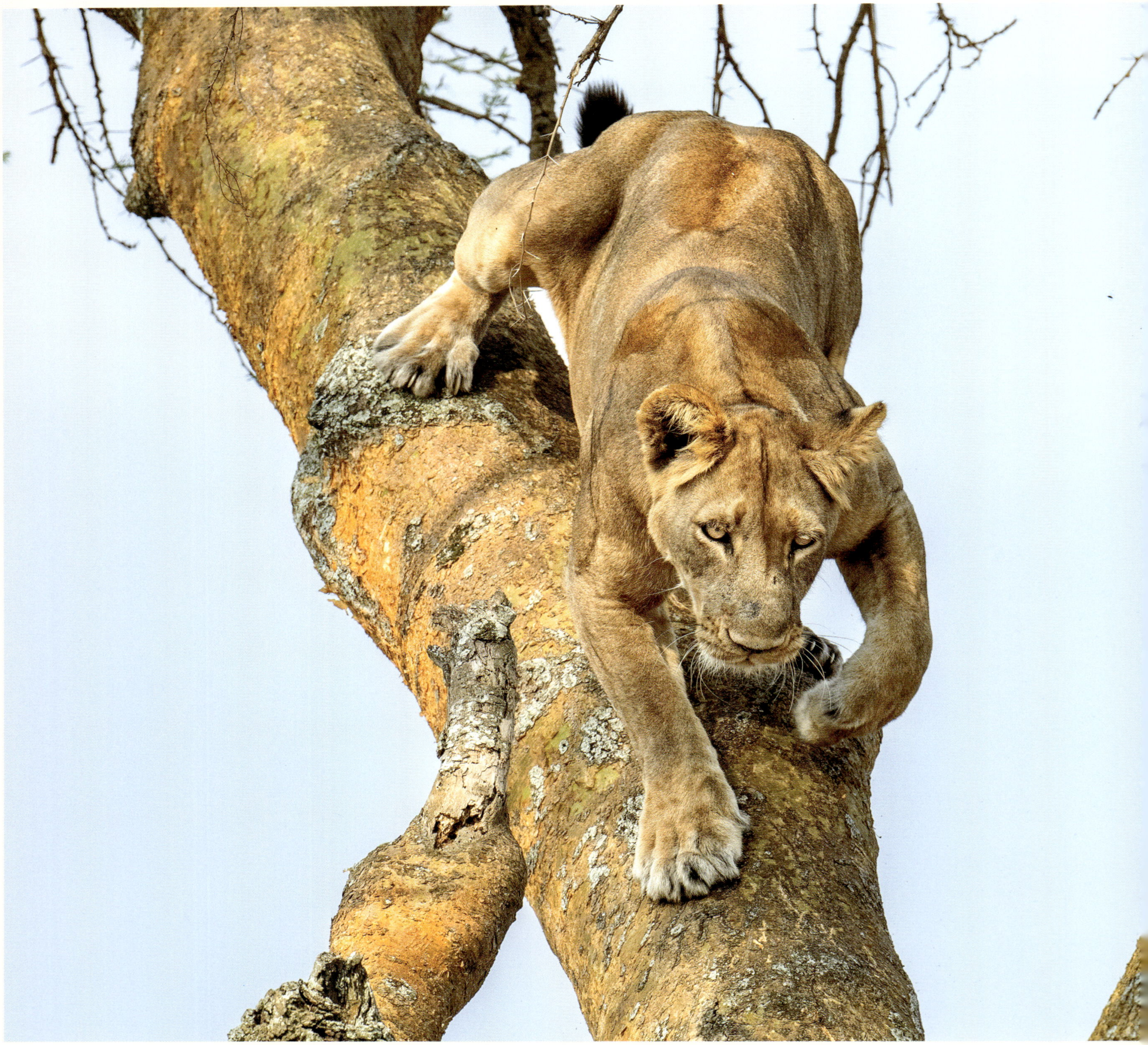

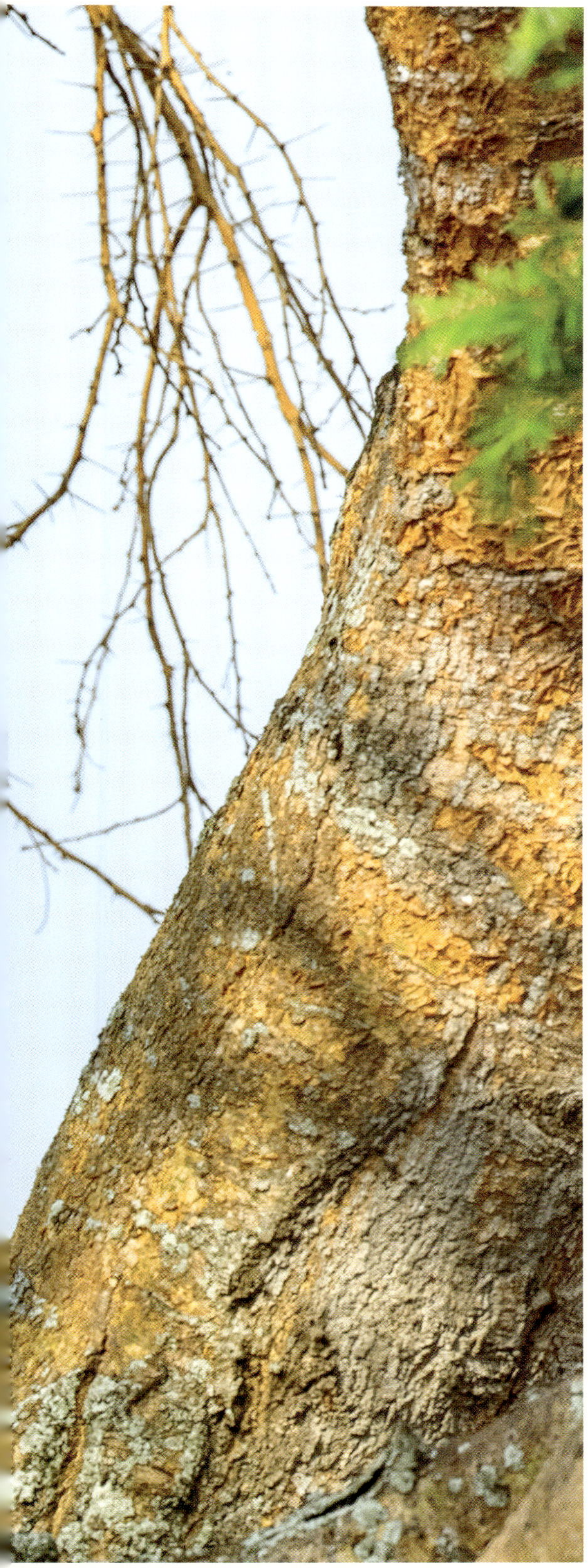

CHAPTER EIGHT

African Lions

Savanna Dwellers and Tree Climbers

With numbers dropping by an order of magnitude in a decade, conservationists are now relocating lions, removing poachers' snares, and upping protections to save these big cats.

A tree-climbing lioness carefully navigates her descent in Uganda's Queen Elizabeth National Park.

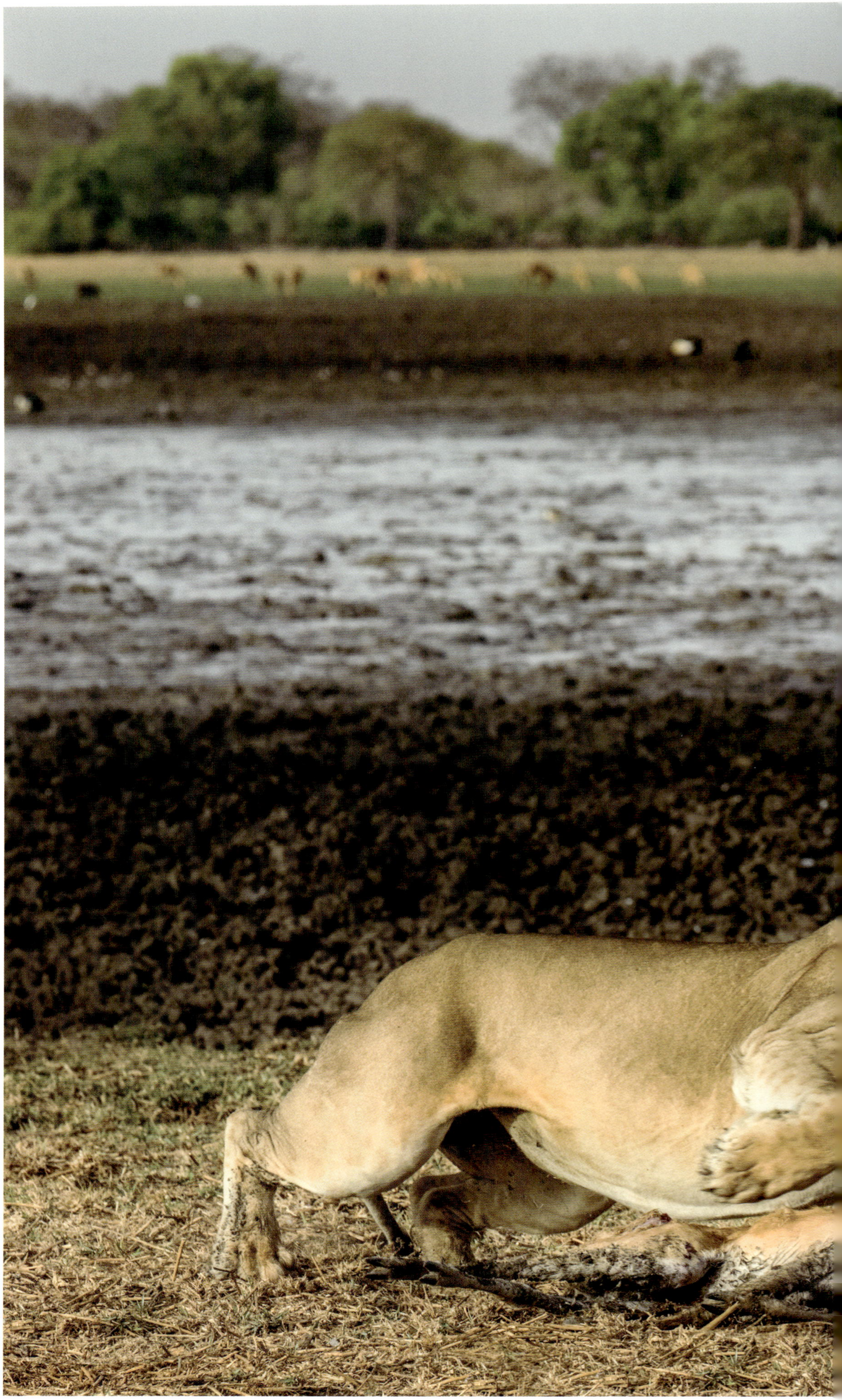

A lion and lioness tussle over a carcass in Zakouma National Park, Chad. Wildlife is rebounding here after rampant slaughter from 2002 to 2010 by heavily armed Janjaweed militias that funded their activities through wildlife trafficking.

PAGES 234–235: Lions are the kings and queens of their realms, admired as a symbol of power, strength, and courage throughout history. Here, a male lion sunbathes, regal in the tall grass in Kenya's Masai Mara National Reserve, home to roughly 850 to 900 lions.

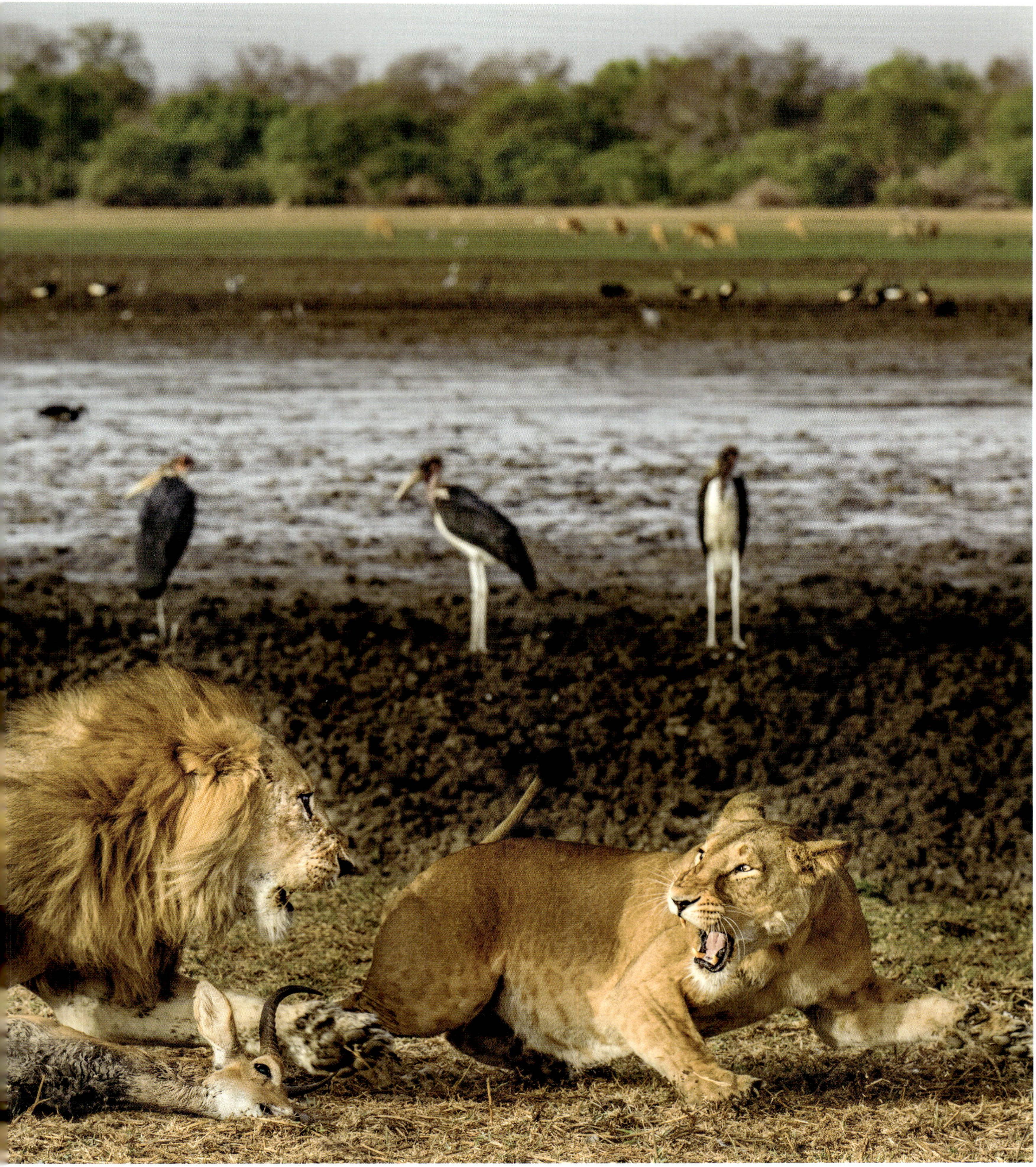

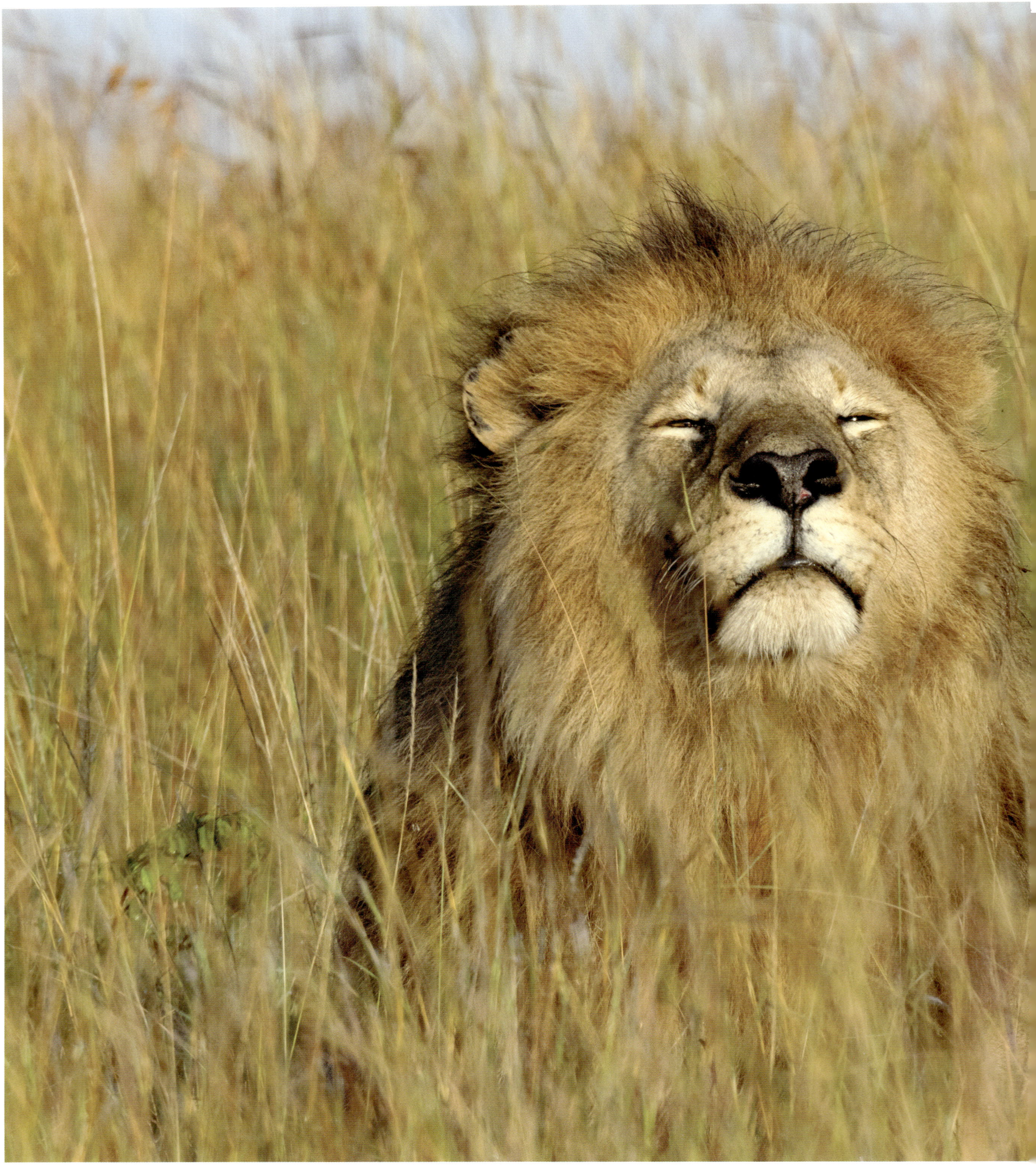

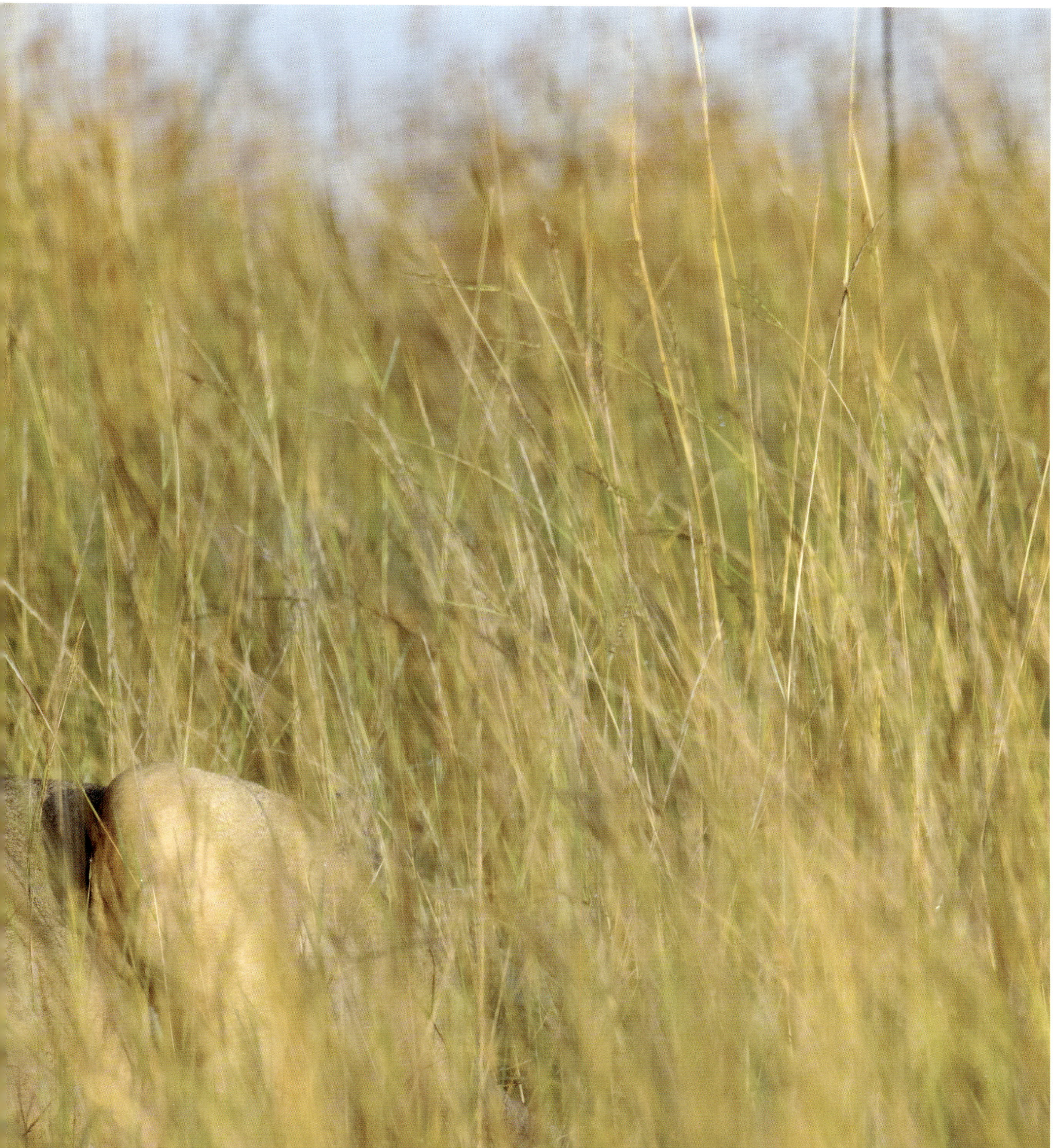

I'd come to Phinda, a game reserve in eastern South Africa, to photograph a leopard research project. On my second night there, I encountered a muscular male lion with an impressive mane resting on the ground beneath a star-studded sky. My assistant shined a light at him, illuminating the cat within a golden circle, and then shut it off so I could make a long exposure to bring up the stars. That was my first encounter with the king of the beasts.

It was then that I learned of their precarious state. They've lost at least 85 percent of their homelands to Africa's ever growing human population. Poachers kill them for their body parts, used in traditional Chinese medicine, and are wiping out their prey for the bushmeat market. Unscrupulous trophy hunting is also taking a toll. The stats are grim: About 23,000 lions are left in Africa today, an order of magnitude fewer than the roughly 200,000 that resided there a century ago.

After that initial encounter in 2009, I was committed to sharing the plight of this iconic cat, working in Botswana and going back to South Africa. Then, an assignment in Chad's Zakouma National Park gave me hope: Lions, elephants, and other wildlife were rebounding after a 30-year killing spree by the Janjaweed (which translates to "devils on horseback"), a terrorist group that funded their activities through poaching.

Years later, I spent time with one of Africa's most unique lion populations: the tree-climbing lions of Uganda's Queen Elizabeth National Park. Every day around 7 a.m. these cats find a tree—a large fig, acacia, or candelabra—to gain a strategic view of their territory and to escape the heat, catch a breeze, and get away from tsetse flies and other biting pests on the ground. It was surreal watching 300- to 350-pound lions bound up trees and perch on a branch out in the savanna.

I spent nearly a month with what we called the "fig tree pride," working with scientist and filmmaker Alex Braczkowski. During that time, we encountered a young female lion named Naturinda who was seriously injured and alone. She'd been trapped in a poacher's snare that sliced her face and blinded her left eye, and though a veterinarian freed her, he hadn't stitched her up. I knew she wouldn't survive without treatment. In a place with just 71 lions, the population can't afford to lose a single female—and the cubs she would bear. The woman who ran our lodge had political clout, and I convinced her to call the vet. He treated Naturinda, and she survived.

The region is still riddled with snares, many of them set by poachers who slip back and forth across the border, selling Ugandan-caught bushmeat in the Democratic Republic of the Congo. Just 40 members are left in the fig tree pride.

But the biggest threat is Uganda's exploding human population, which numbered 50 million in 2025 and is projected to hit 125 million over the next three decades. Humans continue to encroach into protected areas. For example, cattle farmers graze their livestock unattended in and near the national park, and some are inevitably eaten by lions. Animosity is rising.

Many people here make just a few dollars a day. Milk and meat from their livestock may provide a family's only income, and losing one to a lion is devastating. Locals increasingly retaliate by poisoning carcasses with insecticides, usually aldicarb or carbofuran, killing entire prides when they come back to eat.

Conservationists are trying to compensate farmers for losses and educate local communities about the importance of lions that keep nature in balance by preying on antelope that eat the same grasses their herds rely on.

In 2018, my friend Andrea Heydlauff told me about an amazing project that she was working on with the nonprofit African Parks. They were capturing lions in South Africa, where they were doing well, and rewilding them into areas where they'd been wiped out.

Researchers identified a pride in Phinda that they could move to Liwonde National Park in Malawi. There hadn't been a breeding population there in decades.

In Phinda, I met up with the team: a veterinarian, park guards, biologists, park officials, and conservationists. I climbed into a truck with the vet, and we approached the pride. When we were close enough, he raised his rifle and shot the dominant female with a tranquilizer dart. Then, one by one, he anesthetized them all.

The rest of the team descended, putting a specially designed blindfold on each of the six lions to protect their eyes and faces during transport. The vet did a quick exam, took blood samples, and then the team hefted the sleeping lions onto canvas stretchers and loaded them into pickup trucks. It was a short drive to a large, wood-fenced holding area. They would stay there until their bloodwork came back. Lions can carry contagious diseases, and tuberculosis is a particular concern. Ultimately, 12 lions were released in Liwonde. The park is now a haven for wildlife.

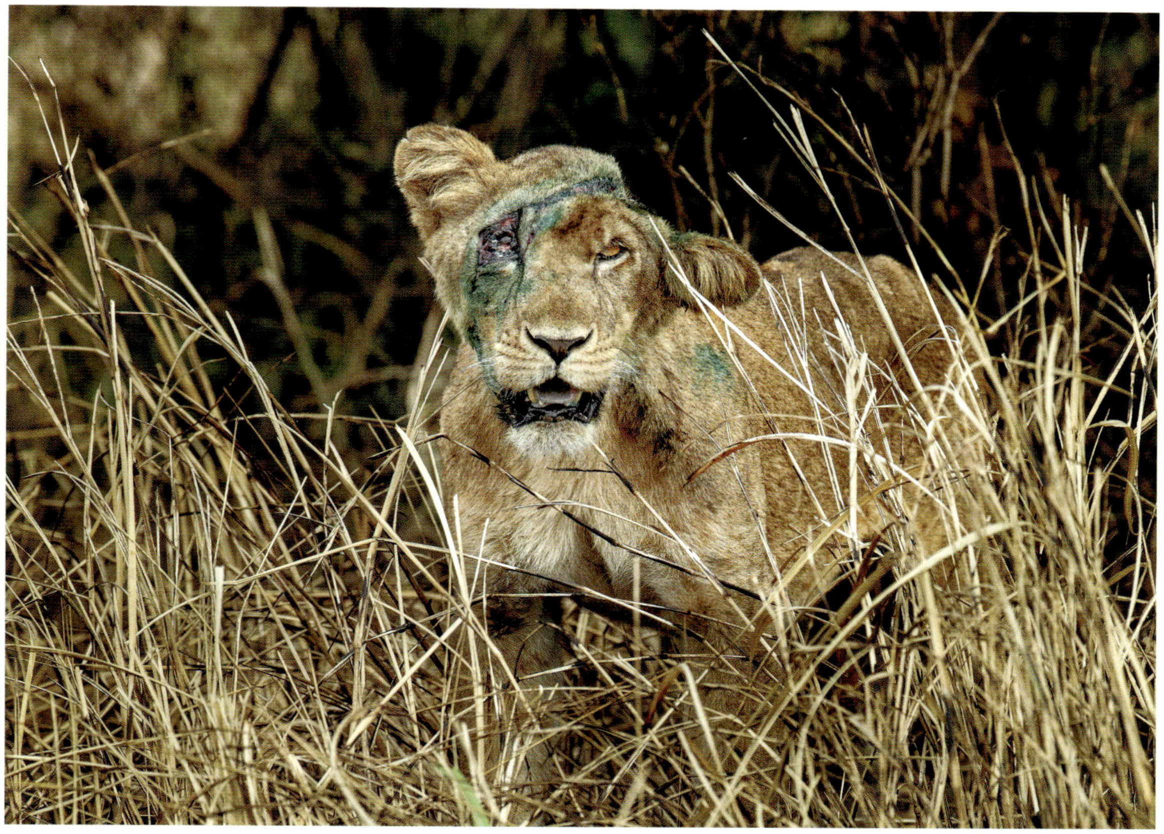

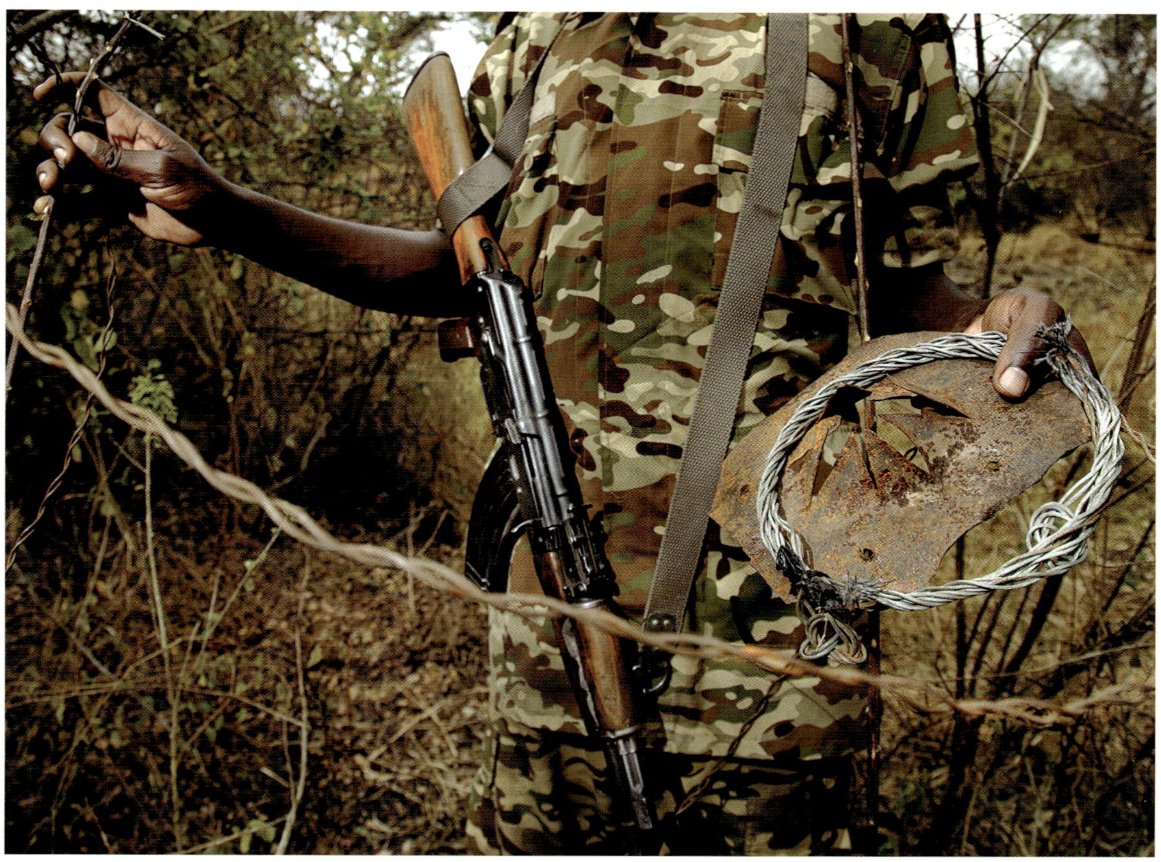

ABOVE: Poaching and loss of habitat are the top threats facing Africa's lions. This young lioness, caught in a snare in Queen Elizabeth National Park, was blinded and required treatment by a wildlife veterinarian.

RIGHT: Rangers fight a constant battle against poachers in the park who set snares that indiscriminately kill all kinds of wildlife. Some poachers are local; others slip across the border from the Democratic Republic of the Congo.

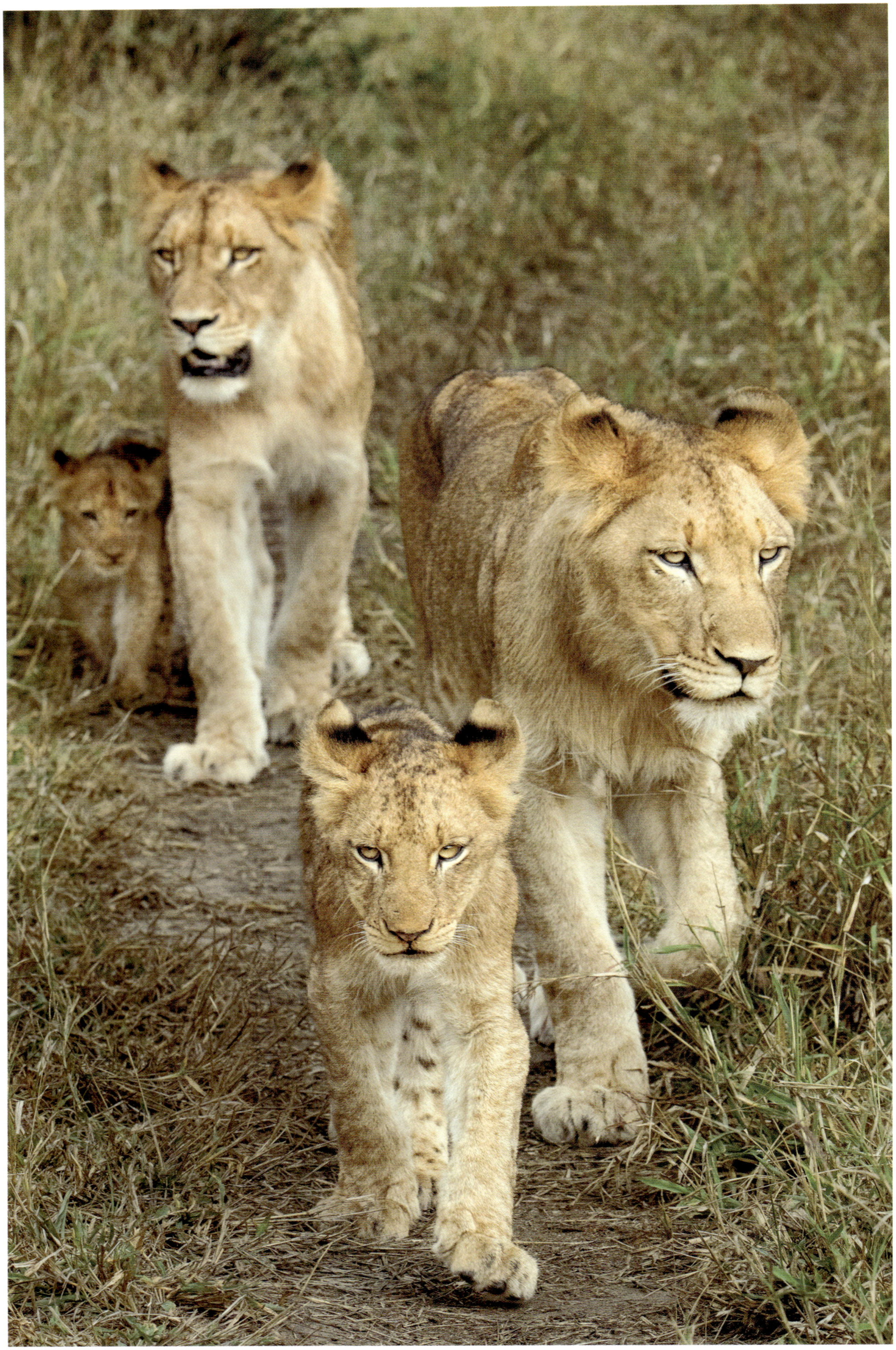

Two females with their cubs are part of a pride in South Africa's Sabi Sand Nature Reserve. Lions are the only social big cat, living in prides that can include as many as 40 animals.

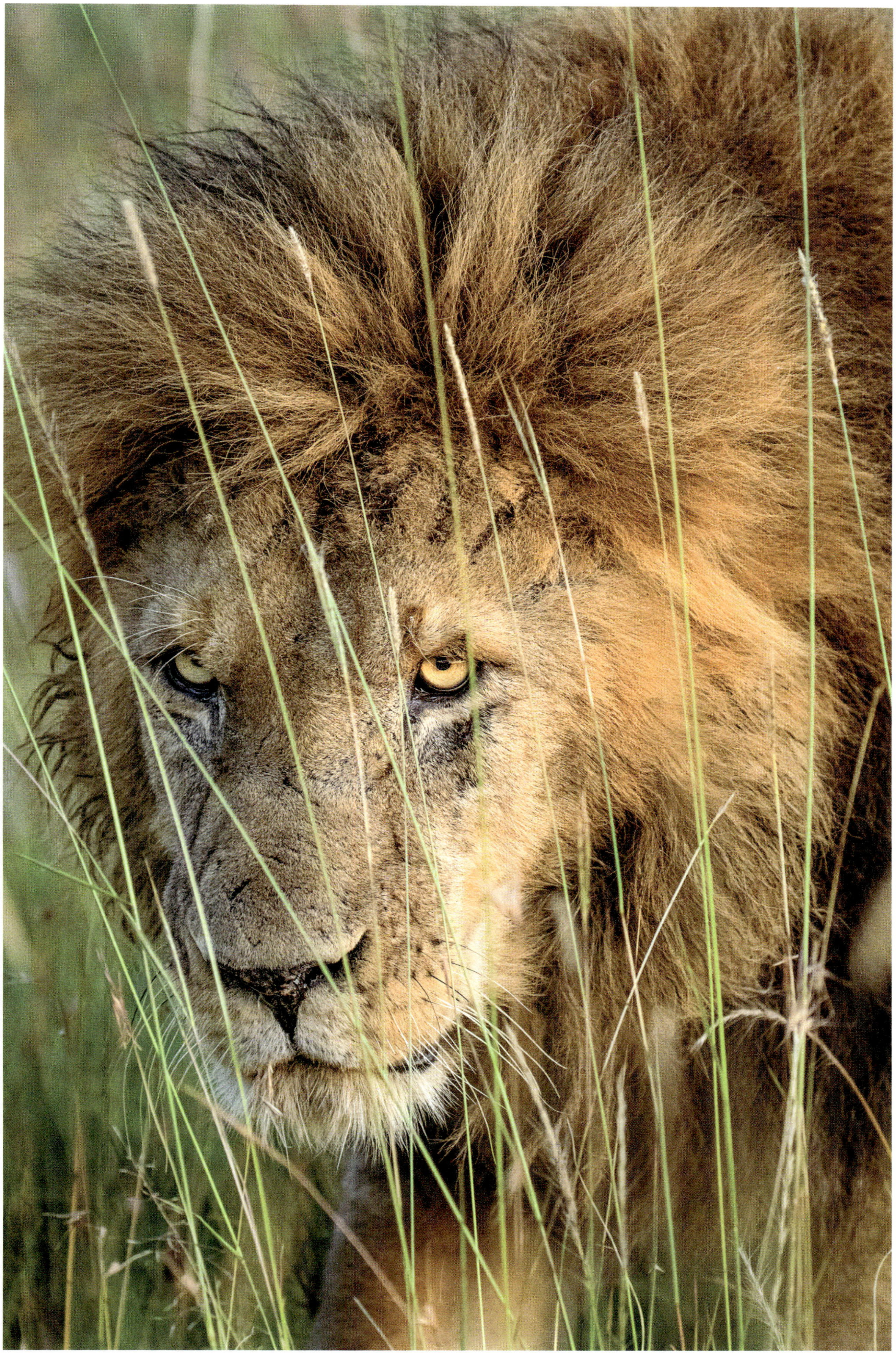

This lion, nicknamed Scarface by local biologists, roams Kenya's Masai Mara National Reserve. As the reigning male, he fathers cubs and protects the pride, fighting off intruders.

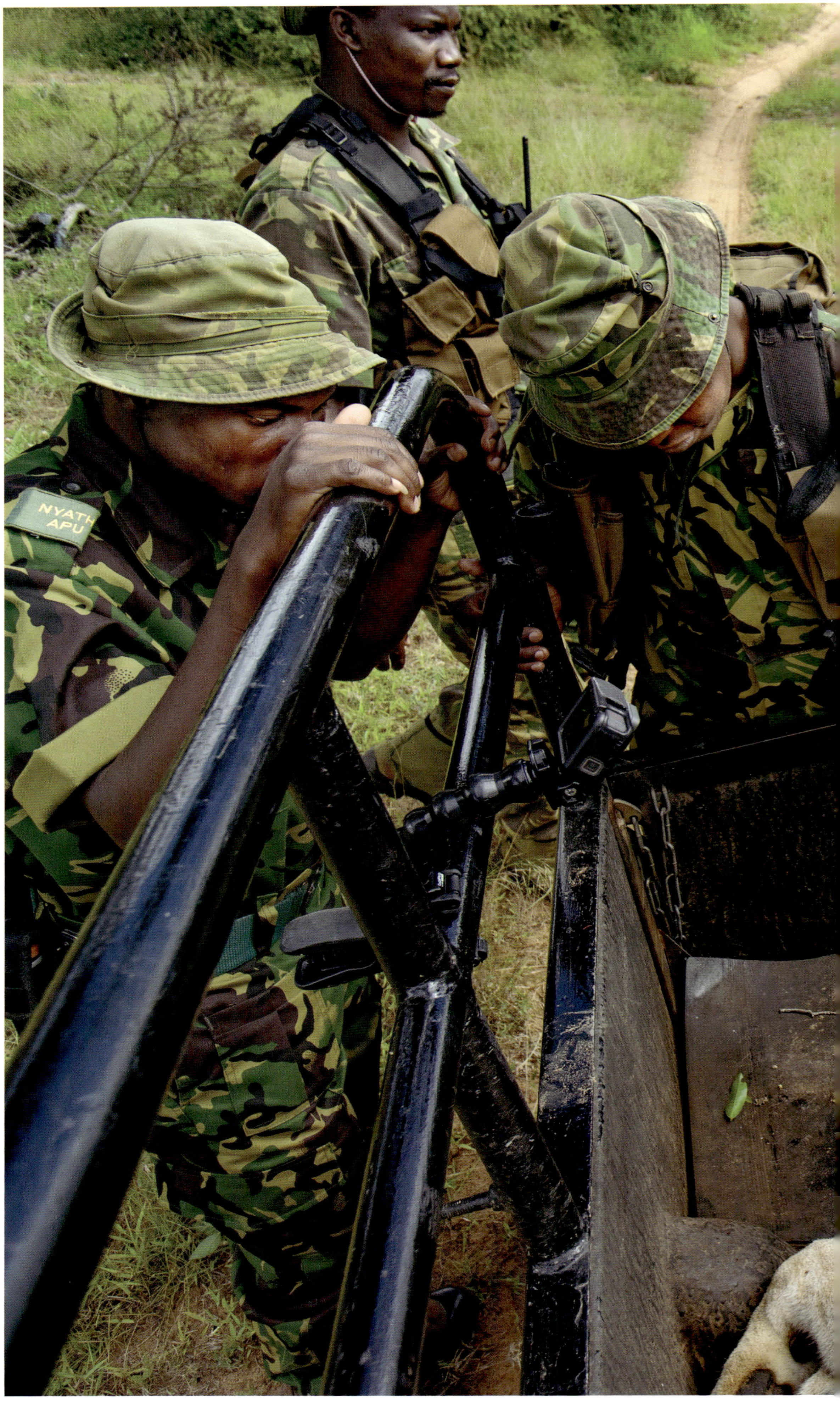

A team readies sedated lions for transport. This South African pride was captured in Phinda for relocation to Liwonde National Park in Malawi, which hadn't had a breeding population of lions in decades.

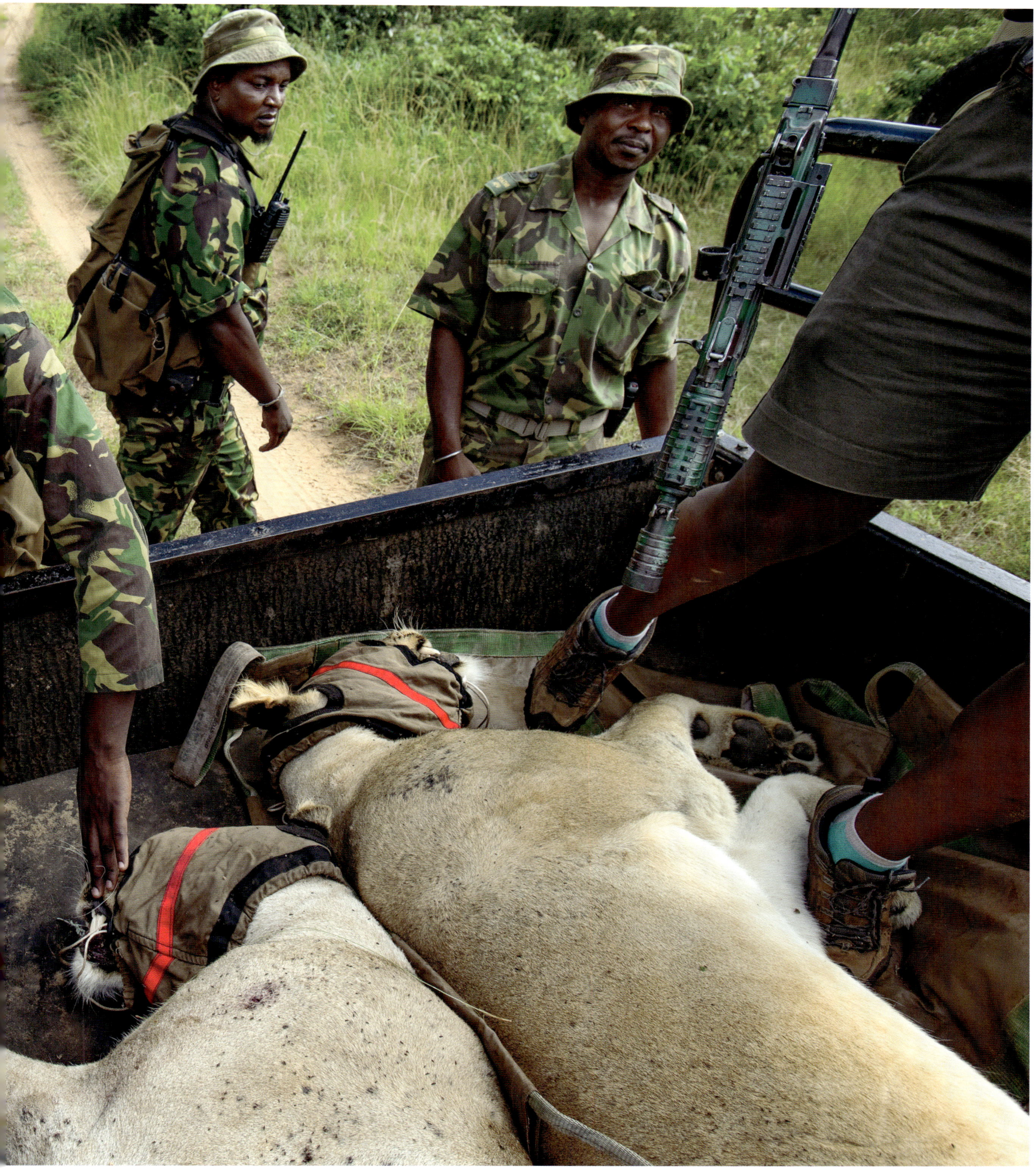

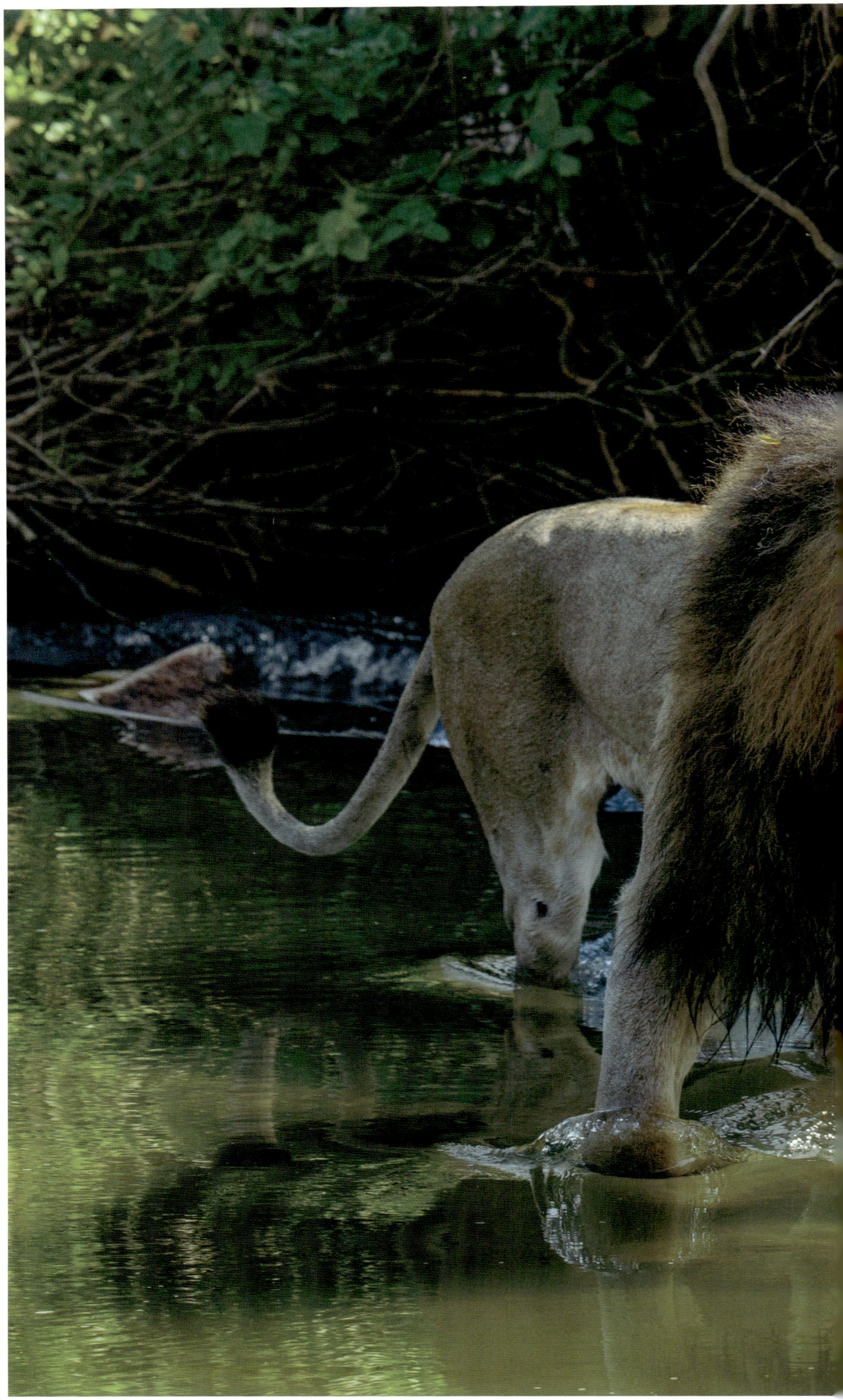

A lion and lioness cross a river in South Africa's Sabi Sand Nature Reserve, a private protected area.

PAGES 244–245: A lioness waits for her turn to feed on a giraffe under a canopy of stars in Sabi Sand. While females do most of the hunting, males eat first.

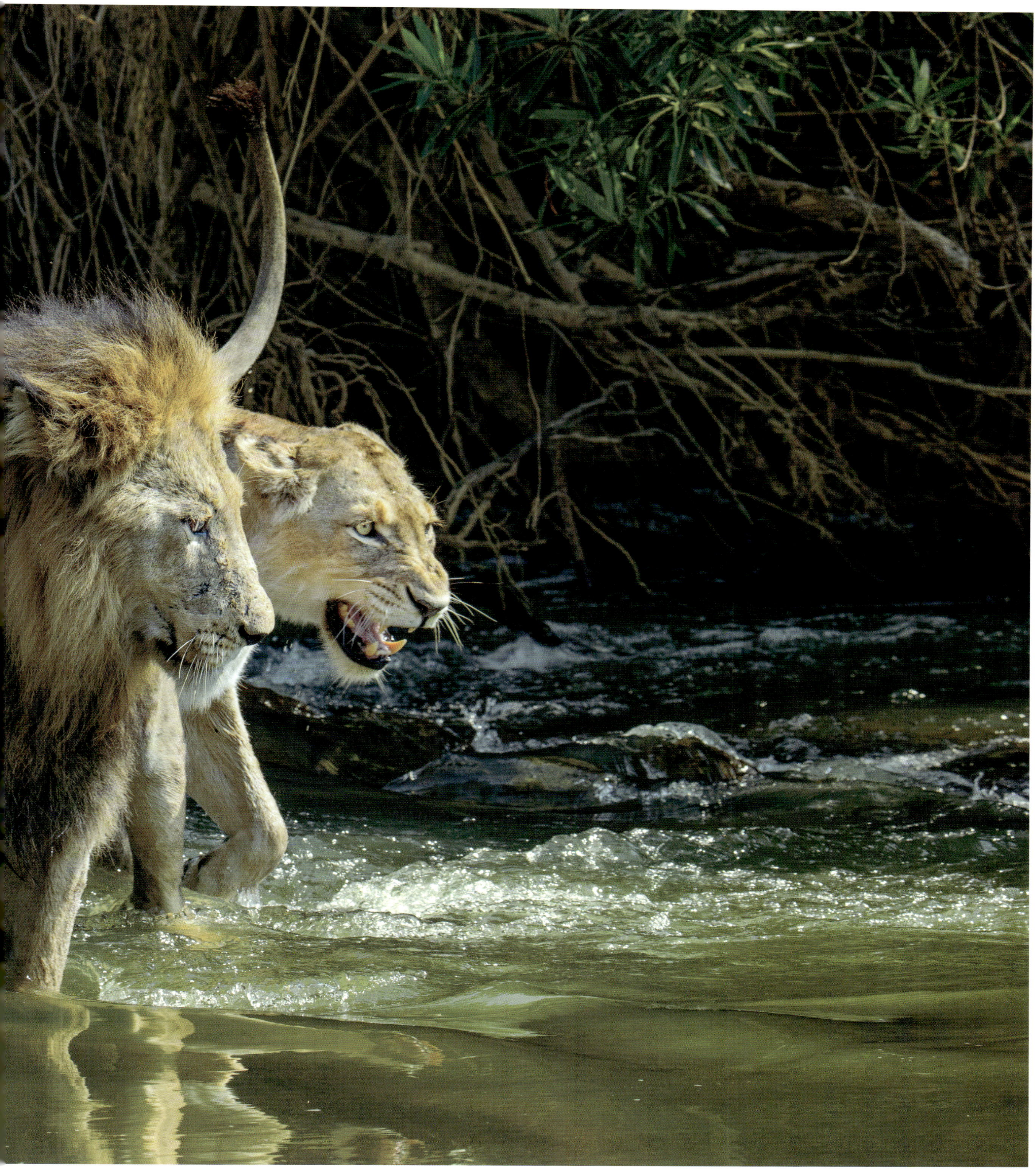

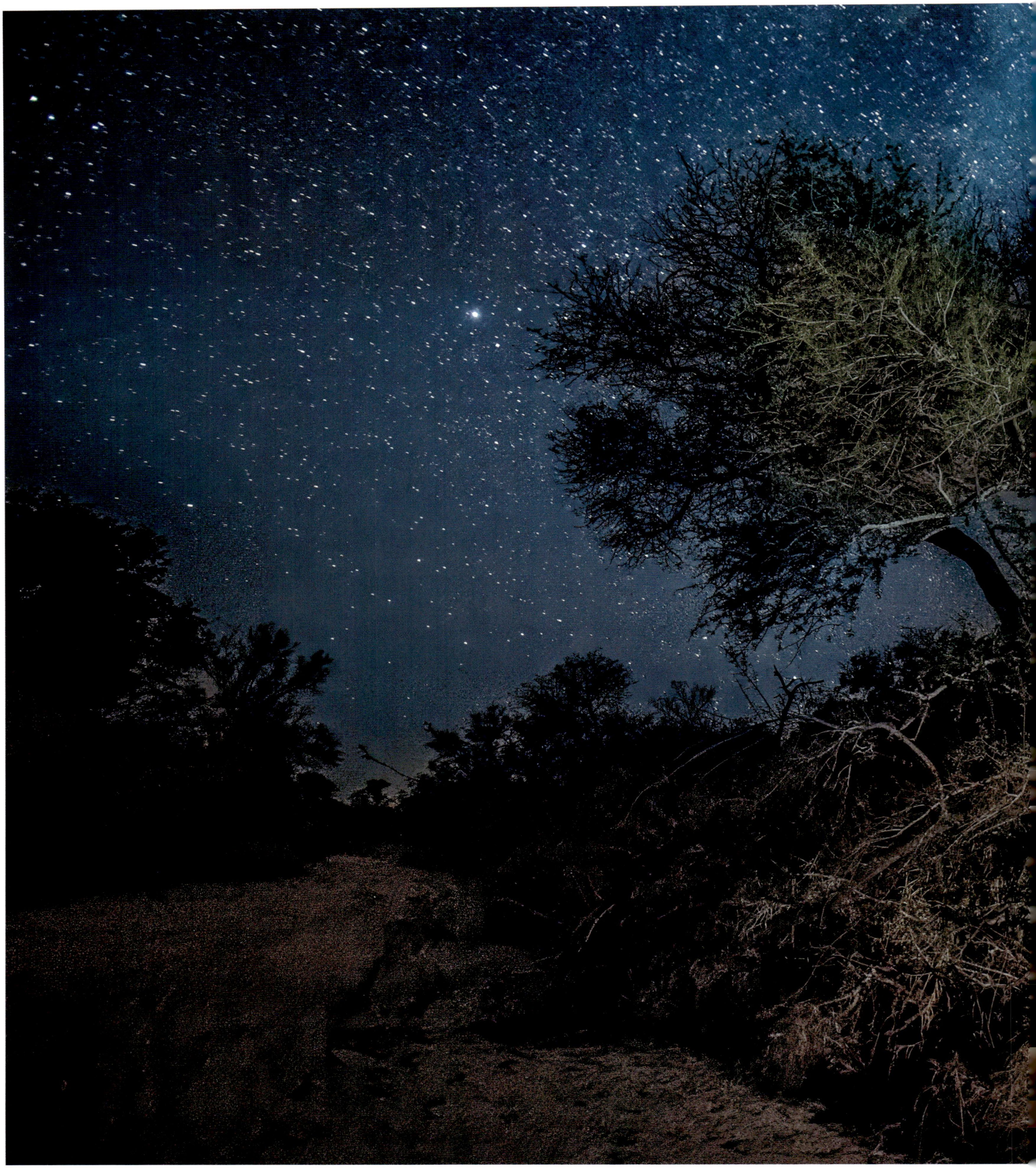

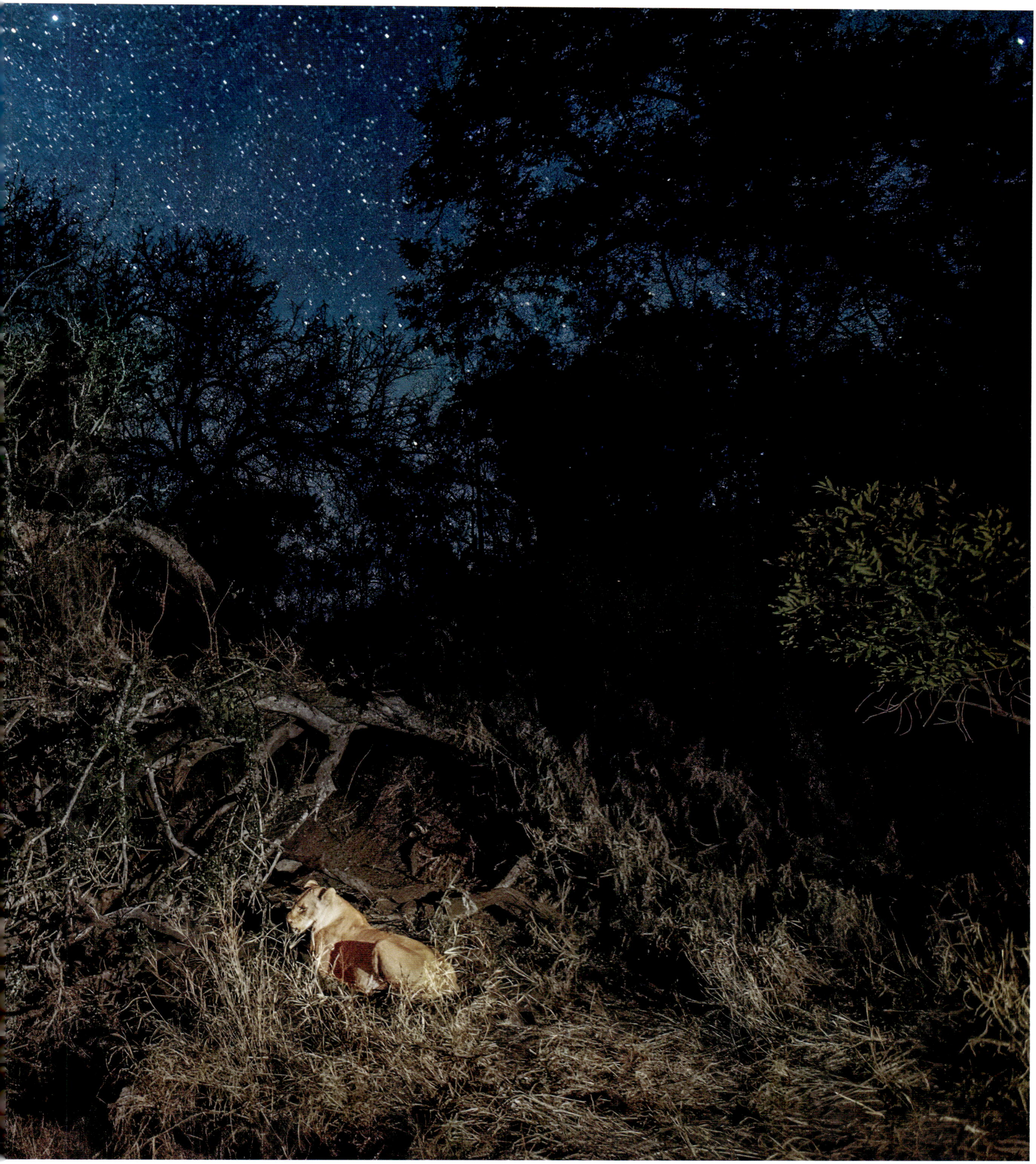

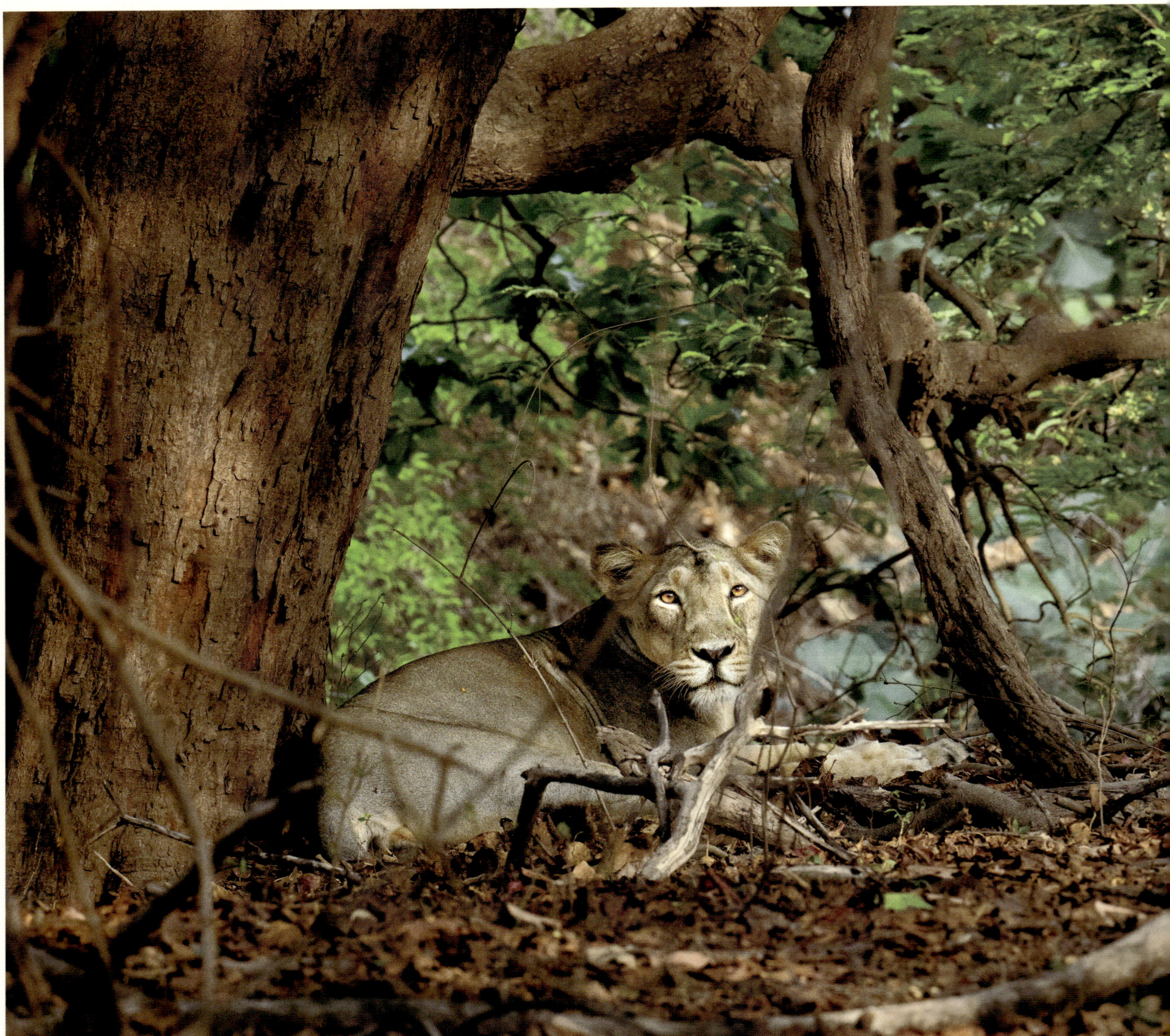

CHAPTER NINE

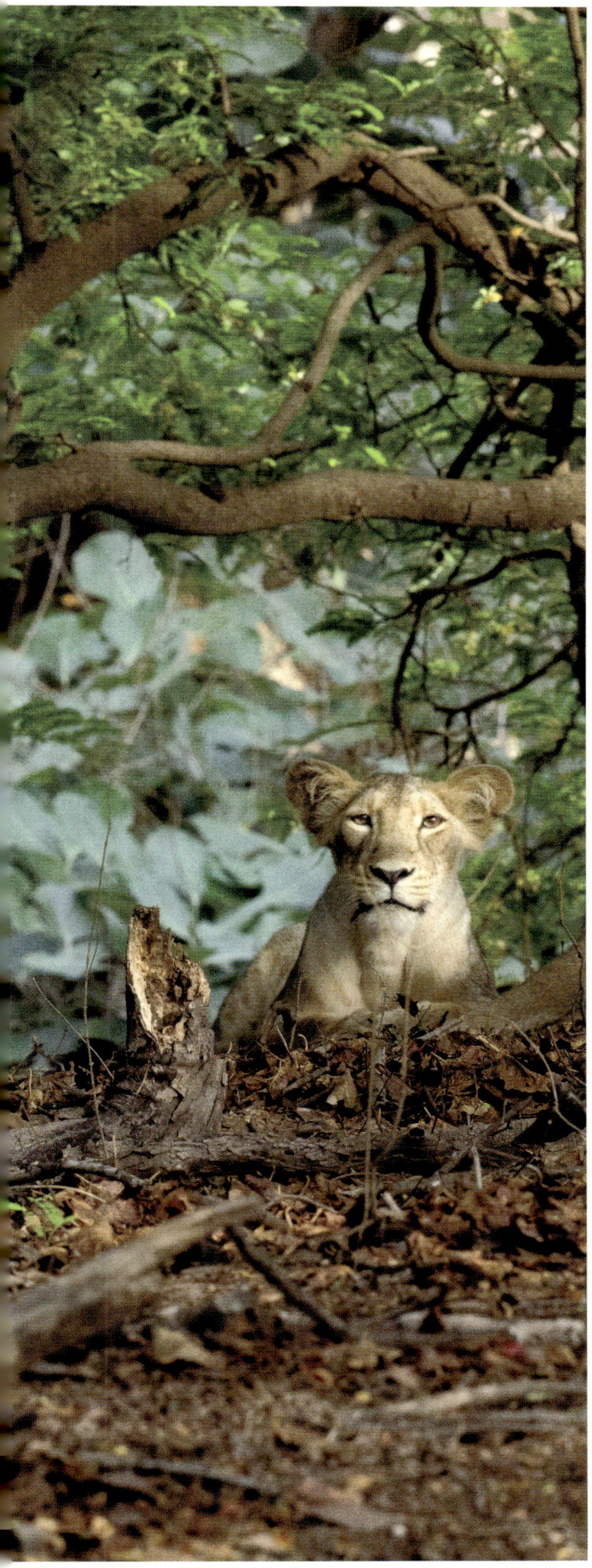

Asiatic Lions
The African Lions' Little-Known Cousins

With Herculean efforts,
India has brought this subspecies
back from the precipice of extinction.
Today, the world's most populous country
seeks coexistence with lion prides.

Few people realize that African lions have a cousin in India, the Asiatic lion. Here, two Asiatic lions rest in Gir National Park. This dry deciduous forest grows lush during monsoon season but becomes parched and bare during the scorching late spring months.

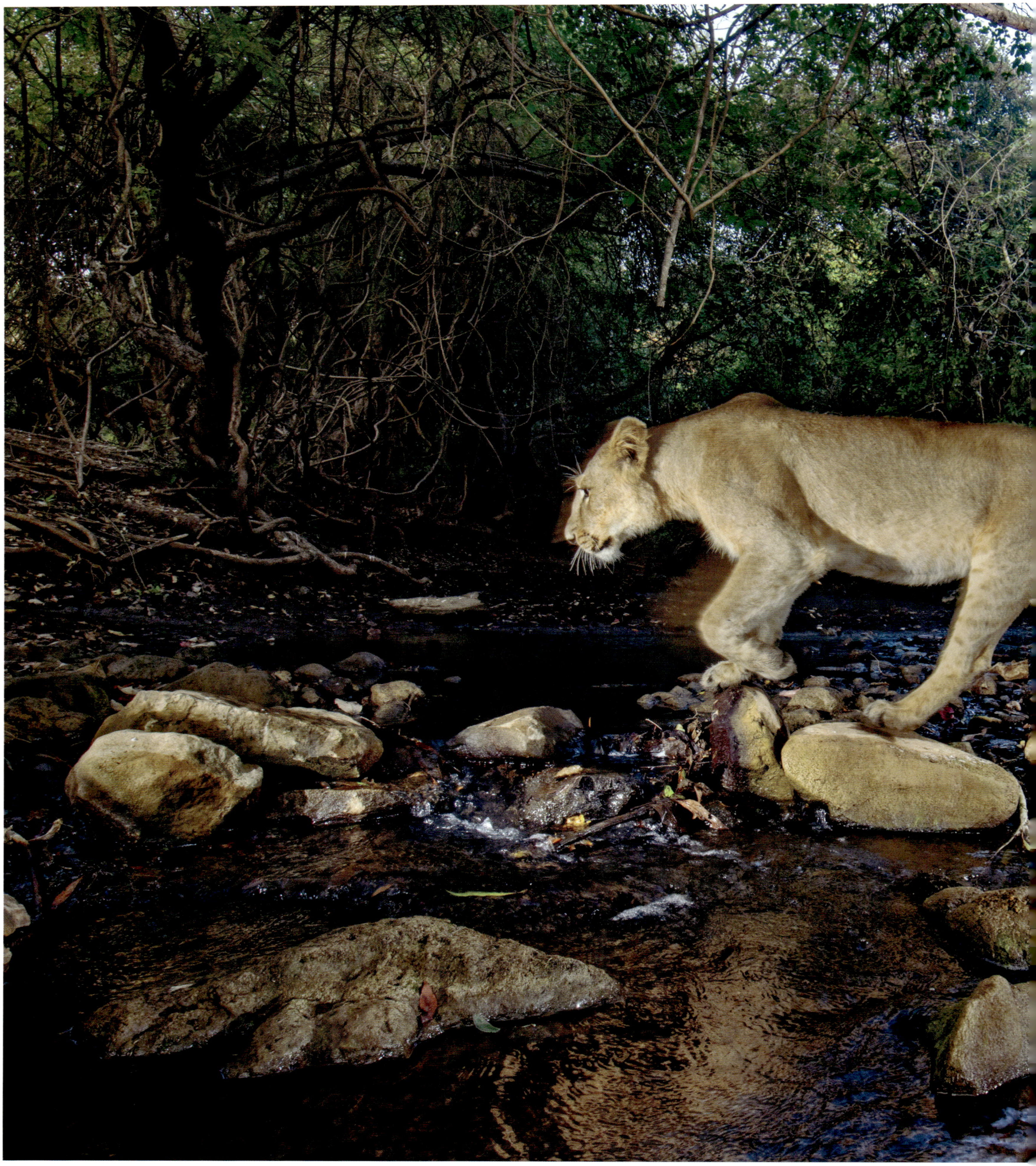

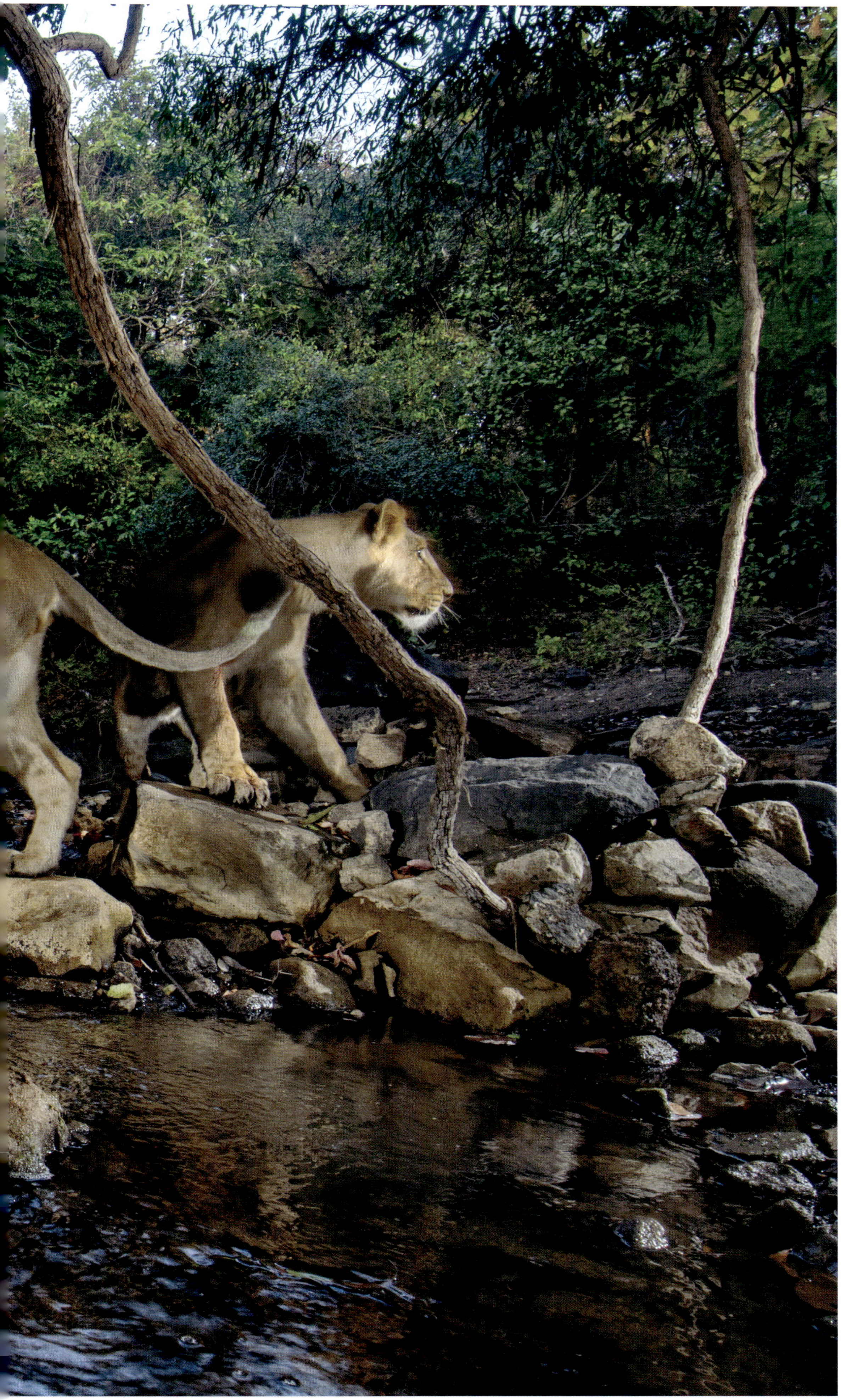

A camera trap captured this image of a lioness and a yearling cub crossing a stream. Asiatic lions once roamed from Greece, across the northern Mediterranean, and through Mesopotamia and Persia to the Indian subcontinent, but the world's last population now lives in India's Gujarat state.

PAGES 250–251: This pair stayed together, mating, for three days before parting ways. Females couple with multiple partners so that males are confused about paternity and don't harm cubs that aren't theirs.

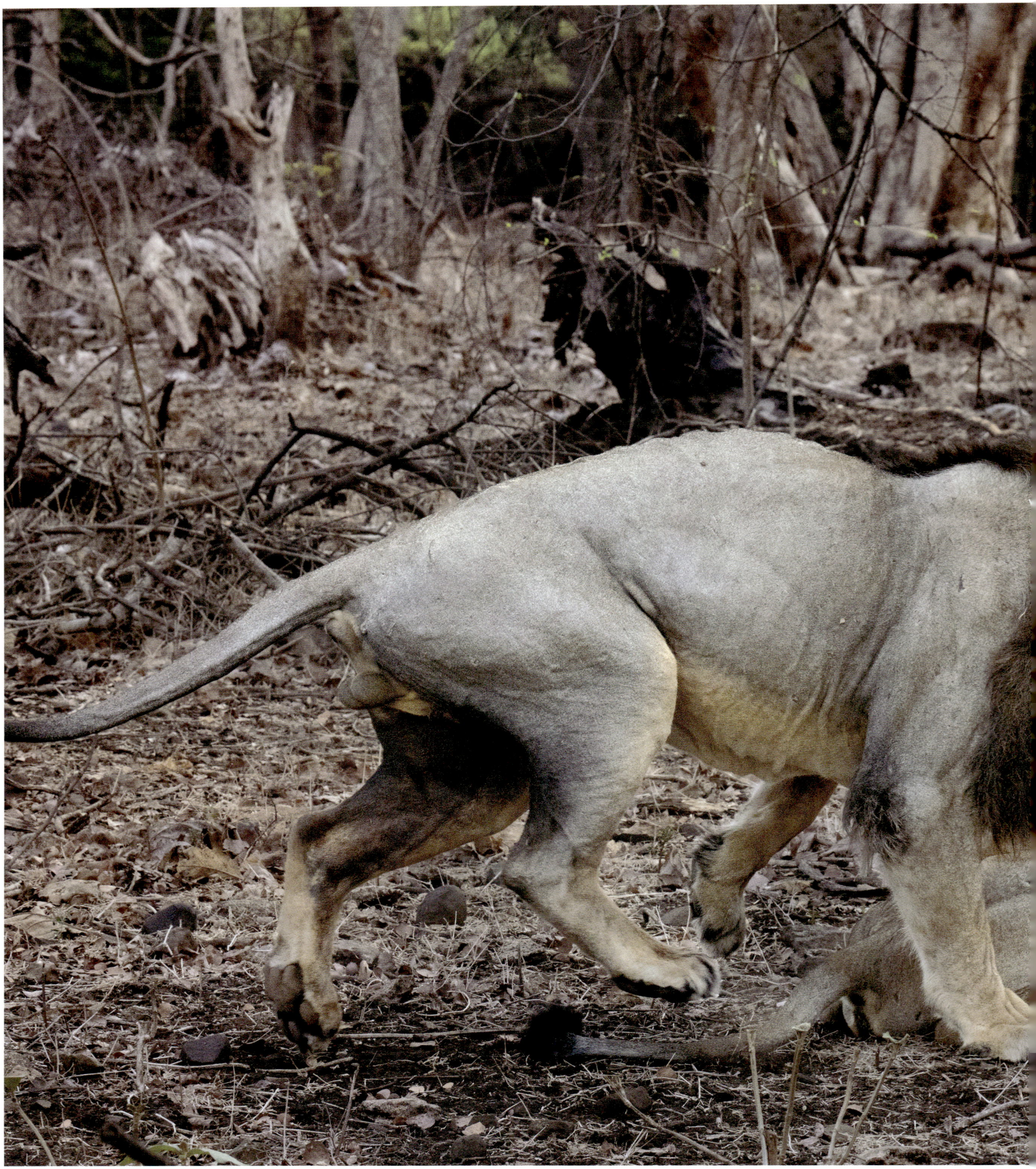

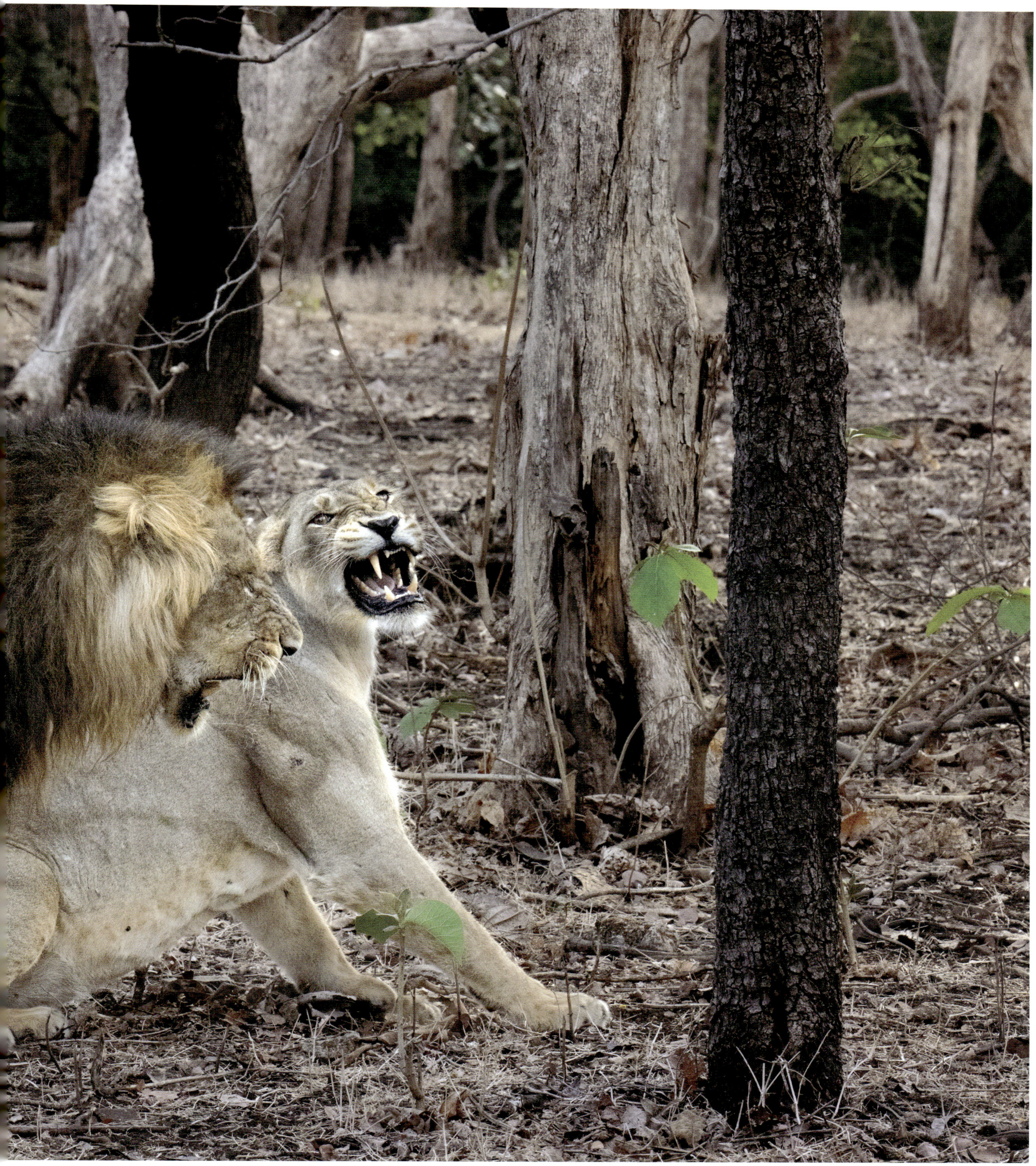

The lion was seriously injured.

He looked like a prizefighter, with old scars, bloody scratches, and a patch of fur missing beside his left eye. An open wound gaped along his spine. He'd been rushed by ambulance to the Lion Hospital from India's Gir National Park, where he'd fought with the resident male in a turf war. He was furious. He lunged against the cage bars and roared at ear-splitting decibels.

I was allowed to photograph from a distance as the medical team transferred him into a squeeze cage, cranking the sides inward to immobilize him. He lowered his head, his roars softening. Dr. Paresh Vadher, the head veterinarian, quickly inserted an IV into the lion's tail, injecting pain meds, antibiotics, vitamins, and Ayurvedic medicines. His team treated the wounds they could reach without being mauled, then wheeled him off to a large, quieter enclosure away from the other inpatient lions and leopards.

Dr. Paresh peeled off his surgical gloves and gave us an update. The cat would remain here under observation for at least two weeks, then hopefully he'd be released, rejoining the world's last remaining population of Asiatic lions that live only here, in India's Gujarat state.

But this isn't another tragic tale of a magnificent animal's demise. On the contrary, it's a resurrection story in a time marked by relentless wildlife declines.

A century ago, just a handful of Asiatic lions remained, yet they've survived. Extraordinary efforts have brought them back from near extinction. The government established sanctuaries and a national park, adopted 21st-century monitoring, built wildlife hospitals, and increased community programs. A massive guard force eliminated lion poaching—an extraordinary achievement that I've never seen anywhere else, and I've worked in more than 100 countries.

Lions have also benefited from residing in Prime Minister Narendra Modi's home state of Gujarat: The central government allocated $43.7 million to protect *Panthera leo* between 2021 and 2024. But the most important factor in their rebound may be the local people's remarkable capacity to coexist with lions.

With protection, numbers hit 300 in 2008, and the most recent census in 2025 counted 891. Their future, though, is not yet secure. Saving a large apex predator in the world's most populous country—home to 1.5 billion—is complicated. More than half of Gujarat's cats now live outside protected areas, wandering through villages and cities. They prey on livestock. Sometimes, they attack people.

They need more protected space. A single cataclysm—a cyclone, wildfire, flood, or virus—could derail the lions' miraculous recovery. "We have all the eggs in one basket," says Yadvendradev Jhala, former dean of the Wildlife Institute of India, who's spent 40 years studying Asiatic lions.

Mohan Ram calls the cat's rebound "a great conservation success story." He serves as the Gujarat Forest Department's deputy conservator of forests, and he's masterminded conservation initiatives here since 2019, a model, he says, for others trying to protect carnivores around the world. But with that success comes a looming question: What will it take to ensure the lions' long-term survival? Experts say it will require ongoing innovation and cooperation. For Gujaratis, who cherish these lions and work hard to save them, it may require sacrifice: letting some of them live elsewhere.

Few people realize that African lions *(Panthera leo melanochaita)* have an Asiatic cousin. These two "modern lions" diverged from prehistoric cave lions some 70,000 years ago. Asiatic lions *(Panthera leo leo)* once roved from North Africa's coastal forests through northern Greece and the Mediterranean and across Mesopotamia and Persia, reaching India sometime between 4,000 and 10,000 years ago. Regardless of where they've lived, they've been revered, reviled, and relentlessly persecuted.

The "King of the Beasts" has captured the human imagination throughout history, symbolizing strength, power, valor, and nobility. That's why elaborate hunts have been a royal pastime for five millennia, dating to the Assyrians, and why they continued into modern times with India's princely rulers and the British Raj. Like all predators, lions were hunted en masse and nearly eliminated.

But these cats also graced Persia's imperial seals, and they adorn cathedrals, city gates, and grand public buildings. Buddha was depicted as a lion. This feline appears in Homer's *Iliad,* Aesop's *Fables,* the Old Testament, and today on Broadway in *The Lion King.*

Lions have held deity status among Hindus for millennia, says conservationist Rohit Vyas. In India's 3,500-year-old Vedas, Narasimha (a centaur-like half-man, half-lion) appears as an incarnation of Vishnu. The warrior goddess Durga's celestial mount is a lion. Meeting a lion is considered auspicious, a visit from the divine. The lion became the Republic of India's national animal in 1948 (though it was later displaced by the tiger).

A lioness gazes at a potential meal. With restoration efforts in Gir National Park and other wildlife sanctuaries in Gujarat, sambar deer, spotted deer, and other favored prey have proliferated, so lions now kill less livestock near protected areas.

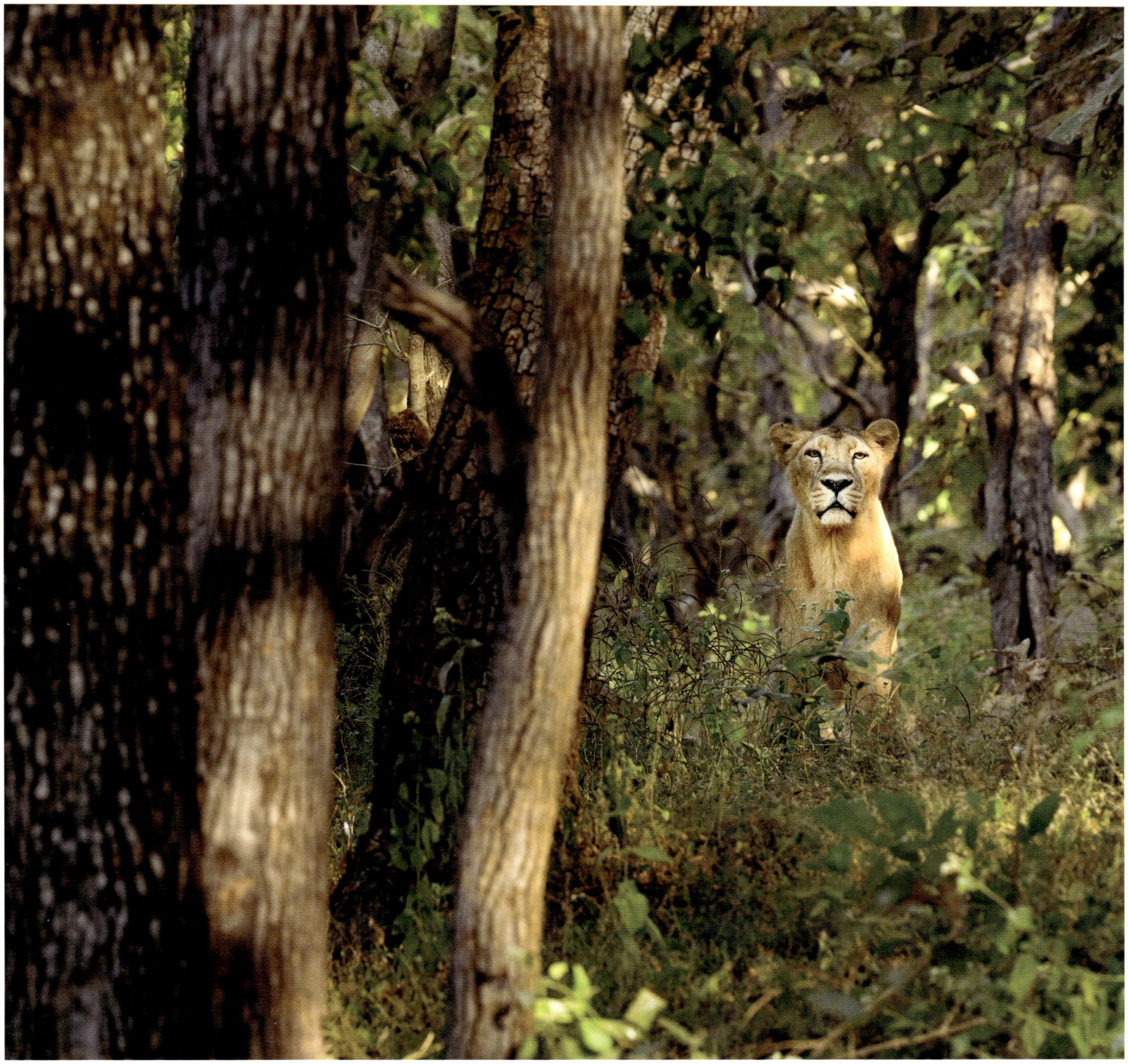

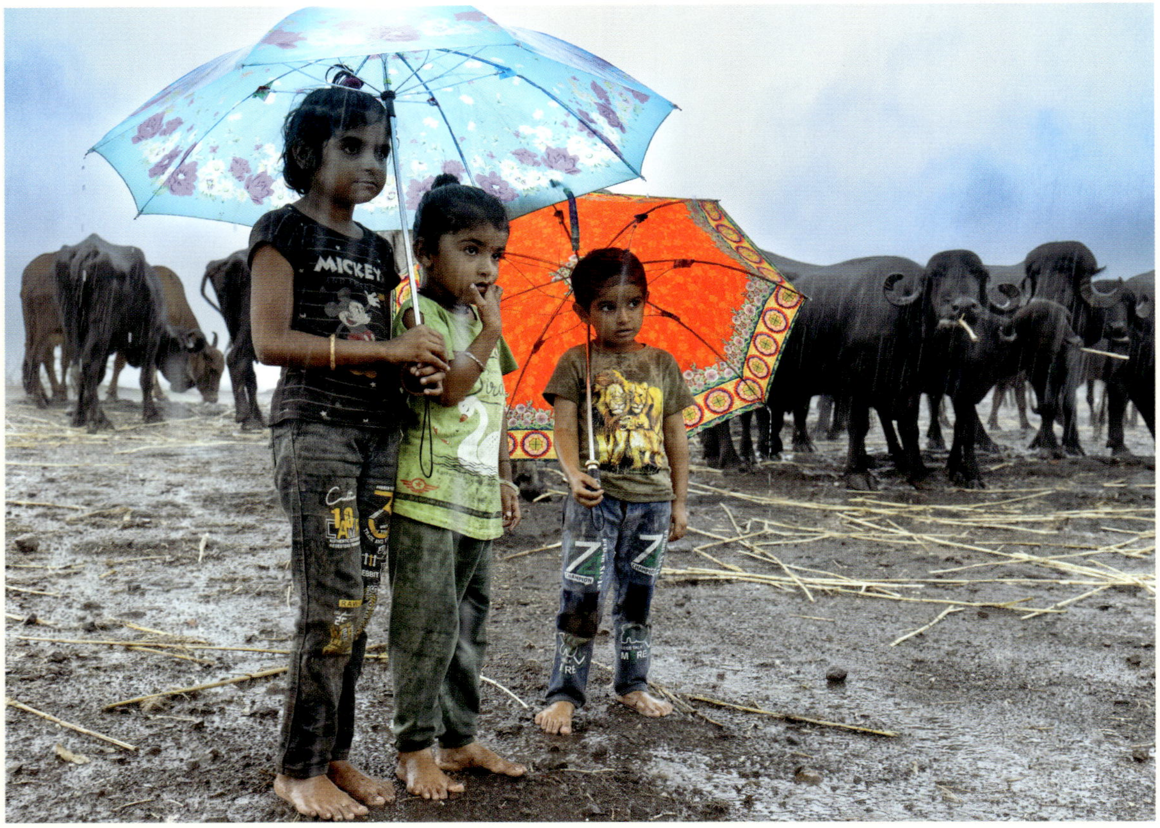

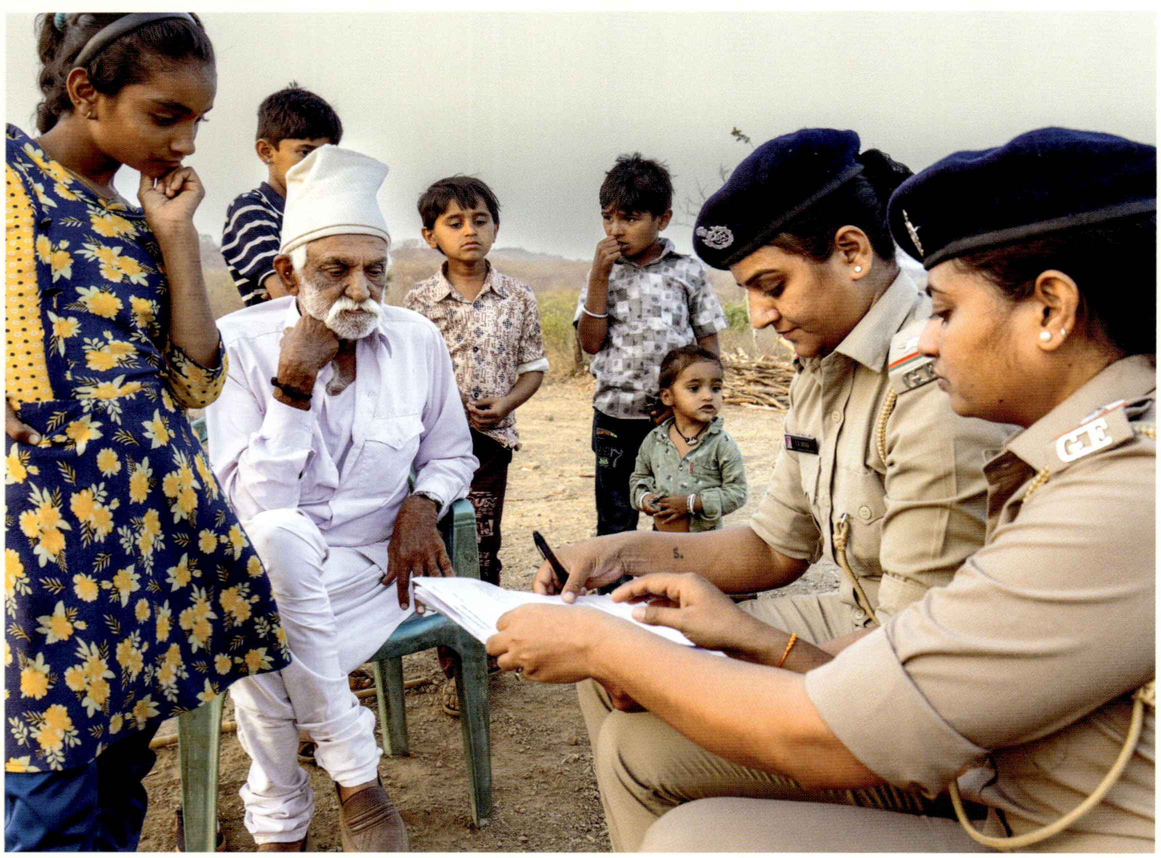

ABOVE: Three Maldhari children stand in the rain. These Indigenous herders have lived in Gir Forest for generations, though many were relocated when the national park was created in 1975. Today, about 4,300 remain in the sanctuary.

LEFT: Rangers with the Gujarati Forest Department meet with Bharabhai Vejabhai, the head of a Maldhari household of 60 people. He lost a buffalo to a lion the night before this photo was made and was filing paperwork to receive government compensation.

This deep reverence is a major factor in the lion's extraordinary rebound. "Here, people have learned to coexist with them," says Yadvendradev Jhala. But it's a fragile balance.

It took five years for Sharon and me to get journalist visas and the high-level permissions I needed to set up camera traps in the national park. We'd first proposed the story to *National Geographic* in 2017. The paperwork took forever, and then the pandemic hit. Finally, in May 2024, I got a call from my producer, Nimit Pandya: We had our papers.

Though it was the hottest time of year, far from ideal working conditions, I had to establish footing, build relationships to make the story work, and get there quickly: Monsoon season loomed, and the park would be closed for four and a half months.

Dawn blanched the sky as we drove into Gir National Park, revealing parched, golden grasslands, a landscape woven with spiky acacia trees, nearly bare teak, and huge banyan trees with riotous aerial roots.

We scanned for movement in the nascent light, looking for lions. Sagar Manjariya, the park's lead tracker, walkie-talkied other guards for intel. I photographed herds of spotted deer that regarded us warily as we passed and a troupe of acrobatic langur monkeys that bounced through branches above. A flock of green parrots. Wild boars. Scampering mongooses. A tall, antlered sambar, its rump more horse than deer. Parading peacocks. The wind was strung with birdsong: There are 338 avian species in Gir, as well as 41 mammals, 47 reptiles, and some 631 plants and trees. It's a biodiverse wonder.

And then I saw them: A lioness camouflaged in flaxen grass with her tiny three-month-old cub frolicking nearby. We killed the engine. She was vigilant, but regal, mostly ignoring us. At first glance, she looked like her African cousins, but as I photographed, I noticed differences. She was slightly smaller, with distinctive belly folds that resembled stretch-marked skin following weight loss.

Her baby leaped on her head, gnawed her ears, licked her face, and tugged on her flicking, tufted tail. Finally she swatted back. She was old, about 10. She'd lost one of her two cubs, so she was her offspring's only playmate. They rolled and chased for a delightful half hour. Then the lioness froze, listening. She roared in rhythmic blasts that reached a thunderous crescendo before quieting to short, guttural grunts. Perhaps she was calling to her pride. She sauntered into the forest and disappeared, her cub bounding close behind.

Two males were nearby, the only lions in this group who'd been named: Veeru, the king, and his brother, Jai. The males slept through the afternoon heat, but as I sprawled in a thicket to photograph them, I could see that their manes were visibly scantier than those of their African counterparts. Their tails had larger tufts, and they sported the Asiatic lion's trademark: a slim, dark Mohawk running from their crown to their mid-back.

But the most striking difference is how these gregarious cats organize their communities. In Africa, a group of males live with an extended family of females and cubs—all of them ruled by one dominant male. But in Gir, the two sexes live mostly separate lives. Two or more lionesses form a pride, raising and protecting their cubs in a communal crèche. So much of my work covered solitary cats. It was different shooting images of these affectionate families, greeting each other nose-to-nose or nuzzling faces before plopping down together.

The males buddy up in a coalition of two to five relatives. They protect—and procreate with—a few prides, but one male reigns supreme. The females also mate with multiple partners, but for them, it's a strategy to protect their cubs, Mohan Ram explains. With males confused about who fathered a litter, it "ensures that their progeny isn't killed by any of [them]," he says.

It took two weeks to get the go-ahead for camera traps. Sagar helped us choose locations to set up the cameras and protected us while we were out of the jeep, on the ground, in lion territory. (One time a pride walked through an area we'd left maybe 10 minutes before.) Camera traps are finicky. We had unrelenting technical problems, but we got three up and received permission to stay for two weeks after the park shut down for monsoon season. I got my first pictures: a mom and cubs drinking at a water hole, cats mating, and portraits of Maldharis. About 4,300 of these seminomadic herders live in the wildlife sanctuary and have resided there, with lions, since the late 1800s. Then the rains came, and I went home.

I returned for six weeks in November and then came back a third time in March, with funds raised through our nonprofit, Big Cat Voices. There'd been a baby boom, and we spent time with a veritable herd of lions: four females and their 10 cubs. We often saw Veeru and Jai sleeping through blast-furnace, 106°F afternoons.

About half of these cubs may survive to adulthood, staying with the family until they're two or three years old. Then females absorb into a pride, but males must set off to find their own territory, traveling anywhere from a few miles to a few hundred. And that's where challenges arise.

Lions need substantial territory, and Gir's protected areas are full. For the first time, most Asiatic lions—507, to be exact—roam outside preserves. So a big part of this story was how to manage lions that roam through a human-dominated mosaic of towns and cities, farmlands, industrial complexes, and lands sliced and diced by roads and railways. Lions are broaching new territory, where people haven't lived with them in generations and may not have the same tolerance or caution as their forebears did. We learned that managing them requires a 21st-century conservation paradigm.

With multiple railway lines running through Gir, a shocking number of lions were dying on the tracks. The Supreme Court of India ordered action, so I documented guards patrolling railroads near wildlife areas. We visited Gir's Hi-Tech Monitoring Unit, a mother ship of computers, hard drives, and large-screen monitors where staff tracks collared cats around the clock and monitors tourist safari

(Continued on page 259)

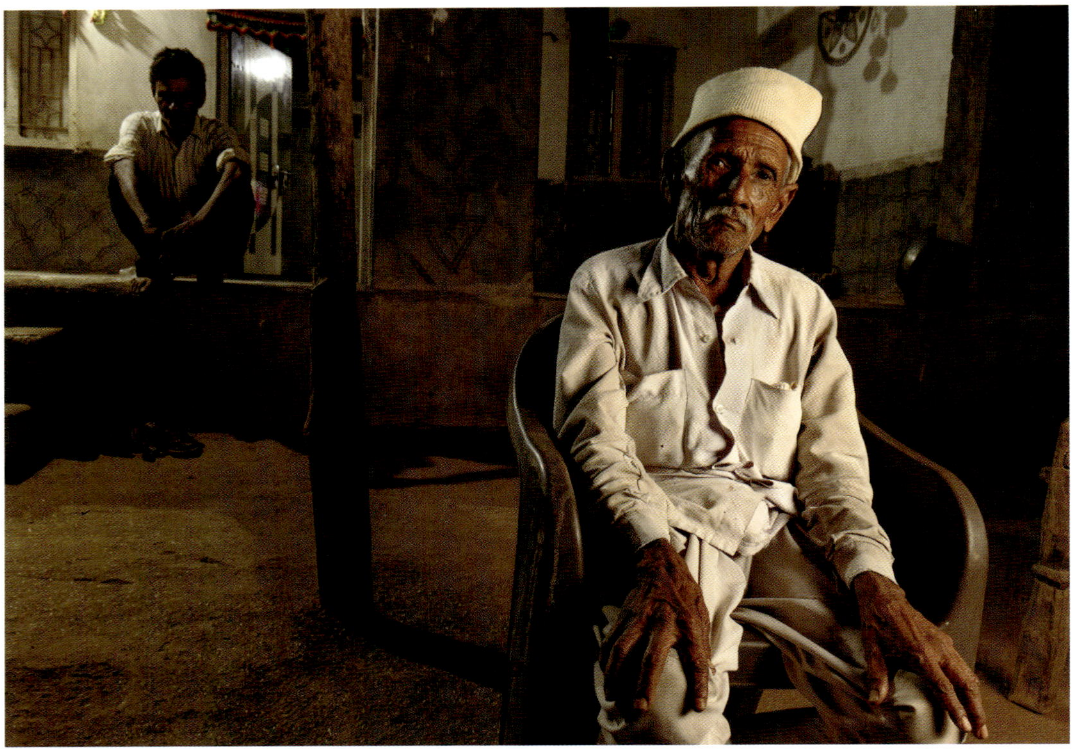

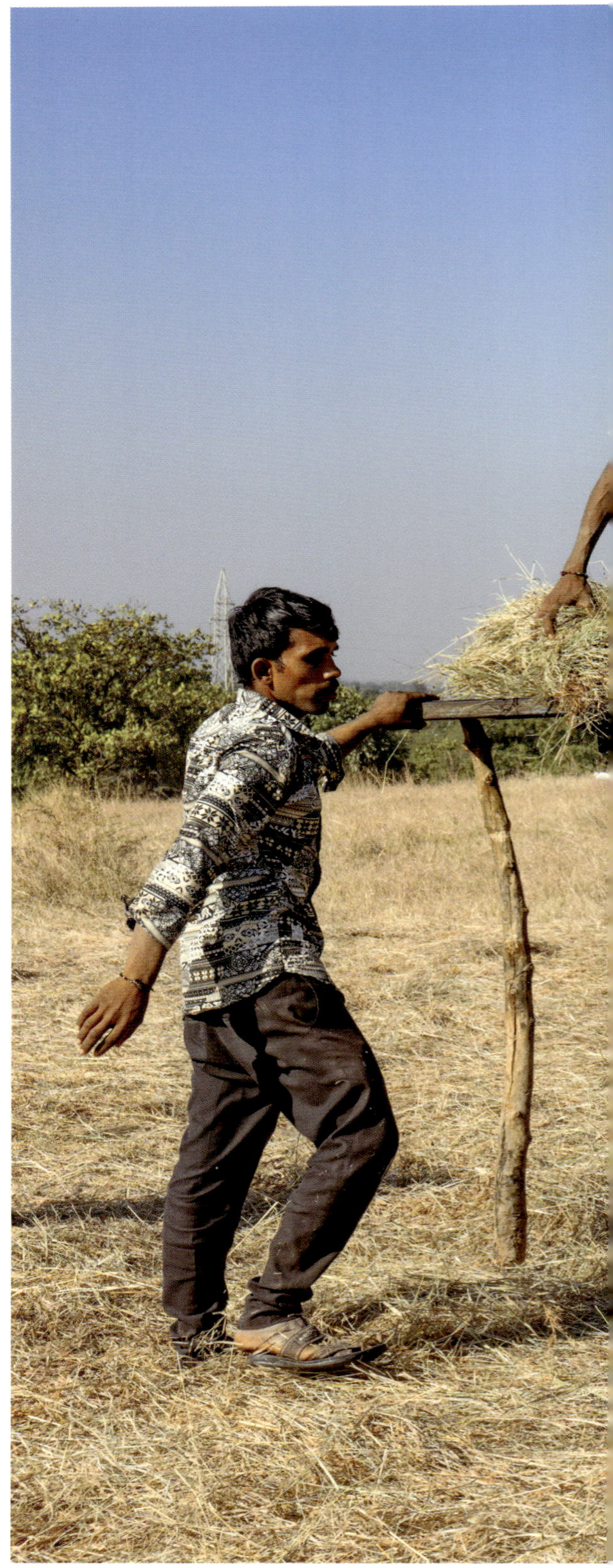

ABOVE: Tab Bhai Makwana was ambushed by a lion on his farm just outside Gir's wildlife sanctuary 33 years ago, leaving him permanently crippled and in constant pain. Attacks are rare – but lions are apex predators.

RIGHT: Healthy native grasses keep predators and prey within park boundaries. Here, workers bale these nutrient-rich grasses, which the Forest Department gives to local farmers to feed their livestock during monsoon season, promoting tolerance for lions and other wildlife.

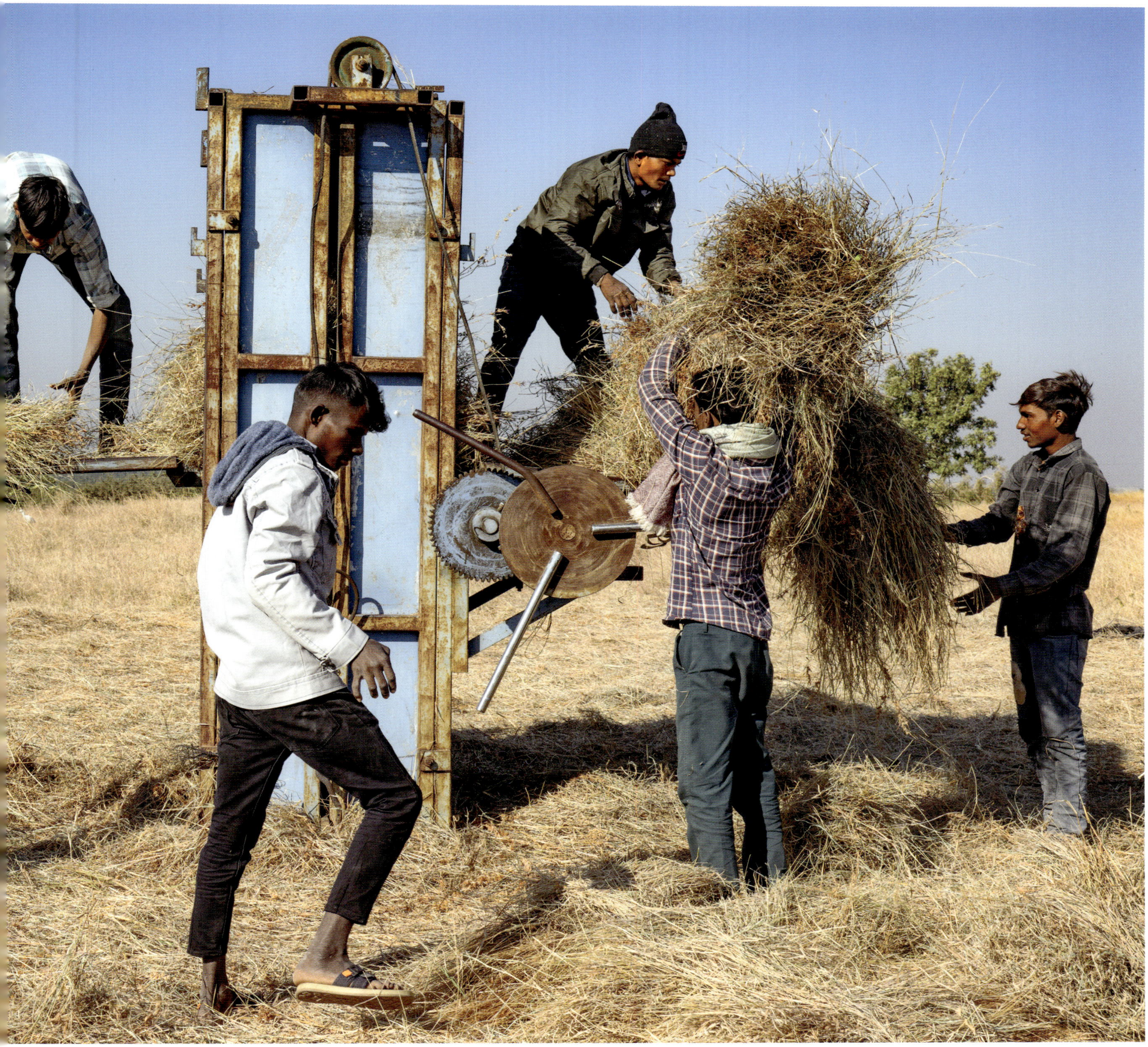

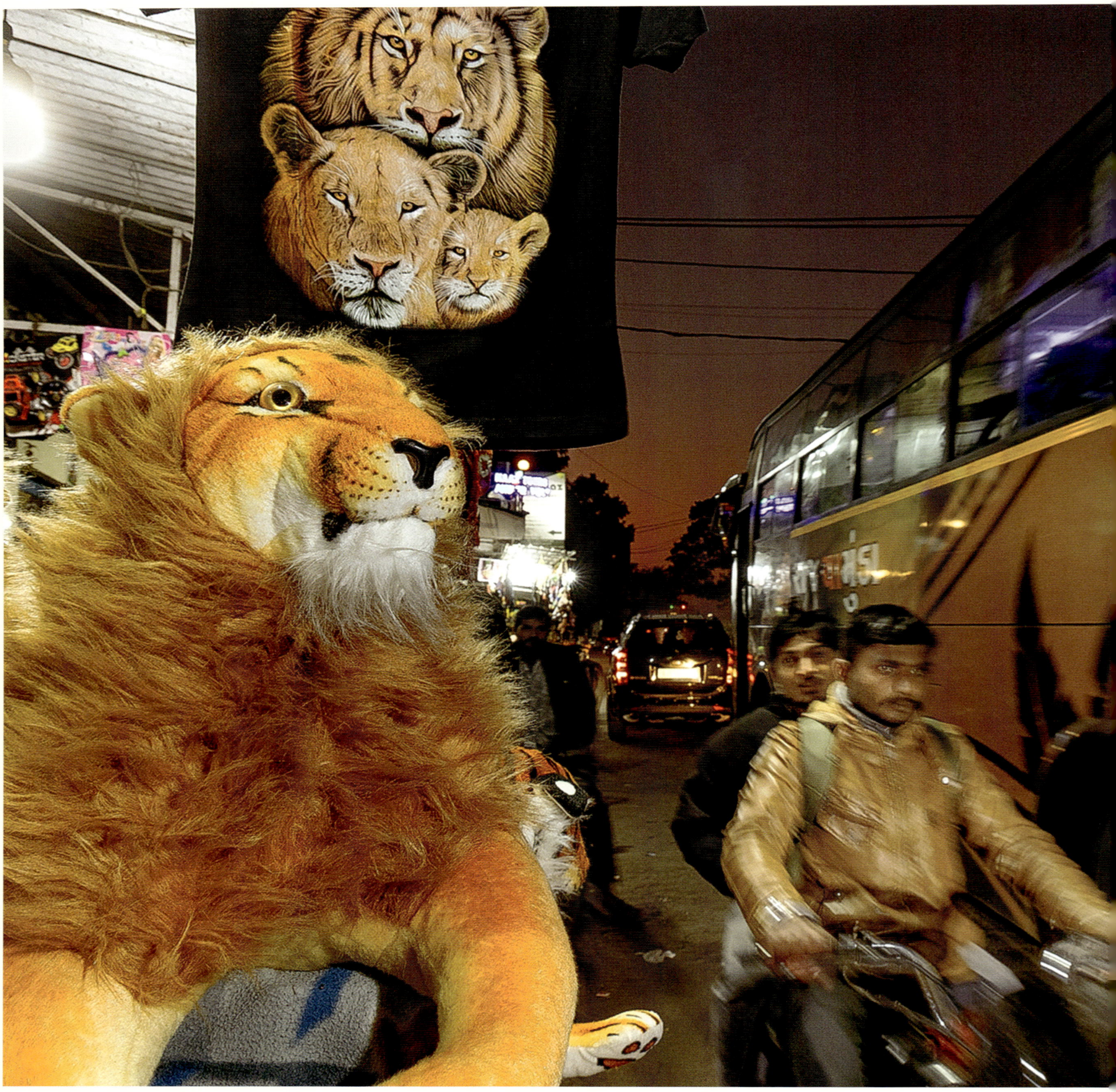

(Continued from page 255)
vehicles in the park and cars that drive on main roads where wildlife regularly crosses. We went to a Maldhari settlement; there, guards filled out paperwork for a man who lost a buffalo to a lion the night before. The forest department pays for livestock taken by lions.

But humans are in danger, too. On average, lions in India maul 14 people and kill one every year. Victims' families receive government compensation, and forest guards trap killer cats and move them to zoos or fenced safari parks.

Animal attacks—by sharks, alligators, big cats—are headline clickbait. These events are horrific, but the risk is overblown. For context, 24,000 people a year die from lightning strikes worldwide and car crashes claim 1.35 million.

I photographed Tab Bhai Makwana, a survivor of a lion attack. Years ago, a lioness ambushed him from behind, sinking her claws deep into his back and ripping off one buttock. "It was like a hit and run, the worst experience of my life, mentally and physically," he said. He was permanently crippled and lives in pain, but as I rose to leave, he told me, "I still have immense respect for the big cats."

Makwana's neighbors welcomed us into their homes so we could photograph them, too. No one could remember the last time a lion had hurt someone nearby. They were pragmatic: Lions are a fact of life. But they avoid walking through fields in the murky light of dawn or dusk and steer clear of mothers with cubs.

Jhala notes that "the cost-benefit of having lions in your neighborhood" will ultimately determine the limits of these lions' rebound. On the plus side, lions prey on deer, wild pigs, and other animals that devour crops. The forest department provides solar streetlights, constructs ponds, digs wells, vaccinates livestock, and has built 10,000 metal machans (observation towers) for farmers to safely watch over their fields.

More broadly, preserving the huge tracts of land needed by a wide-ranging predator protects the entire spectrum of life that they live beside. "Under the umbrella of this flagship species, we are conserving this ecosystem," Mohan says.

These lions need more space. A decades-long debate still rages over relocating a pride to a distant location. In 2013, the Supreme Court of India mandated such a move to Kuno National Park in neighboring Madhya Pradesh, but that ruling was ignored. "The Gujarati government has, up until now, been opposed to starting a second population," Jhala tells me.

I still can't get that basic question out of my head: What will it take to ensure that these magnificent cats still walk the Earth when my granddaughter is my age? Hope lies in the incredible efforts India has made, and in the lion's biology and resilience. Cats are prolific, and they thrive with just the basics: food, water, and habitat. When you add protection, strong laws, and monitoring, as they've done in Gir, even endangered populations like Asia's last lions can bounce back.

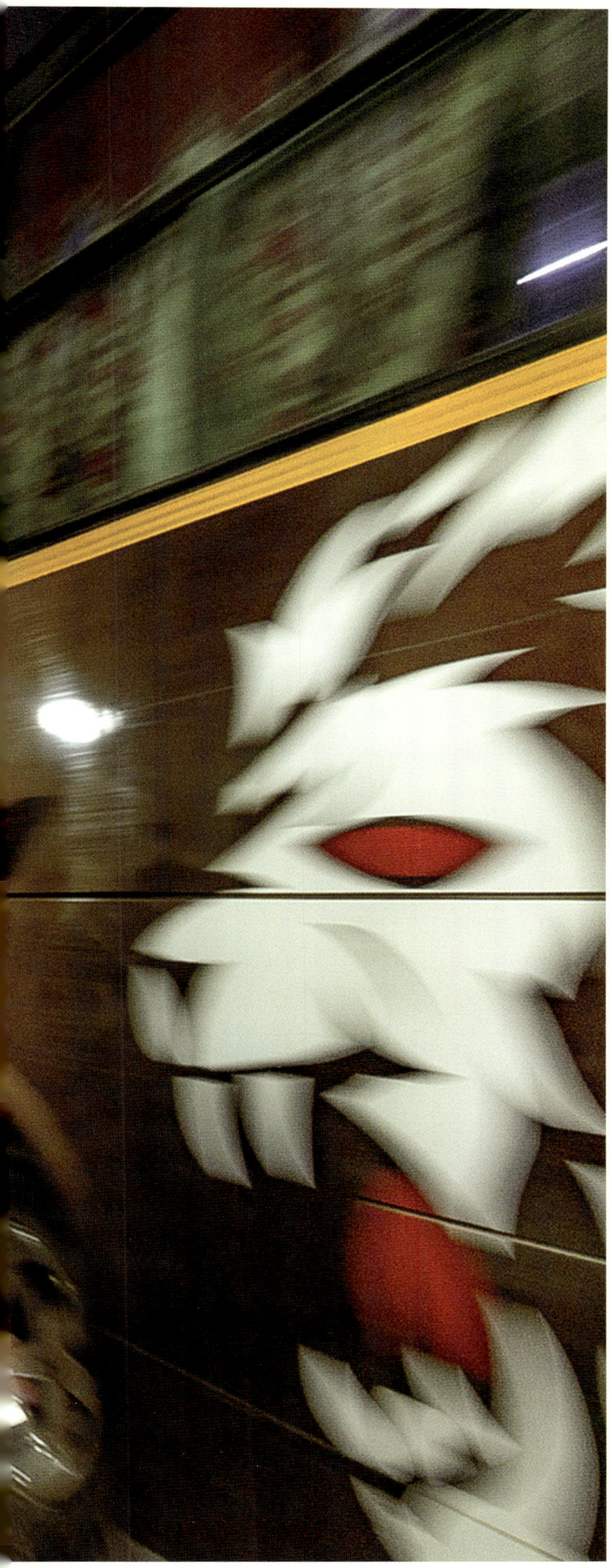

Tourists leave on day safaris from Sasan Gir, a town now flourishing from ecotourism income. In 2024, 855,000 visitors came to see Gujarat's lions.

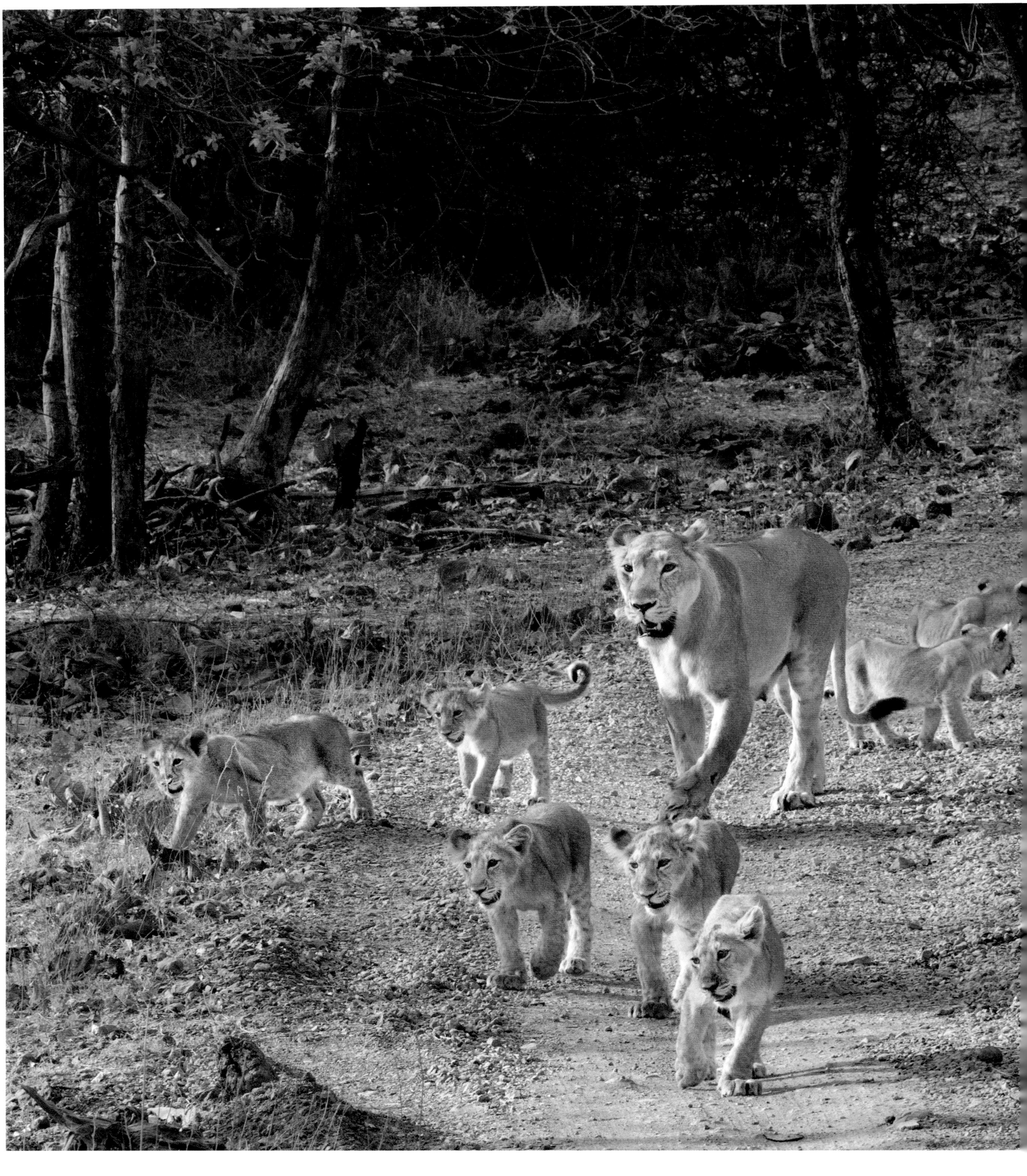

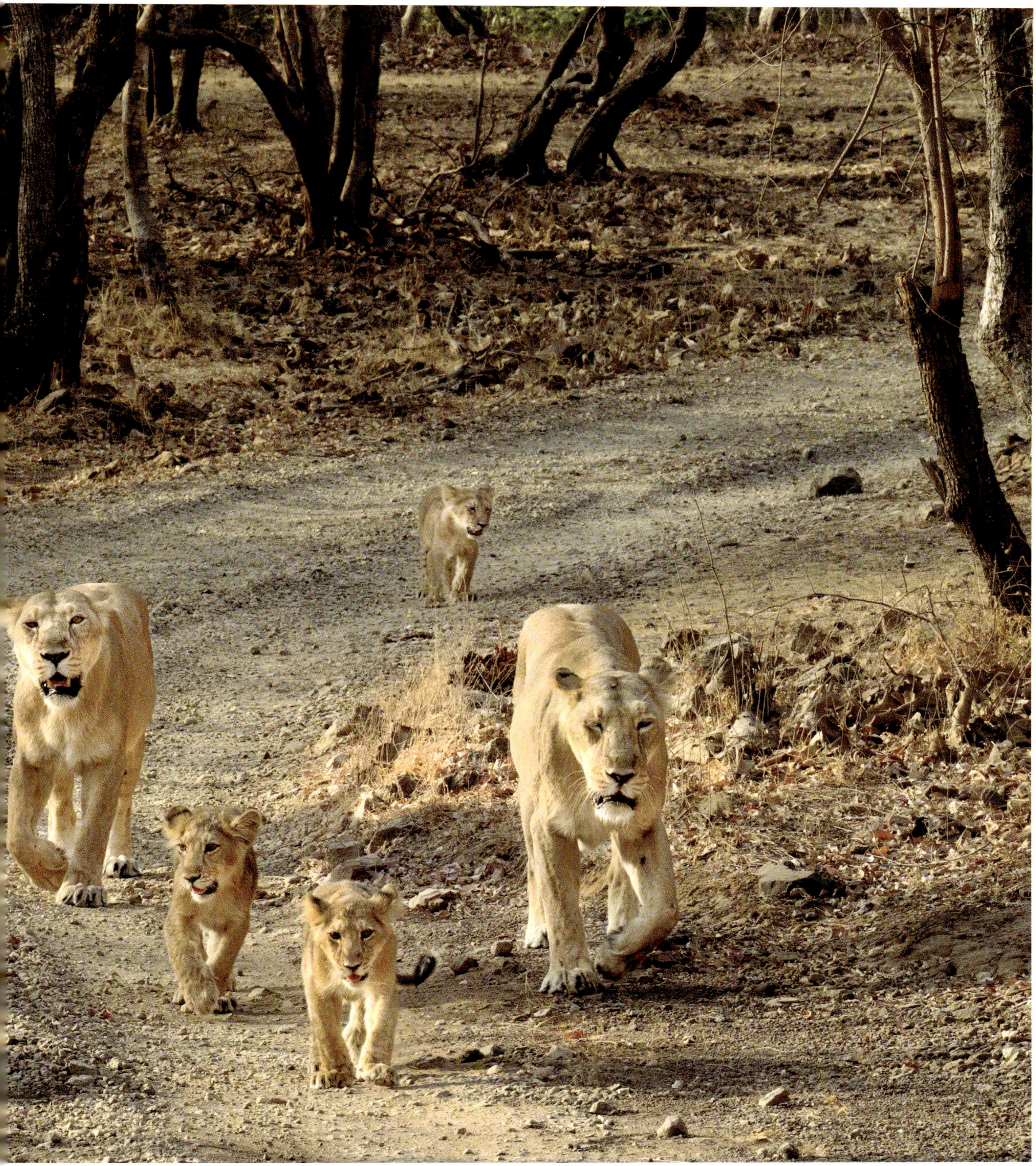

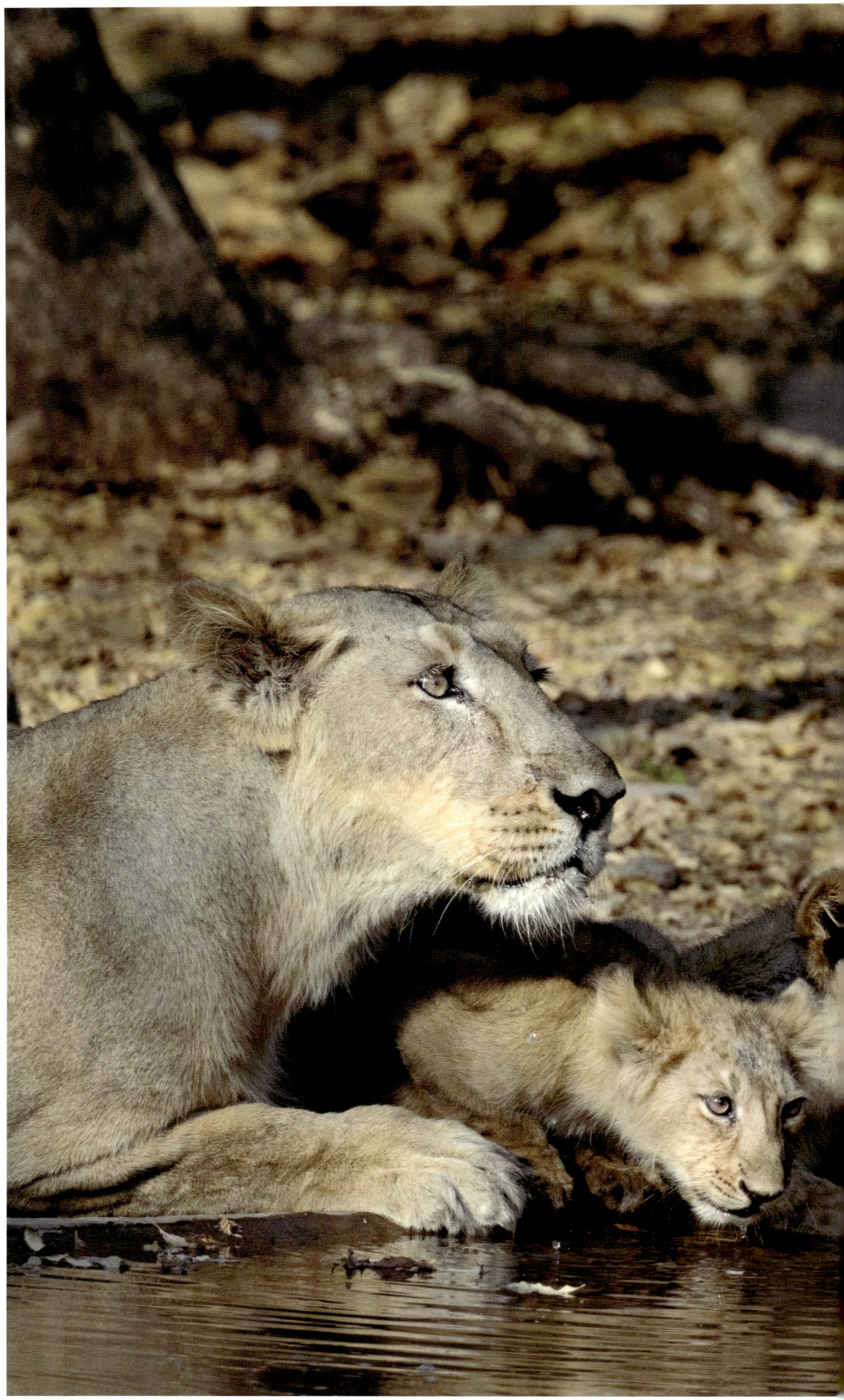

Lions gather at an artificial water hole with a solar-powered pump in Gir National Park. Since this forest simmers under extreme heat, the Forest Department has built water holes for the animals and wells for locals to foster coexistence between the species.

PAGES 260–261: Lionesses with three to five cubs are representative of a healthy ecosystem in Gir National Park, with plenty of prey and water. Only about 50 percent of cubs survive to adulthood, however.

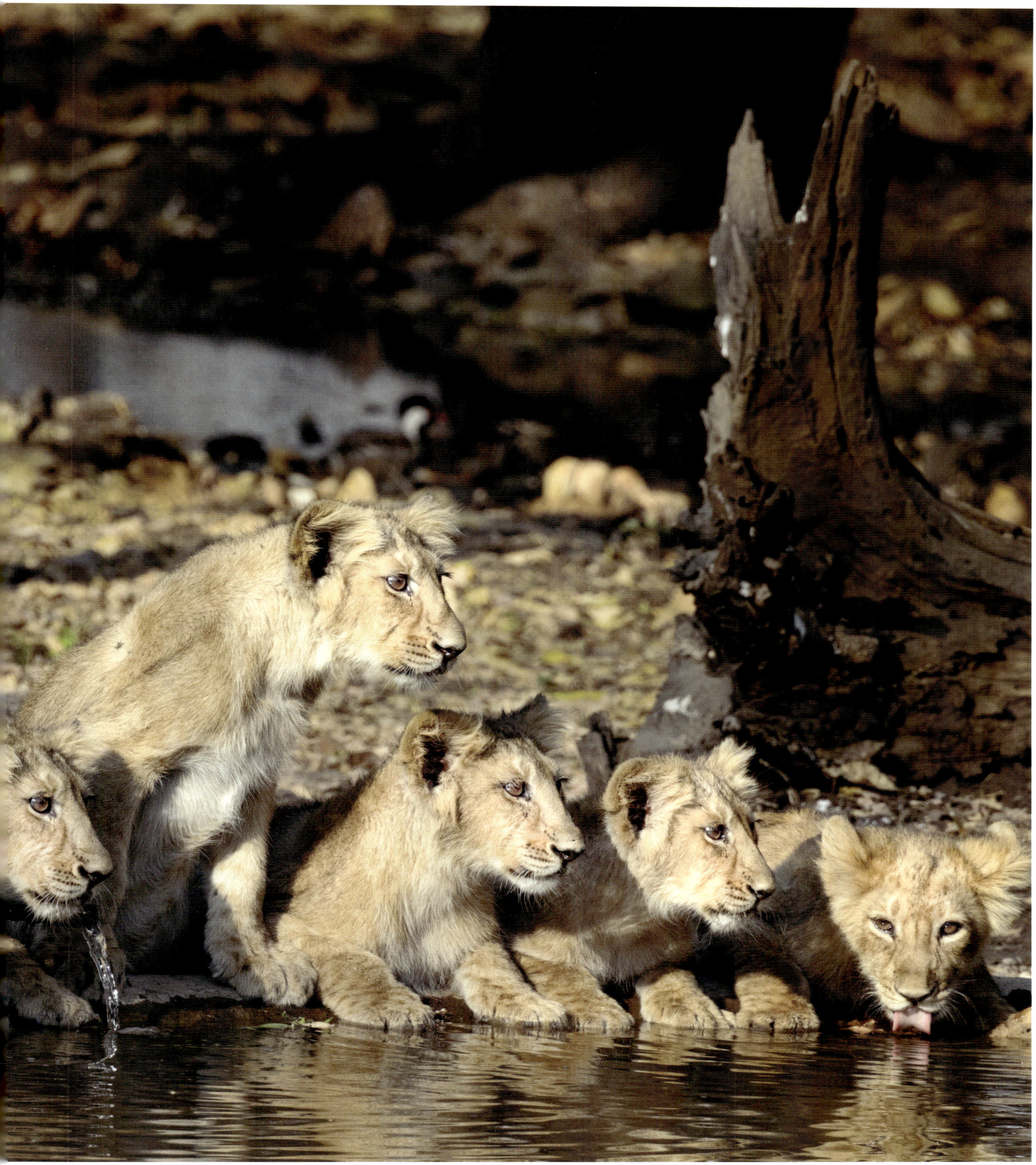

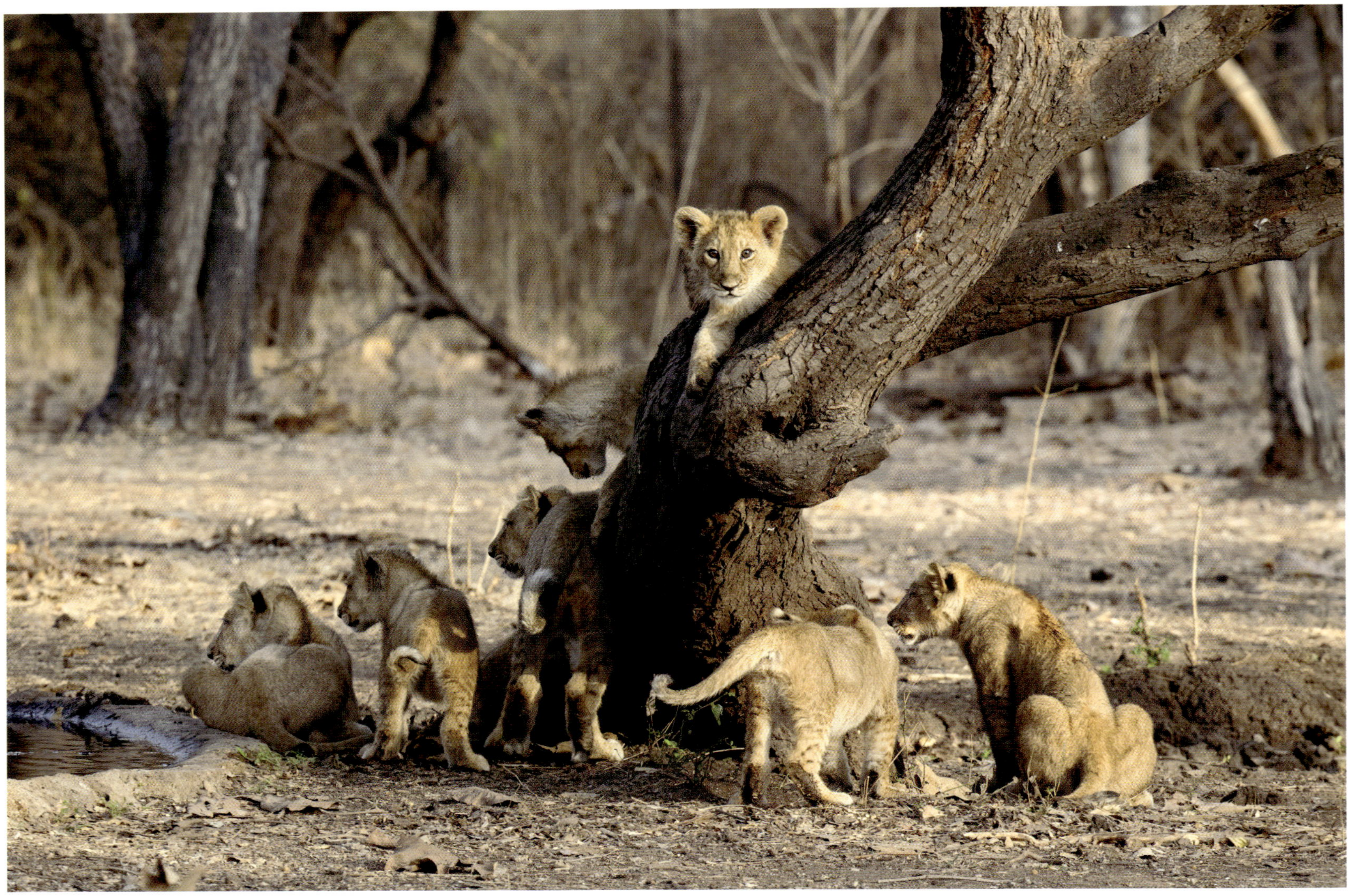

When cubs play, ambushing their brothers and sisters, it's practice for a future where they'll need to hunt to feed themselves. When we arrived in Gir National Park in March 2025, there had been a baby boom: A pride of four females had given birth to 13 cubs that year.

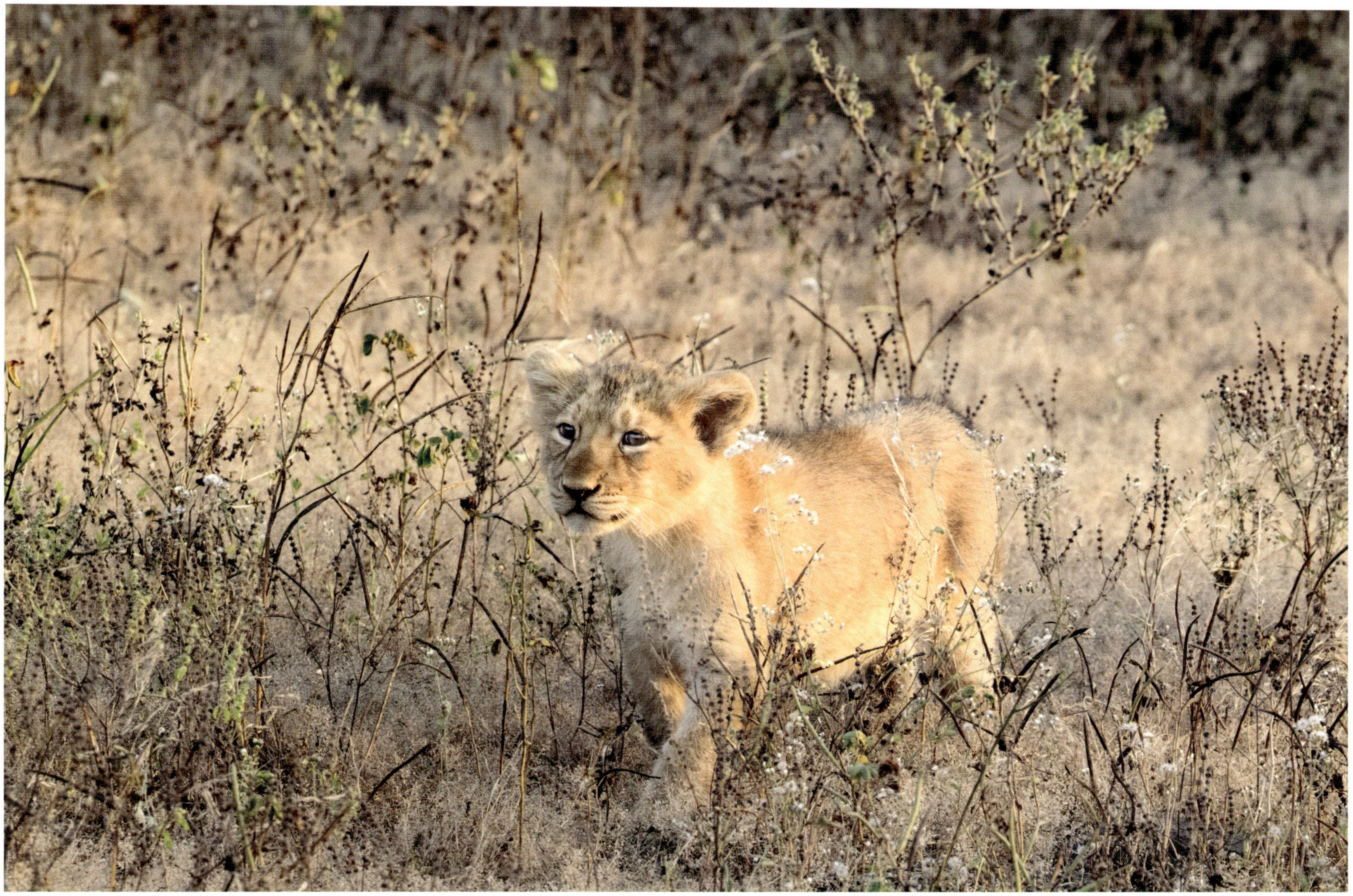

A cub follows its mother through the underbrush. While cubs still face a plethora of threats, including snakes, leopards, rogue male lions, disease, and injury, they are well protected by park guards: There is zero poaching in Gir's protected areas.

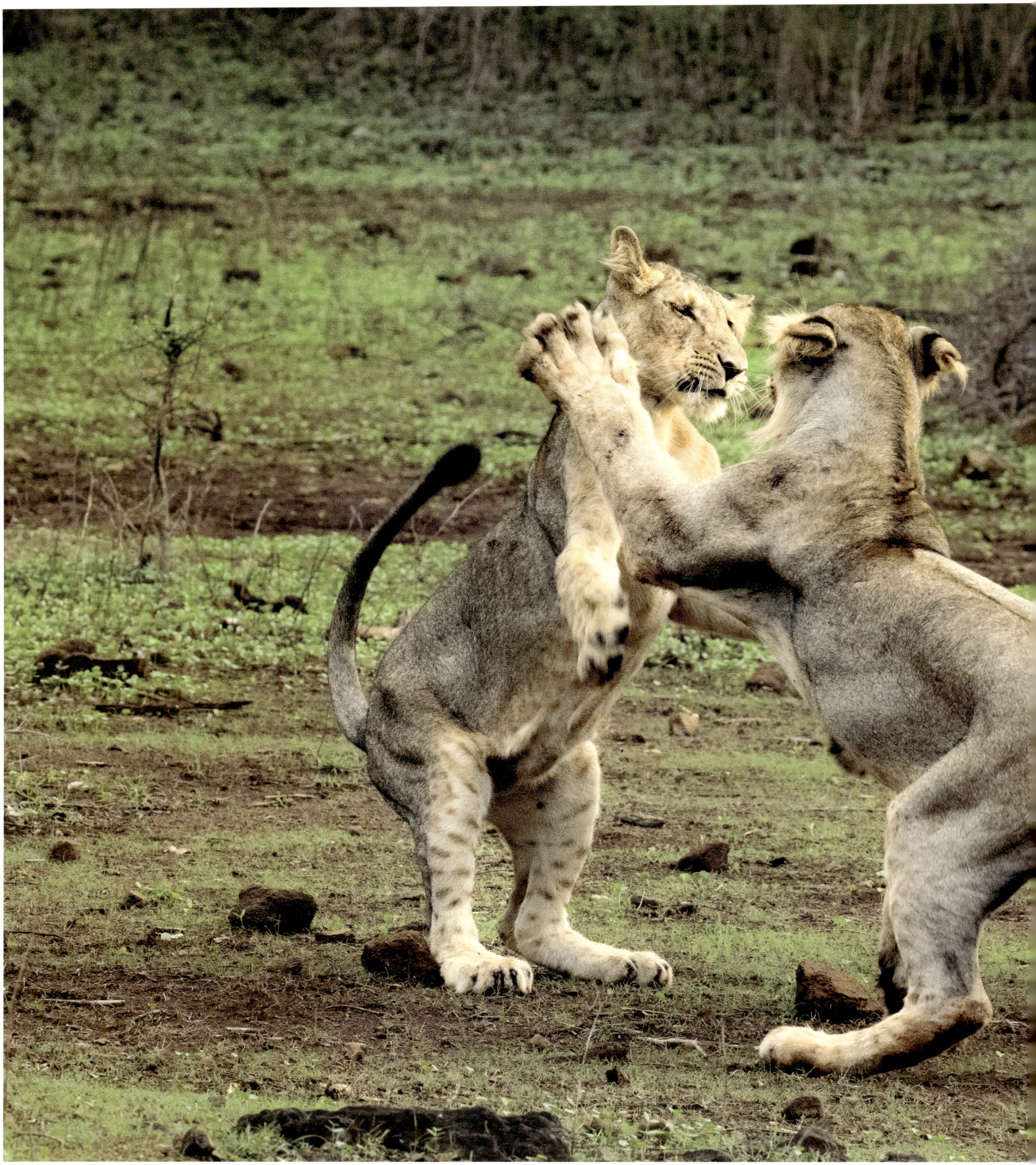

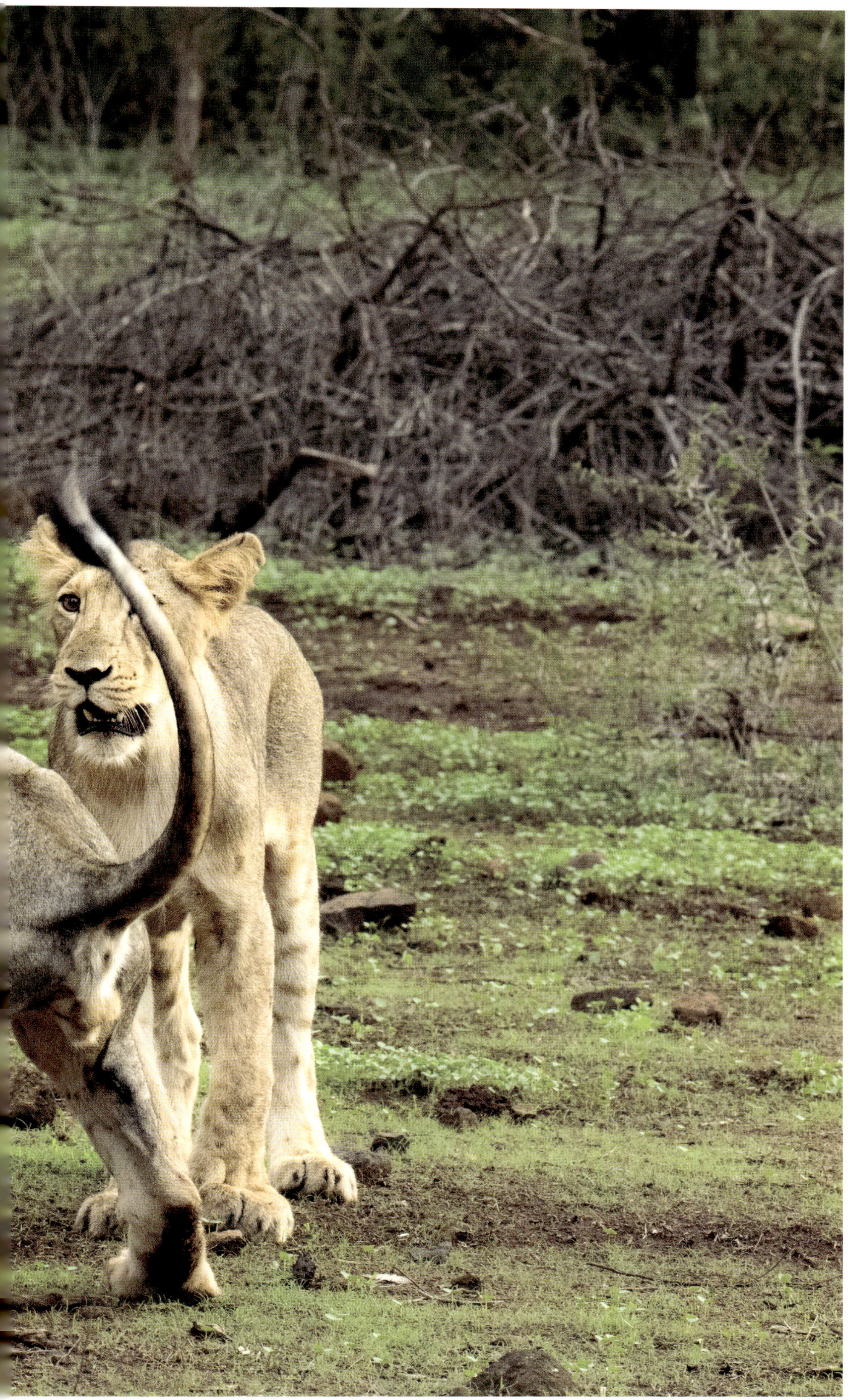

These nearly grown cubs will soon disperse; at about the age of two, they leave their family. Protected areas are full, so the cats are now moving deeper into a human-dominated landscape to find their own territory.

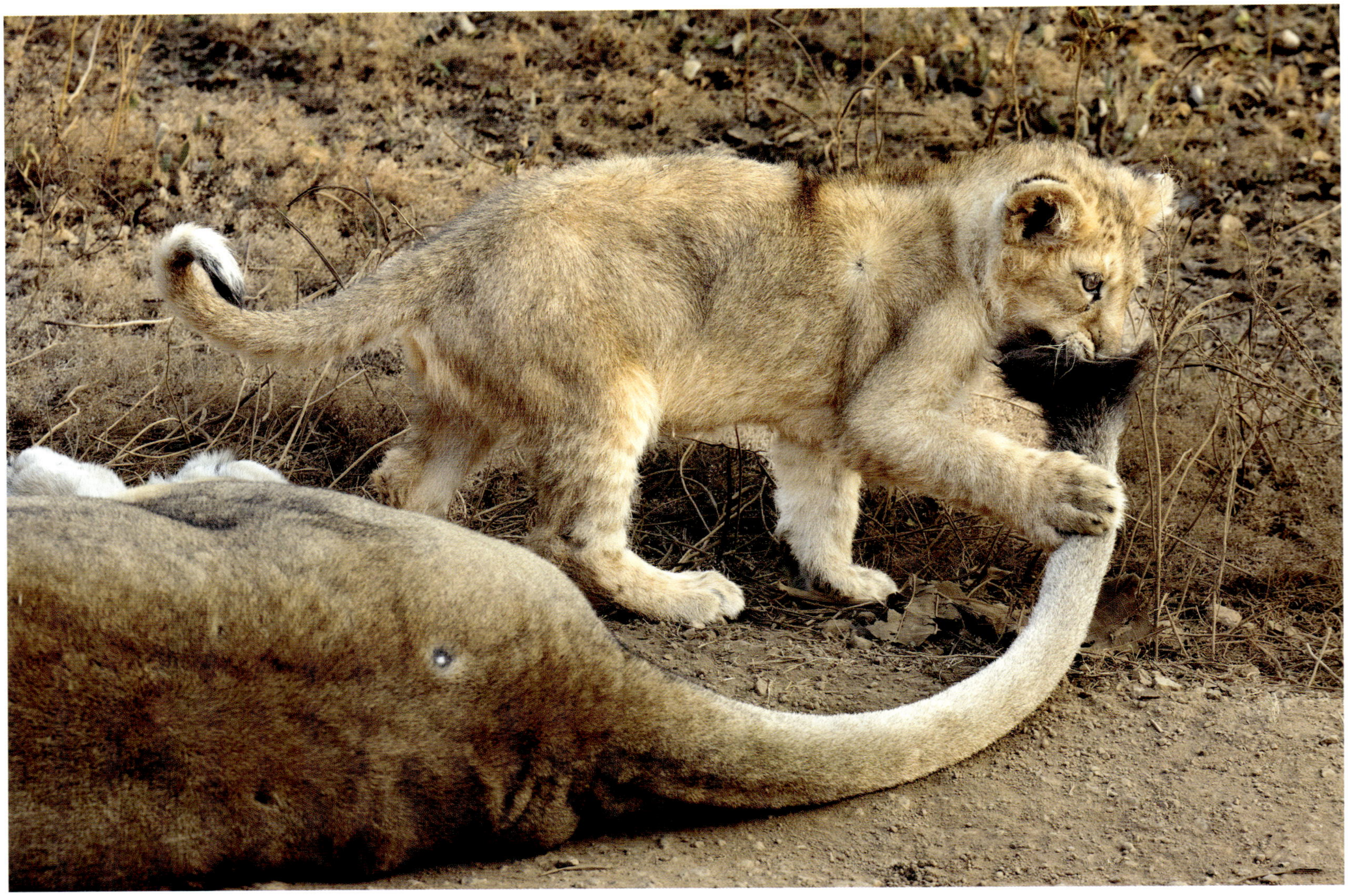

This cub, born in Gir National Park, was part of an exceptionally large pride of 17 females and young cats.

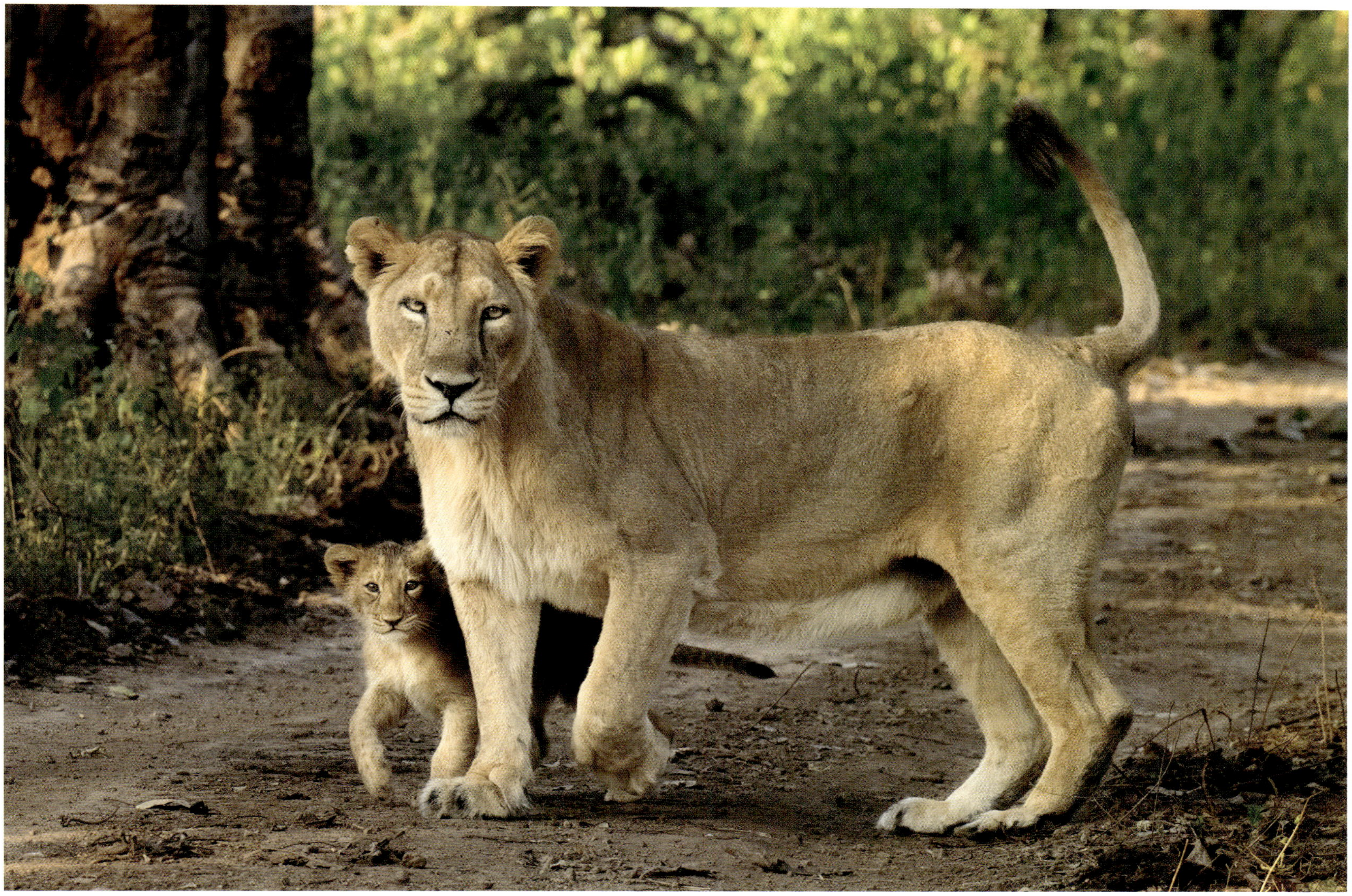

A lioness leads her sole surviving cub down a road in Gir National Park. Though the cub has no siblings, it has numerous cousins: Female Asiatic lions form prides, raising and protecting their cubs in a communal crèche.

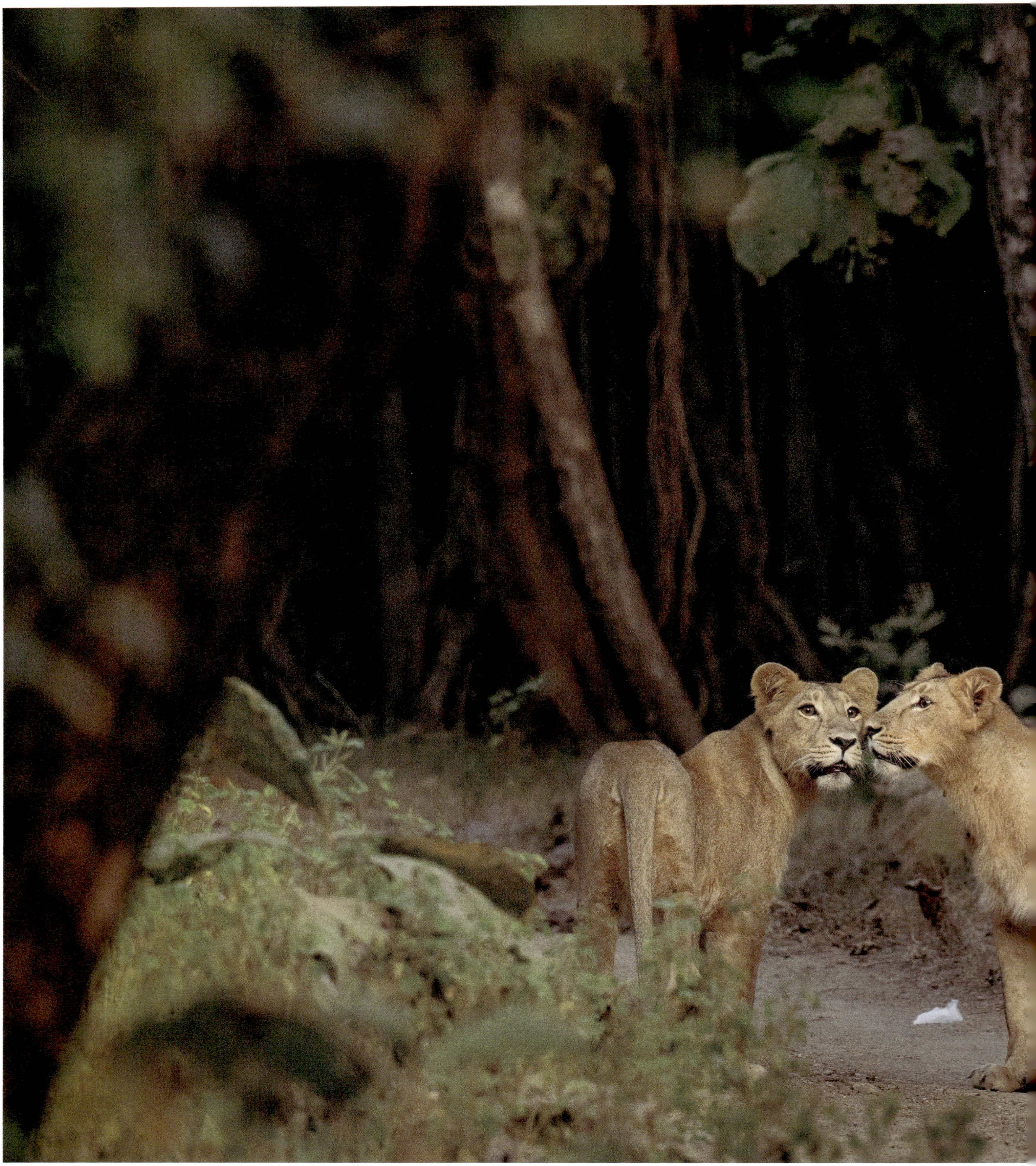

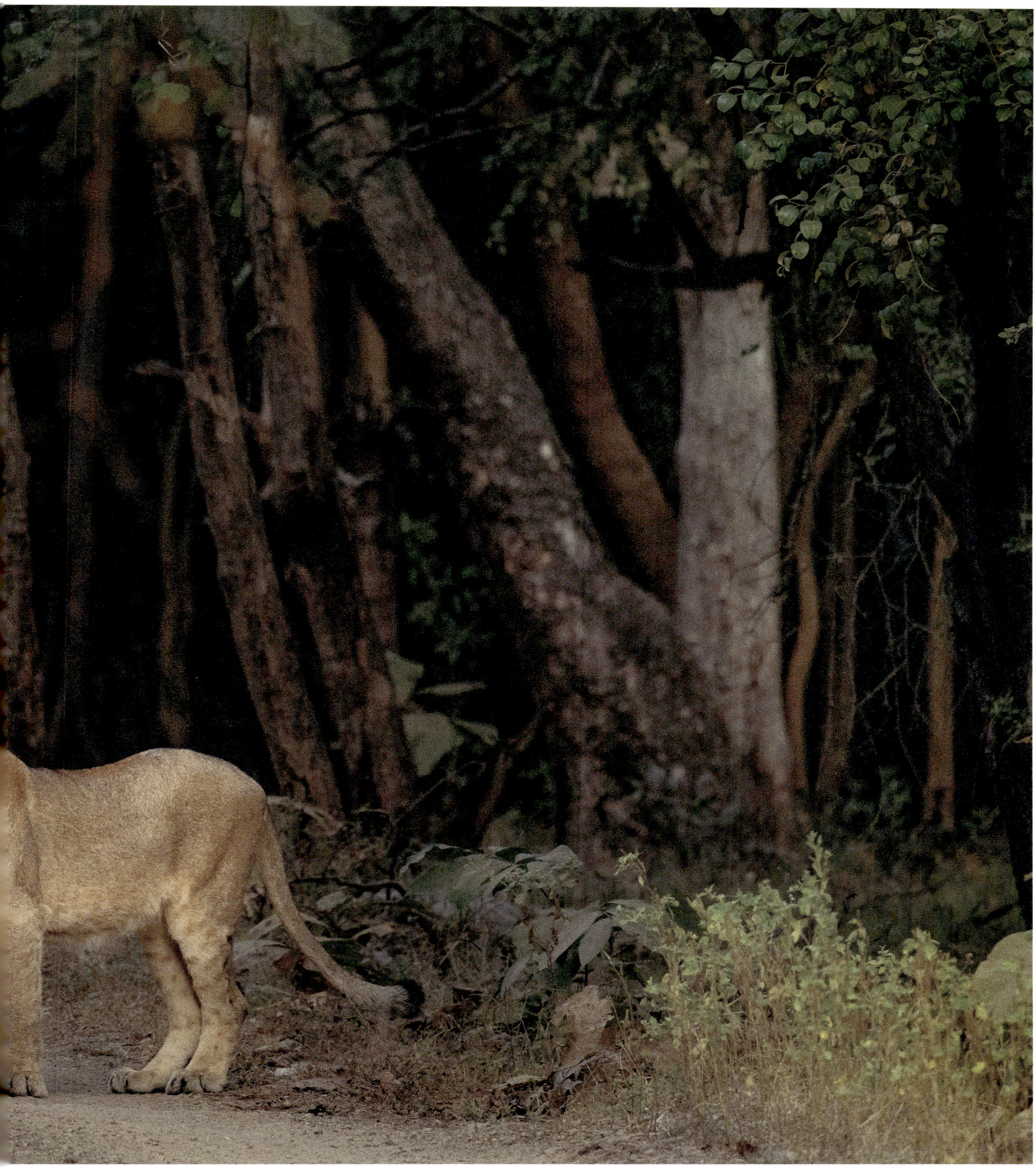

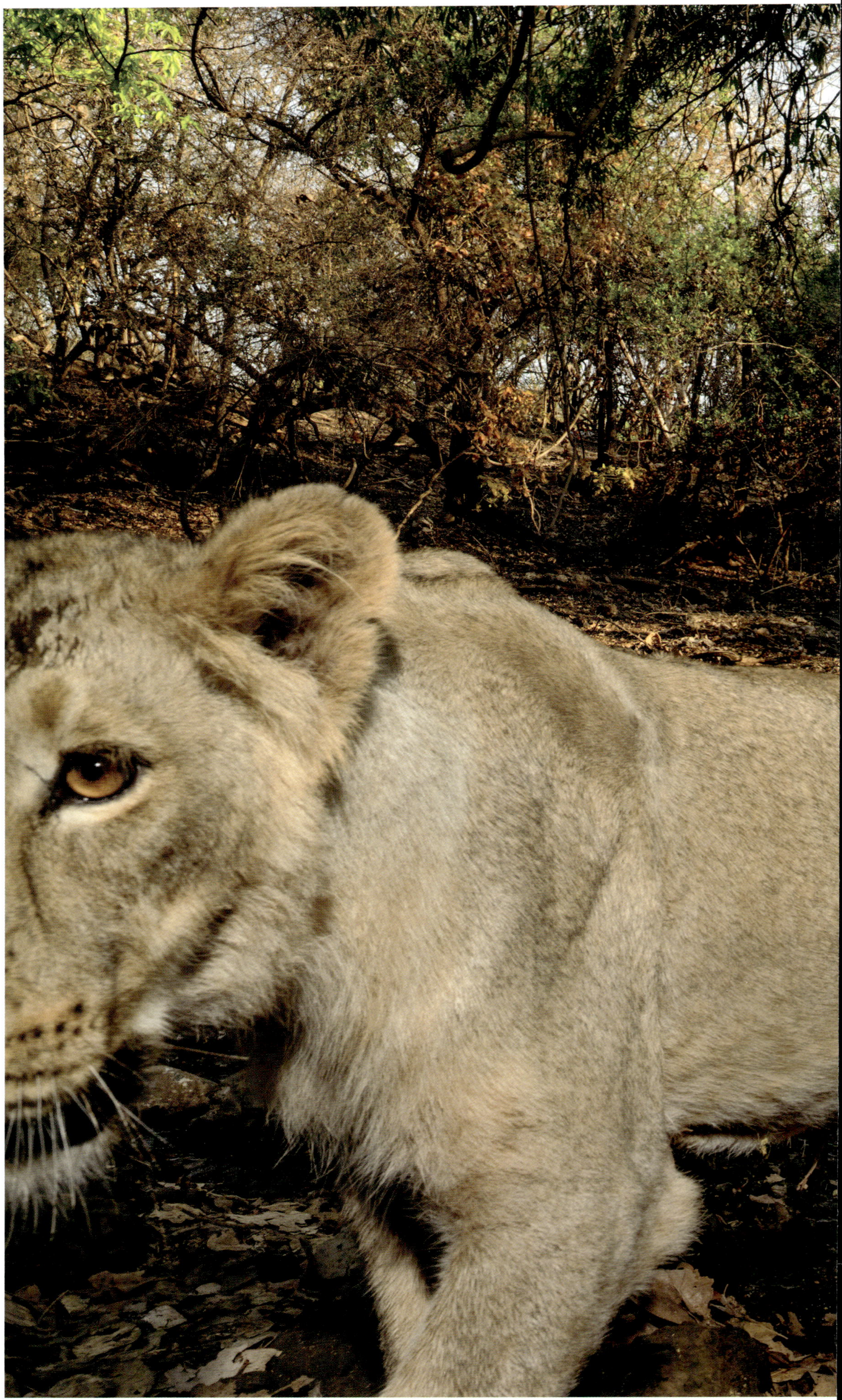

A camera trap captures an image of two subadults and a cub crossing a stream in Gir National Park. Lions need substantial territory. When these cubs reach young adulthood, they will likely have to journey beyond the protected area to find their own homeland.

PAGES 270–271: Two or more lionesses form a pride, and members of a pride greet each other nose-to-nose or by nuzzling faces before plopping down together.

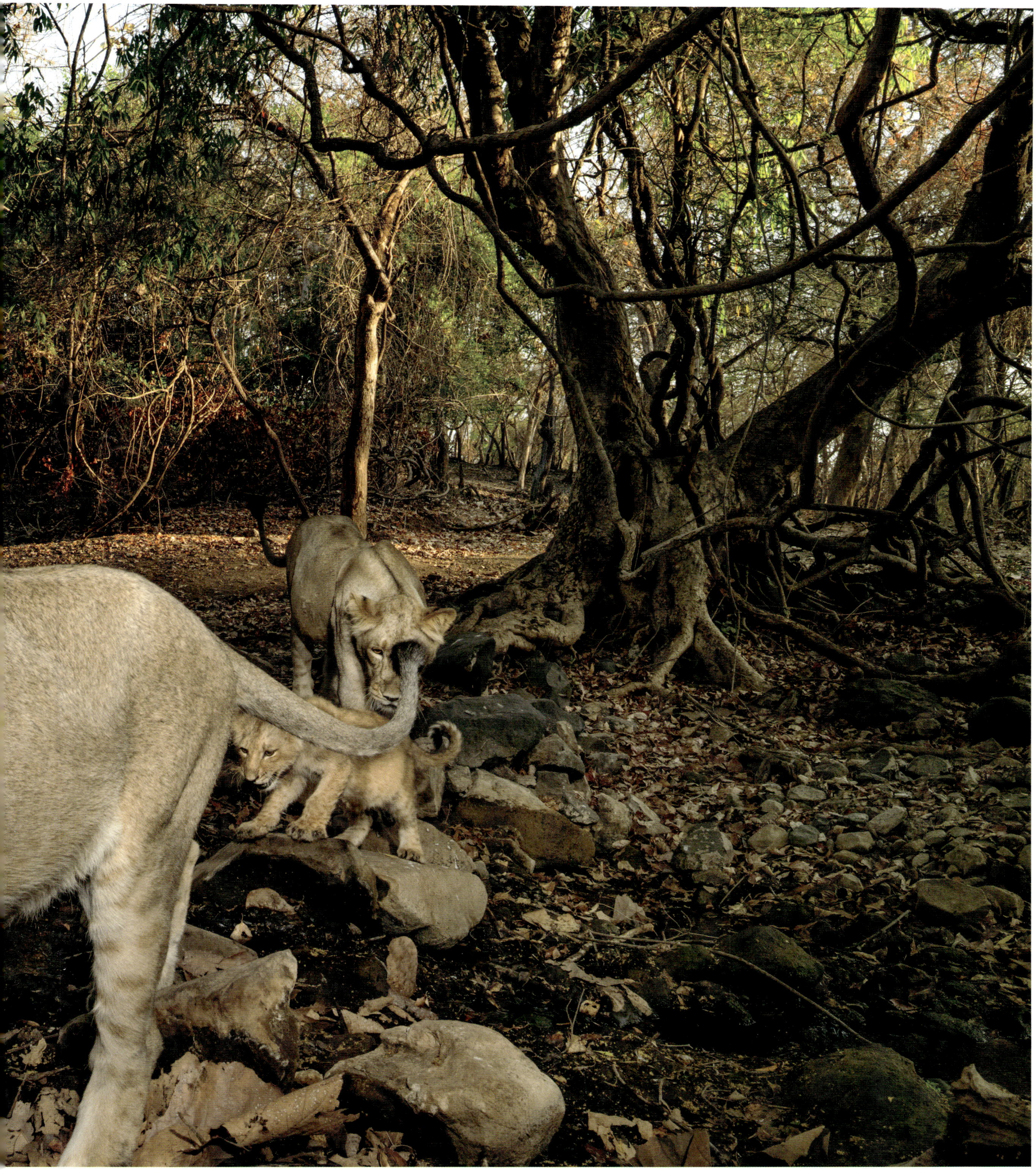

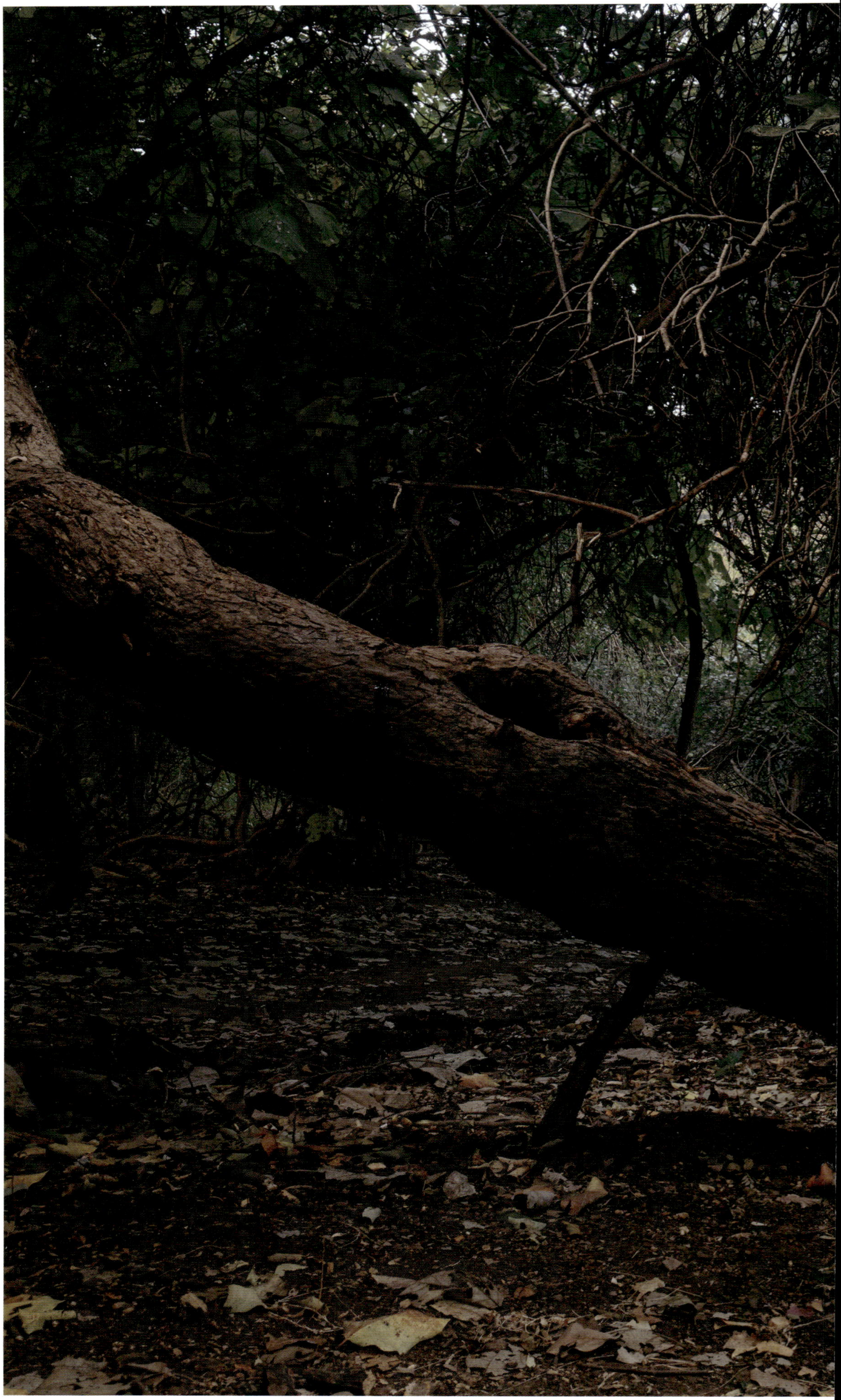

A young male lion gazes at a flash from a camera trap. He is beginning to acquire an Asiatic lion's features, distinct from those of an African lion: a dark Mohawk, a scantier mane, and a fluffier tuft on his tail.

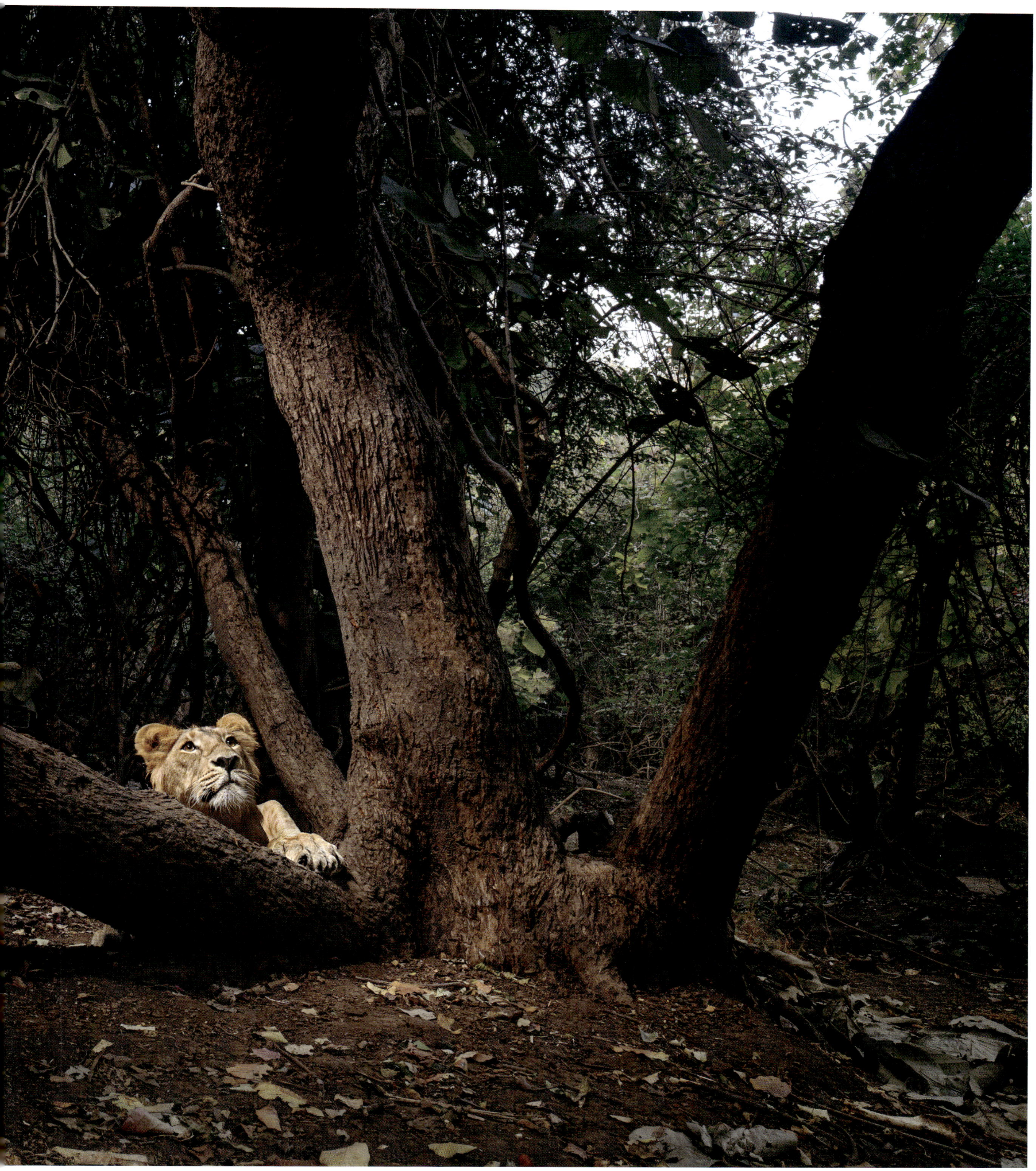

EPILOGUE

When I was born, there were three billion people on the planet. During my lifetime, we've surpassed eight billion. As we've expanded, humans haven't made the wisest choices regarding our future, the survival of other creatures, or protecting the planet. We need to think differently, understanding that voracious consumption is deadly and we can choose to live in ways that heal the damage we've inflicted.

That damage includes the demise of innumerable plants, animals, and entire landscapes. We're amid Earth's sixth mass extinction crisis, compounded by the ever increasing impacts of changing climate and extreme weather events. Together, they threaten fauna, flora, and human communities worldwide.

Every living thing is woven into an interconnected system, and we are rapidly losing species, pulling threads from this fabric of life, the impact varying with each creature's role in its world. Losing apex predators—including big cats—puts an ecosystem at very real risk of collapse.

Each of the world's wild felines are displaced, persecuted, hunted, and exploited by humans. Most of the cats in this book are endangered. Some are critically endangered, a paw's breadth from the precipice. The Javan tiger went extinct in the 1980s.

But we can turn this around. With many proven strategies for coexistence, there is the possibility of a bright future. Big cats are resilient, adaptable, and—importantly—prolific breeders.

These majestic cats can bounce back, but efforts that last just a year or even a decade won't save felines or any other species over the long term. Renowned field biologist George Schaller once told me he'd learned long ago that conservation has no victories. "It's a never-ending process that each of us must take part in," he said. And there are dedicated, brave, innovative people working to save wildlife across the planet, sometimes at great personal risk.

As Jane Goodall constantly reminded us all, hope is crucial to changing the world for the better, but it "requires action and engagement." She also noted: "Hope does not deny all the difficulty and all the danger that exists, but it is not stopped by them. There is a lot of darkness, but our actions create the light."

But we don't have much time. Legendary marine biologist, National Geographic Explorer, and conservationist Sylvia Earle (honorarily nicknamed "Her Deepness") emphasizes, "What we do in the next five years will determine the future of life in the next 10,000 years." "Never before has there been greater urgency," she says, "and never again will there be a better time to act to safeguard the habitability of Earth."

Sharon and I will be out in the field, telling stories about big cats and other animals for the rest of our lives. Our goal in producing this book is to reveal these cats' stunning beauty and their unique behaviors. To document why their existence is so precarious. And to show the many ways that individuals, organizations, and media coverage have given voice to the plight of wild felines.

These photographs have helped conserve habitat, sparked protections, encouraged economic incentives to save these animals, and created safe passages through our urbanized world—and, in the case of P-22, launched a love affair with a cat that made an entire metropolis care about urban wildlife.

Our hope is that these images inspire the awe that we feel when we're in the presence of a cougar, a jaguar, a leopard, a tiger, or a lion in some of the world's remaining wilderness. But our greatest hope is that these pictures and stories inspire you to act, to do something to help save wildlife, wildlands, and this planet we all call home. Every action has impact. Photography can change the world—but every one of us can make a difference.

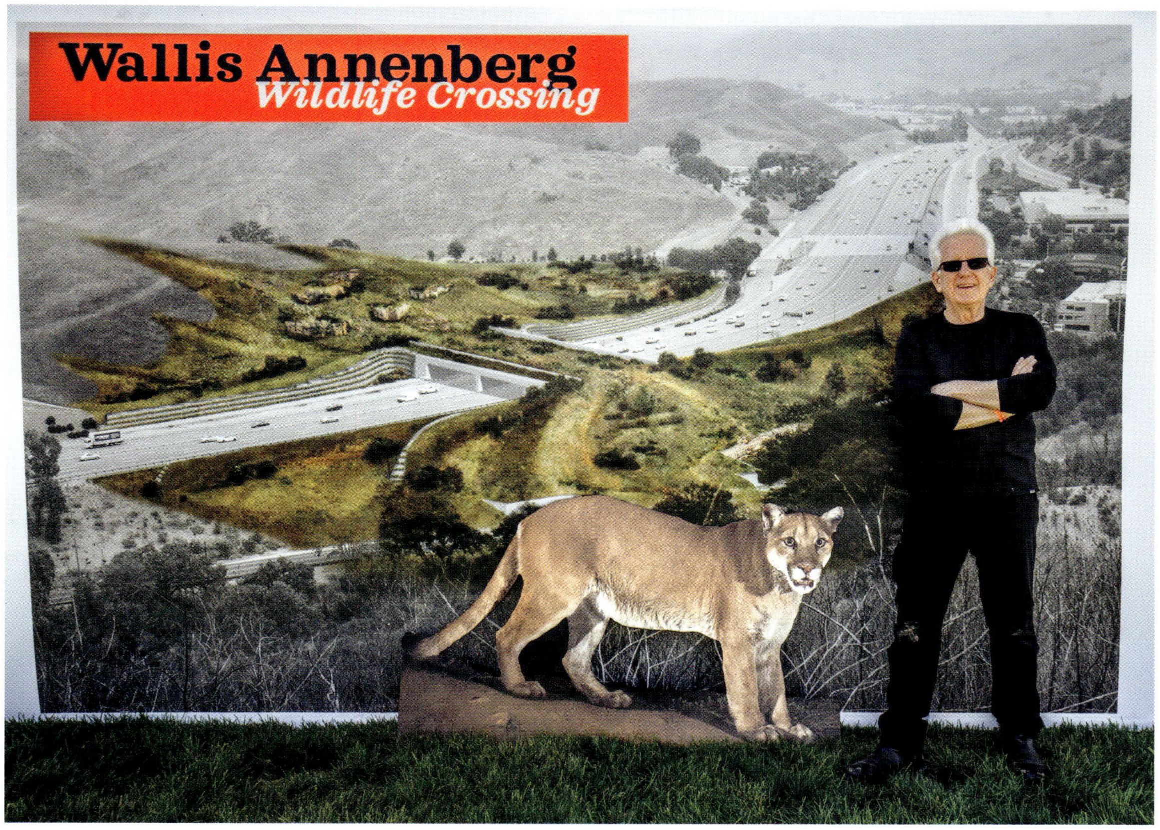

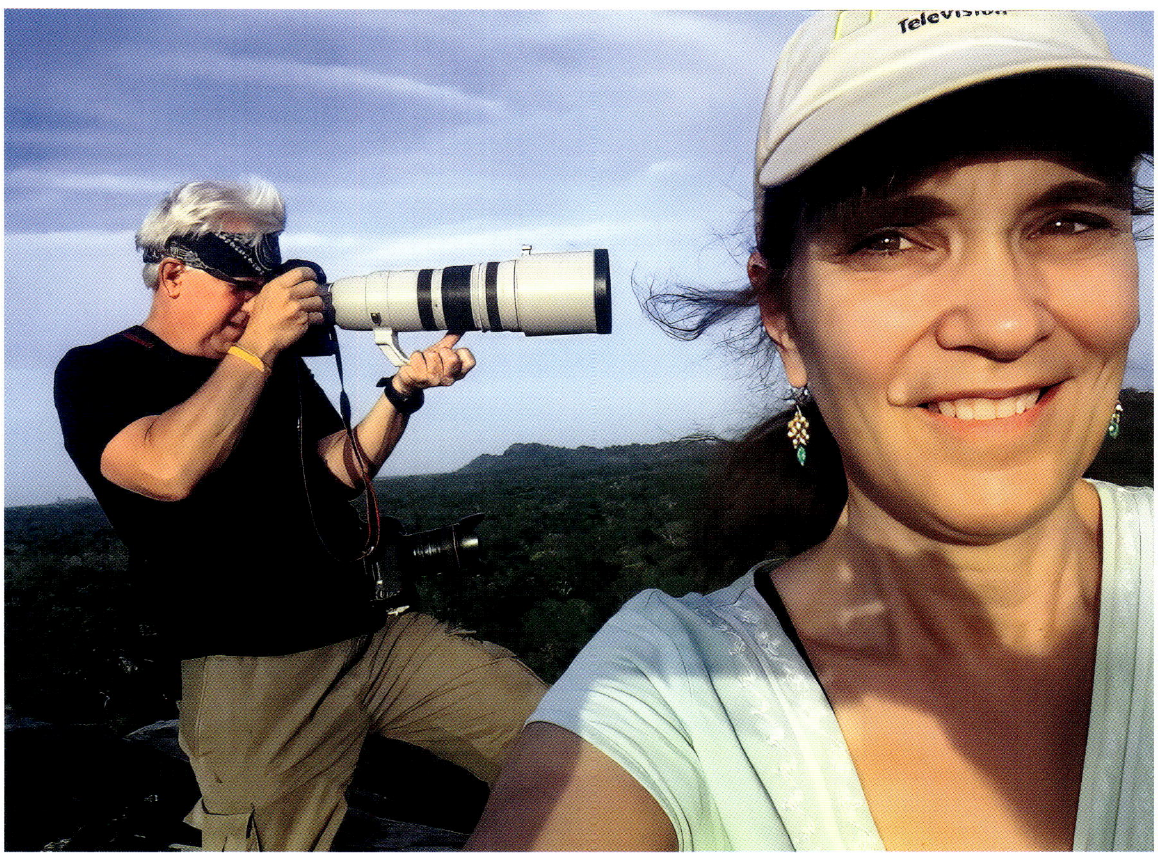

ABOVE: My photo of P-22 under the Hollywood sign made him a celebrity in Los Angeles and inspired the world's largest wildlife crossing.

RIGHT: We took this selfie on top of Elephant Rock in Sri Lanka while working on a leopard story. We collaborate often, interspersing photo, writing, video, and audio skills in ways that make us stronger storytellers.

ACKNOWLEDGMENTS

From Steve: Sharon and I would not be able to do our work without the knowledge, generosity, and guidance of so many, from local fixers, conservationists, scientists, assistants, and colleagues to editors, friends, and family.

The images on these pages span more than 25 years, produced on four continents, so we have an extensive list. First off, I want to thank Kathy Moran, my editor on these stories and my guiding light. Thank you, Doug Chadwick, for years of collaboration and for your brilliant wordsmithery.

Sharon and I are grateful for the support of *National Geographic* magazine editors Chris Johns, Bill Allen, David Griffin, Susan Goldberg, Sarah Leen, Sadie Quarrier, John Echave, Nathan Lump, Alexa Keefe, Debra Adams Simmons, and Oliver Payne. Special thanks to extraordinary photo engineers Kenji Yamaguchi, Tom O'Brien, and Eric Flynn. Thank you to Gary Knell, former National Geographic Society and Partners CEO.

We're grateful to the National Geographic Society for generous funding that allowed the field time needed to produce in-depth and investigative wildlife stories: Rebecca Martin, Kaitlyn Yarnall, Alex Moen, and Jill Tiefenthaler. Thank you to Janet Han Vissering for support of our films and nurse Karen Barry for keeping us healthy. Thanks to Deb Benson, Katherine Potter, Julie Gerdes Becnel, and everyone at National Geographic Live.

A big shout-out to Bertie Gregory, my brilliant collaborator and dear friend.

Working with Alan Rabinowitz in the field in Latin America and Asia changed my life, and I will be forever grateful. And thank you, George Schaller, for all that you've taught me.

On every story, scientists shared their expertise with us, which is always critical to our success on challenging stories, often in remote locations: Howard Quigley, Ullas Karanth, Achara and Saksit Simcharoen, Jeff Sikich, Beth Pratt, Seth Riley, Tom McCarthy, Vidya Athreya, Raghu Chundawat, Joanna Van Gruisen, Valmik Thapar, John Goodrich, Fernando Tortato, Krithi Karanth, Rodney Jackson, Rafael Hoogesteijn, Sandra Cavalcanti, Ricardo Boulhousa, Munawar Kholis, Tony Lynam, Andrea Heydlauff, Hariyo Wibisono, and Firoz Ahmed. Thank you!

So many brilliant conservationists spent vast amounts of time helping us understand the complexity of big cat conservation: Bittu Sahgal, Belinda Wright, Nitin Desai, Debbie Banks, Sybelle Foxcroft, Carson Barylak, Scott and Tanya Smith, Emily McCormack, Pat Craig, Tim Harrison, Meredith Whitney, and many others.

Thank you to those who guided, researched, assisted, or shot video on these stories: Toby Sinclair, Oliver Sinclair, Nimit Pandya, Emilene Ostlind, Drew Rush, Joe Riis, Alex Braczkowski, Fabricio Dorileo, Tashi Tundup, Lala, E.A. Kuttappan, Kunwar, Manju Barua, Gabe DeLoach, Dhruv Cohen, and Sonali Cohen.

Thank you to our exceptional team at National Geographic Books: Susan Tyler Hitchcock, Adrian Coakley, Elisa Gibson, Kathy Moran, Elaine Bradley, Tyler Daswick, Becca Saltzman, Jen Noon Hess, and Mary Stephanos—and especially Lisa Thomas, for the years of brainstorming that produced this book. Thank you, Veronica Sharon, for the untold ways (small and large) you facilitated my work for more than a decade, kept us organized, and helped produce this book and *Tigers Forever*.

Elizabeth McGovern and Michael McGovern: How can we thank you for bringing our nonprofit, Big Cat Voices, to life and for helping us do this work every day? Vishal Subramanyan: You helped make BCV fly.

Lastly, I want to thank my mother, who taught me to dream big, and my father, who gave me my first camera when I was seven and taught me photography.

From Sharon: So many dear friends have bolstered us through physically and emotionally challenging assignments—and we are immeasurably grateful: Laurie Fabiano, Joe Della Fave, Brian Skerry, Amy Weinberg, Jackie Bonnanno, Nancy Celini, Glenn Scherer, Pamela Hassell, Chad Cohen, Deanne Fitzmaurice, Stacy Gold, Deborah Kaplan, Alice Keating, Maura Mulvihill, Sergio Pitamitz, Luca Venturi, Ami Vitale, and Randy and Sally Peabody.

Thank you, Nancy Green and Bijal Trivedi, for inspiration and support every day.

Lauren Herzer Risi, Jennifer Turner, Meaghan Parker, Roger-Mark De Souza, Kim Conner, and Lindsay Collins: Thank you for your incredible support of my work at the Wilson Center.

My beloved Nanny and Poppy—Harriet and Bill McKenzie—formed so much of who I am. They bought me an Instamatic camera when I was five, gave me a typewriter and a suitcase when I was seven, and taught me to watch birds and grow flowers. Thank you for stepping in as my parents.

From Steve and Sharon: Sending big love to our family, Nick Ruggia, Michael Wright Ruggia, and Winter Rose Ruggia: We adore you.

Nick: How can we begin to thank you for the magic of this shared life together? For photo assisting, for shooting video, for writerly advice and editing on this manuscript—and for showing us the world through your eyes over the decades we've spent out in the wide world together. We are so incredibly lucky to be able to collaborate on these stories, brainstorming and supporting one another—whether we're working together or independently—speaking for the world's majestic, voiceless big cats and for the wild world.

Finally: Thank you, cougars (especially P-22), jaguars, snow leopards, leopards, tigers, and lions for the incredible honor of sharing your world and telling your stories.

ABOUT THE PHOTOGRAPHER AND AUTHOR

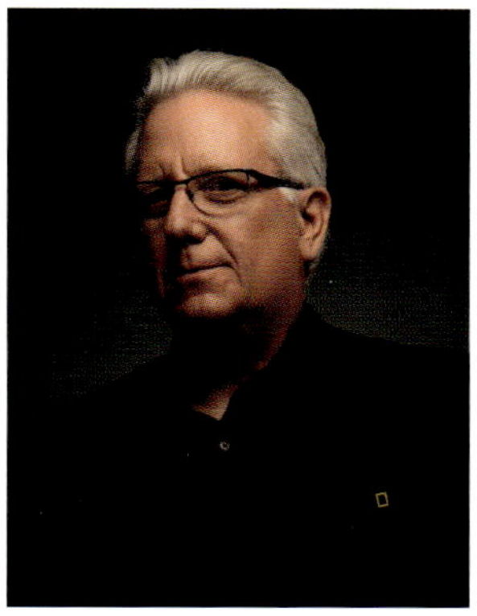

Steve Winter
Steve Winter has been a wildlife photojournalist for *National Geographic* magazine for over two decades, producing stories on some of the world's most elusive and iconic animals—and specializing in big cats. He is a National Geographic Explorer, and his documentary films on jaguars, leopards, tigers, and tree-climbing lions have aired on Nat Geo WILD. His numerous awards include BBC Wildlife Photographer of the Year and BBC Wildlife Photojournalist of the Year, top nature story honors from World Press Photo, and Picture of the Year International's Global Vision Award. His work jump-started construction of the world's largest wildlife crossing in Los Angeles, helped create protected areas, started community programs to build coexistence with big cats, and helped pass a U.S. law, the Big Cat Public Safety Act. Winter has appeared on *60 Minutes, CBS Evening News,* NPR, BBC, and CNN, and he speaks globally on big cats and conservation. In the field, he spends months working in steamy jungles and grasslands—or camping at 15,000 feet in the winter in the Himalaya—to capture images of these animals' stunning beauty and secret behaviors.

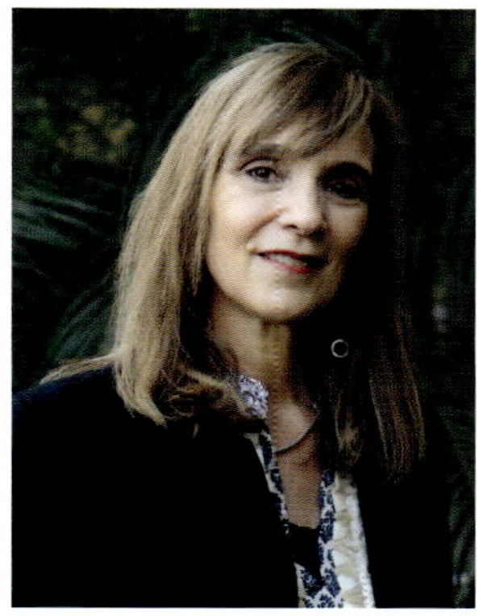

Sharon Guynup
Sharon Guynup is an award-winning journalist and producer. She writes about wildlife, ecosystems, and the threats they face. Her work has appeared in *National Geographic,* the *New York Times, Scientific American, Smithsonian,* and elsewhere. Guynup was named a 2024 Changemaker by New York University and has served as a global fellow with the Wilson Center's Environmental Change and Security Program and China Environment Forum. She created and launched *The State of the Wild* book series for the Wildlife Conservation Society.

Guynup and Winter frequently collaborate. They jointly received the 2024 Eliza Scidmore Award for Outstanding Reporting from the National Geographic Society and India's Sanctuary Nature Lifetime Service Award. They've co-authored two books, *Tigers Forever: Saving the World's Most Endangered Big Cat* (National Geographic, 2013) and *The Ultimate Book of Big Cats* (National Geographic Kids, 2022). Together, they launched the nonprofit Big Cat Voices. They live in Hoboken, New Jersey.

Since 1888, the National Geographic Society has funded more than 15,000 research, conservation, education, technology, and storytelling projects around the world. National Geographic Partners distributes a portion of the funds it receives from your purchase to National Geographic Society to support their mission to illuminate and protect the wonder of our world.

National Geographic Partners, LLC
1145 17th Street NW
Washington, DC 20036-4688 USA

Get closer to National Geographic Explorers and photographers, and connect with our global community. Join us today at nationalgeographic.org/joinus

All photos by Steve Winter except pages 18 and 277 (BOTH) by Sharon Guynup and page 279 (LE) courtesy of National Geographic.

ISBN: 978-1-4262-2417-1

The authorized representative in the EU for product safety and compliance is Disney Trading B.V., Asterweg 15S, 1031 HL, Amsterdam, The Netherlands email: DCP.DL-EU.bookscontact@disney.com

Printed in South Korea

26/QPSK/1